Hugh Weightman

The Medical Practitioners' Legal Guide

The Laws Relating to the Medical Profession

Hugh Weightman

The Medical Practitioners' Legal Guide
The Laws Relating to the Medical Profession

ISBN/EAN: 9783743410954

Manufactured in Europe, USA, Canada, Australia, Japa

Cover: Foto ©berggeist007 / pixelio.de

Manufactured and distributed by brebook publishing software (www.brebook.com)

Hugh Weightman

The Medical Practitioners' Legal Guide

THE

MEDICAL PRACTITIONERS'

LEGAL GUIDE;

OR,

THE LAWS RELATING TO THE MEDICAL PROFESSION.

BY

HUGH WEIGHTMAN, Esq., M.A. Cantab.

BARRISTER-AT-LAW OF THE INNER TEMPLE, AND OF THE
OXFORD CIRCUIT.

HENRY RENSHAW,
356, STRAND, LONDON.

1870.

PREFACE.

It is now forty years since the publication of Mr. Willcock's treatise on "The Laws relating to the Medical Profession," and although that work is still considered a high authority on questions as to which the law has effected no subsequent alteration, it is obvious that time alone must have rendered it as a book of reference practically obsolete. But it is not time only that has thus impaired its value, and as a consequence its usefulness. Modern legislation has completely revolutionized the whole system upon which qualifications for the medical profession were based, or rather it has reduced into a system that which, until a very recent date, was but a confused body of chaotic matter.

The first idea that naturally occurred to the author was the production of a new edition of the original work; but, at a glance, it was evident that such a mode of treating the subject was simply impossible. A very large portion of Mr. Willcock's treatise is occupied with matter which is no otherwise identical with the cognate subjects of the present day than in name. Cases, rules, and above all statutes, which were not in existence at the time when Mr. Willcock's work was published, to say nothing of the since exploded highly penal jurisdiction of the College of Physicians, have entirely changed the whole aspect of the profession both *inter se*, as well as *inter alios*. The intention therefore of publishing a new edition of that book was speedily abandoned; the author felt that it was impossible to do justice to the subject except by attempting an entirely new work. At the same time, it is only right for him to state that some of the abstract ideas contained in Mr. Willcock's treatise have afforded him much assistance, and from the same source much valuable information has been obtained and is hereby duly acknowledged. But the Medical

Acts, the Lunacy Acts, the Sanitary Acts, the Copyright Acts, the Contagious and Infectious Diseases Acts, the Vaccination Acts, the Poor Law Acts, and even the Will Act, which have been passed since the date of the publication of Mr. Willcock's work, have precluded the author from availing himself of any practical materials for his present purpose from that treatise, as, in point of fact, though a work of great merit at the time, it has been entirely submerged by the flood of subsequent legislation.

That the subject of the laws relating to, and regulating the medical profession, is one of sufficient importance to justify a treatise, there cannot be a doubt. The rights, the liabilities, the duties, privileges, and immunities, the legal, as well as moral responsibility involved in connexion with the pursuit of the profession, in all and any of its branches, are based, as indeed they ever have been, upon principles which modern legislation has endeavoured to define for the public weal, and not as is sometimes alleged upon mere abstract, arbitrary rules, the offspring of superstition, intolerance, and ignorance. An endeavour has been accordingly made to obtain for this work at least a character for consistency, by keeping general principles in view, and *that* without trenching on the domain of medical science, or affecting a knowledge to which the author makes no pretensions, his sole aim and object being to furnish the medical practitioner, no matter what his grade, or in what department of the noble art he may practise, a ready means of ascertaining and asserting his own status, and of accurately apprehending his rights, privileges, duties, liabilities, and responsibilities, as recognised, defined, protected, and jealously guarded by the laws of the land.

It is hoped that such a work may be also found useful to members of the legal profession, when applied to for advice in difficulties arising out of medical practice, or incidentally resulting from such professional avocations.

Coroners, clerks to magistrates, to boards of guardians, or those whose duty it is to administer the laws relating to lunacy or the public health, medical officers generally, and even higher functionaries, it is believed will find in this publication that information ready at hand which it is not at all times easy to condense, and which, through a difficulty in acquiring the materials alone, cannot always be obtained when required.

The book is now presented to the medical profession as containing the law on the subject of which it treats, as it exists at the present time, and though not primarily intended for the higher branch of the legal profession, a hope is entertained that even in that sphere it may not be found wholly valueless, and may meet with a favourable reception.

The author would not think it necessary to refer to either the novelty or the difficulty of the task involved in the preparation of the work, were it not that he desires to acknowledge the original idea as most kindly emanating from Dr. Druitt, one of the officers of health of the parish of St. George, Hanover Square, to whom he is also indebted for some very valuable suggestions; and he also is anxious to record that the chapters on "Medical Ethics," and "Joint Medical Education," have been carefully revised by an eminent member of the medical profession, to whom he takes this opportunity of returning his heartfelt thanks; Mr. Clarke, M.R.C.S., of 23, Gerrard Street, Soho, must also be included in the number of those medical friends to whom the author's thanks are due.

In the event of any further legislation affecting the medical profession during the present Session of Parliament, such will be published by way of a supplement, and may be obtained *gratis* by all previous purchasers of the present work, upon application to the publisher.

HUGH WEIGHTMAN.

5, KING'S BENCH WALK, TEMPLE.

May, 1870.

CONTENTS OF THE WORK.

INTRODUCTION.

CHAP. I. ANCIENT ORDERS OF THE MEDICAL PROFESSION.

,, II. MODERN ORDERS OF THE MEDICAL PROFESSION; OR THOSE ENTITLED TO BE REGISTERED UNDER THE MEDICAL ACT (1858).

,, III. RIGHTS, PRIVILEGES, AND IMMUNITIES OF DULY REGISTERED MEDICAL PRACTITIONERS.

SUMMARY OF THE DIFFERENT QUALIFICATIONS ENTITLING THEIR POSSESSORS TO RECOVER AT LAW THEIR FEES ACCORDING TO THEIR RESPECTIVE QUALIFICATIONS.

GENERAL OBSERVATIONS.

,, IV. OF THE LIABILITIES OF LEGALLY QUALIFIED MEDICAL PRACTITIONERS, AND THE DISABILITIES OF UNREGISTERED PRACTITIONERS. ALSO THE MALPRACTICE OF BOTH QUALIFIED AND UNQUALIFIED PERSONS.

,, V. CHARACTER—DEFAMATION.

,, VI. LAW OF COPYRIGHT, ESPECIALLY AS APPLIED TO WRITINGS AND LECTURES OF MEDICAL MEN.

,, VII. SCHOOLS OF ANATOMY.

,, VIII. LAW OF PARTNERSHIP AS AFFECTING THE MEDICAL PROFESSION.

,, IX. THE LAW OF LIFE ASSURANCE AS AFFECTING THE MEDICAL PROFESSION.

,, X. THE PUBLIC HEALTH (COMMON AND STATUTE LAW.)
CONTAGIOUS AND INFECTIOUS DISEASES. VACCINATION.

CHAP. XI. THE LAW RELATING TO IDIOTS, LUNATICS, OR PERSONS OF UNSOUND MIND AND INCAPABLE OF MANAGING THEIR OWN AFFAIRS, AS ALSO THE NATURE AND DEGREE OF INSANITY WHICH WILL RELIEVE A PERSON FROM THE PENAL CONSEQUENCES OF CRIME.

„ XII. CORONERS' INQUESTS.

„ XIII. REGISTRATION OF BIRTHS, DEATHS, AND MARRIAGES.

„ XIV. POOR LAW LEGISLATION AS AFFECTING MEDICAL PRACTITIONERS. PAUPER LUNATICS, INCLUDING CRIMINALS AND PRISONERS FOR DEBT.

„ XV. CHEMISTS, DRUGGISTS, AND DENTISTS.

„ XVI. MEDICAL ETHICS, AND MEDICAL EDUCATION.

SUPPLEMENTARY CHAPTER.—ON TITLE OF DOCTOR OF MEDICINE, OR DOCTOR OF PHYSIC.

APPENDIX.

TABLE OF CONTENTS.

CHAPTER I.

Ancient Orders of the Medical Profession—The Antiquity of the Art of healing—Hippocrates—his voluminous Writings—his personal Character—and his Pharmacopœia—Galen—his Writings—the Basis for Compilers of Medical Works—Translated into Arabic—these Works introduced into England at the Time of the Christian Mission under Pope Gregory the Great—The Monks and secular Clergy at first the only regular Medical Practitioners—Interference with the Medical Practice of the Monks by the Decree of the Council of Tours (A.D. 1163)—Clerical Medical Men—The Rev. W. Clark, M.D., F.R.S., F.R.C.P., Professor of Anatomy at University of Cambridge—Three Regular and many Irregular Orders of the Medical Profession—Celebrated Regulations of the College of Salerno—First attempted Interference with Women practising Physic—The Act, the Foundation of all Statute Law regulating the Practice of Physic—Establishment of the College of Physicians, London—The Knights Hospitallers, and the Order of St. Lazarus—The Barbers' Company—Barber-surgeons—Apothecaries or Grocers—Empirics—Jews—Female Medical Practitioners—Touching for the King's Evil . . . . pp. 1—14

CHAPTER II.

Modern Orders of the Medical Profession ; or those entitled to be Registered under the Medical Act (1858)—Royal College of Physicians, London—History of the College—Eligibility for its Fellowships—The College of Physicians, Edinburgh—King and Queen's College of Physicians, Ireland—College of Surgeons of England—College of Surgeons of Edinburgh—The Faculty of Physicians and Surgeons of Glasgow—The College of Surgeons, Ireland—Society of Apothecaries, London—Apothecaries' Hall, Dublin—Universities—Doctor by Doctorate—Doctors of Foreign and Colonial Universities or Colleges—Person practising Medicine before August, 1815 . . . . pp. 15—28

CHAPTER III.

Rights, Privileges, and Immunities of duly-registered Medical Practitioners—Right such to recover their Fees, and what Fees, whether in Medicine, Surgery, Pharmacy, or Midwifery—Summary of the above—General Observations
pp. 29—35

CHAPTER IV.

SECTION I.—Of the Liabilities of legally qualified Medical Practitioners, and the Disabilities of unregistered Practitioners—Unregistered Persons not entitled to the Expenses of Medical Witnesses, nor to make *post-mortem* Examinations—Appeal from a Conviction for assuming the Title of Doctor of Medicine
pp. 36—37

SECTION II.—Malapraxis—Negligence—Civil Liability of Medical Practitioner not terminated with the Death of the Patient—Doctrine of legal Malice—Ignorance—Competent Skill—Culpable Want of Skill—Improper Use of the *Vectis* or *Lever*—A Person negligently supplying a wrong Drug, in consequence of which Death ensues, he is thereby guilty of Manslaughter . . . . pp. 37—42

CHAPTER V.

Character—Defamation—Maliciously imputing to a Medical Practitioner a want of Qualification, Attention, Skill, or Capacity—Distinction between Slander and Libel—For Libel two Remedies are given—Malice expressed and implied—Falsity of the Imputation—Consequence of the Slander—Cases in which no special Damage need be proved, and why—By what other Means than by Writing a Libel may be expressed—Motives under which a public Officer acts in doing a Duty incumbent upon him do not make his Conduct Actionable—Challenging public Criticism by Advertisements—Privileged Occasion—Publication, what amounts to—Justification in a Civil Action and in a Criminal Proceeding—Repeating a Libel to another—Selling Copies—Privileged Communications—Trials for Libel or Slander in the County Courts . . . . . pp. 43—52

CHAPTER VI.

Law of Copyright as applicable to the Writings and Lectures of Medical Men—Copyright before Publication—The Right of an Author or Proprietor of a Manuscript to its first Publication—Power of the Court of Chancery to restrain the Publication of Medical Lectures delivered orally—Right to restrain the Publication not only of Etchings and Photographs, but also of a descriptive Catalogue of them—No Copyright before Publication allowed in a Work of an immoral or irreligious Character—Copyright after Publication—How far it is now extended—Lectures delivered in any University, Public School, and some other Places are not protected—What amounts to an Infringement of Copyright—International Copyright, and the Acts and Conventions relating thereto—The different Countries with which International Copyright has been arranged—Copyright in a foreign Print, and our own Engraving Acts—Copyright in the Pharmacopœia . . . . pp. 53—62

CHAPTER VII.

Schools of Anatomy—The Act regulating them—What led to this Enactment—Burke and Hare—Stealing Shrouds from a dead Body—Resurrectionists—Disinterring a Body an indictable Offence—Selling the dead Body of a Felon for the Purpose of Dissection, where Dissection formed no Part of the Sentence, a Misdemeanour at Common Law—Penalties under the Anatomy Act—Actions of Trespass for violating Graves or Burial Grounds, maintainable by whom . . . . pp. 63—67

CHAPTER VIII.

Law of Partnership as affecting the Medical Profession—What Agreement amounts to a Partnership—Between an outgoing and incoming Surgeon—A common Stock or Plant not essential to constitute a Partnership—One Partner finding Property and another Skill—One Partner indemnifying others against Loss—Partners *in futuro*, as between Master and Pupil, or Principal and Assistant—Partnerships between qualified and unqualified Medical Practitioners—Appeal from a summary Conviction under the Penal Clause of the Medical Act, the Appellant's Name not being found in the Medical Register coupled with other Circumstances—Authority of each Partner to hire or discharge Servants, Clerks, or Assistants, engaged in the Business—What Class of Servants are entitled to more than a Month's Warning, or a Month's Wages—Yearly Hirings when determined—A Master has no Right to dismiss an Apprentice merely because he misbehaves, except real Danger was by his Misconduct occasioned to his Master's Business—How far Clerks, Assistants, &c. remunerated by a percentage on their Earnings are to be regarded as Partners—A Partnership though not *inter se*, yet may exist *quoad* third Persons—A Master justified in dismissing a Clerk or an Assistant for claiming to be a Partner—Partnership Contract need not be in Writing—A *quasi* Partnership—In some Cases, under the Statute of Frauds, Agreement *must* be in Writing . . . . pp. 68—77

CHAPTER IX.

The Law of Life Assurance as affecting the Medical Profession—Medical Referees—The most perfect *bona fides* must be observed—Wilfully false Statements made by the Medical Referee, or any Collusion with the Proposer, although the Referee may have no Interest in the Contract, render him liable to an Action—Claim of Medical Referees against Insurance Office for a Fee for their Certificates—Cases decided on the Subject—The ordinary Medical Adviser—The Meaning of the Term "usual Medical Attendant"—Need the "usual Medical Attendant" be a regular qualified Practitioner?—What constitutes moral Fraud and legal Fraud—What amounts to a wilful Suppression of the Fact of other Offices having been applied to and refused the Life—What Evidence of the "usual Medical Attendant" will be sufficient to rebut the Imputation of Fraud in not having disclosed that the Deceased was of a Consumptive Family . . . . pp. 78—81

CHAPTER X.

SECTION I.—The Public Health (Common and Statute Law)—Contagious and Infectious Diseases—The ancient Writ of "De leproso amovendo"—The exposing a Child infected with Small-pox in a Public Way a Misdemeanour at Common Law—The Plague—Inoculating for the Small-pox—The "Nuisance Removal Acts" and the "Diseases Prevention Acts"—"Sanitary Acts," "Workshops Regulation Act," and "Factory Acts" epitomized and explained . . . . pp. 82—88

SECTION II.—The "Contagious Diseases Acts" abstracted and epitomized
. . . . pp. 88—90

SECTION III.—The "Vaccination Acts"—Effect and Operation—Appeals from Summary Convictions under them . . . . pp. 90—92

CHAPTER XI.

The Law relating to Idiots, Lunatics, or Persons of Unsound Mind, and incapable of managing their own Affairs; as also the Nature and Degree of Insanity which will relieve a Person from the Penal Consequences of Crime—Medical Certificates generally—Townley's Case—Provisions of Lunacy Acts—Form of Medical Certificate in the Case of Private Persons (not Paupers)—Evidence of Medical Witnesses in open Court—The Nature of Insanity recognised by Law—McNaughten's Case—Answers of the Judges to the House of Lords—Feigned Insanity—Juries in Criminal Cases—Civil Cases—Certificate of Medical Man, accompanied by Affidavit, for the Issue of a Commission "De lunatico inquirendo," or for furnishing Information to the Lord Chancellor—Meaning of the Term "Unsound Mind"—Case of W. F. Windham—Legal Definition of the Word "Lunatic"—Habitual Drunkenness—Medical Men witnessing or drawing-up the Wills of Patients—A Will made by Lunatic during a Lucid Interval—Fraud and Imposition upon Weakness of Mind—Will in favour of a Medical Attendant in whose House the Testator resided—Judgment of Sir W. Wynn as to the Effect of an Inquisition of Lunacy—Nature of Wills—Memorandum of Form of Attestation—A Person not Insane may yet be incompetent to make a Will—Devise, Legacy, or beneficial Interest given by a Will to the attesting Witness of it, or to the Wife or Husband of such Witness—Further Consideration of the Lunacy Statutes—Liability of a Medical Man at the Suit of a Private Person confined under his Certificate—The Prosecution of Medical Men for signing Certificates without examining the Patient—Common Law Right of a Medical Man to restrain a dangerous Lunatic—Death of Patients in Asylums—What Statements must be drawn-up and signed by the Medical Attendant, and to whom sent—Cases cited on the Subject of Restraint . . . . pp. 93—113

CHAPTER XII.

Coroners' Inquests—The Office of Coroner—Its Antiquity and Dignity—Qualification—Qualities which, according to Sir E. Coke, a Coroner should possess—"Medical Witnesses Act" and "Sanitary Act (1866)"—Duty of Coroner to pay the Medical Fees immediately after the Inquest—Penalties imposed upon Medical Practitioners for disobeying Coroner's Order for their Attendance—Importance of the Evidence of Medical Witnesses—Cases illustrative of it—The Cause of Death the Subject of the Inquest—Concealment of Birth only has no Connexion with the Cause of Death—In such Case the Coroner has no Jurisdiction—Death by Poison—By Prussic acid—By Strychnia—By Arsenic—Circumstantial Evidence not reliable—Doubtful Cases—M. Raspail, the French *Savant*—Great Caution required in Medical Witnesses—Medical Witness has no Privilege, but is bound to reveal all Matters disclosed to him by his Patient—Office of Coroner not determinable by the Demise of the Crown—Chosen for Life, but may be removed by the Lord Chancellor for Inability or Misbehaviour—Aldermen and Councillors Disqualified for the Appointment of Borough Coroner . . . . pp. 114—125

CHAPTER XIII.

Registration of Births, Deaths, and Marriages—Registrars not necessarily Medical Practitioners—In the Case of a Death, the Persons upon whom the Duty devolves of giving the requisite Information to Registrars—In Case of Inquests the Coroner shall inform Registrar of the Finding of the Jury as to the Cause of Death—Suggestions as to the Duties of Officers of Public Institutions, and especially Medical Officers . . . . pp. 126—129

CHAPTER XIV.

Poor Law Legislation as affecting Medical Practitioners.

SECTION I.—Paupers in Workhouses or Poor Law Districts—Medical Officers of such Districts or Workhouses cannot, as a Matter of Right, claim the Appointment of Public Vaccinators—Salary of Medical Officers for a Poor Law District or Workhouse is based upon a different Contract from that of a Public Vaccinator—Remedies for the Recovery of such Salaries respectively—Up to what Date is the Salary payable—In Case of Suspension by Board of Guardians, and final Dismissal by Poor Law Board—Penal Liability of Medical Officers rendering Relief or Assistance to Facilitate the Conveyance of a Poor Person out of one Parish with the Intent to burden another Parish with the Chargeability of such Person, and in Consequence of which such Person does actually become so chargeable—Case cited—Vaccination by Public Vaccinator not to be considered Parochial Relief . . . . pp. 130—132

SECTION II.—Pauper Lunatics, including Criminals and Prisoners for Debt—Medical Certificate of Medical Practitioner as distinguished from that of the Medical Officer—Amendment of incorrect Certificate within what Time, and with what Sanction—Pauper Lunatics not in an Asylum, &c. to be personally visited every Quarter by the Medical Officer—his Report—Object of such Visitation—Special Remuneration to the Medical Officer in such Cases—Pauper Lunatics wandering Abroad away from Control, how to be dealt with—Duties of Medical Officer in advising Leave of Absence of any Patient, and in the Case of the Death of a Pauper Lunatic—Notice and Statement of the Death and other Particulars signed by Medical Officer, and transmitted to whom—Legal Proceedings for furnishing a defective Certificate—Action for Assault and False Imprisonment brought by an alleged Lunatic against a Medical Officer, among others, for confining him as a dangerous Lunatic and in a State of Destitution—Vagrant Acts relating to the Detention of dangerous Lunatics—The Formation of the present System of providing Public Lunatic Asylums at the Public Expense, when such a System was first made compulsory upon Counties and Boroughs, and under what Provisions—Letter of Commissioners of Lunacy to the Lord Chancellor as to the Authority of preliminary Orders and Certificates, whether a Person of unsound Mind be pronounced dangerous or not—The finding of a Jury in no Case essential to justify the Confinement of such Person—Power of restraining at Common Law—Power under the Lunacy Act—Acts relating to Prisoners in Convict Prisons becoming Insane, and to Criminal Lunatics—Visits of two or more Commissioners in Lunacy, one of whom shall be a Physician or Surgeon, to Asylums for Criminal Lunatics, and their Report to Secretary of State—Statement to be filled up and transmitted to the Medical Superintendent with every Criminal Lunatic—Acts relating to Prisoners for Debt becoming Insane—Form of Medical Certificate in such Cases—Act passed in Consequence of the Proceedings in the Townley Case, in the Case of a Prisoner under Sentence of Death, where there is reason to believe that he is Insane—Form of Certificate — Act relating to Prisoners in the Navy who shall become Insane . . . . pp. 133—142

CHAPTER XV.

SECTION I.—Chemists and Druggists—The Pharmacy Acts—Pharmaceutical Chemists, Assistants, Apprentices, and Students—Examination—Definition of Chemists and Druggists under the Amended Act (1868)—Qualifications for Registration—Penalty for contravening the Provisions of the Act—Not to affect any registered legally qualified Medical Practitioner—No Member of the Medical Profession is to be registered under this Act—Registration under

CHAPTER XV.—*continued.*

Pharmacy Acts not to entitle any one to practise Medicine or Surgery—Poisons and the Pharmacopœia—Restrictions on the Sale of Poisons—Penalty for any Infringement—Exceptions both as to Articles and Persons—The making or dealing in Patent Medicines—Medical Council authorized to publish a Book, to be called "British Pharmacopœia," as often as they deem necessary—"General Council of Medical Education and Registration of the United Kingdom" created a Body Corporate, with the exclusive Right of publishing, printing, and selling such Pharmacopœia . . . . pp. 143—146

SECTION 11.—Dentists—Persons desirous of being examined may be so examined by the College of Surgeons of England for the Purpose of testing their Fitness to practise as Dentists, and may be granted Certificates accordingly . . p. 146

CHAPTER XVI.

Medical Ethics—Conventional Rules—Revised Rules of Professional Conduct, whether in Hospital, Private, or General Practice—Difference between the Position of the Medical Man in Private and in Public Practice—Difficult Position, especially of the latter—Conjoint Medical Examination Board for each of the three Divisions of the United Kingdom—One uniform Qualification for Practice—Arguments *pro* and *con.*—Is it desirable to abolish all Distinctions in the Departments of Medical and Surgical Science ? . . . . . . pp. 147—152

SUPPLEMENTARY CHAPTER TO THE BODY OF THE WORK.

Dissertation on the Title of Doctor of Medicine, or Doctor of Physic—Draft of an Act of Parliament of 9 Henry V.—Case of Dr. Bonham (6 Jac. I.)—Remarks of Lord Chief Justice Coke on the relative Positions of the "Doctors of Physic of the College in London" and the "Doctors of the Universities"—"Medicus," a more ancient Designation than that of either Doctor or Bachelor of Medicine
pp. 153—155

Appendix . . . . pp. 156—416

TABLE OF CASES CITED BY NAME.

	PAGE
Abernethy v. Hutchinson	53, 57, 58
Addison v. The Mayor of Preston	131
Allen v. Worthy	92
Allison v. Haydon	23, 30, 34
Alphonso, Dr.	22
Amor v. Fearon	75
Anderdon v. Burrows, M.D.	112
Apothecaries' Company v. Latinga	34
Archer, Dr.	19
Re Ashmore	106
Attorney-General v. College of Physicians	16, 29
Attorney-General v. Pearson	134
Austen v. Colpepper	46, 47
Aveneys v. Mudie	61
Ayre v. Craven	47
Barry v. Batlin	104
Bannatyne v. Bannatyne	104
Battersby v. Lawrence	34
Beaver v. Hickes	42
Beverley's case	104
Blogg v. Pinkers	30
Bohn v. Bogue	59
Bond v. Pittard	106
Bonham's case, Dr.	21, 153
Re Bosanquet	106
Branwell v. Halcomb	59
Bromage v. Prosser	50
Brown v. Kennedy	45
Brown v. Smith	44
Browning v. Budd	104
Cartwright v. Cartwright	104
Cassell v. Stiff	60, 61
Cazenove v. British Equitable Assurance Company	80
Chandelor v. Lopus	79
Ex parte Chuck	69
Re Clark	72
Clay v. Roberts	47
Cockcroft v. Rawles	104
Collins v. Carnegie	16, 45
College of Physicians v. Rose	25
In re John Parsons Cook	122
Cooke v. Cholmondeley	104
Courtenay v. Wagstaff	70
Cox v. Midland Counties Railway Company	30

	PAGE
Creagh v. Blood	105
Cuckson v. Stones	75
Re Davies	107
Davis v. Curtis	105
Dawkins v. Paulet	47
Dayley v. Roberts	47
Dees v. London, Brighton, and South Coast Railway Company	31
De la Rosa v. Pietro	30
Dixon v. Smith	51
Dry v. Boswell	68
Du Bost v. Beresford	46
Duckworth v. Johnson	39
Duffit v. James	31
Duplex v. Economic Life Office	80
	Appendix, 158
Dyce-Sombre v. Troup	105
Edmondson v. Davis	73
Elliott v. Semple	134, 135
Ellis v. Kelly	37
Everett v. Desborough	81
Filcher v. Stafford	92
Filleal v. Armstrong	76
Flower's case	46
Folkes v. Chad	98
Forrester v. Waller	54
Foulds v. Jackson	106
Fromont v. Coupland	69
Gale v. Leckie	70
Gardiner v. Childs	70
Gaze v. Gaze	106
Geddes v. Wallace	70
General Provincial Life Assurance Company, ex parte Daintree	81
Gibbon v. Budd	29
Gilpin v. Elderby	69
Gore v. Sir Geo. Grey and others	137
Gray v. Russell	59
Greenham v. Gray	68
Greenholt, Dr.	22, 38
Haffield v. Mackenzie	30
Hall v. Semple	110
Hall v. Warren	104
Hanley v. Hewson	31
Harrington v. Churchward	69
In re Harrison	15, 34
In re Haynes	66

TABLE OF CASES CITED BY NAME.

	PAGE
Hesketh v. Blanchard	75
Hill v. Featherstonhaugh	31
Hooper v. Economic Life Office	79
Hopkinson v. Smith	72
Huckman v. Fernie	80
Hunter v. Sharpe	48
Hutton v. the Waterloo Life A.C.	80
In re Jackson	72
James v. Moody	104
Jefferies v. Duncombe	46
Keigwin v. Keigwin	106
Kemble v. Church	104
Kenner v. McMullen	31
Keyse v. Burge	68
Lawrence v. Smith	55
Letch's case, Dr.	20, 22
London and South Eastern Railway Company	39
Lovegrove v. Nelson	69
McPherson v. Daniels	50
Maitland v. Goldney	44
Major v. Knight	104
Medical Officer of Tadcaster Union	132
Metzner v. Bolton	73
Millar v. Taylor	53, 55
Moor v. King	105
Morgan v. Hallen	31
Morrison v. Musprat	80
Mortimer v. Prowett	73
Mountain v. Bennett	104
Murray v. Bogue	58
In re Palmer	72
Parker v. Ibbetson	73
Parker v. Parker	107
Pasley v. Freeman	79
Peacock v. Peacock	74
Pedgrift v. Barrington Chevalier	71
Philbrick v. Whetham	79
	Appendix, 156
In re Philpot	106
Prince Albert v. Strange	54, 55
Proud v. Mayall	23, 34
Rawlings v. Bell	79
Rawlinson v. Clark	68
Reede v. Bentley	70
Reid v. Holinshead	69
Regina v. Dawson and another	111
,, v. Palmer	122
,, v. Pinder	96
Rex v. Ball	100
,, v. Burdett	50
,, v. Carlile	50
,, v. Crutchley	117, 118
,, v. Cundick	66
,, v. Dodd	50
,, v. Enoch	118

	PAGE
Rex v. Ferguson	41
,, v. Freeman	121
,, v. Hart	49
,, v. Harvey	50
,, v. Hunt and another	49
,, v. Langley	44
,, v. Long	40, 44
,, v. Philips	44
,, v. Poulton	117
,, v. Reeves	118
,, v. Searle	98
,, v. Sellis	117
,, v. Senior	40
,, v. Simpson	41
,, v. Spiller	41
,, v. Spilling	42
,, v. Tessymond	42
,, v. Trilloe	113
,, v. Van Butchell	40
,, v. Webb	42
,, v. Williamson	40
,, v. Wright	98
,, v. Wright	118
Scott v. Miller	72
Scott v. Wakem	111
Seare v. Prentice	38
Shrewsbury v. Blount	79
Simpson v. Relfe	34
Slater v. Baker	33
Smart v. West Ham Union	131
Southee v. Denny	46, 47
In re C. Stanger, M.D.	18, 19, 22
Sweet v. Benning	58
Sym v. Fraser and another	112
Taylor v. Ashton	81
Tench v. Roberts	72
Thorley v. Lord Kerry	50
Tinsley v. Lacy	59
Todd v. Kerrick	73
In re Townley	95, 140
Turner v. Reynall	60, 74
Tutty v. Alewyn	46
Venables v. Wood	70
Waring v. Waring	105
Watson v. Poulson	79
Webb v. Rose	54
Watton v. Hardisty	78
Williams v. Jones	72, 76
Willis v. Lowe	106
Wilson v. Whitehead	70
In re W. F. Windham	102, 103
Wise v. Wilson	74
Wright v. Greenwood	30
Wright v. Woodgate	50
Young v. Genger	31

STATUTES REFERRED TO.

	PAGE
3 Edw. I., c. 10.	114
14 Edw. III., c. 1, s. 8	114
28 Edw. III., c. 6	114
3 Henry VIII., c. 11	7, 9, 16, 18
5 Henry VIII., c. 6	9
14 & 15 Henry VIII., c. 5	7, 15, 16, 18, 154, 155
32 Henry VIII., c. 42, ss. 2, 3	15, 18, 34, 146
1 Jac. I., c. 72	66
2 Jac. I., c. 31	83
28 Car. II., c. 3, ss. 4, 17	76, 77
8 Anne, c. 19	56
9 Anne, c. 2	83
9 Geo. II., c. 5	66
18 Geo. II., c. 15	23, 24
31 Geo. III., c. 34	25
39 & 40 Geo. III., c. 94	141
43 Geo. III., c. 96	135
55 Geo. III., c. 194	25
6 Geo. IV., c. 78	83
9 Geo. IV., c. 40	135
2 & 3 Wm. IV., c. 75	63, 67
3 & 4 Wm. IV., c. 103, ss. 11, 12	86, 88
5 & 6 Wm. IV., c. 65, ss. 4, 5	56, 57
5 & 6 Wm. IV., c. 76	125
6 & 7 Wm. IV., c. 86	126
6 & 7 Wm. IV., c. 89, ss. 1, 3, 5, 6	36, 115, 116
1 Vict., c. 26, s. 9	105, 106, 107
1 Vict., c. 68, s. 2	116
1 & 2 Vict., c. 14, ss. 2, 3	95
1 & 2 Vict., c. 49	128
1 & 2 Vict., c. 59	59
3 & 4 Vict., c. 29	90
3 & 4 Vict., c. 54	94, 140, 141
4 & 5 Vict., c. 32, s. 2	90, 132
5 & 6 Vict, c. 22	137
5 & 6 Vict., c. 29	137, 138
5 & 6 Vict., c. 45, ss. 17, 23	56, 61
6 & 7 Vict., c. 26	138
6 & 7 Vict., c. 96, ss. 2, 6	49, 52
7 & 8 Vict., c. 12, ss. 2, 8, 11	59, 60, 61
7 & 8 Vict., c. 15, ss. 8, 10, 11, 12, 13, 22, 53	86, 87, 88
8 & 9 Vict., c. 29	88
8 & 9 Vict., c. 126	136
8 & 9 Vict., c. 100, ss. 3, 23, 43, 50, 54, 55, 56, 57, 59, 60	107, 108, 110, 111, 112
9 & 10 Vict., c. 66, s. 6	131
9 & 10 Vict., c. 93, ss. 1, 2	88
10 & 11 Vict., c. 29	88
13 & 14 Vict., c. 54	88
15 & 16 Vict., c. 12, ss. 1, 9, 18	60, 61
15 & 16 Vict., c. 24	106
15 & 16 Vict., c. 56	143, 144
16 & 17 Vict., c. 70	101, 103, 108
16 & 17 Vict., c. 96, ss. 7, 10, 11, 12, 13, 19, 25	97, 110, 112, 136

STATUTES REFERRED TO.

	PAGE
16 & 17 Vict., c. 97, ss. 7, 66, 67, 68, 69, 70, 73, 75, 76, 79, 87, 122	95, 96, 97, 112
16 & 17 Vict., c. 100	90, 92
16 & 17 Vict., c. 104	88
17 & 18 Vict., c. 114	26
19 & 20 Vict., c. 108, s. 23	51
21 & 22 Vict., c. 90, ss. 17, 26, 30, 31, 32, 34, 35, 36, 37, 40, 47, 48, 54, 55	15, 17, 23, 27, 29, 30, 31, 35, 36, 70, 85, 146
22 Vict., c. 21, s. 6	27
23 Vict., c. 7, s. 45	16, 31, 35
23 & 24 Vict., c. 66	16
23 & 24 Vict., c. 75, ss. 14, 15	138
23 & 24 Vict., c. 77	84
23 & 24 Vict., c. 86, ss. 2, 12	102, 103
23 & 24 Vict., c. 116, s. 6	125
24 & 25 Vict., c. 134	139
25 & 26 Vict., c. 91, s. 2	62, 146
25 & 26 Vict., c. 111, ss. 21, 22, 24, 27, 47	96, 97, 133
27 & 28 Vict., c. 29, ss. 2, 3	140, 141
27 & 28 Vict., c. 48, s. 6	88
27 & 28 Vict., c. 119, s. 75	141, 142
29 & 30 Vict., c. 35, s. 15	88
29 & 30 Vict., c. 90, ss. 15, 22, 28, 37	84, 85, 116
30 & 31 Vict., c. 84, ss. 18, 31, 32	90, 92
30 & 31 Vict., c. 142, ss. 5, 10	51
30 & 31 Vict., c. 146	86
31 & 32 Vict., c. 115	85
31 & 32 Vict., c. 116, s. 1	69
31 & 32 Vict., c. 121, ss. 3, 18	143, 144
32 & 33 Vict., c. 96, s. 7	89, 90
33 Vict., c. 117, ss. 1—15	143, 144

STATUTES PRINTED *IN EXTENSO* IN THE APPENDIX.

6 & 7 Wm. IV., c. 89, Medical Witnesses at Coroners' Inquests	166
7 Wm. IV., and 1 Vict., c. 68, Coroners' Inquests	168
8 & 9 Vict., c. 100, Lunacy	170
16 & 17 Vict., c. 96, ,,	211
16 & 17 Vict., c. 97, ,,	224
18 & 19 Vict., c. 105, ,,	281
19 & 20 Vict., c. 87, ,,	280
25 & 26 Vict., c. 86, ,,	286
25 & 26 Vict., c. 111, ,,	292
15 & 16 Vict., c. 56, Pharmacy, Pharmaceutical Chemists Acts	305
31 & 32 Vict., c. 121, ,, ,, ,,	309
16 & 17 Vict., c. 100, Vaccination Acts	317
30 & 31 Vict., c. 84, ,,	321
29 & 30 Vict., c. 35, Contagious Diseases Acts	330
32 & 33 Vict., c. 96, ,, ,, ,,	345
18 & 19 Vict., c. 116, Public Health Acts	355
18 & 19 Vict., c. 121, ,, ,,	357
23 & 24 Vict., c. 77, ,, ,,	376
29 & 30 Vict., c. 90, ,, ,,	380
21 & 22 Vict., c. 90, Medical Acts	397
22 Vict., c. 21, ,,	408
23 Vict., c. 7, ,,	409
23 & 24 Vict., c. 66, ,,	411
31 & 32 Vict., c. 29, ,,	413

INTRODUCTION.

This work is based for the most part on the Medical Acts (21 & 22 Vict., c. 90; 22 Vict., c. 21; 23 Vict., c. 7; 23 & 24 Vict., c. 66; and 31 & 32 Vict., c. 29), but many other Acts of Parliament, such as those relating to Anatomy, Copyright, the Public Health, Lunacy and Insanity, Vaccination, the Poor Laws, Medical Witnesses, and other subjects incidental to the guidance of the medical practitioner in the discharge of his duties, both public and private, are referred to, abstracted, or epitomized, according to their relative importance or bearing upon the main object of this treatise. That object is to furnish the medical practitioner with a compendious and ready means of apprehending the nature of his duties and responsibilities in relation to the laws of the land, and of referring to the authorities which have decided disputed cases, explained and expounded apparently inconsistent or ambiguous enactments, and have thus pioneered the way, possibly for further improvements, in the organization of the profession, or in the protection, remuneration, and security legally furnished to the individual practitioner. All the Acts of practical utility will be found in the Appendix, a list of which is prefixed. The substance of most of the cases cited is given in an abstract form, as the reports containing them might not be accessible to the general reader. At the same time a table of cases is also added, and in the remarks and conclusions made or arrived at by the author, he trusts to be acquitted of presumption, as they are principally founded upon observations and opinions expressed by those who are entitled to the utmost deference and respect. In perusing this work it will be necessary for the general reader to bear in mind that the term "unqualified" person, medical man, or practitioner, is used as distinct from "unregistered," as, although the Medical Act of 1858, in defining the term "legally qualified medical practitioner," or "duly qualified medical practitioner," or "any words importing a person recognised by law as a medical practitioner or member of the medical profession," construes such to mean "a person registered under that Act," such construction only applies "when" such words are "*used in any Act of Parliament*" (s. 34). A person may be "qualified," though not in the above sense "legally" or "duly" qualified as a medical practitioner. He may labour under many disabilities through want of registration, as will be seen in the course of this work, but he cannot, in such case, be proceeded against for penalties as if he had no qualification whatever, much less can he be said to be "wilfully and falsely pretending to be, or taking or using the name or title of a physician," &c., "implying that he is registered under the Act," within the meaning of the penal section (40).

For instance, an unregistered Fellow, Member, or Licentiate of the College of Physicians could not be proceeded against by that or any other body for practising physic, so long as he did so within the terms prescribed by his licence or the reasonable bye-laws of the College; and the same remark applies to Fellows and Members of the College of Surgeons practising surgery though not registered, and also to the Licentiates of the Apothecaries' Society practising physic under their licences, and indeed to all practising under the authority of any licensing corporation and not infringing the privileges of others, although not "registered." In a word, an "unqualified" person is a person not possessing any qualification for registration, whereas a person may possess every professional qualification for registration, but yet for want of such registration he would not be a "legally qualified medical practitioner" in the sense in which such words "are used in any Act of Parliament," unless he were also "registered."

In the list of qualifications, mentioned in Schedule (A) of the Act, there is no reference made to "Members" of any college of physicians, but that omission, which is simply explained by the fact that at the time of the passing of the Act, "Members" had not then been created as distinct from "Licentiates," is supplied by the 22 Vict., c. 21, s. 4, which enacts that the term "Member" shall be added after the term "Fellow" to the qualifications described in the first and second heads of Schedule A. Those columns, with that addition, therefore, now read thus:—

1. Fellow, Member, Licentiate, or Extra-Licentiate of the Royal College of Physicians of London.

2. Fellow, Member, or Licentiate of the Royal College of Physicians of Edinburgh.

The word or term "Member" is directed to be in addition to, not in substitution of, any other qualification described in those two heads. And by the 23 Vict., c. 7, s. 1, Licentiates in Surgery of any university in Ireland, legally authorized to grant such licences, are put upon the same footing as respects registration as is prescribed in the Schedule of the first Medical Act in respect of the registration of any Master in Surgery of any university of the United Kingdom. In referring to that Schedule the reader will be pleased to remember these "additions." A further Act of the 23 & 24 Vict., c. 66, provides that in any new charter which shall be granted to the Royal College of Physicians of London or of Edinburgh, or to the King and Queen's College of Physicians in Ireland, nothing shall in any way affect the rights, powers, authorities, qualifications, liberties, exemptions, immunities, duties, and obligations granted, conferred, or imposed upon such corporations respectively, but so much of the Act of the 14 & 15 Henry VIII., c. 5, as relates to the Elects of the Royal College of Physicians of London, and their powers and functions, is declared to be repealed, and their office and name from the date of that Act (1860) are wholly to cease and determine, it being recited in the preamble that "the main function of the Elects—viz., that of examining and granting letters testimonial in accordance with the provisions" of the Act of Henry VIII., forbidding any person "to practise in physic through England until he be examined by the President and three of the said Elects, and have from them letters testimonial," had "been virtually superseded by the Medical Act, and that the Elects have ceased to grant letters testimonial in accordance with the provisions contained in the statute of Henry VIII., and it is therefore expedient that these provisions should be repealed." It is of course provided that such repeal shall not affect the rights of those already possessing such testimonials, and all trusts to be executed by the Elects are thenceforth to vest in and be executed by the Censors of the College. As also by the Medical Act (1858), if the Royal College of Surgeons of Edinburgh, and the Faculty of Physicians and Surgeons of Glasgow, agree to amalgamate so as to form one united corporation under the name of the " Royal College of Surgeons of Scotland," her Majesty is empowered to grant them such new charter or charters as may be necessary for effecting that union, such changes might still further create confusion, except they were borne in mind when a reference is made to Schedule A of the original Act. With these few explanatory remarks, and remembering that the first three Acts are directed to " be construed together as one Act," and that the last Act on this subject relates entirely to the Colonies, the term " colony," however, not to include the *Channel Islands and the Isle of Man*, it is sufficient to say that the Medical Act of 1858 will in this work be spoken of generally as *The Medical Act*, and sometimes as *The Medical Act* (1858), *or The Act*, and the others, when specially referred to, will be particularized according to their respective descriptions.

H. W.

5, KING'S BENCH WALK, TEMPLE.

THE LAWS

RELATING TO

THE MEDICAL PROFESSION.

CHAPTER I.

ANCIENT ORDERS OF THE MEDICAL PROFESSION.

THE antiquity of the art of healing, as embodied in the science of medicine, and reduced to principles which have ever since, more or less, regulated its practice, can at least be traced to the time of Hippocrates, the second of that name, who was born about the first year of the eighteenth Olympiad—that is, B.C. 160, or 2330 years ago. He was, in many respects, the most celebrated physician of ancient or modern times, and, probably from a belief that no mere mortal could possess the powers and skill which he undoubtedly evinced, a divine origin was attributed to him, and according to his genealogy given by John Tzetzes, with a minute particularity worthy of Sir Bernard Burke himself, he was seventeenth (though, according to Soranus, nineteenth) in descent from Æsculapius, and through his mother, whose name was Phaernarete, was even said to be a descendant of Hercules. Without vouching for the truthfulness of this pedigree, the Divine art may well be supposed to have, at least in some sense, a Divine origin.

Hippocrates was born in the isle of Cos, where his birthday was celebrated with sacrifices on the 26th day of the month of Agrianus. He was instructed in medical science by his father, Heracleides, and by Herodicus, and was also said to have been a pupil of Gorgias of Leontini, and after practising his profession at home died at Larissa, in Thessaly, about the year B.C. 357, in which case he must have been 103 years of age. His ancestors had practised medicine for probably a century previous, and his two sons Thessalus and Dracon, and a son-in-law, Polybus, followed the same profession, and are supposed to have been the authors of some of the works in the Hippocratic collection. These facts tend to explain the ponderous and voluminous collection of writings under the above title, of which it may be safely asserted that it is quite impossible any one man could have written all thus collected under his name. It appears from different parts of this collection that by far the greater proportion of the several hundred substances used medicinally by him belong to the vegetable kingdom. His personal character seems to have been that of a pattern physician, entertaining, as he evidently did, a keen sense of the moral responsibilities and obligations of his profession, and throughout his writings his endeavour is to impress upon his readers the duties of care, attention, and kindness towards the sick; and some of his moral reflections and apophthegms have acquired quite a proverbial notoriety (as for example "Life is short, and art is long"). This is not the place, however, for a biographical sketch even of so celebrated a character; and even an allusion to his vegetable pharmacopœia is only made to remind the reader that, in an age when chemical science had not as yet emerged from its very germ, the simple process of this great man consisted chiefly in watching the operations of nature, and in more accurately discovering the properties of the vegetable kingdom with a view to applying them to the alleviation of human maladies. Not, however, confining himself to this knowledge, in which the most savage state of society can to some extent participate, in surgery he is the author of the frequently-quoted maxim that "What cannot be cured by medicine is cured by the knife; and what cannot be cured by the knife is cured by fire."

Galen, who was born about A.D. 130, ranks next among the medical sages of antiquity, and though his admirers have not gone the length with those of Hippocrates of assigning to him a Divine genealogy, he himself is often devoutly lavish in the praises of his father Nicon, not only for his knowledge of astronomy, grammar, arithmetic, and various other branches of philosophy, but also for his patience, justice, benevolence, and virtue. But even Galen was not above claiming special revelations from the gods. He on one occasion pleaded to the emperor, M. Aurelius, that it was the will of Æsculapius that he should be left at Rome instead of following the fortunes of the imperial warfare on the Danube, and whether he believed or invented the idea of a vision from the god on this occasion, it is certain that he more than once mentions his receiving what he conceived to be Divine communications during sleep, in cases where no personal object can be discovered.

Rome was henceforth the great theatre of Galen's success, and afforded, under imperial patronage, an advantageous field for his great learning and abilities. Here he had opportunities of witnessing the development of Christian practice, if not of doctrine, in the early period of the history of that faith, and in one of his works he speaks of the Christians in the highest terms, praising their temperance and chastity, their blameless lives and love of virtue, in which, as he says, they equalled or surpassed the philosophers of the age.

The works of Galen (Kühn's edition) extend to twenty-one volumes, and embrace the subjects of Anatomy and Physiology, Dietetics and Hygiene, Pathology, Diagnosis and Semeiology, Pharmacy, *Materia Medica*, Therapeutics—including Surgery, Commentaries on Hippocrates, and various miscellaneous and philosophical works, so that it would appear medical science was not even at that now remote period, based upon mere superstition, nor yet is it to any considerable extent found appealing, for lack of human knowledge, to supernatural power, charms, incantations, sorcery, or witchcraft. To the compilers, both among the Greeks and the Romans, of medical works, like Aëtius and Oribasius, Galen's works formed the basis; while, as soon as they had been translated into Arabic, in the ninth century, chiefly by Honain Ben Ishak, they were at once adopted throughout the East as the standard of medical perfection. Opinions differ as to the value afforded by the Arabic writers in explanation and illustration of Galen's works. These Arabic writings, though extant in various European libraries, have never yet been published; although an immense number of European writers have from time to time edited, translated, or illustrated Galen's works, but, unfortunately, no critical edition of Galen's works in the original has been undertaken, the existing commentaries and abridgments having been made from Latin translations of the Arabic, itself a translation from the Greek, abounding with interpolations undistinguished from the original text.

Thus, to a great extent, the writings of Galen became obscured, and in a measure vitiated, by new commentaries, which were, for the most part, mere sophistical speculations founded on Aristotelian logic.

These reminiscences should teach us to deal leniently with the failings of the ancients, remembering that in science, art, and literature, much that was valuable has been lost or obscured, and that Greece, Rome, Egypt, Arabia, and Assyria, of old, were perhaps more advanced in civilization, which goes hand in hand with science in all its branches, than can now be strictly proved, or than we are in candour prepared to admit.

Without entering upon the controversy relating to the introduction of Christianity into Britain, it may with safety be affirmed that its organization into a regular system in this country by Pope Gregory the Great, brought into England many of the Roman writings, and among others the valuable medical works of Galen. For these the monastic libraries at once afforded a repository, and at the same time thus furnished food to the monastic craving for learning, which, whether sacred or profane, found favour in these institutions; and it would, indeed, be a monstrous moral contradiction that the monastery that produced a Bede should have been indifferent to the cultivation of a science which from the time of the founder of Christianity had gone hand in hand with its sublime code of humanity, in healing the sick, cleansing the leper, restoring sight to the blind, making the lame to walk, and even in the case of recovery from suspended animation, to all appearance raising the dead. On the institution of the Universities, medicine necessarily became an object of considerable importance.

In the middle of the seventh century was founded the medical school at Salerno, an institution soon followed by similar seminaries throughout Europe, and the degree of "Medicus" seems to have been known at Oxford soon after the Conquest (Exon. Domesday Book, 5), and in the fourteenth century the degree of doctor of physic was by no means uncommon. (See Wood's Hist. of Oxford, vol. ii. p. 18. Gutch, vol. ii. p. 765). The monks, and even the secular clergy, were necessarily the only regular physicians and surgeons, as none but the clergy paid the slightest attention to learning—or rather, it might be said, that almost all the learned of that day went into holy orders, and the Latin being the language of the Catholic Church, was assiduously cultivated, so that all books, even on philosophy and science, were written in it.

The doctors of that day, as has been just now remarked, were for the most part monks, who seem to have practised physic in all its branches, both prescribing and administering medicine, and performing the operative parts of surgery. Various canons, such as those of the Council of Rheims (A.D. 1131) and the Council of Lateran (A.D. 1139) were passed to regulate this practice, until at length at the Council of Tours (A.D. 1163) it was decreed that none of the *regular* clergy should leave their monastery for the purpose of devoting themselves to this practice. The words of the canon on that point are as follows:—" Inde nimirum est quod se in angelum lucis more solito transfigurans, sub obtentu languentium fratrum consulendi corporibus et ecclesiastica negotia fidelius pertractandi, *regulares* quosdam ad legendas leges et *confectiones physicales* ponderandas de claustris suis educit . . . . statuimus ut nullus omnino post votum religionis, post factum in aliquo religioso loco professionem, ad physicam legesve mundanas legendas permittatur exire." This canon is couched in terms of great respect towards the persons against whom it was aimed, for at the same time that they are forbidden to undertake duties incompatible with their spiritual character as wholly devoted to certain religious occupations, they are complimented as being the most worthy members of their profession, and actuated by the best feelings of human nature—viz., a desire to alleviate the sufferings of their brethren. The canon, however, was only directed against the *regular* or monastic clergy, and not against the secular clergy, who were not bound by any special vows of obligation to monastic rule. From this time the monks confined themselves to the prescribing of medicines, to be compounded and administered by others, and wholly abstained from the manual operations of surgery; but the secular clergy continued to practise physic in all its branches as before, without any interference. It appears that about this time Richard Fitz-Nigel, who died Bishop of London, had been apothecary to Henry II. (Ang. Sacra, t. i. p. 304.)

But even in the present day there are two members of the Episcopacy—viz., Dr. Bickersteth, Bishop of Ripon, and Dr. McDougall, retired Bishop of Labuan, who, having been originally members of the medical profession, have, since their elevation to the Episcopal Bench, been elected, and still remain, Fellows of the College of Surgeons of England. Before quitting the subject of clerical medical practitioners it is perhaps only due to the memory of one of the last clerical-physicians of the present day to say that the University of Cambridge has but recently lost one of its most distinguished members in the person of the Rev. W. Clark, M.D., F.R.S. and F.R.C.P., who in the year 1808 was seventh wrangler, and afterwards became Professor of Anatomy at the University, and indeed to whom the University is indebted for the greater part of its present anatomical collection, and almost for the formation, and certainly for the arrangement, of the anatomical museum. Dr. Clark, who was elected a fellow of Trinity College, was in the year 1826 presented by that College to the rectory of Guiseley, near Leeds, which preferment he held for thirty-three years, and during the whole of that time he also occupied the chair of Anatomy at Cambridge, having held the latter appointment for a period of no less than forty-nine years, viz., from the year 1817 to 1866. It is believed that at one time he also combined the private practice of a physician, but this was probably before he obtained preferment in the Church. The case is somewhat curious, both as evidencing the legal right even in the present day to pursue at the same time the two professions, and as illustrative of the varied qualifications and attainments that may be centred in one individual, as we certainly have here an instance of mathematical, anatomical, medical, and general scientific acquirements, united in the same person, and subsidiary perhaps, only to theological and divinity scholarship. Nor need such an one be one whit the less an efficient clergyman because he has profoundly studied human and comparative anatomy, to say nothing of physiology, the structure, disposition, functions, and development of the organs of the human body, the masterpiece of the works of the Creator.

The effect of the canons on the subject of physic, however, was to create a distinction of orders, and hence arose a division among the duties of medical practitioners. The department of surgery fell by degrees into the hands of barbers and smiths, who had hitherto been employed as assistants in surgical operations, and from this time there grew up three regular, and unfortunately many irregular, orders of the medical profession, which may be distributed thus:— First, the scholars; secondly, the surgeons of all sorts; thirdly, the grocers or poticaries; fourthly, the empirics; fifthly, the alchymists; sixthly, the sorcerers; and lastly, the witches or herbalists; to whom may be added the astrologers, who, however, rather foretold the occurrence of ailments and distempers than attempted to avert them. The physicians were almost uniformly of the order of Galen, compounding their own prescriptions, and, at all events, affecting a great show of learning, and being often men of ecclesiastical rank and wealth, obtained for their branch of the profession a considerable social status, with which the surgeons and apothecaries could never compete.

In the year 1237 were made the celebrated regulations by the College of Salerno, by which every person was required to spend three years in the study of philosophy, and five years in the study of physic, and to obtain a license after examination by two doctors, before he could enter upon practice in that faculty. And similar regulations appear to have been adopted soon after by the French and English Universities. The laity were never expressly excluded from the practice of physic, provided they obtained the necessary degree; but as most who remained long enough at college to graduate in medicine received holy orders, there are not many laymen to be found among the earlier medical practitioners, who were often beneficed clergy, and even bishops and other dignitaries among the secular clergy. The canon of the Council of Lateran, before referred to, which directs that the physician of the body shall first call in the physician of the soul, and postpone all attempts to cure the body until spiritual medicaments had been applied to the soul, gave the clerical physicians a considerable advantage over their lay competitors. The English Universities had no power to prevent any persons from practising, though they had not graduated in physic. On this account, therefore, in the 9th year of Henry V., an attempt was made by the Universities of Oxford and Cambridge to obtain legislative authority for excluding every one from the practice of physic who had not taken the degree of Bachelor of

Medicine, under the penalty of 40*l.* and imprisonment, *whether man or woman*. This measure, however, never really obtained the force of an Act of Parliament; but in the 3rd year of Henry VIII. an Act was passed which is generally received as the foundation of all statute law on the subject. Under this, in order to banish all supernatural obstacles, the bishop of the diocese proceeded to exorcise the fiend, and in order to provide for sufficiency of learning two doctors are required to examine the qualifications of the medical candidate. The third clause of the Act saves the rights and privileges of the Universities of Oxford and Cambridge. But by the 14 and 15 Hen. VIII., c. 5, reciting the Charter of 10 Hen. VIII., the powers of examination were transferred from the persons appointed for that purpose by the former Act, and confided to the College of Physicians, instituted by the charter previously granted by the king as therein recited. Under this statute the University graduates who might desire to practise in London were included; but the rights and privileges of graduates of Oxford or Cambridge to practise throughout the rest of England were reserved.

The appointment of physicians and surgeons to attend upon royalty was from a very early period made by the king, with the assent of the privy council.

Lord Coke, in his remarks upon the College of Physicians, alludes to such appointments thus:—

" Rex adversa valetudine laborans de assensu concillii assignavit Johannem Arundel, Johannem Saceby, et W. Hatcliffe, medicos: Robertum Warren et Johannem Marshall chirurgos ad libere ministrandum et exequendum in et circa personam suam: viz., quod licite valeant moderare sibi diætam suam, et quod possint ministrare potiones, syrupos, confectiones, laxitivas medicinas, clysteria, suppositoria, caput purgea, gargarismata, lealnen, epithimota, fomentationes, embrocationes, capitis rasuram, unctiones, emplastra, cerera ventos. cum scarificatione, vel sine emovodorum provocatione, &c. Dantes singulis in mandatis quod in executione præmissorum sint intendentes, &c." And he adds, " Upon this, four things are to be observed—1, that no physick ought to be given to the king without good warrant ; 2, that this warrant ought to be made by the advice of his councell ; 3, they ought to minister no other physick than that which is set down in writing ; 4, that they may use the aid of those chirurgeons named in the warrant, but of no apothecary, but to prepare and do all things themselves. &c. And the reason of all this is the precious regard had of the health and safety of the king, which is the head of the commonwealth. The science of physick containeth the knowledge of chirurgery."— Coke's Institutes, 4th Pt. 251.

It has been already remarked that the monkish physicians were at first the only regular surgeons, and it is not unlikely that they sometimes employed the barbers and farriers in the less agreeable parts of the occupation. But in the eleventh century arose two singular bodies of medical practitioners, the Knights Hospitallers, and the order of St. Lazarus. The former, in addition to their military duties, undertook the care of the sick and wounded in the field of battle, and accompanied the Christian armies in the double capacity of warriors and surgeons. The latter were lazars, or lepers, themselves, who being excluded from the community, formed a separate society, and at length becoming numerous and useful, obtained the order of knighthood, into which they admitted many who were not affected with the complaint. They afterwards formed two classes, the healthy and the diseased, of which the former, like the sons of Æsculapius, accompanied the armies to battle, but declining the privilege claimed by the heroes of the Barber-Surgeons' Company to stand therein "unharnessed and unweaponed," signalized themselves far more by their martial achievements in wounding and slaying than in the feats of anointing and healing; but in reference to the origin of the order, it was a rule that no one should attain the rank of its master, who had not been himself a leper.

The operative part of surgery being totally abandoned by the monks, in pursuance of the canon of Tours, that practice fell into the hands of the barbers and smiths, but the more regular and dignified surgeons were the gentlemen of the Barbers' Company. The familiar assistants of the ancient clerical surgeons, and provided with the recipes of their former employers, it was natural that they should assume the business relinquished by their superiors. They readily associated the knife and lancet with the razor and the scissors, and at once established as rigorous a doctrine in surgery as the scholastics had previously introduced concerning physic. The first recorded instance of the co-operation of physician and surgeon is the unfortunate case of Richard I., who seems to have been attended, and horribly maltreated by both, when wounded at the siege of Chalus.

> " Interea regem circumstant undique mixtim,
> Apponent medici fomenta, secantque chirurghi,
> Vulnus, ut inde trahant ferrum leviore periclo."
>   *Vide* Pasquier, " Recherch. de la France,"
>   lib. 9, c. 31, p. 825.

With the exception of a few surgeons about the court, the body seems to have held a very low place in the social scale, but yet they were by no means numerous.

The barbers thus practising surgery associated themselves, and formed one of the guilds or companies of London, and at length, in the first year of Edward IV., obtained a charter of incorporation, bestowing upon them certain privileges in the exercise of their mystery.

Notwithstanding this, another body of pure surgeons sprung up, who, although unauthorized, became competitors of such influence, that the barber-surgeons were compelled to acknowledge them, and admit them into one fraternity under the provisions of 32 Hen. VIII., c. 42, a previous Act, 5 Hen. VIII., c. 6, having relieved the barber-surgeons from the duties of watch, ward, quests, and other obligations. All persons were, however, still bound under the 3 Hen. VIII., c. 11, to obtain a license after due examination in order to practise as surgeons within the city of London and seven miles of the same, under a penalty of 5*l.* for every month of such practising. Under the 32 Hen. VIII., c. 42, s. 3, those surgeons using "barbery or shaving within the city of London, suburbs, or one mile circuit of the same city, were prohibited from occupying any surgery, letting of blood, or any other thing belonging to surgery, drawing of teeth only except." And it was further enacted that "whosoever useth the mystery or craft of surgery shall in nowise occupy nor exercise the feat or craft of barbery or shaving, neither by himself nor by none other for him, to his or their use."

And by sec. 4, "no person should presume to keep any shop of barbery or shaving within the city of London, except he were a freeman of the said corporation and company." By these enactments the two callings became at length severed, the above prohibitions having been enacted to prevent the spread of infectious and contagious disorders through persons so suffering resorting to the barbers' shops for the purpose of "being washed or shaven." According to Guido de Cauliaco, who wrote about the middle of the fourteenth century, the method of surgery practice was a most bloody and barbarous one, and yet it was not only the orthodox one, but subsequently received the sanction of the English Parliament, the most cruel and superstitious practices being commanded with the utmost particularity, even so late as 33 Hen. VIII., c. 12.

We read in all the authorities (the last mentioned Act included), of searing irons, of red-hot knives, for cutting and searing at the same time, of the application of a yet bleeding capon to the mutilated seared stump after amputation, and other absurd and equally torturing modes of conducting an operation, such as boring as many holes as possible in a wounded skull, under the name of trepanning.

The apothecaries, or rather the grocers, having in their latter capacity laid in stores of wine, oil, honey, pepper, lard, &c., and also few herbs, very readily superadded ointments and such other herbs as were deemed applicable for medical purposes. Some who had added to this store the prescriptions of the physicians, which they were bound to file as their vouchers, acquired a reputation for the knowledge of simples, and took upon them to doctor their customers. It was not until the introduction of chemical medicines, and a variety of mixtures of the physicians, had rendered the medical department of their trade unintelligible to ordinary grocers, that the pharmaceutics appeared as a separate class, and claimed a superiority over the dealers in cheese, butter, sugar, plums, figs, and home-made wines. The two crafts of grocers and apothecaries, like the barbers and surgeons, formed one of the ancient companies of the City of London, and though separated by the charter of 13 Jac. I. into two distinct corporations, the apothecaries even from this period were regarded as a trading company in the city; and whoever thought proper so to do was at liberty to sell physic throughout the rest of the kingdom, provided he complied with the terms imposed on all traders, in serving an apprenticeship of seven years.

It is, however, to the empirics—viz., those who were neither regular physicians, surgeons, nor yet apothecaries, but irregular practitioners, under the various names of astrologers, alchymists, sorcerers, and even witches, that alone is the faculty indebted for the slightest degree of improvement. These men, despising or ignorant of the medical theories of the schoolmen, and neglecting the dogmas of the surgeons, marked out each for himself a peculiar line and mode of practice. They thus sometimes hit upon an improved remedy or a real discovery in science. The alchymists especially numbered among them many men of real genius. Such a master mind as that of Roger Bacon, whilst concurring with others in the delusive search for the philosopher's stone, almost unconsciously surprised himself at progressive discoveries in chemistry. Some new result valuable to science, and peculiarly conducive to the advancement of medical knowledge, emanated from the vain endeavour to discover the elixir vitæ, or that all-purifying essence which was to transmute all baser ores into the pure and unalloyed residuum of the precious metals. Fortunate thus was it for science that the " auri sacra fames " and a belief in the discovery of the " tree of life " encouraged the observation of nature, and an investigation of its most secret resources, and that thus there arose, out of the study of an absurd philosophy, the discovery of the true philosopher's stone, the adaptation and application of natural forces and elements to the promotion of the happiness and the alleviation of the sufferings of the human race.

A despised race, who had learnt among their brethren scattered throughout Arabia, Greece, and Persia, the arts of healing, at an early period of our history alone possessed any valuable knowledge of medicine, yet pursued their course in secret and under the foulest suspicions. Stigmatized by various opprobrious names simply because they were Jews, their most wholesome drugs were esteemed the bane of Christians, and their most scientific operations in surgery as fatal to every follower of the Cross, and yet their practice was immeasurably superior, both scientifically and operatively, to that of the most approved schools of Christian Europe. A translation of some of the Arabian authors has since divulged the mysteries of the Jewish physicians, and their only remaining excellence henceforth consisted in the simplicity of their practice, and their disinclination to adopt desperate and dangerous remedies.

One class of practitioners alone remains to be considered, and though last, not the least, either in professional reputation as physicians, or as clinical dressers of their wounded patients.

In the pages of Tacitus we read of the competition of the German dames with the venerable Priests of the Mistletoe in the science of physic; and as these ladies were believed to be endued with Divine wisdom, and were more attentive in the choice of their medicines and to the comfort of their patients, they obtained a reputation by no means inferior to that of the regular physicians. Later, in the age of chivalry, the high-born damsel was not only not above, but she deemed it an essential part of her education, to acquire some proficiency in the healing art. Nor does her attention appear to have been confined to the knowledge of the medicinal properties of herbs. She was both physician and surgeon—skilled not only in composing soothing drinks and making unctuous ointments, but, as our ancient poets describe her, occupied in setting broken and dislocated limbs, and dressing wounds, incurred perhaps in championing her virtues, or sought as honourable tokens of her knight's devotion. The kindness and attention of such a surgeon would doubtless do much to alleviate the sufferings of any knight, and probably the simple remedies comprised in the "one touch of nature," were the best and most nutritious regimen. It remains to be seen whether modern female aspirants to medical practice can rival, not to say surpass, the high-souled damsels thus described in the Fairy Queen.

> "Where many grooms and squires ready were
> To take him from his steed full tenderly;
> And eke the fairest Alma met him there,
> With balm, and wine, and costly spicery;
> To comfort him in his infirmity:
> Eftsoones she caused him up to be conveyed,
> And of his arms despoil'd easily;
> In sumptuous bed she made him be laid,
> And all the while his wounds were dressing by him stayed."
> *Faëry Queen*, 1, 2, c. 11, st. 49.

> "So prosper'd the sweet lass, her strength alone
> Thrust deftly back the dislocated bone;
> Then cutting curious herbs, of virtue tried,
> While her white smock the needful bands supplied,
> With many a coil the limb she swathed around,
> And nature's strength returned, nor knew the former wound."
>
> *Ibid.*

It was not until the year 1422 that any attempt was made to interfere with women practising physic, by subjecting them to a fine of 40*l.*, and imprisonment, but the Bill containing this prohibition never had the effect of an Act of Parliament. (Petyt's MSS. v. 33, p. 140.) Comparatively modern times, and a profession of Faith that wholly repudiates all belief in the continued existence of miracles, at length set the seal to credulity, and capped the climax of superstition. Whatever mysterious efficacy in Druidical times may have attached to the mistletoe, when duly impregnated with the moon's propitious beam, cropped by sacred hands with a consecrated golden sickle, and reverently placed on white and holy cloth, it lay garnered in the woodland sanctuary—and whatever may at one time have been the supposed influence of the "inconstant moon" on human intellect or human affections, these, and all other superstitions, pale before the National belief which, under the name of the Service for the "Healing of the Sick," disgraced the Book of Common Prayer until a comparatively recent period, attributing to that "Divinity which doth hedge a king" a remedial power, by the mysterious imposition of whose royal hands, accompanied by the profanity of invoking the aid of sacred writ by way of incantation, the sick recovered.

The following graphic description of the practice of touching for the king's evil is thus referred to by Macaulay, in the third volume of his "History of England"—the reign of William and Mary.

"The days on which this miracle was to be wrought were fixed at sittings of the Privy Council, and were solemnly notified by the clergy in all the parish churches of the realm. See the Order in Council of Jan. 9, 1683. When the appointed time came several divines in full canonicals stood round the canopy of state. The surgeon of the royal household introduced the sick. A passage from the 16th chapter of the Gospel of St. Mark was read. When the words 'They shall lay their hands on the sick, and they shall recover,' had been pronounced, there was a pause, and one of the sick was brought up to the king. His Majesty stroked the ulcers and swellings, and hung round the patient's neck a white riband, to which was fastened a gold coin. The other sufferers were then led up in succession; and as each was touched, the chaplain repeated the incantation, 'They shall lay their hands on the sick and they shall recover.' Then came the epistle, prayers, antiphonies, and a benediction. The service may still be found in the prayer-books of the reign of Anne. Indeed it was not until some time after the accession of George I. that the University of Oxford ceased to reprint the Office of Healing together with the Liturgy. Theologians of eminent learning, ability, and virtue, gave the sanction of their authority to this mummery, and, what is stranger still, *medical men of high note believed*, or *affected to believe*, in the balsamic virtues of the royal hand. We must suppose that every surgeon who attended Charles II. was a man of high repute for skill; and more than one of the surgeons who attended Charles II. has left us a solemn profession of faith in the king's miraculous power. One of them is not ashamed to tell us that the gift was communicated by the unction administered at the coronation; that the cures were so numerous and sometimes so rapid that they could not be attributed to any natural cause; and the failures were to be ascribed to want of faith on the part of the patients; that Charles once handled a querulous quaker, and made him a healthy man and a sound churchman in a moment; that if those who had been healed lost or sold the piece of gold which had been hung round their necks, the ulcers broke forth again, and could be removed only by a second touch, and a second talisman. . . . . . The crowds which repaired to the palace on the days of healing were immense. Charles II., in the course of his reign, touched near a hundred thousand persons. . . . . . . In 1682 he performed the rite eight thousand five hundred times. In 1684 the throng was such that six or seven of the sick were trampled to death. James, in one of his progresses, touched eight hundred persons in the choir of the Cathedral of Chester. The expense of the ceremony was little less than 10,000*l*. a year, and would have been much greater but for the vigilance of the royal *surgeons*, whose business it was to examine the applicants, and to distinguish those who came for the cure from those who came for the gold." See the preface to a Treatise on Wounds, by Richard Wiseman, Sergeant Chirurgeon to His Majesty, 1676. And further information is to be found in the Clarisma Basilicon, by John Browne, Chirurgeon in ordinary to His Majesty, 1684. See also, The Ceremonies used in the time of King Henry VII. for the healing of them that be diseased with the King's Evil, published by his Majesty's command, 1686; Evelyn's Diary, March 28, 1684; and Bishop Cartwright's Diary, August 28, 29, and 30, 1687.

William III. seems to have not only discouraged, but to have denounced the royal practice. "It is a silly superstition," he exclaimed on one occasion when he heard that his palace was besieged by a crowd of the sick. " Give the poor creatures some money, and send them away." On one single occasion was he importuned into laying his hand on a patient. " God give you better health," he said, " and more sense."

Carte, in his History of England (pub. 1747), mentions a case of one Christopher Lovell whom he saw at Bristol in 1717, who had recently been touched in Paris by the King in November, 1716, from which moment his humour dispersed and his sores healed, till, in the beginning of January, he was in perfect health. Dr. Lane, *an eminent physician*, and Mr. Samuel Pye, *a very skilful surgeon*, both of Bristol, mentioned the case as one, if not miraculous, at least as one of the most wonderful events that had ever happened.

It would therefore appear that superstitious beliefs are not peculiar to what some affect to call the Dark Ages, nor yet are they confined to the remote ages of antiquity, neither are they accompanied and sanctified by religious rites solely of a Druidical, heathenish, or barbarous character.

CHAPTER II.

MODERN ORDERS OF THE MEDICAL PROFESSION; OR THOSE ENTITLED TO BE REGISTERED UNDER THE MEDICAL ACT (1858).

Up to the passing of the Medical Act (1858), 21 & 22 Vict. c. 90, three orders of the medical profession were distinctly recognised,— physicians, surgeons, and apothecaries. Chemists and druggists also had, in the courts of law, met with a legal recognition as persons who might make and vend medicines, and even compound them, according to the prescriptions of a physician, or the directions of an apothecary; and they, as well as the dentists, have since received a statutory *status* by the Medical and Pharmacy Acts, but not in any sense as medical men.

The following institutions confer the qualifications entitling persons to be registered as practising medicine, or surgery, or medicine and surgery, as the case may be :—

1. The Royal College of Physicians of London.

And here it is desirable to give a short history of this College.

The Royal College of Physicians of London, or, as it is styled under the new charter proposed to be granted in pursuance of the 47th section of the Medical Act, "The Royal College of Physicians of England," enjoyed and still enjoys, under the statute 32 Hen.VIII., c. 40, s. 3, the right of practising physic in all its branches, among which surgery is specially enumerated. The law, therefore, permits those constituting its body both to prescribe and compound medicines, and as well to perform as to superintend operations in surgery.

The exclusive power of the College is as follows :—By the 14 & 15 Hen. VIII., c. 5, reciting a royal charter of 10 Hen.VIII., and which is still in force, except as it may be inconsistent with their proposed new charter and the Medical Acts, no one is permitted to practise medicine (which includes surgery) within the London precinct—*i.e.*, within the City, and seven miles round it, without a license from the College, under a penalty of 5*l.* a month for every whole month of such practising.

This right, at first rigidly exercised, at length fell almost into abeyance. The College was unsuccessful, in the year 1829, in a suit against one Harrison, reported in 9 Barnewall & Cresswell, 524, since which no action for such unlicensed practising has been brought.

The College has no special power to interfere with practice in the country, but, as the 3 Hen. VIII., c. 11, distinctly forbids any one to practise beyond the London precinct, whether as a surgeon or physician, unless duly licensed by the authorities of that day, and as the 14 & 15 Hen.VIII., c. 5 (reciting the charter of the College), expressly forbids any one thenceforth to practise "through England until he be examined and approved at London by the President and three of the elects of the College of Physicians (except he be a graduate of Oxford or Cambridge)," any infringement, even of this enactment, though no penalty is imposed by the Act, was a misdemeanour at law, as was observed by Lord Denman in Collins v. Carnegie (1 Adolphus & Ellis, 695). His Lordship said: "The statute, indeed, imposes no penalty upon unlicensed practitioners beyond the seven miles, but the prohibitory words are strong enough to make the practice unlawful." Such, therefore, might have been the subject of an indictment. This restriction no longer exists even against unqualified practitioners; the power of the elects thus to examine and grant letters testimonial having been repealed by the 5th section of the Medical Act (1860). The College can, however, grant licenses without restricting their licentiates from compounding and supplying for gain the medicines they prescribe, and such is not an invasion of the privileges of the Society of Apothecaries: Attorney-General v. The Royal College of Physicians, 1 Johnson & Hemming's Rep. p. 561. The former division of the College was, until recently, into fellows, licentiates, and extra licentiates, but now, under a bye-law of the 8th August, 1859, licentiates who shall have been admitted licentiates before the 1st of October, 1859, and extra licentiates under the bye-laws enacted February 16th, 1859, and graduates in medicine who shall be admitted licentiates before the 1st of March, 1860, under the same bye-laws, shall, from and after October 1, 1859, be styled members. The members are alone eligible to the fellowship, but they must be of at least four years, standing, and not less than thirty years of age, and have distinguished themselves in the practice of medicine, or in medical or general science, or literature. All persons who have been either licentiates or members during a period of four years on the whole previously to October 1, 1863, shall be considered members of four years' standing. They are not eligible, however, if they practise pharmacy, or practise physic or surgery in partnership, so long as that partnership continues; nor if they make any engagement with any person for the supply of medicine from which profit is derived, nor if engaged in trade. The members are entitled to the use of the library and museum, and shall be admitted to all lectures and enjoy other privileges, but they shall not be entitled to any share in the government, nor to attend or vote at general meetings of the corporation. And any person not engaged in the practice of pharmacy, who shall have satisfied the College of his knowledge of medical and general science and literature, may be proposed to the College to receive a license to practise physic as a member, provided he is twenty-five years of age. No candidate to be admitted to examination who is engaged in trade, or practises in any of those ways previously enumerated with respect to members. Those licentiates who are not restricted from supplying medicines to their patients (under a resolution of April, 1860) are entitled to register under the Medical Act as licentiates. Such licentiates are not members of the corporation. They must not, under the bye-laws, compound or dispense medicines, except for patients under their own care, nor are they permitted to assume the title of Doctor of Medicine, or use any other name, title, designation, or distinction implying that they are graduates in medicine of an University, unless they are such graduates in fact; nor yet shall they, by virtue of the license, represent themselves as fellows or members of a College of Physicians. All classes, however, whether they be termed fellows, members, licentiates, or extra licentiates, are now, after registration, entitled to practise throughout the Queen's dominions. The following, though somewhat of a digression, may yet be found interesting, and, perhaps, instructive; and as those styled members are now acknowledged as eligible for fellowships, it is hoped, in the words of Mr. Willcock, " that the jealousy which has for a long time disturbed the harmony of this highly-respectable profession will gradually abate," and let us add, " may totally cease."

It should be remembered that—as is truly said by Mr. Willcock— " This College was instituted in a reign during which patents of monopoly and exclusive privileges were daily granted, and more frequently for the purpose of replenishing the exchequer, and enriching individuals by the profits of the monopoly, than in consideration of the public welfare." (p. 36.)

It may here be observed that the order of *candidates*, as mentioned by Mr. Willcock, is not founded in any Act of Parliament or charter, but was created by the bye-laws. They were not, therefore, as Mr. Willcock admits, " officers, or even *members* of the corporation, but mere licentiates; and the admission into this class amounted merely to a declaration of present intention, which created no kind of obligation on the body to elect them into the order of fellows." (pp. 43, 44.)

" It is evident," says Mr. Willcock, " that the charter so far incorporated all persons of the same faculty of and in London (the words of the original charter are, " omnesque homines ejusdem facultatis, de et in civitate prædictâ,") that every person on the

23rd of September, in the tenth year of the reign of Henry VIII., falling within that description, was entitled to be admitted into the association. Such of them as had availed themselves of this privilege, and others subsequently admitted, are the persons described by Act 32 Henry VIII. as commons and fellows." (s. 2.) "But as to the persons who should afterwards enjoy that distinction, the original charter, and all subsequent statutes, are silent." (Willcock's Laws Relating to the Medical Profession, p. 34.)

This opinion of Mr. Willcock's is based upon a legal decision on the point contained in re C. Stanger, M.D., 7 Term Rep. 283-295 (A.D. 1797), which was an application for a mandamus, commanding the College of Physicians to examine the complainant as to his qualification and fitness to be admitted into the said corporation, as a member or fellow thereof. The affidavit in support of the rule, referred to 3 Hen. VIII.,c. 11, and 14 & 15 Hen. VIII., c. 5. The applicant was a doctor of physic of Edinburgh, and in 1796 had obtained a license from the College to practise in London and its precinct. The affidavits on the part of the College disclosed that a bye-law was made by the College in 1637, ordaining that no person should be admitted a fellow unless he had performed all his exercises and disputations in one of our Universities without dispensation; and another was passed in 1751, which declared that the meaning of the above words was, that no person should be admitted into the class of candidates who was not a doctor of physic of Oxford or Cambridge, or having obtained that degree in the University of Dublin, has been incorporated into the University of Oxford or Cambridge; that Charles II., by a letter to the College in 1674, signified his pleasure and directed the College not to admit any person who had not had his education in either the University of Oxford or Cambridge. The question was whether the bye-law was a reasonable one, it being admitted that no notice should be taken of the letter from Charles II.

Lord Kenyon (C.J.) considered that it was, but he added—" If indeed, this (*i.e.* the taking a degree as before described) had been a *sine qua non*, and it had operated as a total exclusion of every other mode of gaining access to the College, it would have been a bad bye-law; but these bye-laws point out other modes of gaining admission into the College. If Dr. Stanger has all those requisites that qualify a person for that high station, any one of the fellows may now propose him: he may apply to the honourable feeling of the College, to the very same tribunal to which this mandamus, if it were granted, would refer him: for at all events he must submit to their examination and determination." Such being the opinion also of the other members of the Court, the rule for a mandamus was discharged.

In 1769, a similar application had been made to the Court of King's Bench by Dr. Edward Archer, to whom had been granted in 1752, after the proper examination, a diploma to practise in London and within seven miles thereof. The Court refused the application, upon the ground that a licentiate could show no title to be admitted, as of right, a fellow, and a member of the corporation. (5 Burrows, 2740-2743.)

As was stated by Lord Mansfield in re Archer (*ante*), the statute in question is only a *private* Act, and the College is therefore to a certain extent a private body: in it however were included, originally, as before mentioned, all men of the same faculty of and in the city of London, and this was for a long time thought to afford sufficient grounds for contending that such a right of incorporation continued to exist, but in *re* Dr. Stanger, before referred to, Mr. Justice Lawrence explained that "the intention of the Crown was to incorporate the six persons named in the charter, and all men practising physic at that time *de et in civitate prædictâ*, and all those persons were entitled to admission; but the Crown did not intend to give any right to those who might thereafter become *homines facultatis*, but intended that the succession should be continued by the powers incidental to all corporations to elect. Had the charter incorporated *nominatim*, every man authorized to practise physic in London, and given no directions as to the succession, they would have been authorized to continue themselves by election as they have done, and the charter has done the same thing in substance by incorporating the same persons by a general reference to their character and situation." (*Ib.*) "The president and fellows," says Mr. Willcock, "constitute the community of this College, being the only members of it." (p. 37.)

"The licentiates of the College," says the same authority, "who may practise within the precincts of London, are of three orders: fellows, candidates, and mere licentiates, of whom the last are alone generally denominated licentiates." (p. 32.)

"The licentiates (*i. e.*, the last named or mere licentiates) are those who have only a license to practise physic within the precinct of London." (*Ib.* p. 38.)

"The second class are those who have received that license, but whose license also states that they are admitted to the order of candidates." (*Ib.* p. 38.)

"The third class are those who have received that license, but whose license shows that they are admitted to the order of fellows. This license has often been called a *diploma, but as it confers no degree* the word is not properly applied, according to its more strict signification." (*Ib.* p. 39.)

The common law has given every man a right to practise in any profession or business in which he is competent, and by the incorporation of the clause in the charter of Henry VIII., into the 14 and 15 Hen. VIII., c. 5, a statutory authority, viz., the College of Physicians, was thenceforth made the judge of that competency. It became, therefore, their duty to admit every one to practise in the faculty who satisfied them of his competency, learning, and medical knowledge, and, it may be added, integrity ; and if the College refused to examine any applicant the Court of Queen's Bench might issue a mandamus to compel the admission to examination of such applicants. See the observations of Lord Mansfield in Dr. Letch's case : 4 Burrows, 2186-2195 (E. T. 1767.) "There can be no doubt," said his Lordship, "that the College are obliged, in conformity to the trust and confidence placed in them by the Crown and public, to admit all that are fit, and to reject all that are unfit."

Bye-laws have from time to time been made with a view of subjecting the admission to the Order of Fellows to very arbitrary rules ; and the validity of many of these bye-laws may fairly be questioned, if only upon the ground of their unreasonableness.

The charter, subsequently embodied in the statute, merely incorporates the men of the faculty in and of London—and this is the only qualification required by the charter—and the incidental power possessed by every corporation of continuing itself, as intimated by Mr. Justice Lawrence (*ante*) where no special mode is expressly pointed out, it is the duty of the body to exercise as often as they may think expedient.

In this corporation, therefore, upon the first principle of corporation law, the original fellows had, and their successors have, an unfettered right to elect such persons as they may think proper to perpetuate the corporation, provided only that they be *men of the faculty of and in London ;* and the persons so elected are entitled to be admitted into the body, and if their right is disputed, to enforce it by the aid of the Court of Queen's Bench.

The statute, in fact, rendered all men of the faculty of and in London *eligible* to the fellowship. If a bye-law goes the length of saying that all men of the faculty of and in London are *not eligible*, it virtually affects to repeal that clause of the Act. It is directly in teeth of the statute. Neither is an examination of already approved members required by the statute, and may indeed be well deemed superfluous. If the legislature had deemed such an examination necessary as a preliminary to eligibility for the fellowship they would have mentioned it, as they have mentioned as necessary an examination for licentiates : they deemed an examination necessary for those from among whom the president was to be elected ; therefore they required it : and they further deemed it expedient that country practitioners should be examined, and therefore they made it a requirement. The examination is to secure the health of the people against the ignorance or rashness of uneducated practitioners, but this can have no reference to the distinguishing office of fellows, who have already been examined for the membership. No doubt there are peculiar qualifications required for that office, such as a reputation for superiority among the other members of the profession, and that knowledge of the world, and general acquaintance with the manners of medical men, which might render them particularly capable of judging of the rules necessary for the regulation of the conduct of medical practitioners. The only reasonable ground of admission of one man above another is that the existing fellows have long observed his conduct, in the course of his professional career, and that from such observation they are induced to consider him peculiarly qualified to become an example to his professional brethren, and to take a share in the "oversight and correction of the men of the faculty within the precinct of London."

The original act of incorporation gave the fellows, no doubt, an unrestricted right of electing whom they pleased, limited only by the profession of the elected, but it gave them no power to place a limit upon the *eligibility* of candidates, by restricting it to any persons in particular, or by excluding others not excluded by the statute. As no person can acquire any inchoate right of admission into the order of fellows, so no person can be excluded by any unreasonable bye-law, neither can the existing fellows be restrained by the bye-laws from electing into their number any persons they may so wish to distinguish. No academical degree of any kind is necessary for election to a fellowship, although at one time the statutes of the college required that every candidate, even for the license, should have received the degree of doctor of medicine, after two years' residence in some academy for the purpose of studying physic, but this is virtually delegating the powers of judging of the candidate's competency and proficiency to a body unconnected with the College, which cannot be done, so that it might be difficult to sustain such a bye-law. At all events it has long since been abandoned, and any persons who have been elected members are now acknowledged to be eligible for the fellowships.

It is unnecessary to refer to the highly penal, and indeed quasi-criminal jurisdiction, at one time claimed and exercised by the College over those practising without its license, except to mention it as a *thing of the past*. In Dr. Bonham's case, 8 Coke, 107-11 b (M. T. 6 Jac. 1.) the College assumed, not only to punish by fine a doctor of physic of Cambridge University, for practising in London without license of the College, but to commit him to prison, without bail or mainprize, for disobedience. But Chief Justice Coke observed "that none can be punished for practising physic in

London but by forfeiture of 5*l*. by the month—and if any practise physic there for less than a month, he shall forfeit nothing." And it was further resolved by the Court, consisting of three judges, "that the clause which gives power to the censors to fine and imprison doth not extend to the clause, 'quod nemo in dicta civitate, &c., exerceat dictam facultatem, &c.,' but extends only to punish those who practise physic, &c, 'pro delictis suis non bene exequendo, faciendo, et utendo facultate medicinæ, by fine and imprisonment,' so that the censors have not power by the letters patent and the Act to fine and imprison any for practising physic in London, but only for ill and not good use and practice of physic."

Several cases occurred in early times in which the power of imprisonment was exercised by the College for malpraxis, but in almost, if not all these cases, the defendants upon being brought up by *habeas corpus* were discharged upon some technicality, and in the case of Dr. Alphonso, 2 Buls. 259 (M. T. 12 Jac. I.), and Dr. Greenoelt, 1 Lord Raymond, 213, 214 (E. T. 9 Wm. III.), the defendants were pardoned by the king. It might be a matter of regret that this summary jurisdiction over malpractitioners should have fallen into desuetude, were it not that the courts of criminal jurisdiction are now sufficiently alive to their duties in dealing with all such cases; but with regard to civil cases and vexatious interference, some words of reproof, that fell from Lord Kenyon in Dr. Stanger's case, it might be as well even now to remember. "By what fatality," said the learned judge, "it has happened, that almost ever since this charter was granted this learned body have been in a state of litigation, I know not; indeed I cannot see what these parties are contending for that is worth the expense and anxiety attending this litigation." And again, in the case of Archer and Fothergill, before referred to, Lord Mansfield, speaking of the College of Physicians, said, "I have foreseen the labyrinth and maze of litigation that this learned body would be involved in by persisting rigidly on both sides in pursuing the points of their dispute, and contesting about a feather. I have read over all their constitutions, statutes, and bye-laws, and many of them appear to be narrow, if not illegal."

As under the new medical Acts the powers of the College are most materially abridged, it is to be hoped that it will be found not inconsistent with their dignity to accept the advice tendered by Lord Mansfield in Dr. Letch's case, the report of which thus concludes, "His lordship concluded with a recommendation to the College to settle all other matters *amongst themselves*, without coming to this Court: at the same time intimating to them a caution against *narrowing* their grounds of admission so much, that even if a *Boerhaave* should be resident here, he could not be admitted into their fellowship." More than a century has elapsed since these words were pronounced, it is to be hoped not without effect.

2. *The College of Physicians of Edinburgh.*—According to a clause in their charter, A.D. 1681, and ratified by an Act of the Scottish Parliament on the 16th June, 1685, it would appear that the fellows and licentiates of this college are entitled to practise medicine only, and not surgery. The rights of the surgeon-apothecaries of Edinburgh are expressly reserved by the charter. The fellows, members, and licentiates are all entitled to registration, and, by the 47th section of the Medical Act, "within twelve months of the granting a new charter" to the College of Physicians of London, "any such fellow, member, or licentiate, who may be in practice as a physician in any part of England, and who may desire to become a member of such newly-styled College of Physicians of England, may do so, and be entitled to receive the diploma of such College accordingly, on the payment of a fee of 2*l.*

3. *King's and Queen's College of Physicians, Ireland.*—This College, and Trinity College, Dublin, form conjointly the "School of Physic" of Ireland. Whether the qualification of an Irish physician extends to the practice of surgery and pharmacy is uncertain. The fellows and licentiates are entitled to registration. This College is entitled to the same privileges under the 47th section, as above, as is the College of Physicians of Edinburgh.

4. *College of Surgeons of England.*—The qualification conferred by this College is to practise surgery only, but surgery includes medical treatment as ancillary to the removal of surgical complaints. (Allison *v.* Haydon, 3; Carrington and Payne, 246; Proud *v.* Mayall, 3; Dowling and Lowndes, 531.) As midwifery is no branch of the medical art, and any one may practise it, from the very necessity of the case, it is questionable how far the bye-law which excludes fellows of the College from practising it is lawful. Fellows and members are entitled to registration.

By 18 Geo. II., c. 15, members of this College might, even before the Medical Act, practise in all parts of the empire. Under their charter (Charles II.) confirmed by the above Act, they still possess an exclusive right of practising surgery within the London precinct, as against those who are not qualified so to do. The penalty is 5*l.* a month, and every person not qualified in surgery is liable to the same penalty for practising surgery in any part of England or Wales. It is needless to say even this penalty is never enforced.

Licentiates in Midwifery of this College, by becoming registered, possess now for the first time a parliamentary recognition, and a legal right to recover their fees, but they have no exclusive privileges.

5. *College of Surgeons, Edinburgh.*—The parliamentary ratifica-

tion of the patent of William and Mary gave to the fellows and licentiates of this College the right of practising pharmacy in addition to surgery. If irrespective of this Parliamentary title the right incidentally flows from their qualification of surgeon, they may be able legally to practise medicine in England, notwithstanding the Apothecaries Act. The exclusive powers of this corporation which still exist as against unqualified persons, extend over the counties of Edinburgh, Linlithgow, Haddington, Fife, Peebles, Selkirk, Roxburgh, and Berwick. Fellows and licentiates are entitled to registration.

6. *The Faculty of Physicians and Surgeons of Glasgow.*—This body is, by its charter granted by James VI. in 1599, and ratified in 1672, notwithstanding its name, only a College of Surgeons. The exclusive rights of this College were given up in 1850, when by 13 Vict. c. 20, s. 2, fellows and licentiates were " to enjoy the same status and privileges as if the said faculty had been specially authorized by law to grant licenses and diplomas in surgery, conferring the same status and privileges as those conferred by any corporation or Royal College in Scotland." If the effect of this clause be to place the faculty on the same footing as the College of Surgeons of Edinburgh, Glasgow Surgeons will be entitled to practise medicine also throughout her Majesty's dominions. Fellows and licentiates are entitled to registration.

7. *Fellows or Licentiates of the Royal College of Surgeons, Ireland.*—This corporation has no exclusive privileges, except that its members alone could be appointed surgeons to a county infirmary. Whether these surgeons could combine the practice of medicine with surgery is somewhat doubtful, but the better opinion is that they could not without a license from the Apothecaries Hall, Dublin. The fellows and licentiates of this College are entitled to registration.

8. *Society of Apothecaries, London.*—Although throughout the Medical Act, pharmacy, as a branch of practice, is nowhere mentioned, the only two branches of practice enumerated in section 31 being "Medicine" and "Surgery," yet Licentiates of the Society of Apothecaries, London, and Licentiates of Apothecaries Hall, Dublin, are mentioned in schedule A as persons duly qualified to practise medicine if registered accordingly. The apothecaries, or rather the grocers, had their own share of medical practice, such as it was, long before the distribution of the medical faculty into physicians and surgeons. The grocers, or potecaries, as they are indifferently termed, formed one of the ancient companies of the city of London; and it was only in the 13th year of James I. that they were formed into two distinct corporations. Even from this period the apothecaries were at liberty to sell physic throughout the rest of the

kingdom, provided they had served an apprenticeship of seven years, and thereby become freemen of the Company. They had from a very early period occasionally prescribed the medicines which they sold, thus trespassing on the province of the physician, as it was thought, until the House of Lords in the case The College of Physicians v. Rose, decided in favour of their practice, since which they have continued to do so unmolested. It was, in consequence of this decision, thought at length desirable to provide more effectually for the education of this branch of the faculty, and therefore by 55 Geo. III., c. 194, every one practising as an apothecary in any part of England or Wales without a certificate from the Company is liable to a penalty of 20*l*., and this clause has operated as an exclusion against all persons (except physicians) from practising medicine accordingly, the penalty having been actively enforced. The exclusion will be powerless in future against Irish apothecaries, and possibly against Scotch surgeons; but remains in full force against all unqualified persons practising as apothecaries in England and Wales. The licentiates of the Apothecaries Company, upon being registered as such, are now enabled to practise in all parts of her Majesty's dominions.

9. *Apothecaries Hall, Dublin.*—This company was first incorporated by royal charter in 1745, having up to that time been an unincorporated civic guild. In 1791 they obtained an Act of further incorporation, the 31 Geo. III., c. 34. They claim the following privileges: the right to charge for professional attendance as well as for medicine; *the exclusive* right to keep shop for dispensing and compounding medicine under a penalty of 20*l*., and many other rights and immunities, which so far as they relate to the selling, compounding, or dispensing medicines, are expressly reserved to them by the 55th section of the Medical Act, and therefore their penalty still remains in force against unqualified persons. If, as is stated, the Irish apothecary may legally practise surgery, his registered qualification will in such case permit him to practise physic and surgery throughout the empire. Licentiates are entitled to registration.

10. *Universities.*—The degree M.D., or M.B., L.M., or Master in Surgery of any University in the United Kingdom. The first three of these degrees clearly give the possessors a right of prescribing in medicine. Whether the Mastership in Surgery limits its possessor to purely surgical practice might be a question, were it not that at Cambridge at least (and this may also apply to the Irish diploma or License in Surgery) the candidate must have passed all the examinations required for the degree of Bachelor of Medicine, besides having attended lectures on surgery and attended during three years hospital surgical practice, and have been house-

surgeon or dresser for at least six months at such hospital. As such a practitioner, having already passed the examinations for the degree of M.B., usually, when arrived at sufficient standing, proceeds to take it, the practical solution of the question would scarcely be likely to occur, as under sec. 30 he could have his further qualification inserted in the register in addition to his previous surgical one. As the restriction of the College of Physicians, England, has been neglected, and indeed is inconsistent with the general right to practise under this Act, these graduates may be said to have the right of practising throughout the Queen's dominions, London and its precincts even, by courtesy not excepted. By the Medical Graduates Act, 1854 (17 and 18 Vict., c. 114), every M.B. and M.D. of the University of London is entitled to practise physic as fully in all respects as every M.B. or M.D. of Oxford and Cambridge is entitled by virtue of his degree, or under any power, license, or authority conferred by either of the last-named Universities, such privileges however *not to extend to the practice of surgery, pharmacy, or midwifery*. By sec. 53 of the Medical Act, the Medical Graduates Act remains in full force, notwithstanding the surrender of the therein recited charter of the University of London, and the granting of the existing charter, or the future determination of it, or any future charter. A medical graduate of London, if he possesses no further registered qualification, can only *prescribe* medicine in the same way as any M.D. or M.B. of Oxford or Cambridge, and does not possess the full privileges of the Royal College of Physicians of England in treating cases of surgery, or in compounding his own medicines. Neither can he practise in midwifery, except as it is apprehended any one may practise it, as being the superintendence and assistance sometimes required, even in a natural and healthy function, and from the necessity of the case. Licentiates in midwifery of the College of Surgeons, London, are specially recognised by the Act, and though they exclude no one else, they have a Parliamentary right to recover as such, which it would appear the M.B. or M.D. of London would not have, being expressly interdicted by the above statute.

Medical graduates of Durham have also a right to practise medicine.

The five universities of Scotland all have the power of granting M.D. degrees, and therefore such graduates have the right of practising medicine; as also Trinity College, Dublin.

None of these universities have any exclusive power, and their medical graduates upon being registered are now entitled to practise in medicine throughout her Majesty's dominions.

11. *Doctor of Medicine by Doctorate, granted before the passing of the Act, by the Archbishop of Canterbury.*—This limitation does not interfere with the power of the Archbishop to continue to grant

such degrees as a purely *honorary* distinction, but they will not be any longer recognised as a qualification for practice. They have hitherto but seldom been granted to others than those who had already achieved a reputation in a lower walk of the profession, and who perhaps were too far advanced in life to commence afresh an academical course of study. They are at least more respectable than many foreign degrees, and may still be bestowed, perhaps upon retiring successful practitioners, as a mark of honourable recognition of a useful and prosperous career. Those possessing them at the time of the passing of the Act are rendered eligible for registration.

12. *Doctors of Medicine of Foreign and Colonial Universities or Colleges practising as Physicians in the United Kingdom before the* 1st *of October*, 1858.—All such who were practising as physicians in the United Kingdom before 1st of October, 1858, who shall produce certificates to the satisfaction of the Council of their having taken the degree of Doctor of Medicine after regular examination, or who shall satisfy the Council, under sec. 46, that there is a sufficient reason to dispense with any of the provisions of the Act in their particular cases, are entitled to be registered. The Act gives to the Council this dispensing power in favour of "persons practising Medicine or *Surgery*, within the United Kingdom, on foreign or colonial *Diplomas* or degrees, before the passing of this Act;" but in schedule A no mention is made of those practising "surgery on foreign diplomas," but this defect is cured by the words "and also" being used in sec. 46. This dispensing power is also extended to the cases of those who have been surgeons, or assistant-surgeons in the army or navy, militia, or East India Company, or are acting as surgeons in the public service, or the service of any charitable institution; and, as far as the Council shall deem expedient, in favour of medical students who had commenced their professional studies before the passing of the Act.

And by 22 Vict. c. 21, s. 6, any person not a British subject, having any foreign University Degree or Diploma of Doctor of Medicine, who shall have passed the regular examinations entitling him to practise medicine in his own country, may still act as medical officer or resident physician of any hospital established exclusively for the relief of foreigners in sickness. Such person, however, is not to act in any other medical practice.

The following persons, in addition, are entitled to registration :— By the 17th section of the Act, "Any person who was actually practising medicine in *England* before the 1st August, 1815, shall on payment of the requisite fee be entitled to be registered, on producing to one of the registrars a declaration to the effect that he was practising as a medical practitioner at a certain place before the day above-mentioned. This section is a continuation of the exemption in the Apothecaries' Acts, whereby all persons in practice as apothecaries on 1st of August, 1815, are relieved from the penalty imposed on every one else (saving the rights of physicians) for practising as apothecaries in England and Wales without a certificate, and such persons alone are within the scope of this section. As 55 years have all but elapsed since August, 1815, there cannot in the course of nature be many now on the register under this section.

CHAPTER III.

RIGHTS, PRIVILEGES, AND IMMUNITIES OF DULY-REGISTERED MEDICAL PRACTITIONERS.

By the 31st section of the Act, "Every person registered under the Act shall be entitled according to his qualification or qualifications to practise medicine or surgery, &c." . . . "and to demand and recover in any court of law, with full costs of suit, reasonable charges for professional aid, advice, and visits, and the cost of any medicines, or other medical or surgical applications rendered or supplied by him to his patients."

Although at common law a physician could maintain no action for his fees unless by special contract, now by the above section a general right of action is given to all registered practitioners; and a physician may now, if registered, maintain an action for attendance since the passing of the Act, without proof of any express contract or implied understanding with the patient that he should be paid. Gibbon v. Budd. 32 Law Journal (Exch.) 282. 2 Hurlstone and Coltman, 92.

It appears from the report of the case that the Royal College of Physicians has availed itself of the proviso contained in the same section, and has passed a bye-law that no Fellow of the College shall be entitled to sue for his fees, but this does not include *members*. If surgery be included in the qualification of a member of any college of physicians, as in the case of the English College, he would, it appears, be entitled to charge also for his services in surgical cases as well as in pharmacy, when, as in the case of licentiates of the same College of Physicians of England, he may even compound and sell the medicine he prescribes, such privileges being reserved to the College, and where no restriction is contained in the licenses granted by it.—*Attorney-General* v. *Royal College of Physicians* (*ante*).

ENGLISH SURGEONS, ETC.

The various statutes which relate to the capacity of medical practitioners to sue are superseded, though not repealed, by the Medical Act.

It is almost needless to observe that at this distance of time an action is not likely to be brought for medical attendance, &c., given before the Act passed, otherwise it would not be necessary to prove registration in order to recover in such action. Wright *v.* Greenroyd. 31 Law Journal (Q. B.) 4. And further, it has been decided in Turner *v.* Reynall, 14 Common Bench. N. S. 328, agreeing with the Irish Court of Exchequer in Haffield *v.* Mackenzie, 10 Irish C. L. Rep. 289, that it is sufficient that the plaintiff be registered at the time of trial, although after the attendance, and even after action brought. If two medical practitioners are in partnership, and one alone is duly registered, they can jointly maintain an action for services rendered by the firm. Per Erle, C. J., and Byles, J., in Turner *v.* Reynall, supra. Section 32, which gives no right of recovery of any charge of any kind unless the claimant prove that he is duly registered, is not confined in its operations to actions against the patients themselves, but extends to a case where a third person had guaranteed payment for medicine, &c., or is primarily liable for it as supplied on his credit. So a medical practitioner, engaged by another to attend his patients in his absence, cannot recover the price of his services without proof of his registration —De la Rosa *v.* Prieto, 33 L. J. (C. P.) 262—but an unregistered assistant may recover his salary from a registered practitioner. S. C. per Cur. Even in the case of a promissory note given to plaintiff for medical attendance, in order to show consideration plaintiff must prove registration. Blogg *v.* Pinkers, Ryland and Moody, 125. Where the register specifies the department of practice, in respect of which the plaintiff is registered or qualified to practise questions may arise similar to those under the former Acts, *e.g.*, as to whether a person registered as qualified to practise as a surgeon can recover for attendance as a physician or apothecary. See Allison *v.* Haydon, 4 Bingham, 619 (in which it was held that a certificated surgeon could not recover for attending a patient in a fever without a certificate from the Apothecaries Company). See also sections 30 and 31, under which a man registered with an inferior qualification, and having since obtained a higher one, may have that higher one added to his inferior qualification, and may then sue according to his newly-inserted qualification, in addition to his former one, provided his new and additional qualification is inserted in the register before action brought. Turner *v.* Reynall (*ante*).

It is as well that it should be known that the superintendent of a railway station cannot, without express authority, make the company liable for a surgeon's bill for attendance upon a person injured by an accident on the railway: Cox *v.* Midland Counties Railway, 3 Exch. Rep. 268; but then, should the sufferer succeed against the company, the surgeon's bill would be paid out of the damages obtained, as such is always allowed in damages (*ib.* per Baron

Parke); but money paid to a medical man for a report on the plaintiff's condition would not be allowed in damages: Deer v. London, Brighton, & South Coast Railway, Jan. 28, 1870 (Q.B., *coram* Justices Blackburn and Hannen). If the defendant has received no benefit in consequence of the plaintiff's want of skill, it stands to reason the latter cannot recover: Kennen v. McMullen, Peake, N. P. 59; Duffit v. James, cited 7 East. 480. But, on the other hand, the remuneration of a practitioner who has used due skill and diligence does not depend on his effecting a cure. If the operation performed by a surgeon *might* have been useful though it has failed in the event, he is entitled to recover; but otherwise if it could have been useful in no event. Per Baron Alderson, in Hill v. Featherstonhaugh, 7 Bingham, 574.

APOTHECARIES.

As licentiates of the Society of Apothecaries may be registered under schedule A as entitled to practise medicine, it is necessary briefly to consider the extent of their qualification and the amount of remuneration to which they are entitled. It was for some time considered, in consequence of a suggestion which proceeded from Lord Tenterden, in 1830: *Hanley* v. *Henson*, 4 Car. & P. 110, that the charge of two shillings and sixpence each for attendances, besides his charges for medicine, was the legal charge for an apothecary, but it seems that all that Lord Tenterden submitted to the jury was the reasonableness of such a charge, and that, subject to the opinion of a jury as to what is reasonable or unreasonable in each particular instance, an apothecary can make what reasonable charges he pleases for attendance in a medical case (Morgan v. Hallen, 3 Nevill & Perry, 498). Of course the custom of the profession would be taken into consideration in this as in every disputed demand made under any registered qualification, subject to the reasonableness of the charge.

The production of the certificate of a licentiate of the Apothecaries Society shall be sufficient proof that the party named therein has been, from the date of the certificate, duly qualified to practise: 14 & 15 Vict. c. 99, s. 7. This and proof of his registration under the new Act (extended by 23 & 24 Vict. c. 7, to 1st January, 1861), or, it is apprehended, registration alone, the registrar having satisfied himself of the qualification (s. 26) will be sufficient to entitle him to recover. If he possesses only a suburban license and has infringed the law of his own society by practising without further permission within the London precinct, he may be fined by the society; but that is a matter of only fiscal regulation between the society and its members: Young v. Genger, 6 C. B. 541.

Summary.

The different qualifications entitling their possessors to recover at law their fees according to their respective qualifications:—

Fellows of the Royal College of Physicians of London are bound by a bye-law of their College by which they are not entitled to sue for their fees.

Members, Licentiates, and Extra Licentiates of the College may sue for their fees in medicine and surgery; and, where not otherwise prohibited (as members), they may also recover for pharmacy and medicine compounded by themselves, and supplied to their patients.

Fellows, Members, and Licentiates of the Royal College of Physicians of Edinburgh are entitled to sue for their fees in medicine.

Fellows and Licentiates of the King's and Queen's College of Physicians of Ireland are entitled to sue for their fees in medicine. Whether they could recover them in surgery and pharmacy is uncertain. The question must mainly depend on their charter, granted by William and Mary in 1692.

Fellows and Members of the Royal College of Surgeons of England may sue for their fees in surgery cases, and also for medical treatment, if ancillary to the removal of the surgical complaint.

Licentiates in midwifery of this College can also sue for their fees in midwifery cases, and also, perhaps, for medical treatment, if ancillary to midwifery cases.

Fellows and Licentiates of the Royal College of Surgeons of Edinburgh can sue for their fees both in surgery and pharmacy (by their charter), and in medicine as ancillary to their attendance on surgery cases.

Fellows and Licentiates of the Faculty of Physicians and Surgeons of Glasgow can sue for fees both in surgery and pharmacy, but the body is, by its charter, notwithstanding its name (granted by James VI. in 1599), only a College of Surgeons, yet the former observations respecting medical fees are applicable to this institution also.

Fellows and Licentiates of the Royal College of Surgeons of Ireland can sue for their fees in surgery, but it is thought not in pharmacy. The above remark on medical fees applies here likewise.

Licentiates of the Society of Apothecaries, London, can sue for fees for medical attendance, as well as for the price of their medicine. Their exclusive privilege, extending over England and Wales, will enable them still to enforce a penalty of 20*l.* as against all unqualified persons.

Licentiates of Apothecaries' Hall, Dublin, can sue in the same way as Licentiates of the Society of Apothecaries, London, and if, as is stated, they can legally practise surgery, they may, in addition, sue for surgical fees, but there is no evidence of this latter right.

Their penalty of 20*l.* can still be enforced within their jurisdiction against all unqualified persons.

Doctors, Bachelors, and Licentiates in Medicine of Oxford or Cambridge University can sue for their fees in medicine.

Masters in Surgery of either University, or those possessing the like diploma in Surgery of any University in the United Kingdom entitled to grant the same, can sue for their fees in surgery, and probably in medicine; but the latter is doubtful except as ancillary to surgical treatment.

Doctors and Bachelors of Medicine of the University of London can sue for their medical fees alone, but are expressly forbidden by the Medical Graduates Act to practise surgery, pharmacy, or midwifery.

Medical Graduates of Durham University can sue for their fees in medicine.

So can *those of Trinity College, Dublin.*

And *all the Doctors of Medicine of the five Scotch Universities.* These Universities do not grant the lower degree of Bachelor of Medicine.

Doctors of Medicine by Doctorate, granted before the passing of the Act by the Archbishop of Canterbury, can sue for their medical fees.

Also *Doctors of Medicine of any foreign or colonial University or College*, registered with the sanction of the Medical Council, can sue for their fees in medicine, if they had, previous to the passing of the Act, been practising in the United Kingdom as physicians.

Likewise *Surgeons, who, previous to the passing of the Act, had practised in the United Kingdom on foreign or colonial diplomas*, may sue, under similar circumstances, for their surgical fees, including ancillary medical attendance.

And, *those who have held appointments as Surgeons or Assistant-Surgeons in the Army, Navy, Militia, or the East India Company, or are acting as Surgeons in the public service, or in the service of any charitable institution*, may sue for their fees as surgeons, in like manner as above, and under similar circumstances.

"*Any person who was actually practising medicine in England before the 1st day of August*, 1815," can sue for his fees as an apothecary.

There is a large class known in non-professional language as "general practitioners," who, upon becoming registered in the double qualification, if they possess it, of surgeon and apothecary, enjoy the right of both classes of practitioners, and, indeed, any physician, it is presumed, if not forbidden by the bye-laws of his College, or any University medical graduate, may be registered in all three qualifications, should he possess them. If, however, he should claim them in right of his medical degree or diploma, it would be incumbent upon him to prove his right by virtue of the chartered constitution of his University or College.

Medical graduates, if elected members of the College of Physicians of London, of course enjoy all the privileges of that position, viz., of practising, if they think fit, in every branch of physic, though not in pharmacy, when by virtue of their mere medical degrees, they would be under limitations, varying according to the charters of their respective Universities. If further elected to the fellowships, they could no longer be able to recover their fees, being bound by the bye-law to that effect.

General Observations.

Surgeons, being expressly excepted in the Apothecaries Act, have nevertheless, as before mentioned, been declared as entitled by law to administer medicines as ancillary to the removal of surgical complaints. Allison v. Haydon, 3 Carrington & Payne, 246; and Proud v. Mayall, 3 Dowling & Lowndes, 531. "The mere fact of the surgeon having supplied medicine, does not necessarily show that he practised as an apothecary, for a surgeon may lawfully do that if the medicines have been administered in the cure of a surgical case. But on the other hand, if a surgeon takes upon himself to cure a fever, he steps out of his lawful province, and is not authorized to administer medicine in such a case." Per Mr. Justice Cresswell in Apothecaries Company v. Latinga, 2 Moody & Robinson, 495. See also Simpson v. Ralfe, 4 Tyrwhitt, 325. In all the cases the judge has always decided the nature of the practice, leaving it for the jury to say in what character the practitioner had acted. Thus in Allisan v. Haydon, Chief Justice Best ruled that typhus fever was a medical, not a surgical case. "The pestilence, syphilis, and such other contagious diseases," have been declared (in England) by 32 Hen. VIII. c. 42, to be surgical cases; also "letting of blood and drawing of teeth," in a word, "wounds, ulcers, fractures, dislocations, tumours" (charter of 5 Car. I., confirmed by 18 Geo. II., c. 15). In Battersby v. Lawrence (1 Carrington & Marshman, 277), which related to a case of dropsy, the judge remarked, "It is said that the case was a surgical one; but not so altogether." Small-pox, according to the legislative definition before mentioned, would appear to be a surgical as well as a medical case, it being both contagious and infectious. But after all, the real question is, as put to the jury by Lord Tenterden in the Coll. of Physicians v. Harrison, A.D. 1828, (MSS. at the College of Physicians,) "Has the party complained of practised as a physician? Has he written prescriptions which are signed by him as a physician would sign them? Has he practised physic, or has he practised as a surgeon only?"

Such are the respective rights and privileges of all duly registered medical practitioners. The Medical Act also enacts that henceforth no one " shall hold any appointment as a physician, surgeon, or other medical officer, either in the military or naval service, or in emigrant or other vessels, or in any hospital, infirmary, dispensary, or lying-in hospital, not supported wholly by voluntary contributions, or in any lunatic asylum, gaol, penitentiary, house of correction, house of industry, *parochial or union workhouse*, or *poorhouse, parish union*, or other public establishment, body, or institution, or to any friendly or other society for affording relief in sickness, infirmity, or old age, or as a medical officer of health, unless he be registered under this Act" (s. 36). But by s. 4 of 23 Vict., c. 7, "No person authorized to be registered, who shall be acting as a medical officer under an order of the Poor Law Commissioners, or Poor Law Board, shall be deemed to have been disqualified to hold such office, or any appointment mentioned as above (s. 36), unless he shall have failed to be registered on or before 1st January, 1861." And further, " No certificate required by any Act then in force or hereafter to be passed, from any physician, surgeon, licentiate in medicine and surgery, or other medical practitioner, shall be valid unless the person signing the same be registered under this Act" (s. 37) Medical Act. By the 35th section, " Every person who shall be registered, shall be exempt, if he shall so desire, from serving on all juries and inquests whatsoever, and from serving all corporate, parochial, ward, hundred, and township offices, and from serving in the militia."

CHAPTER IV.

SECTION I.

OF THE LIABILITIES OF LEGALLY QUALIFIED MEDICAL PRACTITIONERS, AND THE DISABILITIES OF UNREGISTERED PRACTITIONERS—ALSO OF THE MALPRACTICE OF BOTH QUALIFIED AND UNQUALIFIED PERSONS.

THE sections 35 & 36, quoted in the last chapter, imposing as they do disabilities on the unregistered and conferring corresponding privileges on the registered, practically compel the registration even of those who may be prohibited by their own bye-law from suing for their fees, as otherwise they would be ineligible for the appointments therein mentioned, neither could their certificates be received (s. 37). Moreover, unregistered persons are not entitled to the expenses of medical witnesses in courts of justice, or to make *post mortem* examinations, as under the provisions of 6 & 7 Wm. IV., c. 89, (Medical Witnesses Act) such persons must be "legally qualified," and by the 34th s. of the Medical Act the word "legally," or "duly qualified," "when used in any Act of Parliament," shall be construed to mean a person registered under that Act. By the 40th section, any person who shall wilfully and falsely pretend to be, or take, or use the name or title of Physician, Doctor of Medicine, Licentiate in Medicine and Surgery, Bachelor of Medicine, Surgeon, General Practitioner or Apothecary, or any name, title, addition, or description implying that he is registered, or that he is recognised by law as a Physician, or Surgeon, or Licentiate in Medicine and Surgery, or a Practitioner in Medicine, or an Apothecary, shall upon a summary conviction for such offence pay a sum not exceeding 20*l*.

In a case upon appeal from a decision of the Pinner Bench of Petty Sessions under this section, Baron Bramwell said, "Assuming that the respondent was not a Doctor of Medicine, the question is whether he has assumed the title 'wilfully and falsely.' The section must be read as pointing to wilful falsity. No one can doubt that the respondent had a foreign diploma of some kind, and had on the strength of that called himself 'Dr. Kelly.' Is it to be said that he has wilfully and falsely pretended to be, or taken or used the name or title of 'Doctor of Medicine'?" Ellis *v*. Kelly,

30 Law Journal (M. C.) p. 35. In this case the respondent was registered as a Member of the College of Surgeons, England, and Licentiate of the Society of Apothecaries, London, but having a foreign diploma as a Doctor of Medicine of Erlangen, had previous to, and subsequent to the passing of the Medical Act called himself "Dr. Kelly." Judgment was given in his favour, with costs. It would seem then that a registered practitioner, clothed with a foreign medical degree, may assume its academical title, although not registered as such, if he is guilty of no wilful falsity. This remark would also apply to the degree of Doctor by Doctorate, granted by the Archbishop of Canterbury after the passing of the Medical Act, although such is no longer a qualification for registration.

SECTION II.

Negligence, in which though there is no criminal or dishonest conduct, but a gross want of that attention which the case of the patient requires, may be the subject of an action at law, and in point of fact is a serious misdemeanour, and offence at common law, "because," as Blackstone observes in his Commentaries, "it breaks the trust which the party had placed in his physician, and tends to the patient's destruction." Such was the decision in Dr. Graenvelt's case, reported in 1 Lord Raymond, 213, 9 Wm. III. Ignorance is also a great misdemeanour at common law, whether in a licensed or unlicensed practitioner, and at the same time the injury arising to any individual from his being a victim to this practice is a private wrong, for which he may recover adequate damages, but where a practitioner is a duly qualified practitioner, of course there is a considerable difficulty in fixing him with *crassa ignorantia*, however he may have erred in judgment. The circumstance of qualification affords a presumption in favour of the practitioner which juries will consider in such cases. There is no doubt that owing either to the death of the only person who could maintain the action, by the unskilful treatment, perhaps, of the medical practitioner, or the difficulty of producing the necessary evidence, which is obviously very great, or the hitherto uncertainty of medical practice, there has rarely occurred an opportunity of trying the point of the liability of the physician towards his patient for want of skill and attention, but the whole tenor of the books is that such an action can be supported. A physician, *as such*, is usually called upon on each distinct occasion, when his presence and aid are specially required, to counsel and direct those already in charge of the case.

But the surgeon and apothecary, when undertaking the exclusive care of the patient, have been universally held responsible for injuries arising from want of care and attention. Seare *v.* Prentice, 8 East 347-358. In fact, the rule is obvious that in all cases where damage accrues to another by the negligence, ignorance, or misbehaviour of a person in the duty of his trade or calling, an action will lie, as if a farrier kill a horse by bad medicines, or prick him in the shoeing. Butler's Nisi Prius, 73.

"Again, ye are to distinguish of other manslayers, as of physicians, &c., for physicians and chirurgeons are skilful in their faculties, and probably do lawful cures, having good consciences, so as nothing faileth to the patient which to their art belongeth: if their patients die they are not thereby menslayers or mayhemors; but if they take upon them a cure, and have no knowledge or skill therein, or if they have knowledge, if nevertheless they neglect the cure, or minister that which is cold for hot, or take little care thereof, or neglect due diligence therein, if their patients die, they are menslayers or mayhemders." Mirror, c. 4, s. 16. But to recur to the authoritative cases. In Seare *v.* Prentice, (an action against a surgeon,) Lord Ellenborough, C. J., said "that an ordinary degree of skill is necessary in a surgeon who undertakes to perform surgical operations, in the same manner as it is necessary for any other man to have it in the course of his employment, and I am ready to admit that a surgeon would be liable for *crassa ignorantia*, and would be justly responsibile in damages for having rashly adventured upon the exercise of a profession, without the ordinary qualification of skill, to the injury of a patient."

Again in Slater *v.* Baker and Stapleton, 2 Wils. 359, an action against a surgeon and an apothecary for ignorance and unskilfulness, tried before Chief Justice Wilmot, upon motion to set aside the verdict and damages which the plaintiff had obtained, the Court, without even hearing the counsel for the plaintiff, said: "He who acts rashly acts ignorantly: and although the defendants in general may be as skilful in their respective professions as any two gentlemen in England, yet the Court cannot help saying that in this particular case they have acted ignorantly and unskilfully, contrary to the known rule and usage of surgeons."

The liability of the medical practitioner is no longer terminated with the death of his, in such cases, unfortunate patient, for by the 9 & 10 Vict., c. 93, s. 1 & 2, whenever the death of a person is caused by some wrongful act, neglect, or default which would (if death had not ensued) have entitled the injured party to an action, then the person who would have been liable if death had not ensued, shall be liable for damages, though the death was caused by an act amounting to felony: and the action shall lie for the benefit of the wife, husband, parent, and child of the deceased, and in the name of the executor or administrator, and the word "parents" is explained in the Act to include father, mother, grandfather, grandmother, stepfather, and stepmother, and the word "child," to include son, daughter, grandson, granddaughter, stepson, and stepdaughter. The escape from liability is thus rendered next to impossible; and even the small weekly wages of a deceased son who contributed them towards the expenses of his parents' house, is evidence upon which the jury may act, though it is not shown that the sum contributed did more than cover the expense of the child's maintenance. Duckworth v. Johnson, 4 Hurlstone & Norton, 653. Proof must be given of a pecuniary loss through the death, sustained by the members of the family who sue, but the slightest pecuniary loss will support the action, or even a reasonable expectation of pecuniary advantage had the deceased lived, may be taken into consideration by the jury. London and South Eastern Railway Company, 4 Common Bench (N.S.) 296. See also Franklin v. South Eastern Railway Company, 3 H. & N., 211.

We now come to the consideration of *mala praxis*, both in qualified and unqualified, registered and unregistered medical practitioners.

Although a person may not be registered under the Act, he may nevertheless be *qualified* for registration, but labouring under disabilities, and excluded from the enjoyment of the privileges attached to registration, at the same time neither a quack nor an impostor. For reasons best known to himself, he may not choose to be registered, and yet may not be the subject of any penalties. On the other hand, many persons may still insist upon practising without even so much as a qualification, and may risk all penalties. Malpractices, therefore, may still be very prevalent.

Cases of great difficulty and nicety have hitherto arisen with regard to the question of legal malice, where medicines have been carelessly or unskilfully administered by incompetent persons, and though they may be liable to action, it may at the same time be very difficult to render them amenable to the criminal law. Indeed it may not be always easy to fix them even with civil liability.

The law on the criminal responsibility of every man in such cases, is thus laid down by Lord Hale:—

"If a physician gives a patient a potion without any intent of doing him any bodily hurt, but with intent to cure or prevent a disease, and contrary to the expectation of the physician, it kills him, this is no homicide; and the like of a surgeon. And I hold their opinion to be erroneous that think, if it be no licensed surgeon or physician that occasions this mischance, then it is a felony, for physic and salves were before licensed physicians and surgeons, and, therefore, if they be not licensed according to the statutes, they are subject to the penalties in the statutes, but God forbid that any mischance of this kind should make any person not licensed guilty of murder or manslaughter." 1 Hale's Pleas of the Crown, 429. Upon the latter point Sir William Blackstone appears to agree with Lord Hale. If a physician or surgeon, he says, gives his patient a potion or plaster to cure him, which, contrary to expectation, kills him, this is neither murder nor manslaughter, but misadventure, and he shall not be punished criminally, however liable he might formerly have been to a civil action for neglect or ignorance; but it has been held, that if he be not a *regular* physician or surgeon who administers the medicine or performs the operation, it is manslaughter at the least. Yet Sir Matthew Hale very justly questions the law of this determination, 4 Bl. Com. c. 14. The correctness of Sir M. Hale's opinion has been recognised in many cases since. Thus in R. v. Van Butchell, 3 Carrington & Payne, 632, Baron Hullock said, that it made no difference whether the party was a regular or an irregular surgeon, adding that in remote parts of the country many persons would be left to die, if irregular surgeons were not allowed to practise. The same opinion was expressed by Justice Park, in a subsequent case, in which he observed that whether the party was licensed or unlicensed is of no consequence, except in this respect, that he may be subject to pecuniary penalties for acting contrary to charters or acts of parliament. R. v. Long, 4 C. & P. 398. But whether the party be licensed or unlicensed, if he displays gross and culpable ignorance, or criminal inattention, or culpable rashness in the treatment of his patient, he is criminally responsible. There is no doubt, said Mr. Baron Hullock, that there may be cases where both regular and irregular surgeons may be liable to an indictment, as there may be cases where from the manner of the operation, even malice might be inferred. R. v. Van Butchell. Where a person, who though not educated as a surgeon, had been in the habit of acting as a man-midwife, and had unskilfully treated a woman in childbirth, in consequence of which she died, was indicted for the murder, Lord Ellenborough said that to substantiate the charge, the prisoner must have been guilty of criminal misconduct, arising either from the grossest ignorance or the most criminal inattention. One or other of these was necessary to make out a case of manslaughter. R. v. Williamson, 3 C. & P. 635. And this ruling was cited with approbation by Justice Park in the case above quoted, R. v. Long. But where a person, grossly ignorant, undertook to deliver a woman and killed the child in the course of the delivery, it was resolved by the judges that he was rightly convicted of manslaughter. R. v. Senior, 1 Moody; Criminal Cases, 346. The rule with regard to the degree of misconduct, which will render a person practising medicine criminally answerable, is thus laid down by Mr. Justice Bayley: "It matters not whether a man has received a medical education or not. The thing to look at is, whether in reference to the remedy he has used, and the conduct he has displayed, he has acted with a due degree of caution, or on the contrary, has acted with gross and improper rashness and want of caution. I have no hesitation in saying, that if a man be guilty of gross negligence in attending to his patient, after he has applied a remedy, or of gross rashness in the application of it, and death ensues in consequence, he will be liable to a conviction for manslaughter. R. v. Long. In a case tried at the Lancaster Assizes, March 14, 1829, viz., R. v. Simpson, 4 Car. & Payne, 407, and 1 Lewin, Crim. Cases, 172, prisoner was indicted for manslaughter. It appeared that the deceased, who was suffering from salivation, went, upon the recommendation of an acquaintance, to the prisoner, an old woman at Liverpool, for an emetic to get the mercury out of his bones. She gave him a solution of corrosive sublimate, one dose of which caused his death. Mr. Justice Bayley, in addressing the jury, said: "I take it to be perfectly clear, that if a person, not of a medical education, in a case where professional aid ought to be obtained, undertakes to administer medicines which may have a dangerous effect, and thereby occasions death, such person is guilty of manslaughter. He may have no evil intention, and may have a good one, but he has no right to hazard the consequences in a case where medical assistance may be obtained. If he does so, it is at his own peril. It is immaterial whether the person administering the medicine prepares it, or gets it from another." Mr. Baron Bolland in another case, said: "The law, as I am bound to lay it down, as it has been agreed upon by the judges, is this: if any person, whether he be a regular or licensed medical man or not, professes to deal with the life or health of his majesty's subjects, he is bound to have competent skill to perform the task that he holds himself out to perform, and he is bound to treat his patients with care and attention and assiduity." R. v. Spiller, 5 Car. & Payne, 333. In a case of this kind, before Lord Chief Justice Tindall, his lordship said to the jury: "You are to say whether in the execution of the duty which the prisoner had undertaken to perform, he is proved to have shown such a *gross* want of care, or such a *gross* and *culpable* want of skill, as any person undertaking such a charge ought not to be guilty of, and that the death of the person named in the indictment was caused thereby." R. v. Ferguson, 1 Lewin, C. C. 181. In a case which occurred before Lord Lyndhurst, when Chief Baron, the law on the subject was thus laid down:—"I agree that in these cases there is no difference between a licensed physician or surgeon, and a person acting as physician or surgeon without a license. In either case, if a party having a competent degree of skill or knowledge makes an accidental mistake in his treatment of a patient, through which death ensues, he is not thereby guilty of manslaughter; but, if where proper assistance can be had, a person, totally ignorant of the science of medicine, takes upon himself to administer a violent and dangerous remedy to one labouring under disease, and death ensues in consequence of that dangerous remedy having been so administered, that he is guilty of manslaughter. I shall leave it to the jury to say whether death was accelerated or occasioned by the medicine administered, and if they say it was, then I shall tell them that the prisoner is guilty of manslaughter, if they think that in so administering the medicine, he acted either with a criminal intention, or from any gross ignorance." R. v. Webb, 1 Moody & Robinson, 405.

The next case is that of a person who had for nearly thirty years carried on the business of an apothecary and man-midwife in the county of York, and was duly qualified by law for that profession. His practice was very considerable, and he had attended the deceased on the birth of all her children. It appeared that on the occasion in question he made use of a metal instrument, known in midwifery by the name of a *vectis* or *lever*, inflicting thereby such grievous injuries on the person of the deceased as to cause her death within three hours. It was proved by the medical witnesses that the instrument was a very dangerous one, and that at that period of the labour it was very improper to use it at all; and also, that it must have been used in a very improper way, and in an entirely wrong direction. Mr. Justice Coleridge told the jury that the questions for them to decide were, whether the instrument had caused the death of the deceased, and whether it had been used by the prisoner with due and proper skill and caution, or with gross want of skill or gross want of attention. No man was justified in making use of an instrument in itself a dangerous one, unless he did so with a proper degree of skill and caution. If the jury thought that in this instance the prisoner had used the instrument with great want of skill, or gross want of caution, and that the deceased had thereby lost her life, it would be their duty to find him guilty. The prisoner was convicted. R. v. Spilling, 2 Moody & Robinson, 107. A chemist, likewise, who negligently supplies a wrong drug, in consequence of which death ensues, is guilty of manslaughter. The apprentice to a chemist by mistake delivered a bottle of laudanum to a customer who asked for paregoric; and a portion of the laudanum being administered to a child, caused its death. The apprentice was indicted for manslaughter, and the jury, under the direction of Mr. Justice Bayley, that if they thought him guilty of negligence, they should find him guilty of manslaughter, did so. R. v. Tessymond, 1 Lewis, C. C. 100.

CHAPTER V.

CHARACTER—DEFAMATION.

A REGULAR professional practitioner is perhaps more entitled to protection against defamation and libel than a private person, because it tends to injure him in his profession. He is entitled to recover a pecuniary compensation against any person who, without reference to his private character, has maliciously imputed to him, either by words or writing, a want of qualification, attention, skill, or capacity, or any misconduct by which he is likely to be prejudiced in the profits of his profession.

At a very early period slander was recognised as an individual wrong, and an offence against the public peace. In its primitive legal acceptation it was applied solely to words spoken; libel, or written defamation, among an unlettered people being scarcely known; but the term slander has since been used to embrace written as well as oral defamation. In Bacon's "Abridgment" *slander* is defined to be the publishing of words in *writing* or by speaking; by means of which the person to whom they relate becomes liable to suffer some corporal punishment, or to sustain some damage. "If a man, by letter, *write* slander of another to a third person," an action lies. 1 Anderson, 119. Again, in "Buller's Nisi Prius," *slander* is defined as "the defaming a man in his reputation by speaking or *writing* words which affect his life, office, or trade; or which tend to his loss of preferment in marriage or service, or to his disinheritance, or which occasion any other particular damage." The term will, therefore, be used in this work as comprehending *oral* as well as *written* defamation, except where otherwise defined. By the common law of England reputation is acknowledged as an inherent personal right. Revenge is forbidden, and even the tongue, that "unruly member," is restrained. Character is treated as the safeguard of social position and success in life, and professional reputation, as an absolute individual right, not to be assailed with impunity.

According to Lord Camden, "If the words be true, they are no slander, and may be justified." 2 Wils. 301.

Blackstone says, "If the defendant be able to justify and prove the words to be true, no action will lie, even though special damage have ensued, for there it is no *slander* or *false tale*." Bl. Comm., vol. 3, p. 125. According to this authority also, libels are defined to be "any writings, pictures, or other signs, which immediately tend to injure the character of the individual, or to occasion mischief to the public." Each individual, whatever be his talents, his rank, or his station, is more or less dependent on others for security of person, property, or even life. He must trust his lawyer for the preservation of his property, his physician in time of sickness, and his servants for the faithful discharge of the meaner duties of life. Character, therefore, is the most valuable personal right of each. The right, then, of every man to the character and reputation which his conduct deserves, stands on the same footing with his right to the enjoyments of life, liberty, health, property, and all the comforts and advantages which appertain to a state of civil society.

Slander, or in its more primitive sense *oral* defamation, is an injury for which an action will lie; but an indictment will not lie for mere words not reduced into writing (2 Salkeld, 417, Rex. *v.* Langley, 6 Modern Reports, 125), unless they be seditious, blasphemous, grossly immoral, or addressed to a magistrate whilst in the execution of his duties, or uttered as a challenge to fight a duel, or with the intention to provoke another to send a challenge to fight a duel (R. *v.* Phillips, 6 East. 464).

Libel, or written defamation, is in the eye of the law of a greater and more aggravated nature than slander, by means of the more durable publicity which may thus be given to the defamatory matter, and the deliberation of the defamer in reducing the slander to writing. Two remedies are therefore given in libel: one by criminal procedure against the libeller, the other a civil remedy for damages by action at law.

Actionable slander therefore consists of the maliciously defaming a person in his character, reputation, dignity, profession, trade, or office, by words or signs—as to say of another, "He forged his master's name to a cheque on his master's bankers;" or by *signs*— as by significant motions of the hands and fingers, as to impute an unnatural or other crime to a person. Or to say that a man has poisoned another, or is perjured; or to charge him with having an infectious disease, the tendency of which would exclude him from society; or to call a tradesman a bankrupt, a *physician* a *quack*, or a lawyer a knave. 3 Bl. Comm. 123, Brown *v.* Smith, 23 "Law Journal" (Com. Pleas), 151; 13 Com. Bench Rep. 596.

The communication must be *malicious*, but only in a technical or legal sense, as merely signifying the absence of any legal justification or excuse. In ordinary actions of slander *express* malice, therefore, need not be proved (Maitland *v.* Goldney, 2 East, 426);

it is to be implied from the slander itself (Bacon's Abridgment, Tit. Slander, D). And the action is maintainable without proof of special damage.

It is essential, however, that the imputation should be *false*, but the law allows the *truth* of the words to form a justification in an action for slander. But the plaintiff is entitled to the presumption of innocence in his favour, and is not required to prove the falsity of the alleged calumny.

The consequences of the slander must be to occasion some injury or loss to the plaintiff in law or in fact. But no proof of *special* damage is required in the following cases:—

1. If an indictable offence be imputed.
2. If a contagious or infectious disorder be imputed.
3. If any injurious imputation be made affecting the plaintiff in his office, *profession*, or business.

In these cases, as the immediate tendency of the slander is to produce great and irreparable injury to the party whose character is assailed, even though none can be proved, or at least proved in time, so as to save the party perhaps from absolute ruin, the law considers the very publication of the slander to be injurious, and to confer a substantive right of action. In all other cases the plaintiff must prove the damage which he has sustained from the words.

Where an action is brought for words spoken of a barrister or a physician, it must appear that he practised as such at the time the words were spoken (7 Bac. Abridgment, 269, *ib.* 271), for otherwise the words could not have affected him professionally. It is no slander to say of a person who is unlawfully practising in England as a physician, that he is a quack, or an impostor, or an unqualified person (Collins *v.* Carnegie, 1 Adolphus & Ellis, 695). It does not, however, follow, that in an action for defamation brought by a medical practitioner, that he should not be able to recover damages because he is not registered under the Medical Act. In such case, if his qualification be denied, it must be proved by the production of his diploma, or in any other manner in which before the passing of the Act such qualification might have been proved. As mentioned in a former part of this work, a practitioner may be qualified though not registered, and whatever may be his disabilities as a non-registered practitioner, they do not affect his qualification. A doubt has been raised whether damages are properly recoverable by physicians for words relating to their profession, since their fees are merely honorary (Brown *v.* Kennedy, 33 " Law Journal," Ch. 342); but now by the 31st section of the Medical Act all registered practitioners, physicians included, can recover their fees unless the latter are prohibited by a bye-law of their College; the actual decisions, even before the passing of the Act, leave no doubt upon the subject (3 Wilson, 59); and even if their situation be considered as merely confidential, their right to recover rests upon the same foundation with that of magistrates and others, whose offices are of a similar description. The words are actionable if they impute any want of knowledge, skill, or diligence, in the exercise of an office or vocation; as to say of a surgeon, " I wonder you had him to attend you. Several have died that he has attended, and there have been inquests held upon them." These words were held actionable without the aid of an innuendo (Southee v. Denny, 1 Exch. Rep. 196). So also the words—"He is a bad character. None of the medical men here will meet him," are actionable, as imputing the want of a necessary qualification for a surgeon in the ordinary discharge of his professional duties (*Ib.* p. 202). So also to say of an apothecary, " It is a world of blood he has to answer for in this town; through his ignorance he did kill a woman and two children at Southampton; he did kill J. P. at Petersfield; he was the death of J. P.; he has killed his patient with physick." (Tutty v. Alewin, 11 Modern Rep. 221.)

To say of a physician that he is "no scholar," is actionable, a learned education being considered to be an essential qualification in the medical profession (7 Bac. Abridgment, 269). So where the defendant said of a midwife (Flower's case, Cro. Car. 211), "Many have perished for her want of skill."

In actions for slander in which no indictable offence is imputed, but merely an *immoral one*, the Courts have always regarded them with jealousy, even though the words were spoken of the plaintiff in his profession or business. In most of these cases the Courts have required proof of special damage in order to sustain such actions; the exception being in the case of beneficed clergymen.

Libel, or written slander, is a species of defamation, as has been said, of a more aggravated form, and, for the reasons before mentioned, has always been treated as a more grievous injury than that of verbal defamation; and is therefore the subject of an indictment or criminal information as a criminal offence against society at large, as well as of an action at law for the civil injury inflicted on the individual.

A libel may be expressed by printing, writing, signs, or pictures.

The libel may be contained in any book, letter, placard, pamphlet, or indeed printing or writing of any kind, even on a wall or post: Austen v. Colpepper, Skinner, 123.

It may also be expressed by signs, types, or pictures; as to affix a gallows or other reproachful or ignominious sign at a person's door: Jefferies v. Duncombe, 11 East, 226; or by a picture: 11 Modern Rep. 99, per Chief Justice Holt; or by caricature: Du

Bost v. Beresford, 2 Campbell, 511, per Lord Ellenborough, C.J.; or by exhibiting a person in any shameful or disgraceful position, or in the company of disreputable persons: 5 Coke, 125; Skinner, 123; Raymond, 431; 3 Keb, 378.

In an action by a physician for words imputing adultery to him, the words were alleged to have been spoken of him "in his profession;" no special damage was laid. After verdict for the plaintiff judgment was arrested on the ground that such words, alleged merely to be spoken of the plaintiff "in his profession," were not actionable without special damage; and that, if spoken so as to convey an imputation upon his conduct in his profession, the declaration ought to show how the speaker connected the imputation with the professional conduct. Ayre v. Craven, 2 Adolphus & Ellis, 7; and see Southee v. Denny, 1 Exch. 202 (ante). This case, it seems then, was determined upon a legal technicality and not upon the merits. Upon a distinct finding of a jury, that the words were not spoken with reference to the plaintiff's profession, but that they had a "tendency to injure him morally and professionally," it was held that upon this finding the words were not actionable. Dayley v. Roberts, 3 Bingham's New Cases, 835.

As in the case of oral slander, so in the case of written slander, or libel, affecting a person in his office, profession, trade, or business, an action will lie without proof of actual malice or special damage. It will not be necessary to cite cases here which apply equally to this species of defamation whether the action be for words spoken merely, or deliberately written. It is at the same time evident from the authorities that words written, tending to disparage a man in his profession, will support an action when the same words spoken will not. The law, however, respects communications made in confidence, even though they be false and erroneous, and may tend to injure the party referred to, so long as they are not malicious, nor, under the cloak of private confidence, are intended to defame. But, in the case of a public officer, the motives under which he acts in doing a duty which it is incumbent upon him to do cannot make the doing of that duty actionable, however malicious they may be. Dawkins v. Paulet, 21 Law Times Rep. N.S. 584, Q.B. Although this was an action by a military officer against his superior, the rule is equally applicable to *medical* officers.

It has been held that to write and publish of a physician that he has met homœopathists in consultation is no libel, though it be alleged in the declaration that, in the opinion of the profession, meeting homœopathists in consultation would be a breach of professional etiquette, and injurious to the professional character, reputation, and practice of a physician. Clay v. Roberts, 2 Jurist (N. S.)

580. When a medical practitioner, by way of advertisement, published large portions of his book, occupying from time to time whole columns of the *Times* and other newspapers, the effect of which was to represent to the public that he was in possession of a specific remedy for the cure of consumption, the defendant, in commenting thereon with bitterness and severity, denounced the author as an impostor and a quack; and also (by reason of the plaintiff describing himself as M.D. when he had only an American diploma) as like " scoundrels who utter base and forged coin :" and there being evidence that the plaintiff's publications contained statements that were fallacious, delusive, exaggerated, and alarming, it was held that the subject of the plaintiff's publications and representations was matter of public interest and public concern, and a fair and proper subject for public comment; and that if the defendant in commenting thereon as a public writer really believed them to be fallacious, delusive, &c., he was justified in asserting that they were so; and that if in drawing therefrom inferences of imposture and bad intention, he fell into error, yet if he wrote honestly and with the intention of exercising his vocation as a public writer fairly, and with reasonable moderation and judgment, he was privileged by the occasion. Hunter *v*. Sharpe, 4 F. & F. 983. And it was observed by Chief Justice Cockburn, in his charge to the jury :—" Here is a man challenging public criticism, by bringing forward what professes to be a new system of treatment, and inviting the public to adopt it as the only means of curing the most destructive disease known among us. In doing this he challenges public criticism, and if a public writer, using a reasonable degree of temper and moderation, as behoves any one who makes imputations upon others,—if a public writer, thus discussing the subject in the exercise of his vocation, falls into error as to the facts or the inferences, and goes beyond the limits of strict truth, he is nevertheless privileged. The occasion is a privileged one, and if the privilege is exercised honestly, faithfully, and with reasonable regard to what truth and justice require, then, though he may exceed the limits of what he can logically prove to be the truth, he is protected from liability. It is not, therefore, necessary that the justification should appear to you to be made out, if you think that the defendant, or the writer, was in the reasonable and honest exercise of his vocation as a public writer, even although he was not fully warranted in drawing the inferences he did as to the conduct of the plaintiff, and though it may be that he was not entirely justified by the absolute truth." (*Ib.*)

In the case of words published by writing it is only necessary that they should be calculated to degrade or disparage the plaintiff, and hold him up to ridicule or contempt, in order to make them actionable.

With respect to the evidence in support of a Criminal information or indictment, whatever is a sufficient publication of a libel to entitle a plaintiff in a civil action to a verdict, is equally so in a criminal proceeding, with this addition, that the sending of a libel to the individual reflected on, without exposing its contents to a third person, is a sufficient publication to support an indictment, on account of its tendency to provoke that individual to commit a breach of the peace. 1 Wms. Saund. 132 n. 2 ; 2 Esp. 226; 5 Mod. 163.

And the defendant may be found guilty of the publishing, and acquitted of the composing or printing of a libel, where both are conjointly alleged. R. v. Hunt and another, 2 Camp. 583; R. v. Hart, 10 East. 94.

Until recently, there was a great distinction between justifications in civil and criminal proceedings for libel. In the former, the truth of the defamatory matter was a ground of justification, in the latter it was not. The truth is a justification in a civil action, principally because the very foundation fails on which the claim to damages might otherwise be erected—that foundation being the falsity of the defamatory matter. On the other hand, the *tendency* of the defamation to produce a breach of the peace, is of the essence of the offence as far as the public are concerned ; and, therefore, the truth or falsity of the publication is collateral to the offence, and it is obvious that the imputation may be not the less provoking because it is true.

An important alteration has however been made in the law in this respect by a comparatively recent statute (6 & 7 Vict., c. 96), by section 6 of which Act it is enacted, that on the trial of any indictment or information for a defamatory libel, the defendant having pleaded such plea as thereinafter mentioned, the truth of the matter charged may be inquired into but shall not amount to a defence, unless it was for the public benefit that the said matters charged should be published ; and that, to entitle the defendant to give evidence of the truth of such matters charged as a defence to such indictment or information, it shall be necessary for the defendant, in pleading to the said indictment or information, to allege the truth of the said matters charged, in the manner now required in pleading a justification in an action for defamation ; and further to allege that it was for the public benefit that the said matters charged should be published ; to which plea the prosecutor shall be at liberty to reply generally, denying the whole thereof ; and that if after such plea, the defendant shall be convicted, it shall be competent to the Court, in pronouncing sentence, to consider whether the guilt of the defendant is aggravated or mitigated by the plea in question. So that even now in criminal proceedings

E

the truth of the libel is no defence unless it was for the public benefit that the matters charged should be published. And it may be added that whenever the publication of a libel is criminal as concerns the public, it constitutes also a civil injury reparable in damages at the suit of the party calumniated. If a person, who has either read a libel himself, or who heard it read by another, afterwards maliciously reads *or repeats* any part of it to another, he is guilty of an unlawful publication of it. Hawkins' Pleas of the Crown 62 c. 73 s. 10. But the so reading it without knowing it to be a libel, with or without malice, does not amount to a publication. R. v. Dodd. 2 Sessions Cases 33; Holt's Law of Libel 284, 2nd Edit.

Where a libel is printed, the sale of each copy is a distinct publication, and a fresh offence, and a conviction for one such offence will be no bar to an indictment for publishing another copy. R. v. Carlile, 1 Chitty 451; 2 Starkie on Slander 320, 2nd Edit.

Where a man publishes a writing upon the face of it libellous, the law presumes that he did so with a malicious intent, and it is unnecessary for the prosecution to give evidence of circumstances from which malice may be inferred. R. v. Harvey, 2 Barnewall and Cresswell 257; R. v. Burdett, 4 Barnewall and Adolphus 95.

Where, however, the expressions are ambiguous, or the intentions doubtful, evidence may be adduced for showing the malice which prompted the publication.

In the case of servants' or assistants' characters, such communications are privileged where made to persons who have a right to ask it or expect it. In all such cases malice in fact must be proved. Per Justice Bayley in Bromage v. Prosser, 4 B. & C. 256; McPherson v. Daniels, 10 B. & C. 272 : " Where a man has a right to make a communication you must either show malice intrinsically from the language of the letter, or prove express malice. Per Baron Parke, Wright v. Woodgate, Tyrrwhitt & G. 15. And wherever an action will lie for a libel without laying special damage, an indictment will also lie. And also wherever an action will lie for *verbal* slander without laying special damage, an indictment will lie for the same words if reduced to writing and published; but the converse of this proposition will not hold good, for an indictment may lie for words written which could not be maintained if merely spoken. Thorley v. Lord Kerry, 4 Taunton 355; for instance, if a man write or print and publish of another that he is a scoundrel or villain it is a libel; although if this were merely spoken, it would not be actionable without special damage. 2 H. Blackstone 531.

An instance of special damage is the following :—In an action by a surgeon or accoucheur, for slander imputing to him, in words spoken to D, that a female servant had a child by him; whereby (as special damage) D would not employ him as an accoucheur to attend his wife, and the declaration alleged that the plaintiff was otherwise injured in his business, and that the profits thereby had been diminished: it was held that the plaintiff's damages were not limited to the mere loss of the fee for attending D's wife in her confinement, and that the jury might give damages for the loss of business arising directly from the slander spoken by the defendant to D, but that he was not entitled to such general damages as might be supposed to have arisen from repetition of the slander by other persons. Dixon v. Smith, 5 Hurlstone & Norman 451. 29 Law Journal (Exch.) 125.

It may here be mentioned that by the 19 & 20 Vict., c. 108, s. 23 (County Courts Extension Act), any action for libel or slander may be tried in a county court by consent of both parties to a memorandum to that effect signed by them or their respective attorneys, and specifying the county court in which they agree to try the action.

And by the County Court Act of 1867 (30 & 31 Vict., c. 142), it is enacted by section 5, that in any case of tort, where the action is brought in a superior court, and the plaintiff does not recover damages exceeding 10*l*., he will not be entitled to costs, unless the judge certifies that there was sufficient reason for bringing the action in the superior court. But no original jurisdiction is given to the county court in actions of libel or slander, so that an agreement between the parties is still required in order to try such cases in the county court. But by the 10th section of the same Act, any person against whom any action for libel or slander is brought in a superior court, may make an affidavit that the plaintiff has no visible means of paying the costs of the defendant, should a verdict be not found for the plaintiff, and thereupon a judge of the court in which the action is commenced may make an order that unless the plaintiff shall, within a time to be therein named, give full security for the defendant's costs, or satisfy the judge that he has a cause of action fit to be tried in a superior court, all proceedings in the action are to be stayed; or in the event of the plaintiff not giving such security or not so satisfying the judge, that the cause be remitted for trial before a county court to be therein named; and the county court so named is to have all the same powers and jurisdiction as if both parties had agreed by a memorandum signed by them that the county court should have power to try the action; and the costs of the parties incurred subsequent to the order of the judge of the superior court are to be allowed according to the scale of costs in use in the county courts.

There can be no doubt that this provision will materially diminish, if not entirely get rid of a class of vexatious actions, brought either to compel a compromise, or for the sake of costs, for thus whatever may be the amount of damages claimed the plaintiff is forced into the county court, and cannot therefore either gratify his attorney's avarice or his own vindictiveness by inflicting a high scale of costs upon his opponent, where such unworthy motives have prompted the litigation.

As vindication of character is the main object of legal proceedings for defamation, except where special and serious pecuniary loss has been sustained through the publication of the defamatory matter, the defendant is permitted by 6 & 7 Vict., c. 96, in actions for libel only, tried either in the superior or the county court, to give in evidence (after due notice in writing of his intention so to do, given before trial) in mitigation of damages, that he made or offered an apology to the plaintiff *before* the commencement of the action, or as soon afterwards as he had an opportunity to do so, in case the action were commenced before he had an opportunity of offering such apology. Such a privilege, however, is limited to cases of libel only and does not apply to slander, and exclusively to such libels as are contained in any public newspaper or other periodical, and such plea must be accompanied by the payment of money into court by way of amends, otherwise the plea shall be deemed a nullity; sec. 2. The defendant is not, therefore, allowed to escape altogether a pecuniary penalty, and he must also state in such plea that such libel was published without actual malice and without gross negligence, otherwise it will not avail him.

CHAPTER VI.

LAW OF COPYRIGHT, ESPECIALLY AS APPLIED TO WRITINGS AND LECTURES OF MEDICAL MEN.

THE subject of copyright is far too extensive to be discussed at any length in these pages, but a few observations on the rights of medical men especially, may nevertheless be useful and within the object as well as the limits assigned to this treatise. The term copyright includes two rights—the right exists in every unpublished innocent production of art, literature, or science, and may be termed copyright before publication. It exists *a fortiori* in every such published production, and may be termed copyright after publication. In reference to the first of these, viz., copyright before publication, it may be said generally that when any material has embodied the ideas which spring up in the mind of the author or the artist, the law protects their privacy until the material has itself been, by its owner, given to the world. In the language of Mr. Justice Yates, 4 Burrows, 2378, " A man has certainly the right to consider whether he will make his ideas public, or commit them only to the sight of his friends; in that state a manuscript is in every sense his peculiar property, and no man can take it from him, or make a use of it which he has not authorized, without being guilty of a violation of his property. And as every author or proprietor of a manuscript has a right to determine whether he will publish it or not, he has a right to the first publication, and whoever deprives him of that privilege is guilty of a manifest wrong, and the Court have a right to stop it."

In the case of Abernethy v. Hutchinson, 3 Law Journal (Chancery), 209, 213, 219, Mr. Abernethy had filed a bill to restrain the publication in the *Lancet* of certain lectures delivered by him orally at St. Bartholomew's Hospital. Lord Eldon refused to grant the injunction, because the plaintiff was unable to swear that his whole lecture had been reduced to writing at the date of its delivery. Lord Eldon observed : " Where the lecture is orally delivered, it is difficult to say that an injunction can be granted upon the same principle upon which literary composition is protected, because the Court must be satisfied that the publication complained of is an invasion of the work, and this can only be done by comparing the composition with the piracy." His lordship, however, upon a second argument, granted the relief prayed, but on a different ground, viz., that of a breach of an implied contract between the lecturer and his audience, that the latter would do nothing more than listen to the lecture for their instruction. The earliest case of an infringement of copyright before publication, appears to be Webb v. Rose, cited in Burrows, 2330, which was decided in the year 1732. Sir Joseph Jekyll, Master of the Rolls, there granted an injunction against a conveyancer's clerk, when he threatened to publish his master's conveyancing drafts as legal precedents, not having first obtained the draftsman's permission. Forrester v. Waller, cited as *supra*, is a somewhat similar case; an injunction was granted against printing without his consent the plaintiff's notes of cases. Again, in 1754, a clerk of Sir John Strange made a clandestine abridgment of his master's MSS., with a view to publish them. Lord Hardwicke did not hesitate to restrain such publication.

The Prince Albert v. Strange is almost the only modern case, as it is the most leading case on copyright before publication. As etchings or lithographic sketches or photographic pictures may sometimes form the private collection of a medical practitioner's professional experiences, the result perhaps of much labour and investigation, which he may or may not purpose some day giving to the world, a brief narration of the facts of Prince Albert v. Strange may not be here out of place. Her Majesty and the Prince had, solely for their own amusement, made certain etchings, and from these they had struck off a few lithographic impressions for their own use, and not for the purpose of publication; other impressions they had ordered to be struck off, and some of these latter had been surreptitiously retained by one of the pressmen employed in the operation, and from him they had passed into the possession of the defendant, a London publisher, who declared his intention of publicly exhibiting them, and also of selling to the public a descriptive catalogue of the lithographs. The injunction was granted by the Vice-Chancellor, and on appeal his judgment was affirmed by Lord-Chancellor Cottenham. An extract from the judgment of the Vice-Chancellor may not be uninteresting to those to whom perhaps the doctrine of copyright before publication is not as yet familiar. "Upon the principle," said the Vice-Chancellor, "of protecting property, it is that the Common Law, in cases not aided nor prejudiced by statute, shelters the privacy and seclusion of thoughts and sentiments committed to writing, and desired by the author to remain not generally known. Such then being, as I believe, the nature and foundation of the Common Law as to manuscripts, its operation cannot of necessity be confined to literary subjects. That would be to limit the rule by its example. Wherever the produce of labour is liable to invasion in an analogous manner, there must, I suppose, be a title to analogous protection or redress. To consider, then, the case of mechanical works, or works of art executed by a man for his private amusement, or private use, whatever protection they may have by act of parliament, they are not, I apprehend, deserted by the Common Law. The principles and rules which it applies to manuscripts must, I conceive, be to a considerable extent, at least, applicable to these also. Mr. Justice Yates, in Millar v. Taylor, said that an author's case was exactly similar to that of an inventor of a new mechanical machine; that both original inventions stood upon the same footing in point of property, whether the case were mechanical or literary, whether an epic poem or an orrery; that the immorality of pirating another man's invention was as great as that of purloining his ideas. Property in mechanical works, or works of art, executed by a man for his own amusement, instruction, or use, is allowed to subsist certainly, and may before publication by him be invaded, not merely by copying, but by description, or by catalogue, as it appears to me. A catalogue of such works may in itself be valuable. It may also as effectually show the bent and turn of the mind, the feelings and taste of the artist, especially if not professional, as a list of his papers. The portfolios of the studio may declare as much as the writing table." This extract will perhaps be sufficient to elucidate the principle involved in the decision.

No one can have copyright before publication in a work calculated by publication to do injury to society and to offend against the law.

The law on this point was very fully discussed in Lawrence v. Smith, 2 Merivale, 440. The plaintiff, the eminent surgeon, had delivered at the College of Sugeons, and had afterwards published, a work under the title of "Lectures on Physiology, Zoology, and the Natural History of Man." The bill was filed to restrain the defendant from selling a pirated edition. An *ex parte* injunction having been obtained, the defendant moved to dissolve it, on the ground that the lectures were so hostile to natural and revealed religion, that they ought to have no protection. Lord Eldon dissolved the injunction, and in doing so said, among other things, "Looking at the general tenor of the work, and at many particular parts of it, recollecting that the doctrine of the immortality of the soul is one of the doctrines of the scriptures, considering that the law does not give protection to those who contradict the scriptures; and entertaining a doubt, I think a rational doubt, whether the book does not violate that law, I cannot continue the injunction." On similar grounds, Lord Eldon refused the motion for an injunction to restrain the publication of a pirated edition of Lord

Byron's "Cain," and Sir John Leach, Vice-Chancellor, in 1823, on similar principles, dissolved an injunction which had been obtained *ex parte* against the publication of a pirated edition of a portion of "Don Juan."

It may well be doubted whether the above decisions would be deemed guides in the present day.

The Act, 8 Anne, c. 19, was the first statutory dealing with copyright after publication. As the name implies, the right commences from the moment the eye of the public is allowed to rest upon the subject. This Act, however, only dealt with literary work, and copyright after publication in literary works is now regulated by an Act 5 & 6 Vict. c. 45 (commonly called Talfourd's Act), by which former Acts are repealed, and thenceforth "the copyright in every book which shall be published in the lifetime of the author shall endure for the natural life of such author, and for the further term of seven years, commencing at the time of his death; and shall be the property of such author and his assigns: Provided that if the said term of seven years shall expire before the end of forty-two years from the first publication of such book, the copyright shall, in that case, endure for such period of forty-two years : and that the copyright in every book which shall be published after the death of its author shall endure for the term of forty-two years from the first publication thereof, and shall be the property of the proprietor of the author's manuscript, from which such book shall be first published, and his assigns."

And further : "The copyright which at the time of the passing of this Act shall subsist in any book theretofore published, shall be extended for the full term provided by the Act in cases of books thereafter published, and shall be the property of the person who, at the time of the passing of the Act, shall be the proprietor of such copyright : Provided that, in all cases where such copyright shall belong in whole or in part to a publisher or other person who shall have acquired it for other consideration than that of natural love and affection, such copyright shall not be extended by the Act, but shall endure for the term which shall subsist therein at the time of the passing of the Act, and no longer, unless the author if he be living, or his personal representative, if he be dead, and the proprietor of such copyright shall, before the expiration of such term, consent and agree to accept the benefits of the Act in respect of such book, in which case such copyright shall endure for the full term by the Act provided in cases of books to be published after the passing of the Act."

This Act passed on 1st July, 1842, and a previous Act, 5 & 6 Wm. IV., c. 65, entitled an "Act for preventing the publication of lectures without consent," enacts that from the date of that Act

(Sept. 1, 1835) the author of any lecture, or the person to whom he has sold or otherwise conveyed the copy, in order to deliver the same in any school, seminary, institution, or other place, or for any other purpose, should have the sole right of printing and publishing such lecture.

A penalty is attached to a violation, and the Act declares that if any person shall, by taking down a lecture in shorthand, or otherwise in writing, or in any other way obtain a copy, and shall print or lithograph, or otherwise copy and publish it without the leave of the author, or his assignee, and every person, who knowing the same to have been printed, &c., without such consent, shall sell, publish, or expose to sale any such lecture, shall forfeit such print or copy, together with one penny for every sheet thereof found in his custody, either printed, lithographed, or copied, or printing, lithographing, or copying, or published or exposed to sale contrary to the Act, one moiety to the Crown, and the other to the prosecutor; and that any printer or publisher of any newspaper who shall, without such consent, print and publish in such newspaper any lecture or lectures, should be deemed a person printing and publishing without leave, within the provisions of the Act, and liable to the aforesaid penalty and forfeiture in respect of such printing and publishing.

Lectures delivered in any university, or public school, or college, or on any public foundation, or by any individual in virtue of or according to any gift, endowment, or foundation, are excluded from the protection of the Act 5 & 6 Wm. IV., c. 65. Also lectures of the delivery of which notice in writing shall not have been given two days previously to two justices living within five miles of the place of delivery; sec. 4.

The case of Abernethy v. Hutchinson, before referred to, occurred before the passing of this Act; and though decided upon the ground of breach of trust only, the point involved is expressly provided for under the third section of that Act, thus: "No person allowed, for certain fee and reward, or otherwise, to attend and be present at any lecture delivered in any place, shall be deemed and taken to be licensed or have leave to print, copy, and publish such lectures only because of having leave to attend such lecture or lectures." Mr. Abernethy's lectures were delivered at St. Bartholomew's Hospital, and it might at first sight, therefore, appear that they would not enjoy the protection of the Act, but would probably come within the exception excluded from such protection by the 5th section, as having been delivered in a "public school" of medicine or surgery, "or on a public foundation," or "by an individual in virtue of or according to some gift, endowment, or foundation," but this would, of course, depend upon the definition of the words "public school," &c., and whether the lecturer at an institution supported purely by voluntary contributions (though, doubtless, St. Bartholomew's Hospital is otherwise endowed), can be said to deliver his lectures "in virtue of any gift, endowment, or foundation." As the section concludes by declaring " that the law relating thereto shall remain the same as if this Act had not been passed," it would at all events seem that in such a case the lectures would be protected where, as was there alleged, "no persons had a right to attend the lectures except those who were admitted to that privilege by the lecturer; that it had always been understood by him, and by those who preceded him in the office, and those who attended the lectures, that the persons who so attended did not acquire, and were not to acquire, any right of publishing the lectures which they heard; but that the plaintiff and his predecessors respectively had and retained the sole and exclusive right of printing and publishing their respective lectures, for his and their own respective benefit; that there was an implied contract between the plaintiff and those who attended his lectures, that none of them should publish his lectures, or any part thereof; that the defendants had been furnished with a copy of the lectures, which they had printed, through the medium of some person who had attended the lectures under Mr. Abernethy's above-mentioned permission: and that it was a breach of contract or trust in such person so to furnish the copy, and in the defendant to print and publish the same." Upon these allegations, the Lord Chancellor said "The point to be determined was whether there was such a violation of the contract as to sustain an action; if not whether an injunction could be asked for. . . . . He had no doubt whatever an action would lie against a pupil who published these lectures. One question had been, whether Mr. Abernethy, from the peculiar situation which he filled in the hospital, was precluded from publishing his own lectures for his profit; but there was no evidence before the Court that he had not such right. Therefore the defendants must be enjoined in future." (Extract from Lord Eldon's judgment in the case.)

A lecturer now, even without proof of an implied contract between himself and the audience, has a property in his lectures, though orally delivered, and is entitled to the benefit of the protection afforded by the 3rd section of the Act, provided at least he cannot be brought within the 4th section, as before mentioned.

It is frequently difficult to decide what amounts to an infringement of copyright—whether in fact the use made of a work is fair and reasonable, or is only colourably and evasively different from it. Sir R. Kindersley has defined unfair use of an original work to be the extraction of its vital part. Murray v. Bogue, 1 Drew 369, and Mr. Justice Maule, in Sweet v. Benning, 16 Common Bench 485, said—it was "not so much of kind as of degree." In Bohn v. Bogue, 10 Jurist 420, Sir L. Shadwell, V.C.E, remarked: "In cases of this nature, if the pirated matter is not considerable, that is, where passages which are neither numerous nor long have been taken from different parts of the original work, this Court will not interfere to restrain the publication of the work complained of, but will leave the plaintiff to seek his remedy at law." The dicta of the Vice-Chancellor in the case last mentioned, must not, however, be relied on as an argument that piracy is to be judged solely by the quantity taken from the copyright work. Lord Cottenham has said, in reply to an argument of that description, "When it comes to a question of quantity it must be very vague; one writer may take all the vital part of another's book, though it might be a small proportion of the book in quantity: it is not only quantity but value that is always looked to. It is useless to refer to any particular case as to quantity." Bramwell v. Halcomb, 3 M. and Cr. 738, and see Sir W. P. Wood, V.C. in Tinsley v. Lacy, 111 W.R. 877. Licher's observation is very pertinent to this point. "The thief only takes the wheat, not the straw, which is the bulk of the crop." And said Mr. Justice Story, in Gray v. Russell, 1 Story R. 11, "Extracts non numerantur sed ponderantur. The quintessence of a work may be piratically extracted so as to leave a mere caput mortuum, by a selection of all the important passages in a comparatively moderate space." But this is matter rather for the lawyer than the medical practitioner, or even the medical author, and as "sufficient for the day is the evil thereof," proceedings against another for infringement of a copyright would so inevitably involve the retention of the services of a legal adviser on the occasion, that any mere general information that this treatise might supply would furnish but little practical assistance. It only remains, therefore, to observe that the first International Copyright Act was 1 and 2 Vict. c. 59. It embraced books only, and was repealed by 7 and 8 Vict. c. 12, and this Act was followed by a convention between this country and France. The convention was concluded at Paris, November 3rd, 1851, for an extension to each country of a reciprocal copyright in works of literature and art. The convention was followed by an Order in Council of January 10, 1852, to the following effect. "From and after the 17th of January, 1852, the authors of works of literature or of art, in which the laws of Great Britain give to British subjects the privilege of copyright, and the executors, administrators, and assigns of such authors, &c., shall as respects works published within the dominions of France after the above date of the Order, have the privilege of copyright therein for a period equal to the term of copyright which authors, &c., of the like works respectively first published in the United Kingdom are by law entitled to, provided such books, &c., have been registered, and copies thereof have been delivered according to the requirements of the said recited Act (7 & 8 Vict. c. 12) within three months after the first publication thereof in any part of the French dominions, or, if such work be published in parts, then within three months after the publication of the last part thereof. It seems that a newspaper is not a work published in parts within the meaning of the provision, because it could not have been intended that at any period, however remote, the publisher of such a work might register it and carry back his copyright therein to the earliest period in 1852, when French authors had a copyright in this country. Sir W. P. Wood (V.C.) in Cassell v. Stiff, 2 Ka. & Jo. 279. In order to enable her Majesty to make similar stipulations in any treaty with any other foreign powers, Parliament passed another Act (15 & 16 Vict., c. 12) the first 9 sections of which are incorporated with the 7 & 8 Vict., c. 12, and sec. 18 of the earlier Act is repealed by the latter so far as the same is inconsistent therewith. Under that Act the Queen in Council may now direct that the authors of books published after a specified day in any foreign country, their executors, &c., may (subject to the provisions of the 15 & 16 Vict., c. 12) prevent the publication in the British dominions of any translations of such books not authorized by them for a period (to be specified by her Majesty) not exceeding five years from the first publication of an authorized translation, and in the case of books published in parts for a period not exceeding, as to each part, five years from the first publication of an authorized translation of that part, 15 & 16 Vict., c. 12, sec. 2. The laws which protect British copyright are, by the same Act, conditionally extended to such authorized translations.

But no author, his executors, &c., can have the benefit of the Act in respect of any translation of a book, &c., unless within three months of its first publication he register and deposit a copy in the manner required for original works by 7 & 8 Vict., c. 12, and unless the author notifies in the titlepage (if any) or on some conspicuous part of the book, that it is his intention to reserve the rights of translating it, and unless the authorized translation, or a part thereof, is published either in the foreign country (named in the order by which the translation is protected) or in the British dominions, within one year, and wholly published within three years after the registry and deposit of the original work under the Act; and unless the translation itself be registered and a copy of it deposited within a period (to be named in the order) relating to such translation, and in the manner prescribed by 7 & 8 Vict., c. 12, for the registry and deposit of original works. The above requisitions apply to articles originally published in newspapers or periodicals, if the same be afterwards published in a separate form, but they do not apply to such articles as originally published. (Sec. 8.)

To give effect to these stipulations, the Act 15 & 16 Vict., c. 12, expressly declared that during the continuance of the convention the provisions of the Act should apply to it, and to translations of books, &c., which should, after the passing of the Act, be published in France, in the same manner as if an Order in Council had been issued for giving effect to the convention, and had therein directed that such translations should be protected for five years from the first publication, and as if a period of three months from the publication of such translation was specified in the order as the period for registering and depositing a copy of the translation (S. 11). And further, any article of political discussion which has been published in any newspaper or periodical in a foreign country, may, if the source from which the same is taken be acknowledged, be republished, or translated in any newspaper or periodical in this country. Any article relating to any other subject may, if its source be acknowledged, be also reprinted or translated, in like manner, unless the author has signified his intention of reserving the copyright therein and the right of translating the same in some conspicuous part of the newspaper or periodical in which the same was first published, in which case the same will, without the formalities required by section 8 of the 15 & 16 Vict., c. 12, receive the same protection as is by virtue of 7 & 8 Vict. c. 12, and 15 & 16 Vict., c. 12, extended to books (S. 7.)

On a motion to restrain the infringement of an alleged copyright in a French newspaper, Sir W. P. Wood (V.C.) intimated grave doubts whether the protection just alluded to did not mean the same protection which the author of any book would obtain under an Order in Council made pursuant to the International Copyright Acts. He therefore refused to assist the plaintiff because the requisitions of the Order in Council governing international copyright in French newspapers as to registry had not been complied with. He directed the motion to stand over, with liberty to the plaintiff to bring an action, and liberty to all parties to apply. (Cassell *v.* Still, *supra*.)

The Act 15 & 16 Vict., c. 12, prohibits the importation into any part of the British dominions without the consent of the registered proprietor or his agent authorized in writing, of any copies of any works of literature or art in which there is copyright under the International Copyright Acts, and which have been printed or made in any foreign country except that country in which such work shall have been first published; also all unauthorized translations of it, and extends to such copies and translations sections 17 & 23 of the 5 & 6 Vict., c. 45.

The French law of copyright, taken in connexion with the French convention, seems to admit the work of any alien holding a French copyright to the benefit of the convention. (See 1 Jurist, N.S. pt. 2, p. 523.)

International copyright has since been arranged with Prussia, Saxony, Brunswick, Thuringia, Hanover, Oldenburgh, Anhalt, Hamburgh, Belgium, Spain, Sardinia, and Hesse. In the case of Avoneys v. Mudie, the Court of Exchequer has decided that the proprietor of a foreign print, who claims a copyright therein under the International Copyright Acts, must, in reference to such print, comply with the provisions of our own engraving Acts (10 Exch. Rep. 203). By 25 & 26 Vict., c. 91, sec. 2, the copyright in the Pharmacopœia is vested in the "General Council of Medical Education and Registration of the United Kingdom."

CHAPTER VII.

SCHOOLS OF ANATOMY.

The Act for regulating schools of anatomy is that of 2 & 3 Wm. IV., c. 75. By this Act, the Secretary of State for the Home Department in Great Britain, or the Chief Secretary in Ireland, is empowered to grant a license to practise anatomy to any fellow or member of any College of Physicians, or Surgeons, or to any graduate or licentiate in medicine, or any person lawfully qualified to practise medicine in any part of the United Kingdom, or to any professor or teacher of anatomy, medicine, or surgery, or to any student attending any school of anatomy, on application from such party for such purpose, countersigned by two of his Majesty's justices of the peace, acting for the county, city, borough, or place, wherein such party resides, certifying that to their knowledge or belief such party so applying is about to carry on the practice of anatomy.

Under this Act, inspectors of places where anatomy is carried on are appointed, every such inspector to make a quarterly return of every deceased person's body that during the preceding quarter has been removed for anatomical examination to every separate place in his district where anatomy is carried on, distinguishing the sex, and, as far as is known, the name and age of such persons. Such inspectors to visit and inspect, at any time, any place in their separate districts, notice of which has been given that it is intended there to practise anatomy.

Any executor, or person having the lawful possession of the body of any deceased, and not being an undertaker or other person entrusted with the body only for the purpose of interment, may permit the body of the deceased to undergo anatomical examination unless such deceased person shall have expressed his desire (in the mode indicated in the Act) that his body after his death might not undergo such examination, or unless the surviving husband or wife, or any known relative, shall require the body to be interred without such examination. If the deceased himself has directed (in the manner therein indicated) that his body after death shall be examined anatomically, and such direction be made known before burial to the party in lawful possession of the body, then any person so authorized as beforementioned to make such examination shall be requested, by the person in lawful possession of the dead body, to make such examination, unless an objection is made by certain relatives as before. And any person duly licensed to perform such examinations may receive or possess the body of any such person for the purpose of anatomical examination if permitted so to do by the party having lawful possession of the body, provided a certificate of the illness of which the deceased died, or the cause of death, be delivered with the body, which certificate the party so receiving the body shall within twenty-four hours after the removal of the body transmit to the inspector of the district, or if there be no inspector to some medical man residing at or near the place to which the body is thus removed, and also a return of the day and hour when, and from whom the body was received, the date and place of death, the sex (and as far as is known) the Christian and surname, age, and last place of abode, and shall enter such particulars and a copy of the certificate in a book to be kept by him for the purpose, and shall produce such book whenever required to do so by an inspector.

And no one shall carry on or teach anatomy at any place, or receive or possess for anatomical examination, or examine anatomically any deceased person's body after its removal unless such person, or the owner or occupier of such place, or some person legally authorized to examine bodies anatomically, shall have previously given to the Secretary of State notice of the place where it is intended to practise anatomy, at least one week before the first receipt or possession of a body for such purpose. Such body after undergoing anatomical examination is to be decently buried in consecrated ground, or in some public burial ground, such provision to be made by the party removing or causing the body to be removed as aforesaid, and a certificate of such interment to be transmitted to the inspector within six weeks of the reception of the body. Any person offending against the provisions of this Act is to be deemed guilty of a misdemeanour, and be liable to three months' imprisonment or a fine of 50l.

The medical man who certifies to the cause of death shall not be concerned in examining the body after removal, and no such removal shall take place till forty-eight hours after death, nor until after 24 hours' notice to the inspector of the district, or otherwise, of the intended removal of the body.

A crime hitherto unheard of gave occasion to the passing of the above Statute. Such is unmistakeably recorded in the preamble of the Act itself as follows:—" Whereas a knowledge of the causes and nature of sundry diseases which affect the body, and of the best methods of treating and curing such diseases, and of healing and repairing divers wounds and injuries to which the human frame is liable, cannot be acquired without the aid of anatomical examination. And whereas the *legal* supply of human bodies for such anatomical examination is insufficient fully to provide the means of such knowledge. And whereas, in order further to supply human bodies for such purposes, divers great and grievous crimes have been committed, and lately murder, for the single object of selling for such purposes the bodies of the persons so murdered. And whereas it is highly expedient to give protection, under certain regulations, to the study and practice of anatomy, and to prevent, as far as may be, such great and grievous crimes and murders as aforesaid: Be it therefore enacted," &c., &c. Thus is indelibly recorded in the statute-book of England that in the nineteenth century of the Christian era there were wretches prowling about and deliberately murdering their unoffending fellow-creatures for the sole purpose of themselves living like ghouls upon the dead bodies of their victims. Burke and Hare were the names of the principal offenders, who, having met with the fate they so richly deserved, were the diabolical cause which resulted in the present legal provision for the due supply of the anatomical demand.

The preamble of the statute more than hints at "divers" other "great and grievous crimes" which "had been committed" in order to meet the increased demand for subjects, and possibly the crime of disinterring and carrying away dead bodies for the purpose of dissection is here alluded to; for though, of course, such a crime bears no comparison in atrocity with that of Burke and Hare, yet a violation of the sanctity of the grave has ever been considered, by all nations and creeds alike, to be a gross outrage upon public decency, and a cruel insult to the feelings of disconsolate relatives and friends.

The first charge on record relating to digging up graves was preferred against one Haynes, in the time of James I., but he was charged only with stealing the winding-sheets, he having returned the bodies to the sepulchre after divesting them of these habiliments. In East's "Pleas of the Crown," vol. ii. p. 652, we find a note of this case among others. It is there stated, "There can be no property in a dead corpse, and therefore stealing it is no felony, but a very high misdemeanour."

In the case of Doctor Haudyside, where trover was brought against him for two children that grew together, Lord Chief Justice Willes held the action would not lie, as no person had any property in corpses. But a shroud stolen from the corpse must be laid to be the property of the executors, or whoever else buried the deceased, and not of the deceased himself. Sir Edward Coke, who, as he states, advised upon the case of Haynes, "for the rareness of the case, consulted with the judges at Serjeants' Inn, in Fleet-street; where we all resolved, that the property of the sheets was in the executors, administrators, or other owner of them, for the dead body is not capable of any property, and the property of the sheets must be in somebody: and according to this resolution he was indicted for felony at the next assizes, but the jury found it but petty larceny, for which he was whipped, as he well deserved." (Coke's Inst. P.C. p. 110.) His words in the marginal note of this case are *furtum inauditum.*

In 1788, one Lynn was convicted on an indictment charging him with entering a certain burying ground, and taking a coffin out of the earth, from which he took a dead body, and carried it away for the purpose of dissecting it. The only Act having any relation to such a crime was that of 1 Jac. I., c. 72, which prior to this time had been repealed by 9 Geo. II., c. 5, making it felony to take up a dead body out of any place where it rested, for the purpose of using it in any manner of witchcraft, sorcery, charm, or enchantment; but here the charge was, "for the purpose of using it for dissection." The Court, however, said that "common decency required that the practice should be put a stop to. That the offence was cognizable in a criminal court, as being highly indecent, and *contra bonos mores*, at the bare idea alone of which nature revolted. That the purpose of taking up the body for dissection did not make it less an indictable offence." It will be observed that it was the *disinterring*, and not the *dissecting*, that constituted the offence; and the statute with which this chapter is prefaced provides, therefore, for the decent interment of the body after the anatomical examination, and allows of no further disturbance of the remains. Since then, several persons have been punished for disinterring bodies, under the name of *resurrectionists*, and there is also a case (A.D. 1822), of Rex v. Cundick, 1 Dowling and Ryland, M. C. p. 356, establishing that to sell the dead body of a felon capitally executed, for the purpose of dissection, where dissection formed no part of the sentence, is a misdemeanour indictable at common law.

At length it seems to have occurred to the reprobates before mentioned that it was a readier process, and one attended with less risk of detection, to apply a pitch plaister to a living man's mouth and nostrils than to raise a corpse from what ought to be its last resting-place. It may be observed, in conclusion, that the reception of a dead body otherwise than through the medium of supply pointed out by the statute, with any *indicia* of violence upon it, so as to raise a reasonable suspicion of foul play, might place the recipient in a most painfully perilous position in that respect, besides rendering him liable to be punished by fine or imprisonment under the 18th section of the 2 & 3 Wm. IV., c. 75.

The clergyman of the parish, or the directors of a Cemetery Company, or those who possess a vested freehold interest, and whose burying grounds or graves have been violated by any attempt at disinterring, might bring their respective actions of trespass in any such cases, but under existing circumstances such acts of trespass for the future will be perhaps entirely unknown.

CHAPTER VIII.

LAW OF PARTNERSHIP AS AFFECTING THE MEDICAL PROFESSION.

PERHAPS the most accurate definition of the term "partnership" is that given by Mr. Dixon in his work on the "Law of Partnership—I." He there defines a partnership to be "a voluntary and unincorporated association of individuals standing to one another in the relation of principals for carrying out a joint operation or undertaking for the purpose of a joint profit." Such then may be said to be a partnership in the proper sense of the term.

The usual characteristics of a partnership are a community of interest in profits and losses, a community of interest in the capital employed, and a community of power in the management of the business engaged in.

An agreement, therefore, to share, though not necessarily in equal parts, profits and losses, may be said to be the type of a partnership contract. It admits of no doubt whatever that an agreement to share the profits *and* losses arising from trade, business, or adventure, constitutes partnership in that undertaking, even although the words partners or partnership do not occur in the agreement. (Greenham v. Gray, 4 Ir. Com. L. Rep. 501.) And even an agreement to share profits, nothing being said about losses, amounts *primâ facie* to an agreement to share losses also, (Ib. p. 501; Dry v. Boswell, 1 Camp. 330; Keyse v. Burge, 9 C. B. 440,) for it is but reasonable that the chance of gain and loss should be taken by the same persons; and unless, therefore, an intention to the contrary can be shown, persons engaged in any business or adventure, and sharing the profits derived from it, are partners accordingly. Indeed, whether persons are partners is a question of intention, to be determined by a consideration of the whole agreement into which they have entered. This principle is illustrated by those cases in which managers or clerks are paid salaries proportionate to the business in which they are employed. In the case of Rawlinson v. Clarke, 15 M. & W. 292, Clarke was a surgeon, and agreed to sell his practice and drugs and stock in business to Rawlinson, who was to pay for them partly in cash down, and partly in cash at the end of a year. During this year Clarke agreed to carry on his practice as usual, and to introduce Rawlinson to the patients; and

Rawlinson agreed to allow Clarke during this year one-half of the clear profits of the concern, to be paid at the expiration of the year. It was held that the whole agreement showed that Clarke was only to receive a salary for the service rendered to Rawlinson in helping him to continue the business. Formerly it frequently became a question in a criminal proceeding for embezzlement whether the accused were a partner or merely a clerk, servant, or assistant, as, in case he were a partner in the concern defrauded, he could not be convicted of embezzlement. In the case of R. v. Macdonald, the accused, who was paid by a per-centage of profits, having been convicted of embezzlement, was held by the Court of Criminal Appeal to have been rightly convicted, as there was no partnership; and now by 31 & 32 Vict. c. 116, s. 1, it is enacted that " If any person being a member of any co-partnership, or being one of two or more owners of any money, goods or effects, bills, notes, securities, or other property, shall steal or embezzle such property of or belonging to any such co-partnership, or to such joint beneficial owners, every such person shall be liable to be dealt with, tried, convicted, and punished for the same, as if such person had not been or was not a member of such co-partnership, or one of such beneficial owners;" and the case above quoted is therefore now only referred to, together with that of Harrington v. Churchward, 6 Jurist, N. S. 576, as showing that remuneration arising out of a per-centage of profits does not necessarily constitute a partnership, at least as between the person so paid for his services and his masters. In the latter case, the Vice-Chancellor Wood, now Lord Chancellor, held that the contract into which the parties had entered was a contract of hiring and service only. In the above cases, had there been, as there was not, capital or stock common to those who shared the profits, that additional circumstance would have gone far to show that a partnership was in fact intended. (See Reid v. Holinshead, 4 B. & C. 867; *Ex parte* Chuck, 8 Bing. 469; Gilpin v. Enderby, 5 B. & A. 954.) At the same time it is not essential to the existence of a partnership that there should be any joint stock or capital. If several persons labour together for the sake of gain, and of dividing that gain, they will not be partners the less on account of having no common stock, or plant. Thus in Fromont v. Coupland, 2 Bing. 170, also Lovegrove v. Nelson, 3 M. & K. 1, two persons who horsed a coach and divided the profits were held to be partners, although each found his own horses, and the other had no property in them.

The ordinary agreement between publishers and authors, to the effect that the author shall contribute the manuscript, and the publisher shall, in the first instance, defray the expenses of publication, and repay himself out of the proceeds of the sale of the work, and that then the profits shall be divided, furnishes an instance of a partnership confined to profits only (Gardiner v. Childs, 8 C. & P. 345), although in a recent case, Venables v. Wood, 3 Ross, L. C. on Com. Law, 529, it has been held that authors and publishers are not partners at all; but even if this is the correct doctrine, there are numerous authorities in support of the position that there may be a partnership confined to profits only. (Reede v. Bentley, 3 K. & J. 271, and 4 ib. 656; Wilson v. Whitehead, 10 M. & W. 503 ; Gale v. Leckie, 2 Starkie, 107.)

Again, it frequently happens that one person has property, and another skill, and the agreement is that the profits alone shall be divided. Such cases are not likely to arise among medical men, although there possibly might be cases of a partnership existing where one partner possessed all the capital, stock, drugs, &c., and the other might simply prescribe in cases of some specialty, to which he devoted himself entirely, and yet he might share in the profits of the whole business. There is nothing in the principles governing partnerships to prevent one or more partners from agreeing to indemnify the others against loss, or to prevent full effect being given to a contract of partnership containing such a clause of indemnity. (Bond v. Pittard, 3 M. & W. 357; Geddes v. Wallace, 2 Bligh, 270.) Such agreements are, however, necessarily of a complex character, and frequently lead to much litigation, because if the indemnity were construed to extend to the loss of the very advances themselves, and not merely to the losses beyond the advances, the contract would lose its character as a contract of partnership, and become a mere contract of loan.

It is not unusual for a person who contemplates joining another in business, to agree that such business shall be carried on upon certain terms, not themselves creating a partnership, and to stipulate for an option to become a partner, either at a specified time, or at any time the person having the option may choose. Such agreements, if *bonâ fide*, and not mere colourable schemes for creating a partnership, and at the same time concealing it (Courtenay v. Wagstaff, 16 C. B., N. S. 110), do not create a partnership till such time as the person having the option has exercised it, and elected to become a partner. Such cases may occur as between a master and pupil, or between a principal and assistant.

And this is the more important in the medical profession, because, notwithstanding the medical acts, it has been decided that although, under the Act of 1858, s. 32, no person shall be entitled to recover any charge in any court of law for any medical or surgical advice, attendance, or for the performance of any operation, or for any medicine which he shall have both prescribed and supplied, unless he shall prove upon the trial that he is registered under the

Act, it has been intimated by high authority that if only one member of a firm is duly registered, the requisitions of the statute are complied with. (Turner v. Raynall, 14 C. B., N. S. 328, per Chief Justice Erle.) " Even if Smith (Turner's partner in the case) had not been registered at all, I should not," said his lordship, " as at present advised, have felt disposed to yield to the argument urged on the part of the defendant. . . . . How can it matter to the patient whether the attendance is given and the medicine dispensed by an assistant, under the supervision of his master, or by one who calls himself a partner and takes a share of the profits?" At all events, as registration at the time of trial has been held sufficient to maintain an action (Ib.), it follows that a pupil not yet qualified might be in partnership with a duly registered practitioner, and upon being subsequently qualified and registered the day before trial, the firm might recover in the action. And, indeed, the firm in such case might recover all claims extending over a period of six years before action brought, under similar circumstances. Of course such an unqualified partner would be liable to be sued for penalties, and might be summarily convicted under the 40th section of the Act, if he had wilfully and falsely pretended to be a medical practitioner recognised by law; but no disability of a fiscal or penal nature would interfere with the practice of a qualified, though not registered practitioner, and under the cloak of a partnership he could, at least if registered before trial, recover his charges in the name of the firm: and, if Chief Justice Erle's intimation be accepted by the Courts, it would seem that an unqualified—not merely unregistered—and a qualified practitioner might carry on a partnership practice, and so long as the latter was registered, the firm could recover their charges. The penal clause of the Act may be supposed sufficient to defeat such nefarious practices, but artifices of evasion are not unusually conceived in a spirit of much ingenuity; for instance, in the case of Pedgrift v. Barrington Chevalier, S C. B., (N. S.) p. 246, which was an appeal from a conviction of the Halesworth (Suffolk) Petty Sessions, held on October 26th, 1859, under the penal clause of the statute. Among other things it was proved that on the door of a house in which the appellant and one Mr. Irwin, a registered surgeon, lived, and for which they were jointly assessed to the rates, was a plate in a wooden frame, on which was engraved—

" Mr. Pedgrift,
Mr. Irwin,
Surgeon, Accoucheur, &c."

It appeared that the name of Mr. Pedgrift was on a separate piece of plate from the rest; but there was no division between the names except the line, which was necessarily apparent where the two pieces of plate joined. It also appeared that " surgery " was written on another door, and " Surgeon Accoucheur " on the lamp over the door. The justices considered that the words "Surgeon, Accoucheur, &c.," on the plate, were meant especially to apply to Mr. Pedgrift and Mr. Irwin. Such were the main facts of the case. Indeed, coupled with the fact that the appellant's name was not found in the copy of the "Medical Register" produced, this was the evidence relied upon by the justices in convicting; but Chief Justice Erle said, "This case raises a question upon a statute of very grave public importance. A conviction has been brought before us upon an extremely scanty state of facts; . . . . there is nothing upon the face of the case to show that the appellant was not a surgeon in practice before the passing of the Act—nothing to show that he had not a diploma or other qualification, or that he was not recognised by law as a surgeon in a sense that he had a right by law to practice as such, and might have enforced payment of fees by action. There is nothing, in fact, to negative his having been a duly qualified surgeon before the passing of the Act. I feel bound, under these circumstances, to say that the conviction must be quashed."

By several statutes it is unlawful for any person not duly qualified to act by himself or another as an attorney or solicitor, or to suffer his name to be made use of upon the account, or for the profit of an unqualified person. An agreement for a partnership between a person duly qualified as an attorney or solicitor and one who is not, is clearly illegal. (Williams v. Jones, 3 B. and C. 108; Scott v. Miller, Johns. 220.) The statutes on this subject are very stringent. An offending attorney is liable to be struck off the roll, and for ever after disabled from practising, and the unqualified person to be imprisoned for one year. Nor can the statutes be evaded by any agreement to the effect that the unqualified person shall receive a share of the profits as a salary, and that he shall not be a partner with the other. (Tench v. Roberts, 6 Madd. 145; Re Jackson, 1 B. & C. 270; Re Clark, 3 D. & R. 260; and Hopkinson v. Smith, 1 Bing. 13.) Neither can an attorney's clerk (unless himself qualified) act as an attorney, under cover of his principal's name. (Ib. and Re Palmer, 2 A. & E. 686.) The mere fact of not taking out a certificate is not deemed a disqualification within the meaning of the Act; it simply, like want of registration under the Medical Act, operates as a barrier to the recovery of fees, or costs, and would make many acts void, as well as render the uncertificated person ineligible for sundry appointments. It is also illegal for two persons, one qualified and the other unqualified, to hold themselves out as partners, and to put both their names to bills of costs and other documents in which their names ought not to appear, unless they are qualified solicitors. (Edmondson v. Davis, 4 Espinasse, 14.) Upon the whole, the restrictions and discipline for regulating the legal profession are more stringent and severe than those which apply to the medical; but perhaps a further experience of the working of the present Medical Acts may suggest improvements, not to say remedy imperfections. Whilst upon the subject of partnership, it may not be out of place to refer to the relative rights possessed by principals and assistants. And here it may be mentioned that as every partner may in general be regarded as the agent of the firm, and as such is endowed with authority to do all acts within the scope of the partnership business so as to bind the firm, each partner would, generally speaking, have authority to hire or discharge such servants, clerks, or assistants as might be necessary for the purpose of carrying on the business of the partnership. No general rule can be laid down as to what class of servants are entitled to more than the month's warning to which domestic servants are limited. But the Court of Exchequer has held in Todd v. Kerrick, 8 Exch. Rep. 151, that a governess at 60*l.* a year and board and lodging does not fall within the rule by which a menial or domestic servant may be discharged with a month's notice or a month's wages. The position which she holds, the station she occupies in the family, and the manner in which such a person is usually treated in society, certainly place her in a very different situation from that which mere menial and domestic servants hold. The same may be said of a tutor. But in these and similar cases an arrangement should be made at the time of hiring as to the notice expected or required, or intended to be given. In the absence of any such special agreement about notice, the question of notice in the case of assistants to medical practitioners must be determined partly by the custom of the profession. Thus, evidence has been adduced of a usage of trade enabling a master to dismiss a commercial traveller, at 150*l.* a year, upon giving him three months' notice. (Metzner v. Bolton, 9 Exch. 518.) On the other hand, evidence has been received of a custom in the woollen trade to dismiss an agent at one month's notice. (Parker v. Ibbetson, 27 L. J., C. P. 236.) And in Mortimer v. Prowett, Q. B., N. P., June 18, 1856, evidence was given to show that, where no condition was expressed in the agreement of hiring, it was usual to give a printer one month's or at least a fortnight's notice, a publisher three months, and a sub-editor to the end of the current year.

It has sometimes been contended, from analogy to the rule which prevents a yearly tenancy from being determined before that time of the year corresponding with the period when it commenced, that particular yearly hirings can only be determined in a similar manner.

But this point will in most cases depend upon the particular custom of the trade or business in question.

There is, of course, a great distinction between a contract of apprenticeship and a contract of service. "A person has a right to dismiss a servant for misconduct, but has no right to turn away an apprentice because he misbehaves" (per Lord Denman, in Wise v. Wilson, 1 Carr & Kirw. 662). And yet, as in this case, there may be a mixed compact, "something between apprenticeship and service. This plaintiff's son goes to the defendant (a surgeon) to render him *assistance* in his business, although he is also to pursue his studies; and, as a justification for his dismissal, the defendant has pleaded, not that the plaintiff's son did not perform all things on his part to be performed, but that he did things injurious to the defendant's practice, and so misconducted himself as to be dangerous to the defendant's practice as a surgeon. It is proved beyond all doubt, that on some occasions the plaintiff's son came to the defendant's house intoxicated; but I think," said Lord Denman, "that that alone would not justify the defendant in dismissing him. It is also proved that on several occasions, in consequence of the plaintiff's son coming home late, he could not compound the medicines, and employed the shop-boy to do it. Now, I think this affords matter for serious consideration, *and if you* (the jury) *think that, from this conduct of the plaintiff's son, real danger was occasioned to his master's business*, you ought to find your verdict for the defendant, *as the defendant was then, in my opinion, justified in dismissing him*." It should, however, be mentioned that in this case, although a premium had been paid, the written agreement was not an indenture of apprenticeship, strictly speaking.

Arrangements are sometimes made between masters and clerks, travellers, agents, and possibly assistants, of various kinds, under which such persons, in lieu of receiving a fixed salary, are remunerated by a portion of the sums received by them on account of their masters, or by a per-centage on their earnings, or by a sum calculated with reference to the gross or net profits (as the case may be) of the principal, or in some similar manner. These various modes of payment, which are generally adopted with a view to secure increased exertion, often give rise to a question of considerable importance—viz., how far the persons whose services are so remunerated are to be regarded as partners, and not mere servants, agents, or assistants. Where the question arises *between the parties themselves*, the mere fact that the servant was remunerated by a portion of the profits will not alone constitute him a partner (Peacock v. Peacock, 2 Camp. 45), if it appears from the whole scope of the agreement that the intention of the parties was otherwise. Again, *quoad* third persons, there might be a partnership, but yet as *inter se*, even a division of profits might amount to nothing more than a compensation for trouble. (Hesketh *v.* Blanchard, 4 East, 144.) However, in the following case, before referred to, the parties were held not to be partners in any sense:—C. sold to R., and conveyed to him by deed, his interest in the profession and practice of a surgeon, &c., for 900*l*.—500*l*. to be paid on the execution of the deed, and 400*l*. at the end of the year. C. covenanted not to practise within three miles of the original place of business, and also that during one year from the date of the deed he would reside there, and carry on the business as before, and introduce R. to the patients, and promote the interest of the concern. In consideration thereof R. covenanted to allow C., during the year, a moiety of the clear profits of the concern, to be paid at the expiration of the year. In an action of covenant brought by C. upon the deed, R. claimed to set off certain sums received by C. during the year from patients, as money received to his use. Chief Baron Pollock refused to receive evidence of the receipt by C. of such sums, on the ground that the parties were partners during the first year. The Exchequer Chamber, however, reversed this decision, considering that the parties were *not* partners. It has been held that a master was justified in dismissing a clerk, at a yearly salary, who also, at certain periods, received a portion of the profits (but this, the master alleged, was mere gratuity) for claiming to be a partner, as he thereby disclaimed being a servant. (Amor *v.* Fearon, 9 A. & E. 548.)

Although the question of whether or not a servant was rightly discharged must of course often depend upon the nature of the services for which he was engaged, and the terms of the engagement itself, it is conceived that, whatever be the nature of the engagement, the following acts of either insubordination or impropriety of conduct would justify his discharge notwithstanding:—

I. Wilful disobedience of any lawful order of his master.

II. Gross moral misconduct, whether pecuniary or otherwise.

III. Habitual negligence in business, or conduct calculated seriously to injure his master's business.

IV. Incompetence, or permanent disability from illness.

It is unnecessary to say that proper proof must be adduced of the cause or ground of dismissal, should such right to dismiss be disputed, as, for instance, a few days, or an occasional illness, would not create a permanent disability (Cuckson *v.* Stones, 28 L. J., Q. B. 25); neither would a day or two's protracted absence after a vacation had expired, provided the master was not compelled to hire another person, and especially if it did not appear that the assistant's absence was caused by any act of gross immorality

(Filleul *v.* Armstrong, 7 A. & E. 557), amount to habitual negligence, or conduct calculated seriously to injure his master's business.

As a partnership contract is one that need not be created by any writing whatever, and that even persons who are not partners may nevertheless be liable to third persons as if they were, it is necessary to be careful to avoid even the appearance of a partnership where such does not in reality exist.

A *quasi* partnership (rendering the members of it liable to third parties) may be proved by showing a holding out to the plaintiff as such ; and if in attempting to establish a *quasi* partnership a real partnership should be shown to exist, the liability of the persons sought to be charged will only be established the more completely. It is, therefore, always desirable to reduce to writing the terms of any agreement between principal and assistant, so that no doubt may exist as to their relative positions whether as between themselves or others—and such agreement should be properly stamped. And in an agreement for a partnership it is always better to define in writing the limits within which such co-ordinate powers are to be exercised. Besides, in some cases the agreement, in order to be valid under the Statute of Frauds, *must* be in *writing*. The 4th section of that statute enacts " that no action shall be brought whereby to charge any person upon any agreement that is not to be performed within the space of one year from the making thereof, unless the agreement upon which such action shall be brought, or some memorandum or note thereof, shall be in writing, and signed by the party to be charged therewith, or some other person thereunto by him lawfully authorized." To illustrate the effect of this prohibition ; an agreement that a partnership shall commence between the parties at some time distant more than a year from the making of the agreement is not binding unless some note or memorandum thereof is in writing and duly signed. (See per Justice Holroyd, in Williams *v.* Jones, 5 B. & C. 108.)

But there is yet another prohibition contained in the former portion of the same section—viz., " that no action shall be brought whereby to charge any person upon any contract or sale of lands, tenements, or hereditaments, *or any interest in or concerning them*, unless such agreement or some memorandum or note thereof shall be in writing, and signed by the party to be charged therewith, or some other person thereunto by him lawfully authorized." Should, then, *any interest in or concerning* the premises where the business is carried on be contracted for, as the word " lands " includes " houses," and the word " tenements " is still more comprehensive, the agreement could not be enforced unless it or some note or memorandum of it were in writing and signed according to the requirements of the statute (s. 4).

And further, no contract for the sale of any goods for the price of 10*l.* sterling or upwards shall be allowed to be good, except the buyer shall accept part of the goods so sold, and actually receive the same or give something in earnest to bind the bargain or in part of payment, or that some note or memorandum in writing of the said bargain be made and signed by the parties to be charged by such contract or their agents thereunto lawfully authorized (s. 17). If, therefore, in a contract for the sale of a practice any goods, such as drugs, surgery fittings, or furniture, for the price of 10*l.* or upwards, are included, the above requirements must be strictly observed.

CHAPTER IX.

THE LAW OF LIFE ASSURANCE AS AFFECTING THE MEDICAL PROFESSION.

IT is proposed to consider in this chapter the duties of medical referees both with regard to the person proposing the insurance and the insurance office with which such proposer is in treaty. Medical referees are too prone to consider themselves as acting on behalf of the proposer, and in practice give the most favourable answers which they conscientiously can to facilitate the issue of the policy. In a recent leading case in the Exchequer Chamber, the judges on that occasion, as well as in the Court below, have definitely decided that "In cases where the representations of the referees are made the basis of the policy, the answers of the referees are binding upon the insured, so as to make the falsehood of their representations an answer to an action on the policy; but in the present case," said the Court, "considering that the parties insuring were only required to state their belief as to the matter under discussion, and that the life and the referees were really acting in fraud, both of the plaintiffs and the defendants, and that they were not at all in the capacity of persons negotiating the contract, we can see no pretence for making them the agents of the parties insuring, so as to render their fraud that of the plaintiffs." And again —" The insurer and the insured being equally ignorant of material facts to influence their contract, if the insurer asks for information, and the insured does his best to put the insurer in a situation to obtain the information, and to form his own opinion whether the information is sincere, can it be permitted that when the insurer, without any blame being attributed to the insured, has allowed himself to be deceived, he shall be able to say to the insured, 'You warranted all the information I received to be true, and having received your premiums for many years, now the life drops, I tell that you I was incautious and the policy I gave you a nullity?'" (Wheetton *v.* Hardisty, 3 Jurist, N. S. 1170; 5 Jur. N. S. 14.)

In the explanation of this case it should be mentioned that the plaintiffs were the directors of a reversionary interest company, who, having failed to effect an insurance upon the life in question in the European office, had insured the same life—viz., that of a tenant for life—in the office of the defendants. The plaintiffs had been no parties to the false statements of either this tenant for life, his usual medical attendant, or his private friend, upon which statements the defendants appear to have relied; but the papers of the European office were handed to the defendant's office by the plaintiffs, and the medical officers of the defendant's office never even saw the tenant for life, but accepted the proposal in the names of the plaintiffs, and in the form filled up; in lieu of the usual statements as to health and habits, the secretary of the reversionary company wrote, "See papers received from the European Company," and added, "We (the plaintiffs) do hereby declare that we believe that the above particulars and statements are true."

No doubt, the most perfect *bona fides* is to be observed in life insurances as in other contracts; and should the referee make a *wilfully* untrue statement, or collude with the party making the proposal, he may, although not personally interested in the contract, render himself liable to an action if loss should ensue (Pasley *v.* Freeman, 3 Term Rep. 51); for if a man will wickedly assert that which he knows to be false, and thereby draws his neighbour into a heavy loss, even though it be under the specious pretence of serving his friend, he makes himself liable for the damage, and that even if no fraud was intended by him. (Watson *v.* Poulson, 15 Jur. 1111 Exch.) A statement false in fact, but not known to be so by the party making it, or not made with intent to deceive, but, on the contrary, believed to be correct, will not render him liable in an action, although it may have induced the insurers to undertake the insurance, and they may have suffered pecuniary loss thereby, since to support an action of deceit the scienter must be proved. (Shrewsbury *v.* Blount, 2 M. & G. 475; Rawlings *v.* Bell, 1 C. B. 951; Chandelor *v.* Lopus, 1 Smith's Lead. Cas. 77.)

How this liability might be enforced in practice, and in what cases, it would be beyond the province of this work to inquire. It is sufficient here to state the general liability in such cases, principally to guard against its occurrence.

Medical referees of parties proposing insurances on their lives have of late years laid claim to a fee for their certificates, and to this there can be no objection; but in claiming the fee of the insurance office, and not of the patient, they have pushed the proposition too far. The point has been raised in cases tried in three different Small Debts Courts, and has signally failed. (Philbrick *v.* Whetham, 4th Oct. 1850, Colchester County Court; Hooper *v.*

The Gresham Life Assurance Company, June 1851, Shoreditch County Court; and Duplex v. The Economic Life Office, Aug. 1852, in the Sheriffs' Court, before Mr. Russell Gurney.) As there are no authorized Reports of County Court cases to which to refer the reader, the substance of these cases will be found set out in the Appendix. As a matter of policy, it would seem to be to the companies' interest to pay the fees, since of course it is optional to the medical man to reply to the letter of application without the fee; or if, after declining to give information without payment, the company continue to request the information, they would be bound to pay for it.

All insurance offices are naturally desirous to consult with the medical man who has been last in attendance on the life to be insured (Morrison v. Musprat, 4 Bingham, 60); and where the reference was made to a person who had been the ordinary medical adviser, but no mention was made of the person attending at the time of the insurance, the policy was vacated. (Everett v. Desborough, 5 Bing. 503; Hutton v. The Waterloo L. A. C., 1 F. & F. 735; Cazenove v. The British Equitable A. C., 6 C. B., N. S. 437.) At the same time, a reference to a person who, immediately before the time at which the policy is effected, has been casually consulted for trifling maladies, will not satisfy the requisition for a reference to the *usual* medical attendant, although the person who had formerly attended may have retired from practice at the time. (Huckman v. Fernie, 3 M. & W. 505.) "Who is your usual medical attendant?" involves a question of considerable importance. "Now," said Lord Abinger (Chief Baron), in the case just cited, "let it be considered for a moment what is the grammatical sense of that question. It is in the present tense. Suppose a person goes to effect a policy on his life who had no medical attendant in the last year: if the answer to the question were, 'I have no such medical attendant,' must not that question of necessity be followed by another question, which is, 'Who was your former medical attendant?' The terms and nature of the question show that it was designed to extract from the person who is the medical attendant best able to give an account of her constitution at that time; and if she has no medical attendant in the precise grammatical sense of the question, it appears to me that she is bound to mention who is the medical attendant who could give that information." Also, "the word 'usual,'" said the same authority, "implies having attended more than once."

Again, in another case, the plaintiff, wishing to insure the life of a man named House, referred the agent of the company to him for the necessary information for filling up the declaration, and he stated, "I have never had occasion for a doctor: sometimes I have taken Harvey's quack pills, but Mr. V. knows as much of me as any man." The plaintiff signed the proposal, stating that Mr. V. was the usual medical attendant. It was proved at the trial that Mr. V. had not attended House for nearly twenty years, but that he had occasionally been attended, after fits of intemperance in which he was wont to indulge, by a quack doctor named Harvey: of this the plaintiff was unaware. The Court decided that it was no matter whether Dr. Harvey was a good medical attendant or not, he was the person actually attending him: that the circumstance that the plaintiff was ignorant of the error did not affect the question, but that the policy was void. Hence it seems that it is not necessary that the usual medical attendant should be a regularly qualified practitioner: it is sufficient that he acts in that capacity. And, *à fortiori*, a qualified medical practitioner, though not registered, may be thus designated. (Everitt *v.* Desborough, 5 Bing. p. 503.)

In order to constitute moral fraud in any one, it is not necessary that the statement should be false to the *actual* knowledge of the party making it; if untrue in fact, *and not believed to be true* by the parties making it, and made for a fraudulent purpose, it is both a legal and a moral fraud. (Taylor *v.* Ashton, 11 M. & W. 415.)

Although two sisters of the insured had died young, of consumption, the evidence of the usual medical attendant, that consumption could not be detected in the insured until some months after the proposal to the company was made, is evidence *pro tanto* to rebut any imputation of fraud in that respect; but in reply to the question, "Have you ever been proposed to, or declined by, any other office? If so, name it," the answer filled in by the agent being, "I have been and am still in correspondence with other offices, as the amount to be insured is large,"—the fact being that eight offices had refused, and in one instance only had a proposal been accepted—amounted to a wilful suppression of the fact that other offices had been applied to and had refused the life, and allowing another to act on a belief of facts known to be untrue was fraud. (*In re* The General Provincial Life Assurance Company, *ex parte* Daintree. Before Malins, V.C., January 24, 1870.)

CHAPTER X.

Section I.

THE PUBLIC HEALTH (COMMON AND STATUTE LAW). CONTAGIOUS AND INFECTIOUS DISEASES.

THE doing of any act by which the health of the public might be endangered, whether by the propagation of an infectious disease, or by erecting any establishment which might corrupt the air by noxious exhalations, was ever a great misdemeanour at common law.

Three diseases were formerly regarded with the utmost horror—the plague, the syphilis, and the leprosy; and the strictest precautions were taken to prevent the propagation of them, especially the first and the last named.

It was at one time generally believed that syphilis was so contagious as to render the mere presence of a person suffering from it dangerous to others; and the very ancient writ "De leproso amovendo" would appear to have been directed against syphilis quite as much as against leprosy.

It is thus described in Fitzherbert's "Naturâ Brevium," coupled with a commentary supposed to be written by Lord Chief Justice Hale:—

"Where a man is a lazar or a leper, and is dwelling in any town, and he will come into the church, or amongst his neighbours where they are assembled to talk with them, to their annoyance and disturbance; then he or they may sue forth this writ to remove him from their company."

"But it seemeth, if a man be a lazar or a leper, and will keep within his house, and will not converse with his neighbours, that then he shall not be moved out of his house. But there are divers manners of lepers; but it seemeth that the writ is for those lepers who appear to the sight of all men that they are lepers by their voice, and their sores, and the putrefaction of their flesh, and by the smell of them. But for those who are infected with that disease in their bodies, where it doth not appear outwardly upon their bodies, *quære,* whether such writ lieth to remove them." (Fitzh., "Naturâ Brevium," p. 233.) The writ speaks of "the contagion of the disease," and directs that such diseased person should be "seen and diligently examined by certain discreet and lawful men who have the best knowledge of such disease," and if he be found to be a leper "then without delay" he is to "be carried away, and removed from the communication of his neighbours, to a solitary place, to dwell there, as the custom is, lest by his common conversation damage or peril should in any wise happen to the said men." "Lazars" are said to be included in the writ, and Dr. Johnson derives the term "lazar" from "Lazarus in the Gospel," who is not described as a leper in the common acceptation of that term—*i.e.*, as afflicted with the "mala scabies"—but as "covered with sores;" and open and putrefying sores seem to be the description of leprosy, or contagious disease, according to Hale, against which the writ was directed, rather than against that description of leprosy which "did not appear outwardly upon the body" in the manner before described by the learned commentator.

The exposing, by carrying, a child infected with small-pox along a public way, near to dwelling-houses, &c., to the great danger of infecting persons there passing, is a great misdemeanour at common law. (Rex *v.* Vantandillis, 4 Maule & Selwyn, p. 73.) "There can be no doubt," said Mr. Justice Le Blanc, in passing sentence, "that if a person unlawfully, injuriously, and with full knowledge of the fact, exposes in a public highway a person infected with a contagious disorder, it is a common nuisance to all the subjects, and indictable as such. No person having a disorder of this description upon him ought to be publicly exposed to the endangering of the health and lives of the rest of the subjects."

The first statute relating to the plague is that of 2 Jac. I., c. 31, which declared that all infected persons going abroad with any infectious sore upon them, uncured, should suffer death as felons; but if such persons had not such sore found upon them, they were to be punished as vagabonds. And similar regulations were afterwards incorporated into several acts relating to quarantine, the principal of which are the 9th Anne, c. 2, and 6 Geo. IV., c. 78. Although, until recently, inoculating for the small-pox was not illegal in itself, yet in an indictment charging the defendant, an apothecary, with inoculating for that disease and injuriously causing the infected person to be carried along a public street, &c., Mr. Justice Le Blanc observed that the introduction of vaccination did not render the practice of inoculation for the small-pox unlawful, but that it was at all times unlawful, and an indictable offence, to expose persons infected with contagious disorders, and, therefore, liable to communicate them to the public, in a place of public resort. (R. *v.* Burnett, 4 Maule & Selwyn, 271.)

Modern legislation on the question of the public health might be the subject of almost a work in itself. It is not proposed, therefore, in the present chapter, to do more than briefly refer to such Acts (for the most part) as more immediately relate to infectious and contagious diseases; for, however the whole of this branch of legislation may be deemed within the compass of a purely medical treatise, the object of the present work is, as has been before remarked, necessarily confined to those laws which directly affect, not so much the public, *qua* public, but the medical practitioners, as members of a particular profession.

It will not, therefore, be desirable to allude to such Acts as relate to the establishment of baths and washhouses, however beneficial they may be in a sanitary point of view; nor yet to those which have reference to the supply of water to, and the sewerage, drainage, cleansing, and paving of towns and populous places, however these may promote the health and comfort of their inhabitants; nor yet to the Acts to prevent the spreading of contagious and infectious disorders amongst sheep, cattle, and other animals, or to prevent the adulteration of articles of food or drink, however such may check the supply of unwholesome food. Neither will such Acts as have been passed to abate nuisances arising from the smoke of furnaces or steam-vessels be deemed within the province of this work, even though such exhalations may be sometimes noxious, as well as offensive, the object being to devote the few pages allotted to this subject to a consideration of those laws whose tendency is *directly* to prevent diseases, and not merely *indirectly* to promote the greater salubrity of the population.

The first Act, therefore, which calls for attention is the 23 & 24 Vict., c. 77, amending the existing Acts for the Removal of Nuisances and the Prevention of Diseases, and constituting certain bodies therein named the local authorities to execute the "Nuisance Removal Act," and repealing all former enactments constituting a local authority for the execution of the "Diseases Prevention Act" (except those contained in the "Metropolitan Local Management Act"), and directing that the Board of Guardians, or the overseers of the poor (as the case may be) where there is no board of guardians, shall be the local authority for executing the "Diseases Prevention Act," who are thereby directed to provide carriages for the conveyance of infected persons suffering from contagious or infectious diseases, and two justices at petty sessions, upon proper complaint, may order the removal of a nuisance on any private premises, and the guardians of any union, or parish not within an union, may at any time employ one of their medical officers to make inquiry and report upon the sanitary state of the union or parish. And by the 29 & 30 Vict., c. 90 (styled the "Sanitary Act, 1866"), s. 22, the nuisance authority, which shall mean any authority empowered to execute the "Nuisance Removal Act," s. 15, upon the certificate of any legally qualified medical practitioner, may direct the owner or occupier of any house, or part thereof, to cleanse and disinfect the same, and in default of compliance with such order the nuisance authority shall cause such to be disinfected at the expense of the owner or occupier, or, in the case of the poverty of such persons, then the authority may, with the consent of the owner or occupier, at his own expense, cleanse and disinfect such house, &c., and any articles therein likely to retain infection. The same authority is required to provide a proper place for the disinfection of woollen articles of clothing or bedding free of charge.

An infected person entering a public conveyance without notifying that fact to the owner or driver will be liable to a penalty of 5l.

And, with the consent of the superintending body of a hospital within the district of a nuisance authority, any justice may, upon a certificate signed by a duly qualified medical practitioner, direct the removal to such hospital, at the cost of the nuisance authority, of any person suffering from any dangerous, contagious, or infectious disorder, being without proper lodging or accommodation, or lodged in a room occupied by more than one family. A penalty of 5l. is imposed upon any one suffering from any dangerous infectious disorder, who wilfully exposes himself without proper precaution in any public place or public conveyance; and also upon any one in charge of such person so exposing the sufferer, and any owner or driver of a public conveyance who does not immediately provide for the disinfection of his conveyance after it has, with his knowledge, conveyed any such sufferer; and any person who, without previous disinfection, gives, lends, sells, transmits, or exposes any bedding, clothing, rags, or other things which have been exposed to infection, shall be liable to a like penalty.

And if any person knowingly let any house or room in which any person suffering from any dangerous infectious disorder has been, to any other person, without having first disinfected the same to the satisfaction of a qualified medical practitioner, testified by his certificate, such person shall be liable to a penalty of 20l.

By the 34th section of the Medical Act (1858), the words "legally qualified," or " duly qualified medical practitioner," when used in *any Act of Parliament*, shall be construed to mean a person *registered* under that Act.

By the Sanitary Act of 1866 (29 & 30 Vict., c. 90), s. 37, the sewer authority, or in the metropolis the nuisance authority, may provide for the use of the inhabitants within its district hospitals or temporary places for the reception of the sick. And by the amended Sanitary Act (31 & 32 Vict., c. 115), the same authorities shall have the like power to make provision for the temporary supply of medicine and medical assistance for the poorer inhabitants as it now has to provide hospitals or temporary places for the reception of the sick under the above section of the former Act. But such latter power shall not be exercised without the sanction of her Majesty's Privy Council.

By the Workshop Regulation Act (1867), 30 & 31 Vict., c. 146, " if upon the complaint of any officer of health, inspector of nuisances, or other officer appointed by a local authority, or of any superintendent of police, it appears to any justice of the peace that there is reasonable cause for believing that any of the provisions of the Sanitary Act, 1866, are contravened in any workshop, such justice may, by written order, empower the complainant to enter such workshop at any time within forty-eight hours from the date of such order, and examine any person whom he finds in such workshop touching any matter within the provisions of the Act, or of the Sanitary Act."

By the 3 & 4 Wm. IV., c. 103, ss. 11 & 12, no child under the age therein specified was to be employed in any factory or mill without having personally appeared before some surgeon or physician of the place or neighbourhood of its residence, and submitted itself to his examination, nor unless he certified to having had a personal examination or inspection of such child, and that such child was of the ordinary strength and appearance of children of such age—such certificate to be within three months of its date countersigned by some inspector, or justice.

Under 7 & 8 Vict., c. 15, s. 8, inspectors of factories were empowered to appoint a sufficient number of persons practising surgery or medicine to be such certifying surgeons, and in every such appointment specify the factories or districts for which each surgeon is appointed; but every such appointment, and every order annulling such appointment, might be revoked by the Secretary of State upon appeal made to him. And the inspector of the district is required to make known the names of the certifying surgeons to the occupiers of the factories in that district; but no surgeon being the occupier of a factory, or having a beneficial interest in any factory, shall be a certifying surgeon. A form of certificate is given in the schedule annexed to the Act, and the former counter-signature is dispensed with. By section 10, no surgical certificate given by any than a certifying surgeon shall be of any force unless given by a person duly authorized by an university or college, or other public body having such authority, to practise surgery or medicine, and countersigned, according to the form to the Act annexed, by a justice, not being the occupier of a factory, nor the father, son, or brother of the occupier of a factory, in the presence of the person named therein, and upon proof of his identity.

By section 11, personal inspection of the person named in the certificate is required, and the examination of the person and signature of the surgical certificate must be made at the factory, except special cause be allowed by an inspector; and if a certifying surgeon refuses such certificate, he must specify in writing, according to the form given, his reasons for such refusal.

By section 12, the occupier of a factory and the certifying surgeon may agree in writing for the payment for the examination of persons for whom surgical certificates are required; and if the terms of such agreement are in conformity with the regulations for the guidance of surgeons which shall be made by the inspector of the district, and shall be countersigned by the inspector, all penalties incurred by either party for breach of the contract may be recovered as other penalties under this Act may be recovered, and shall be applied as other penalties under this Act are to be applied, and no such agreement shall be liable to stamp duty.

By section 13, an inspector shall fix the amount of fees to be paid by the occupier of a factory to the certifying surgeon, and also the times when such surgeon shall visit such factory, provided he be required to fix such fees and visits by such occupier. And such fees shall in no case, where the surgeon shall examine more than one person, exceed 1s. for each person, together with 6d. for every half mile beyond one mile from such surgeon's residence, and such fees, including mileage, shall not be less than 1s. nor more than 5s. for any one visit, except he have to examine on any one visit more than ten persons, in which case he shall receive 6d. per head, instead of all other fees; and where the factory is situate within one mile of such surgeon's residence the fee shall not exceed 2s. 6d. for each visit, except he shall at any one visit examine more than five persons, in which case he shall receive 6d. per head, instead of all other fees; and no certifying surgeon shall receive more than 6d. for any certificate signed by him elsewhere than at the factory.

Further, by the 22nd section, if an accident should occur in a factory causing any bodily injury to any person employed therein, so as to prevent such person from returning to his work before nine on the following morning, the occupier of the factory or his agent is required within twenty-four hours of such absence to send notice in writing to the surgeon appointed to grant such certificates, and the surgeon shall send a copy of such notice by post to the sub-inspector of the district, and without delay proceed to such factory and fully investigate the cause of such injury, and within twenty-four hours send to the inspector a report thereof.

By the 53rd section, although the surgical certificate shall be evidence in the first instance of the age of the person named therein, it shall not protect any person knowing such person to be less than the age certified from any penalty for employing or conniving at the employment of such person otherwise than is allowed by the Act; and in any proceeding for employing any person contrary to the Act, a declaration in writing by the certifying surgeon, that he has personally examined such person and believes him to be under such age as set forth in such declaration, shall be evidence in the first instance, until the contrary shall be made to appear.

In 8 & 9 Vict., c. 29 (an Act to regulate the labour of children, young persons, and women in print works), the above enactments are incorporated.

By 16 & 17 Vict., c. 104, the 3 & 4 Wm. IV., c. 103, the 7 & 8 Vict., c. 15, the 10 & 11 Vict., c. 29, and the 13 & 14 Vict., c. 54, as amended by this Act, are to be construed together as one, and all the Factory Acts are, by 27 & 28 Vict., c. 48, directed, by section 6, to be incorporated with it, and to apply to the several manufactures and employments therein mentioned. And finally, by 30 & 31 Vict., c. 103 (1867), entitled "An Act for the extension of the Factory Acts," the 14th section gives power, in the case of blast furnaces and iron mills, to the Secretary of State, by written order, to dispense with surgical certificates given by certifying surgeons, and substitute therefor such other regulations as to proof of the age of children and young persons, and of their bodily health and capacity of working daily for the time allowed by such Acts, as he may think expedient.

Section II.

By the Contagious Diseases Act, 1866, 29 & 30 Vict., c. 35, the Admiralty or the Secretary of State for War may appoint a medical officer for each of the places to which the Act applies, to be visiting surgeon for the purposes of the Act; and also a medical officer to be assistant to any such visiting surgeon.

And the same authorities may appoint a medical officer to be inspector of certified hospitals, and another to be an assistant inspector.

And the same authorities may provide buildings as hospitals, to be certified as such.

And by the 15th section, where an information upon oath is laid before a justice by a superintendent (which includes inspector) of police, to the effect that he has good reason to believe that a woman named in the information is a common prostitute, and either is resident within the limits to which the Act applies, or, being resident within five miles of those limits, has within fourteen days from the laying of the information, been within those limits for the purpose of prostitution—which power is extended by the 32 & 33 Vict., c. 96, amending the former Act and repealing the above section, to women resident within ten miles of those limits, or having no settled place of abode, who have, within the period before named, been either within those limits for the purpose of prostitution, or been outside of those limits for the same purpose, in the company of men resident within those limits—the justice may direct a notice to be served upon such woman; and whether she appears or not in obedience to such summons, upon proof of its due service upon her the justice present, on evidence given upon oath substantiating the charge, may order the woman to be subject to a periodical medical examination by the visiting surgeon for any period not exceeding one year. Time and place for conducting such examination are to be specified in the order.

And further, any woman in any place to which the Act applies may voluntarily, by a submission in writing duly signed, subject herself to such medical examination.

The visiting surgeon is to prescribe the times and places at which such woman is to attend for examination from time to time.

And if any such woman is found to be affected with a contagious disease, or is in such a condition that she cannot be properly examined at that time, she shall be liable to be detained in a certified hospital—in the latter case for not more than five days—to enable the surgeon properly to examine her; and the visiting surgeon shall sign a certificate in triplicate that she is so affected, or that he believes she is, naming the hospital in which she is to be placed, and one of the originals shall be delivered to the woman, and the others to the superintendent of police. And any such woman may, if she thinks fit, place herself in the certified hospital named in the certificate, for medical treatment; but if after certificate she neglects to do so, the superintendent of police, or a constable acting under his orders, shall apprehend her and convey her to the hospital, where she shall be detained until she receives a written discharge by the chief medical officer of such hospital.

The certificate before mentioned shall be both a sufficient authority for apprehending and detaining her.

The inspector of such hospitals may direct the transfer of such woman from one hospital to another. But no woman is to be detained under any one certificate for a longer period than three months, unless the chief medical officer of the hospital, and the inspector of such hospitals, or the visiting surgeon for the place whence she came or was brought, conjointly certify that her further detention for medical treatment is requisite, in which case she may be detained until discharged by the chief medical officer in writing, but so that no woman be detained under one certificate for a longer time in the whole than six months, extended by sec. 7 of the Act of 1869 to nine months. A justice may interpose to discharge the woman, if the chief medical officer refuses to discharge her, if upon reasonable evidence he be satisfied that she is free from such disease; and upon her discharge she shall be sent to the place of her residence, if she desires, without expense to herself. If a woman has ceased to be a common prostitute, and desires to be relieved from the periodical medical examination, she being not under detention at the time in a certified hospital, the visiting surgeon may, upon proof that she has so ceased, by written order direct that she be relieved.

Such are the chief provisions of these Acts, so far as they relate to the duties of the medical officers. The places to which the Acts apply are—

Aldershot.	Sheerness.
Canterbury.	Shorncliffe.
Chatham.	Southampton.
Colchester.	Winchester.
Dover.	Windsor.
Gravesend.	Woolwich.
Maidstone.	The Curragh.
Plymouth and Devonport.	Cork.
Portsmouth.	Queenstown.

VACCINATION.

SECTION III.

The Acts relating to vaccination are the 3 & 4 Vict., c. 29; amended by 4 & 5 Vict., c. 32; extended and made compulsory by 16 & 17 Vict., c. 100; consolidated and amended by the present Act (1867). By this Act (the 30 & 31 Vict., c. 84), the former Acts, together with certain clauses in others incidentally connected with the duties of registrars, are with certain exceptions, repealed. The guardians of every union or parish where the same shall not have been divided into districts for the purpose of vaccination, are directed—unless such union or parish be of so limited an extent as not to require subdivision, in which case it shall be treated as a vaccination district—forthwith to divide it into districts for the performance of vaccination, and they shall enter into a contract with some duly registered medical practitioner for the vaccination of all persons resident within the district. No person to be appointed such public vaccinator, or act as deputy to such, who shall not possess the qualifications required by the Privy Council.

The Privy Council may order to be paid to such public vaccinators, over and above the payments received from the guardians, further payments not exceeding 1s. for each child whom the vaccinator has successfully vaccinated during the time to which the award relates.

Such contracts shall only provide for payment for successful vaccinations—for every such vaccination done at an appointed station at or within one mile from the residence of the vaccinator, or in the workhouse, not less than 1s. 6d.; at any station over one mile and under two from his residence, not less than 2s.; at any station over two miles, not less than 3s. In respect of successful vaccinations performed elsewhere, payment to be according to terms specified in the contract. The parent of every child born in England is required, within three months after its birth, or where, by reason of the death of the parent or otherwise, any other person shall have the custody of such child, such person is required, within three months after receiving such child, to cause it to be vaccinated by the public vaccinator or some medical practitioner. And if the operation is performed by the public vaccinator, the child is to be again brought to him within one week to be inspected, and, if necessary, to be re-vaccinated and again inspected.

If a child has been three times unsuccessfully vaccinated, and is insusceptible of vaccination, or has already had the small-pox, upon the public vaccinator or medical practitioner delivering a certificate to that effect the child shall no longer be required to be vaccinated. Every public vaccinator, having successfully vaccinated any child, is required within twenty-one days after the operation to transmit, by post or otherwise, a certificate to that effect to the registrar of births and deaths of the district within which the birth was registered; but if such is unknown to him, then to the registrar within whose district the operation was performed, and upon request to deliver a duplicate thereof to the parent or otherwise, for which certificate no fee is to be charged. If paid by the child's parent, &c., for performing the operation, he shall not be entitled to payment under his contract also, and *vice versâ*.

Where the operation is privately performed by a private medical practitioner, the parent, or otherwise, shall submit a certificate to be signed by him, and shall within twenty-one days transmit it as aforesaid. The penalty for neglecting to cause such child to be vaccinated, or after vaccination to be inspected, without reasonable excuse, is 20s. upon summary conviction. Any person, as before mentioned, who shall neglect to transmit any certificate required of him within the time specified, and every public vaccinator who shall refuse to deliver the duplicate to the parent, &c., on request, and every medical practitioner who shall refuse to fill up and sign the certificate of successful vaccination, shall be liable to the same penalty; and every person who shall wilfully sign a false certificate or duplicate shall be punishable for a misdemeanour.

By the 32nd section of this Act, any person who shall hereafter produce, or attempt to produce, in any person, by inoculation with variolous matter, or by wilful exposure to variolous matter, or to any matter, article, or thing impregnated with variolous matter, or by any other means whatsoever produce the disease of small-pox in any person, shall be liable to be proceeded against summarily, and if convicted to be imprisoned for one year.

Upon an appeal to the Court of Queen's Bench from a conviction of the magistrates of Huntingdon, the point raised upon the appeal was, Did a previous conviction emancipate the appellant from all liability to future penalties for persistently refusing to have his child (a girl under fourteen years of age) vaccinated; or was he liable to punishment and a fresh conviction every time he was informed against, as for a contempt of the law? According to the old Act, 16 and 17 Vict., c. 100, notice with respect to the vaccination of children had to be given by the registrar before the child reached the age of four months; and by a decision of the Court of Queen's Bench in 1862 (Filcher v. Stafford), it was decided that under that Act there was no power to convict a parent a second time for non-compliance with the law. The Act of 1867, however (30 & 31 Vict., c. 84), was intended to remedy the defects of the old law, and by sec. 31 extended the age of children during which compulsory vaccination could be enforced to fourteen years. It also provided that upon the production of a medical certificate that a child was an unfit subject for vaccination, there was not to be a conviction (sec. 18). In the present case there was such a certificate; but inasmuch as the magistrates came to the conclusion that the practitioner furnishing it had been influenced in giving it by his own conclusions as an opponent of the vaccination theory, they again convicted the appellant. The Chief Justice, in giving judgment, said that although the sections of the new Act were very inconsistent with each other, still upon the whole he was able to come to the conclusion that the magistrates were justified in their decision. It might be that they had decided too hastily that the medical practitioner had allowed his mind to be warped by certain theories respecting the advantages of vaccination; but that involved a question of fact with which this Court had nothing to do. (Allen, appellant, v. Worthy, respondent—in the Court of Queen's Bench, before the Lord Chief Justice, Mr. Justice Mellor, and Mr. Justice Hannen, Jan. 15, 1870.) By this decision convictions under this Act may follow *toties quoties*—a parent is charged with being in continual contempt of the law.

CHAPTER XI.

THE LAW RELATING TO IDIOTS, LUNATICS, OR PERSONS OF UNSOUND MIND, AND INCAPABLE OF MANAGING THEIR OWN AFFAIRS; AS ALSO THE NATURE AND DEGREE OF INSANITY WHICH WILL RELIEVE A PERSON FROM THE PENAL CONSEQUENCES OF CRIME.

It would be quite inconsistent with the object and character of this work to enter upon the abstract question of insanity, or to attempt to give any dogmatic definition of that unhappy malady. Much less would it be consistent with that impartiality which different schools of thought have alike a right to expect in a treatise, not upon the theory of disease, whether mental or bodily, but upon the practical application of the laws towards the general alleviation of human infirmity, that any particular view should be prominently indicated as containing a more accurate or comprehensive definition than another.

Medical practitioners may have to certify to the sanity or insanity of a person, whether with a view to a civil temporary detention of such person, for the protection of himself and his property, or for the purpose of saving him from the infliction of a penalty which he has already, perhaps, most justly incurred, but which from his subsequent loss of reason it would be barbarous to carry into effect. In these cases the medical practitioner must act upon his own judgment and belief, always bearing in mind the heavy responsibility attaching to his decision, both as regards society as well as the individual, and remembering that he ought at any time to be prepared, as he may at any time be called upon, to give the reasons of the decision at which he has arrived.

On the other hand, the medical witness in a court of justice may be required to state symptoms for the guidance of others (that is, of the jury). He may be cross-examined as to his definition of insanity; and if asked his opinion upon a certain state of facts (as in some cases is permitted, as we shall presently see), he should be prepared with a clear and comprehensible definition, avoiding as much as possible professional technical terms, and should be able to give a lucid and sound reason for the opinion he may have formed.

Bearing these remarks in mind, it is proposed now to consider the first class of cases, in which a medical man is requested to examine, with a view to ulterior proceedings, an individual suspected of insanity. He may be requested by the private friends—by some civil authority, from the Lord High Chancellor to the Board of Guardians of the neighbouring Union ; or he may, in the case of a criminal charge, be deputed by the local magistrates, by the Judge at the assizes, or by Her Majesty's Secretary of State, to undertake this always delicate and sometimes most difficult task. Take the following piece of legislation on this subject as relating to public offences. By 1 & 2 Vict., c. 14, s. 2, in all cases of persons then in custody, as apprehended under circumstances that denote a derangement of mind and a purpose to commit crime, under or by virtue of any warrant or commitment made by any justice of the peace ; and after the passing of the Act, in the cases of persons who shall be discovered and apprehended under circumstances that denote a derangement of mind and a purpose of committing some crime for which if committed such persons would be liable to be indicted, it shall be lawful for any two justices of the peace to call to their assistance a physician, surgeon, or apothecary, and if upon such view and examination and due proof they should be satisfied that such persons are insane or dangerous idiots, they are empowered to direct the removal of such persons to the county lunatic asylum, &c.

By the 3rd section.—In the case of any person in custody at the time of the passing of the Act, having been apprehended under the above circumstances, if it appeared upon examination, &c., by the physician, &c., as before appointed, that such person was not insane and might be suffered to go at large with safety, such medical person is thereby required to give a certificate to that effect, to be forwarded to the Secretary of State, who, if he shall think fit, shall order the liberation of such person from custody. This Act is referred to principally as an illustration of the immense importance of a correct judgment in such matters. Here are two magistrates and a Secretary of State trusting entirely to the judgment of, perhaps, an apothecary, as to whether persons already in custody, under circumstances denoting a derangement of mind and a purpose of committing crime, are to be at once let loose upon society, to say nothing of the personal liberty of such persons, as well as that of those not as yet committed, being almost entirely dependent upon the like judgment, requiring much scientific knowledge and medical acumen.

Again : by 3 & 4 Vict., c. 54, if any person while imprisoned under any sentence of death, &c., or under a charge of any offence, or for not finding bail, or in consequence of any summary conviction, shall appear to be insane, any two justices of the peace, with the aid of two physicians or surgeons, are empowered to inquire as to the sanity of such person; and upon such person being certified to be insane, a Secretary of State shall, upon receipt of such certificate, direct the removal of such person to a lunatic asylum, there to be detained in custody until it is duly certified to the proper authorities by two physicians or surgeons that such person has become of sound mind. The provisions of this Act were no doubt grossly abused in the case of Townley, a prisoner under sentence of death, and a great scandal arose to the medical profession. The certificates of the medical men summoned by the justices, attesting insanity, resulted in a reprieve at almost the last moment. Upon some medical commissioners being sent subsequently—*i.e.*, within a month from the date of these certificates—to examine the state of mind of the prisoner, they reported him to be of sound mind; but as, after such an anxious interim, it would have been impossible to execute the capital sentence, it was commuted to penal servitude for life. It is right to add that the prisoner afterwards destroyed himself. These instances of the responsibility attaching to medical evidence in such cases are probably sufficient for the purpose intended.

Although this case led to an alteration in the practice of the law, as will be subsequently seen, the principle involved remains the same.

Such evidence being embodied in certificates is, of course, of a more general character, and less open to criticism than that given orally in open Court.

Before, however, entering upon the subject of evidence in open Court, a brief review of some of the provisions of the Lunacy Acts may here be introduced.

According to 16 & 17 Vict., c. 97, s. 74, no person (not a pauper) can be received into or detained in any asylum without a written order from some person, and an annexed statement of particulars of such person, and two medical certificates signed by *two physicians, surgeons, or apothecaries*, not being in partnership, nor the one an assistant to the other, and each of whom shall separately from the other have personally examined the person to whom the certificate relates, not more than seven clear days previous to the reception of the person into the said asylum, and such order, as before mentioned, may be signed before or after the medical certificates, or either of them. But under special circumstances, to be set forth in the statement, the certificate of one medical practitioner will suffice, provided that within three days from the reception of the patient two other such certificates shall be signed by two other medical practitioners, each of whom shall not be in partnership with or assistant to the other, or the medical practitioner who signed the certificate on which the patient was received. See Schedule (F), Nos. 2 & 3. And any person detaining such patient in an asylum for more than three days without such further certificates shall be guilty of a misdemeanour, and the like if he receives such person without such order and certificate or certificates. Lunatics found so by inquisition are excepted from these regulations (25 & 26 Vict., c. 111, s. 22).

Form of Medical Certificates in the Case of Private Persons (not Paupers).

I, the undersigned, being a physician, surgeon, or apothecary (as the case may be) [*here set out the qualification entitling the person certifying to practise as a physician, surgeon, or apothecary, ex. gra.*, being a Fellow of the Royal College of Physicians of England], and being in actual practice as a physician, surgeon, or apothecary (*as the case may be*), hereby certify that I, on the   day of   at [*here insert the* street and number of the house (if any) or other the particulars*], in the county of   [*in any case where more than one medical certificate is required by this Act, here insert, separately from any other medical practitioner*], personally examined A. B., of   [insert residence, and profession or occupation, if any], and that the said A. B. is a lunatic (or an idiot, or a person of unsound mind), and a proper person to be taken charge of, and detained under care and treatment, and that I have formed this opinion upon the following grounds—viz, facts indicating insanity observed by myself [*here state the facts*].

Other facts (if any) indicating insanity, communicated to me by others [*here state the information, and from whom*].

(Signed)

(Place of abode.)

Dated this   day of   , 18   .

By section 10 of 16 & 17 Vict., c. 96, the medical practitioner certifying is required to specify the facts upon which he has formed his opinion as to the lunacy, idiotcy, or unsoundness of mind of the person referred to in the certificate, and to distinguish such facts observed by himself from those communicated to him by others; and no person shall be received into any registered hospital, or licensed house, or as a single patient, under any certificate which purports to be founded only upon facts communicated by others, and a similar requirement is contained in 16 & 17 Vict., c. 97, s. 75.

And by section 11, power is given, after the reception of the lunatic, to amend the order or the certificate or certificates upon which he was received, if incorrect or defective, within fourteen days next after the reception of such lunatic, provided such amendment

---

* For want of this particularity such a certificate was held to be bad. (Reg. v. Pinder, 24 L. J., Q. B. 148.)

shall receive the sanction of one or more of the commissioners. Of course the definition of "physician, surgeon, &c.," is now regulated by the Medical Acts. (See 25 & 26 Vict., c. 111, s. 47.)

And by the 12th section, no medical practitioner, who, or whose father, brother, son, partner, or assistant is wholly or partly proprietor of, or a regular professional attendant in a licensed house or hospital shall sign any certificate for the reception of a patient into such house or hospital. And no medical practitioner shall, by himself or his agents, board or lodge in any unlicensed house, or take the charge of any person upon any certificate signed by himself, his father, brother, son, partner, or assistant; and no such medical practitioners, having signed any certificate, shall be the regular professional attendant of such person while under care under such certificate; and no such medical practitioners, who, or whose fathers, brothers, sons, partners, or assistants respectively, shall sign the order required for the reception of a patient, shall sign any certificate for the reception of the same patient. A similar provision is contained in 16 & 17 Vict., c. 97, s. 76. And any one pecuniarily interested in the payments made by or on account of any patient, and any "Medical Attendant," as defined by the Lunacy Act, 8 & 9 Vict., c. 100, shall be prohibited from signing any certificate or order for the reception of a private patient into any licensed or other house (25 & 26 Vict., c. 111, s. 24), the definition of "medical attendant" being "every physician, surgeon and apothecary who shall keep any licensed house, or shall in his medical capacity attend any licensed house, or any asylum, hospital, or other place where any lunatic shall be confined."

These regulations, it will be seen, are very stringent, and require to be rigidly adhered to, as medical practitioners giving false certificates, or any one signing a certificate as a medical man when he is not so, shall be deemed guilty of a misdemeanour (section 13); but a medical practitioner doing any act contrary to the provisions of this Act (and not declared to be a misdemeanour) shall for every such offence forfeit 20*l.*

With respect to evidence adduced in open court, especially in criminal cases, the extreme importance of the subject demands the consideration which it deserves. And here, it may be observed, the medical witness should above all things avoid making himself appear in the light of a partisan of any particular medical theory, or as a sympathizer with either side. The facts of the case are all the witness has to consider, even if asked his opinion upon the case itself. The judge and the jury have to decide upon the responsibility of the prisoner, and the medical evidence should be confined to the question whether the accused is insane in the sense in which it is understood by law. The medical witness may be asked whether in his opinion such and such acts amount to insanity, but he has nothing whatever to do with the question of responsibility. It is for the jury to say whether the accused knew the nature of the act he was committing, and whether or not he knew he was doing wrong.

Although, in general, a witness cannot be asked what his opinion upon a particular question is, since he is called for the purpose of speaking as to *facts* only, yet, where matter of skill and judgment is involved, a person competent to give an opinion may be asked what that opinion is. Thus, an engineer may be called to say what, in his opinion, was the cause of a harbour being blocked up. (Folkes *v.* Chad., 3 Douglas 157, 4 Term Rep. 498, S.C.) It is the constant practice to examine medical men as to their judgment with regard to the cause of a person's death who has suffered violence; and where, on a trial for murder, the defence was insanity, the judges to whom the point was referred were all of opinion that in such a case a witness of medical skill might be asked whether, in his judgment, such and such appearances were symptoms of insanity, and whether a long fast, followed by a draught of strong liquor, was likely to produce a paroxysm of that disorder in a person subject to it. Several of the judges doubted whether the witness could be asked his opinion on the very point which the jury were to decide—viz., whether, from the other testimony given in the case, the act with which the prisoner was charged was, in his opinion, an act of insanity. (R. *v.* Wright, Russell and Ryland, 456.) On the authority of the above case, Justice Park, on an indictment for cutting and maiming, allowed a medical man who had heard the case to be asked whether the facts and appearances proved showed symptoms of insanity. (R. *v.* Searle, 1 Moo. & R. 75.) And it seems, in M'Naughten's case, such questions were allowed to be asked. (2 Russell on Crimes, by Greaves, 925 n). A question may arise in these cases whether, where a witness, a medical man, called to give his opinion as a matter of skill, has made a report of the appearances or state of facts at the time, he may be allowed to read it as part of his evidence. The practice in Scotland is as follows:—A scientific witness is always directed to read his report, as affording the best evidence of the appearances he was called on to examine; yet he may be, and generally is, subjected to a further examination by the prosecutor's counsel, or to a cross-examination on the prisoner's part; and if he is called on to state any facts in the case unconnected with his scientific report—as conversations with the deceased, confessions made to him by the prisoner, or the like—he stands in the situation of an ordinary witness, and can only refer to the memoranda to refresh his memory. (Alison's Practice Crim. Law of Scotland, p. 541.)

Without venturing to give a distinct scientific definition of insanity, it may be mentioned that the law does not recognise *moral* insanity; indeed, it would be a highly dangerous doctrine to hold as irresponsible those who, however perverted may be their affectious or moral feelings, have given no indications of disturbed reason. For, after all, in every case of true insanity the intellectual disturbance is more or less present, however difficult sometimes it may be of detection. Moral insanity is not, therefore, admitted as a bar to responsibility in either civil or criminal cases, except in so far as it may be accompanied by *intellectual* disturbance; and even intellectual disturbance, if limited to the affection of insane delusions, will not be a bar to criminal responsibility if the accused, even though acting under the influence of the insane delusion, but not in other respects insane, and the act with which he is charged being at the same time contrary to the law of the land, he had a sufficient degree of reason to know that he was doing an act that was wrong. The accused in such case is assumed to know that the act was contrary to the law of the land, as the law is administered upon the principle that every man must be taken conclusively to know it, without proof that he does know it. Indeed, no actual knowledge of the law is essential in order to lead to a conviction; but if the accused was labouring under such a defect of reason from disease of the mind as not to know the nature and quality of the act he was doing, or, if he did know it, that he did not know he was doing wrong, that is such a degree of insanity and such a quality of insanity as alone will relieve a person from criminal responsibility. (See Answers of the Judges to the House of Lords, M'Naughten's case, *ante*, vide Appendix.)

Insanity is frequently feigned by persons accused of criminal offences in order to procure an acquittal or discharge. The following remarks (the result of a learned judge's experience, as furnished to Dr. Taylor by the writer himself) may be useful to those engaged in the preliminary examination of such persons. " It may be safely held that a person feigning insanity will rarely, if ever, try to prove himself to be sane; for he runs the great risk of satisfying others that he is sane—the conclusion he desires to avoid. But there is no better proof, in general, that the insanity (supposing other evidence to be strong) is real, than keen, eager attempts by the accused to prove that he is sane, and strong and indignant remonstrances against being held to be insane, although that would protect himself against trial and punishment." (Taylor's Medical Jurisprudence, p. 85.)

One question asked by the House of Lords of the judges in M'Naughten's case was, " Can a medical man, conversant with the disease of insanity, who never saw the prisoner previous to the trial, but was present during the whole trial and the examination of all the witnesses, be asked his opinion as to the state of the prisoner's mind at the time of the commission of the alleged crime, or his opinion whether the prisoner was conscious at the time of doing the act that he was acting contrary to law, or whether he was labouring under any and what delusion at the time?" In answer thereto the judges said, " We think the medical man, under the circumstances supposed, cannot in strictness be asked his opinion in the terms above stated, because each of those questions involves the determination of the truth of the facts deposed to, which it is for the jury to decide: and the questions are not mere questions upon a matter of science, in which case such evidence is admissible. But where the facts are admitted, or not disputed, and the question becomes substantially one of science only, it may be convenient to allow the question to be put in that general form, though the same cannot be insisted on as a matter of right."

It may be convenient here to mention a case of feigned insanity, which shows how easily medical men as well as non-professional persons may be deceived by a skilful impostor.

At the Lewes Winter Assizes, Dec. 1856, the prisoner (R. v. Ball), a ticket-of-leave man, was convicted of housebreaking and sentenced to fifteen years' transportation. After the trial, and when in gaol, the prisoner simulated madness so successfully that he effectually deceived three visiting justices and two medical men; and a certificate was about to be signed for his removal to a lunatic asylum, when the deception was discovered by the man having made a confidant of one of his fellow-prisoners. He had previously, in 1851, after having been convicted of robbery at Leicester, whence he was sent to Millbank, feigned insanity, and succeeded on that occasion in deceiving the medical officers, who certified that he was a lunatic, and he was accordingly removed to Bethlehem Hospital, where after two years he obtained a ticket-of-leave.

Doubtless there have been many inconsistent verdicts arrived at by juries, but the general tendency of such a tribunal is to lean to the side of a merciful construction of the acts of an accused, especially where a plea of insanity is set up; but until juries are taken from a more educated class of society they can scarcely be expected to exhibit that nice appreciation of subtle distinctions which the medical witness too frequently endeavours to import into the case. They may be, for the most part, men of good common sense, perfectly capable of judging from the evidence whether the accused knew at the time of the commission of the offence that he was doing that which was wrong, and yet may be wholly incompetent to deal with subtle distinctions between " dementia," " amentia," " mania," " dipsomania," or even " lunacy," or the pathological causes of any such forms of insanity. The object of the law is to present to them the case shorn of technicalities, and to ask them whether they believe, in the case before them, that the prisoner was at the time labouring under such a total defect of reason that he was wholly unconscious that he was doing wrong. In France the juries, at least in capital cases, are taken from the superior classes of society. They are comprised of men of intellect, education, and position; but with us such persons enjoy a privileged exemption, practically, from serving on common and petty juries, being for the most part on the grand jury or special jury panel of the assizes; and at quarter sessions, where now all cases except those of the greatest magnitude are triable, these same classes compose the bench of magistrates who there act as judges; and the quarter sessions grand jury absorbs the class immediately below these in the social scale,—so that the petty jury panel, both at quarter sessions and assizes, is necessarily composed of but one class, viz., the small shopkeepers or farmers, men whose very avocations, be it said without any disparagement to their general good sense, are but ill calculated to prepare their minds for the consideration and decision of abstruse and subtle disquisitions on points of science, after all, more curious than practical.

With respect to civil cases, or the rules which should regulate a medical practitioner's conduct in dealing with idiots, lunatics, or persons of unsound mind and incapable of managing their own affairs.

As this treatise professes to be neither a work on Medical Jurisprudence nor yet one on the general Law of Lunacy, but simply on the laws relating to the Medical Profession, a disquisition on the Law of Lunacy would not be within its province or scope, any more than would a treatise on Forensic Medicine, in itself an exhaustive subject, be compatible with the limits prescribed to the author.

A medical practitioner may sometimes be called upon, not only to certify, but also to make affidavit, as to the state of mind of a person the subject of a petition for the issue of a commission " De lunatico inquirendo," or in the nature of " De lunatico inquirendo."

The costs of this inquiry, which have frequently been ruinously expensive, are not only greatly reduced by 16 & 17 Vict., c. 70, but the process very much simplified by 25 & 26 Vict., c. 86, termed " The Lunacy Regulation Act, 1862," where the property of the alleged lunatic does not exceed 1000*l.* in value, or the income thereof does not exceed 50*l.* per annum; and the 3rd section enacts that " the inquiry to be made under every order or commission of lunacy or issue shall be confined to the question, whether or not the person who is the subject of the inquiry is at the time of such inquiry of unsound mind, and incapable of managing himself or his affairs, and no evidence as to anything done or said by such person, or as to his demeanour or state of mind at any time being more than two years before the time of the inquiry, shall be receivable in proof of insanity on any such inquiry, or on the trial of any traverse of an inquisition, unless the judge or master shall otherwise direct."

By the 12th section, " where by the report of one of the Masters in Lunacy, or of the Commissioners in Lunacy, or by *affidavit* or *otherwise*, it is established, to the satisfaction of the Lord Chancellor, that any person is of unsound mind and incapable of managing his affairs, and that his property does not exceed the amount before-mentioned, the Lord Chancellor, without directing any inquiry under a commission of lunacy, may make such order as he may consider expedient for the purpose of rendering the property of such person, or the income thereof, available for the maintenance or benefit of or for carrying on his trade or business : Provided that the alleged insane person shall have such personal notice of the application for such order as the Lord Chancellor shall by general order direct."

Under this Act medical practitioners may frequently be called upon to assume almost the entire responsibility of furnishing upon affidavit the information required in such cases, and it therefore becomes of the utmost importance to consider the meaning of the term " unsound mind" as applicable to such proceedings.

In re W. F. Windham, a case which excited at the time much public interest, the law was thus stated by the Master in summing up to the jury :—

" What the jury had to decide was, whether Mr. Windham was of such sound mind as to be able to govern himself and his affairs. In the time of Lord Hardwicke it was necessary to find a man absolutely insane; but in 1802, Lord Eldon declared that the Court of Chancery thought itself authorized to issue a commission, provided it were made out that the party was unable to act with any proper and provident management, was liable to be robbed by any one, or (and ?) was labouring under that imbecility of mind which, though not strictly insanity, equally required the protection of the law. Lord Lyndhurst, in 1827, adopted the ruling of Lord Eldon, which might be taken, therefore, as an authoritative declaration of the law of England. Another doctrine laid down by Lord Eldon, and sanctioned by his successors, was, that if a jury found merely the incapacity of an alleged lunatic to manage his affairs, and did not infer from it unsoundness of mind, upon that finding a commission could not go on, though the party might be living where he was exposed to ruin every instant. From all this it resulted that, in order to justify a verdict against Mr. Windham, the jury must be satisfied that he was incapable of governing himself and his affairs by reason of unsoundness of mind. Mere weakness of character; mere liability to impulse or susceptibility of influence, good or bad ; mere imprudence, extravagance, recklessness, eccentricity, or immorality—no, not all these put together would suffice, unless they believed themselves justified, on a view of the whole evidence, in referring them to a morbid condition of intellect."—*Times*, 30th January, 1862.

The distinction between an idiot and a lunatic, or person of unsound mind, was important in the early state of the law, as the custody of the idiot (or *idiota a nativitate*) and his lands were vested in the lord of the fee, and eventually by common consent in the King, as lord paramount and "pater patriae," that he might at the death of the idiot render the same to the right heirs, finding the idiot in the interim necessaries alone. An idiot was deemed a natural fool—one that hath no understanding from his nativity—and therefore presumed by law never likely to attain any (Fitzherbert's Natura Brevium, 233); whereas in the case of a lunatic, or rather a *non compos*—for the word "lunaticus" is not used in the original writ or early statute on the subject, but the definition in the writ is *idiota a casu et infirmitate*, and in the statute "aliquis qui prius habuit memoriam et intellectum non fuerit *compos mentis* suo sicut quidam sunt perlucida intervalla," the King provided for the safe custody of their lands without waste or destruction, and the competent maintenance of their households, with the profits of the same, and the residue, after thus providing for their sustenance, was to be kept for their use, to be delivered up to them when they were restored to their right mind; so that their lands and tenements should in no respect be alienated—and the King took nothing of the profits to his own use. But now for the purposes of the inquiry in the nature of "De lunatico," the distinction is of no moment. Important changes have of late years been made in the law on this subject by 16 & 17 Vict., c. 70, and 25 & 26 Vict., c. 86; one object of which has been the "removing or diminishing the delays and expenses now attending in the execution of commissions in the nature of writs '*De lunatico inquirendo*,' and the proceedings consequent on inquisitions taken thereon."

The 25 & 26 Vict., c. 86, amending the 16 & 17 Vict., c. 70, declares in sec. 2 that the word "lunatic" as used in the Acts is to be construed to mean "any person found, by inquisition, idiot, lunatic, or of unsound mind, and incapable of managing himself or his affairs," so that it is now quite unnecessary to consider within which definition to describe an alleged *non compos*. It is sufficient that he be found to be of *unsound mind, and incapable of managing himself or his affairs*. But upon this very point a medical witness may be summoned to give evidence either *pro* or *con*, and here it may be repeated no amount of intemperance, not even notorious drunken and profligate habits, will satisfy the requirements of the statute, unless such, to use the words of the Master in Mr. Windham's case, can be referred "to a morbid condition of intellect." "Drunkenness alone," said Lord Cottenham, "is not sufficient." But if the mind is completely weakened by *habitual* drunkenness, so that it can no longer be deemed sound, though such unsoundness was originally contracted by this vice, such a state, coupled with an incapacity of managing his affairs (and of the incapacity of managing himself in such case there would be no doubt), would satisfy the statute.

It may sometimes happen that a medical attendant is asked to witness and perhaps, in the absence of a legal practitioner, to draw up the will of a patient. It should be remembered, therefore, that the witness to a will attests the sanity of the testator, and should he afterwards attempt to impeach the testator's sanity his evidence would be regarded with great suspicion, and would require corroboration. A will made by a lunatic during a lucid interval would be valid. (See Beverley's case, 4 Rep. 1236; Kemble v. Church, 5 Haggard, 273.) But then the attesting witnesses would be doubtless required to prove the lucid interval. Lord Hardwicke has observed, in connexion with this subject, that fraud and imposition upon weakness may be a sufficient ground to set aside a will, although such weakness is not a sufficient ground for a commission of lunacy. (2 Vesey, 408.) In Mountain v. Bennett, Cax. 355, a will, after much conflicting evidence as to the testator's mind, was supported principally on the evidence of the attesting witnesses, persons of high character and respectability, who were unanimous as to the testator's sanity and freedom from control.

A suspicion is justly entertained of a will in favour of a medical attendant in whose house the testator resided, (James v. Moody, P.C.C. 16; Major v. Knight, 4 No. Cas. 661; Cockcroft v. Rawles, ib. 237); but it seems that this suspicion goes no further than to necessitate somewhat stricter proof as to the testator's *capacity*, though not as to his *knowledge* of the contents of the will (Barry v. Butlin, 2 Moo. P.C.C. 480); for where the *capacity* be duly proved, the testator will be deemed cognizant of the contents of the will. (Browning v. Budd, 6 Moo. P.C.C. 435.) If a will be rational on the face of it, and it is proved to have been duly executed, it is presumed, in the absence of evidence to the contrary, that it was made by a person of competent understanding. And though an inquisition finding a man a lunatic is *primâ facie* evidence of lunacy during the whole period covered by such inquisition, yet it does not preclude proof that the execution of a will, or any other act, occurred during a lucid interval. (Hall v. Warren, 9 Vesey, 605; Cooke v. Cholmondeley, 2 Mac. & G. 22; and Bannatyne v. Bannatyne, 16 Jurist, 864.)

The principle is ably laid down by Sir W. Wynn in Cartwright v. Cartwright, 1 Phillimore, 100 :—"If you can establish," said the learned judge, "that the party afflicted habitually by a malady of the mind has intermissions, and if there was an intermission of the disorder at the time of the act, that being proved is sufficient, and the general habitual insanity will not affect it; but the effect of it is this, that it inverts the order of proof and of presumption; for until proof of habitual insanity is made, the presumption is that the party, like all human creatures, was rational; but where an habitual insanity in the mind of the person who does the act is established, then the party who would take advantage of the fact of an interval of reason must prove it."

The mere absence, however, of any apparent delusion must not be mistaken for a lucid interval; the disease may nevertheless exist. To constitute a lucid interval, absence of the disease itself, not of the delusion only, must be shown. (Waring v. Waring, 6 Moo. P.C.C. 341, and Creagh v. Blood, 2 J. & Lat. 509; Dyce Sombre v. Troup, 1 Deane, 22.) As the only object of this dissertation has been to show that the law permits a person of hitherto unsound mind to make a will, if at the time he has recovered his senses and is of a disposing mind; and as his medical attendant would be the best authority on that point, and would probably be asked to attest, if not, in the absence of legal assistance, even to reduce the patient's wishes to writing, this is thought to be a fitting place to make a few remarks on the same. The statute 1 Vict., c. 26, s. 9, provides " That no will shall be valid unless it shall be in writing, and executed in manner hereinafter mentioned: that is to say, *it shall be signed at the foot or end thereof by the testator, or by some other person in his behalf, and by his direction; and such signature shall be made or acknowledged by the testator in the presence of two or more witnesses present at the same time; and such witnesses shall attest and shall subscribe the will in the presence of the testator, but no form of attestation shall be necessary.*"

In the first place.—Two witnesses are required.

Secondly.—The signature of the testator must be somewhere near to the end of the instrument, and so as not to be immediately over, or preceding any of the dispositive parts of the instrument, but it need not immediately follow, or be under any of the dispositive parts; whereas formerly the signature might be in any part of the instrument.

Thirdly.—The signature of the testator is to be " made," or " acknowledged" in the *simultaneous* presence of the witnesses. (Moor v. King, 3 Curtis, 243, 2 No. Cas. 45, & 7 Jurist, 205.)

Fourthly.—A form of attestation is expressly dispensed with.

There can be no sufficient acknowledgment by the testator of his " signature" unless the witnesses either saw or might have seen the signature, even though the testator should expressly declare to them that the paper to be attested by them is his will.

When the witnesses either saw or might have seen the signature, an express acknowledgment of the signature itself is not necessary, a mere statement that the paper is his will—(*re* Davis, 3 Curtis, 748;

re Ashmore, *ib.* 756, and 7 Jurist, 1145)—or a direction to them to put their names under his—(*re* Philpot, 3 No. Cas. 2. Gaze *v.* Gaze, 7 Jur. 803)—or even a request by the testator—(Keigwin *v.* Keigwin, 7 Jur. 840)—or by some person in his presence—(*re* Bosanquet, 2 Rob. 577, Foulds *v.* Jackson, 6 No. Cas. sup. 1)—to sign the paper is sufficient.

When the signature is seen, or expressly acknowledged, it is not material that the witnesses are not told that the instrument is his will—(Keigwin *v.* Keigwin, supra)—or are deceived into thinking that it is a deed—(Sugden's R.P.S. 334). But see observations of Sir H. J. Fust in Willis *v.* Lowe, 5 No. Cas. p. 432.

To obviate the inconvenience arising from decisions under the 9th section of 1 Vict., c. 26, it was enacted by 15 & 16 Vict., c. 24 (framed and introduced by Lord St. Leonards when Lord Chancellor), that a will shall be deemed to be valid if the signature be so placed at, or after, or following, or under, or beside, or opposite to, the end of the will, that it shall be apparent on the face of the will that the testator intended to give effect, by such his signature, to the writing signed as his will, and that no such will shall be affected by the circumstance that the signature shall not follow or be immediately after the foot or end of the will, or by the circumstance that a blank space shall intervene between the concluding word of the will and signature, or by the circumstance that the signature shall be placed among the words of the testimonium clause, or of the clause of attestation, either with or without a blank space intervening, or shall follow, or be after, or under, or beside, the names or one of the names of the subscribing witnesses, or by the circumstance that the signature shall be on one side or page or other portion of the paper or papers containing the will, whereon no clause or paragraph or disposing part of the will shall be written above the signature, or by the circumstance that there shall appear to be sufficient space on or at the bottom of the preceding side or page, or other portion of the same paper on which the will is written, to contain the signature, and the enumeration of the above circumstances shall not restrict the generality of the above enactment; but no signature under the said Act or this Act, shall be operative to give effect to any disposition or direction which is underneath, or which follows it: nor shall it give effect to any disposition or direction inserted after the signature shall be made.

The law has not made requisite to the validity of a will that it should assume any particular form, or be couched in language technically appropriate to its testamentary character. It is sufficient that the instrument, however irregular in form, or inartificial in expression, discloses the intention of the maker respecting the posthumous destination of his property. It is therefore far better for a non-legal person, in drawing up a will, to use plain and simple language, such as he would use in any ordinary written communication, than to affect a technical style and the use of legal phraseology, which, from want of an accurate application of the terms, may only tend to confuse the construction, and to render ambiguous the intention of the testator. As the circumstances in every case must more or less vary, it has been thought better not to append any particular form of will, and although no particular form of attestation is necessary, still it is better that the witnesses to the will should subscribe a memorandum of attestation to the effect that—

The *signature of the testator was made, or acknowledged by him in the presence of the witnesses, both being present at the same time, and that they subscribed their names in his presence, and in the presence of each other.*

Of course it is far better that a testator should actually sign, or (where necessary) sign by the hand of another, in the presence of the witnesses, than merely acknowledge the signature before them; but yet a testator, whether speechless or not, may acknowledge his signature by gestures—(*re* Davies, 2 Rob. 337, and Parker *v.* Parker: Milw., Ir. Eccles. Rep. 545.) and even a mark, thus X, where the testator through ignorance is unable to write, is to be preferred to a signature by the hand of another. A person may not be insane, and may yet be incompetent to make a will. Imbecility, through advanced age or through excessive drinking, or a want of a disposing power of mind, so as not fully to comprehend at the time the nature and extent of his property, nor yet the claims of those who might naturally be supposed to be the objects of his bounty, may destroy testamentary power. (See Swinburne, p. 11. ss. 5 & 6.)

By the recent Act, 1 Vict., c. 26, if any person shall attest the execution of any will to whom, *or to whose wife or husband*, any devise, legacy, or beneficial interest shall be thereby given or made, such devise, legacy, &c., be utterly null and void.

As it is not probable that a medical practitioner would be requested to attest the execution of a *deed*, except under the superintendence of a legal adviser, deeds being for the most part drawn up by such, and generally executed under circumstances of more deliberation than wills, it is not thought necessary to introduce that question into this subject. Medical attendants should be careful how they permit themselves to be made confidants of the secrets of their patients, as they possess no privilege in this respect, but if required, are bound to disclose them in a court of law.

Medical visitors, appointed to visit any house or houses licensed for the reception of lunatics under 8 & 9 Vict., c. 100, are entitled to remuneration for their services. But by 23rd section of that Act no person shall act as a commissioner or visitor, or secretary or clerk to the commissioners, or clerk or assistant clerk to any visitors, or act in granting any license, who shall within one year next preceding have been directly or indirectly interested in any such house, or the profits arising from the reception of lunatics within it. And no physician or surgeon (being a commissioner), and no physician, surgeon, or apothecary (being a visitor) shall sign any certificate for the admission of any patient into any licensed house or hospital, or shall professionally attend upon any patient in any such house or hospital, unless he be directed to visit such patient by the person upon whose order such patient has been received into such house or hospital, or by the Lord Chancellor, or by the Secretary of State for the Home Department, or by a committee appointed by the Lord Chancellor; and if any such shall after his appointment be or become so interested, he shall immediately be disqualified; and if after so becoming disqualified he shall continue to act in such capacity, he shall be guilty of a misdemeanour. And if any such medical commissioner or visitor shall sign any certificate for the admission of any patient into any licensed house or hospital (except as aforesaid), he shall for each offence forfeit 10*l*.

By the 3rd section, the "Commissioners in Lunacy," some of whom may be physicians or surgeons, shall not, as long as they remain commissioners and receive any salary as such, accept, hold, or carry on any other office or situation, or any profession or employment from which gain or profit shall be derived, the salary allotted under the Act being 1500*l*. a year and travelling expenses.

By sec. 13, in every hospital there must be a physician, surgeon, or apothecary resident therein, as the superintendent and medical attendant thereof. The provision is different as to licensed houses. In every house licensed for 100 patients or more, the residence of such medical officer is compulsory. In other cases it is optional, but if there is no resident medical man, the house must be visited regularly by a medical attendant, the frequency of such visits being regulated by the commissioners and visitors according to the number of the patients for which the house is licensed, whether below 11, or below 50, or below 100, as the case may be. (*Ib.* ss. 57 & 58.)

A medical visitation book is to be duly kept in every house and hospital by the medical superintendent or attendant in the form prescribed by the Act of 1853 (16 & 17 Vict., c. 70), which superseded the form given in Schedule II in the Act of 1845 (8 & 9 Vict., c. 100, s. 59); 16 & 17 Vict., c. 96, s. 25, and Schedule D.

The medical superintendent or attendant is also to keep a book to be called "The Case Book," the form of which the Commissioners in Lunacy are empowered to prescribe by an order under their common seal (8 & 9 Vict., c. 100, s. 60). On the 20th March, 1863, the Commissioners issued an order accordingly, of which the following is the substance. Such "Case Book" to contain :—

"*First.* A statement of the name, age, sex, and previous occupation of the patient, and whether married, single, or widowed.

"*Secondly.* An accurate description of the external appearance of the patient upon admission, habit of body and temperament, appearance of eyes, expression of countenance, and any peculiarity in form of head; of the physical state of the vascular and respiratory organs, and of the abdominal viscera and their respective functions; of the state of the pulse, tongue, skin, &c.

"*Thirdly.* A description of the phenomena of mental disorder, the manner and period of the attack, with a minute account of the symptoms and the changes produced in the patient's temper or disposition; specifying whether the malady displays itself by any, or what illusions, or irrational conduct, or morbid or dangerous habits or propensities; whether it has occasioned any failure of memory or understanding, or is connected with epilepsy or ordinary paralysis, or symptoms of general paralysis, such as tremulous movements of the tongue, defect of articulation, or weakness or unsteadiness of gait.

"*Fourthly.* Every particular which can be obtained respecting the previous history of the patient; what are believed to have been the predisposing and exciting causes of the attack; what the previous habits, active or sedentary, temperate or otherwise; whether the patient has experienced any former attacks, and if so, at what periods; whether any relatives have been subject to insanity; and whether the present attack has been preceded by any premonitory symptoms, such as restlessness, unusual elevation or depression of spirits, or any remarkable deviation from ordinary habits and conduct; and whether the patient has undergone any, and what, previous treatment, or been subjected to personal restraint.

"*Fifthly.* During the first month after admission entries to be made at least once in every week, and oftener where the nature of the case requires it. Afterwards, in recent or curable cases, entries to be made at least once in every month; and in chronic cases, subject to little variation, once in every three months.

"In all cases an accurate record to be kept of the medicines administered, and other remedies employed, with the results, and also of all injuries and accidents.

"That the several particulars, hereinbefore required to be recorded, be set forth in a manner so clear and distinct as to admit of being easily referred to, and extracted, whenever the Commissioners shall so require;

"And that the present order be in substitution for that of the 9th January, 1846, and that a copy thereof be inserted at the commencement of the 'Case Book.'"

The statutes contain numerous and minute provisions with regard to the admission, treatment, discharge, and other important points con-

nected with the custody, care, and control of the patients received into these establishments. The respective enactments should be carefully studied in the statutes themselves, (to be found in the Appendix,) as a due observance of them is of the utmost importance, as well for the benefit of the insane persons as for the protection of the medical men and others who are instrumental in confining or detaining them. Independent of the penal liabilities which may be incurred under the 56th section of the last-named Act, and under 16 & 17 Vict., c. 96, s. 13, a medical man may be exposed to an action at the suit of a private person confined under his certificate. Hall *v.* Semple, 3 F. & F. 337, is an important case on this point. This case arose in 1862, being an action against a physician under whose certificate the plaintiff had been confined in a lunatic asylum. The following remarks are contained in the summing up of Mr. Justice Crompton: —" If a medical man assumes, under this statute, the duty of signing such a certificate without making, and by reason of his not making, a due and proper examination and such inquiries as are necessary, and which a medical man under such circumstances ought to make and is called on to make, not in the exercise of the extremest possible care, but in the exercise of ordinary care, so that he is guilty of culpable negligence, and danger ensue; then an action will lie, although there has been no spiteful or improper motive, and though the certificate is not false to his knowledge." . . . . . .
" The principal questions," continued the learned Judge, " to which I desire to direct your attention are these—first, whether you think that he signed the certificate, untrue in fact, negligently and improperly, and without making proper and sufficient inquiries. Now I cannot help thinking that in a matter of this kind, a proceeding upon which a man is to be at once consigned to imprisonment as a lunatic, very considerable care is necessary. It will be for your consideration what degree of care is necessary so as to make out by the absence of it a case of culpable negligence. It is not a mere mistake or error in judgment which would amount to such negligence, but you must be satisfied that there was culpable negligence. . . . . . . The main question, as it appears to me, is as to negligence or want of reasonable care; and as to that, the real test, as it appears to me, is how matters would appear to the defendant under all the circumstances; and you should endeavour, as far as possible, to put yourselves in his situation, and see how far appearances were sufficient, without any further inquiries than he made."

The learned Judge again repeated:—" You are not inquiring into an error in judgment, but whether the defendant has been guilty of that culpable negligence which I have explained and described to you—negligence in not making sufficient inquiries, the examination not having been sufficient in his own judgment. It would be dreadful if a man were to suffer merely from an error in judgment. The question is whether there has been a neglect of that duty which a person in a case of this kind owes, not to interfere in a matter which touches the liberty of his fellow-citizen without taking due care and making a careful examination and inquiry. That is the main question."

In answer to questions from the learned judge the jury found specially, on the second count, that on the examination and inquiries, the defendant did not use due care and caution; that the certificate was false in fact; that he signed a certificate untrue in effect, without a proper examination and inquiries, and without probable cause; and that on that account he was guilty of culpable blameable negligence in not ascertaining the sanity of the plaintiff, who they found to be sane in point of fact.

The jury further said, in answer to another question, that the defendant "*bonâ fide* believed in the truth of his certificate, from the examination and inquiries which he made, although he made them carelessly."

Mr. Justice Crompton.—" Did he think that he was acting under the authority of the Act? I assume from what you have found, that you mean he *bonâ fide* believed he was acting under the authority of the Act, and that he was authorized by the Act to do what he did?"

The Jury.—" Yes, we think so. He thought he was authorized, although he did it negligently."

The learned Judge.—" Then that is a verdict for the plaintiff on the ground of culpable negligence and want of reasonable care or probable cause."

Jury.—" Yes."

Verdict for the plaintiff, damages 150*l.*

As to the prosecution of two medical men (Reg. *v.* Dawson and another) for signing certificates without examining the patient, see Sixth Report of the Commissioners, p. 19. But at common law, and apart from the Lunacy Statutes, a medical man may justify measures necessary to restrain a dangerous lunatic. So also in the case of a person suffering from *delirium tremens*, he may justify resorting to measures of restraint during the continuance of the fit, or whilst it is likely to return, 3 F. & F. 328. (Scott *v.* Wakem.)

On the death of any patient, in any licensed house or hospital, a statement of the cause of death, with the name of any person present at the death, must be drawn up and signed by the medical attendant (8 & 9 Vict., c. 100, s. 55), and a certified copy must be sent by the proprietor or superintendent, within forty-eight hours, to the following persons:—The Commissioners; the person who signed the order of admission; the registrar of deaths for the district; and in the case of a house licensed by justices, the clerk of the visitors (s. 55; see also s. 54). The death must be entered in the "Register," and also in the "Book of Admissions," by the proprietor or superintendent (ss. 50 & 51), and in the "Medical Visitation Book," by the medical attendant, (s. 59) and it is further provided by the Act of 1853, that a statement as to the cause of death, and the duration of the fatal disease, shall be entered in the "Case Book," and a certified copy transmitted within two days to the coroner, so that he may hold an inquest if he deems it necessary. (16 & 17 Vict., c. 96, s. 19.)

The various provisions with regard to private lunatics received into public asylums, under the 16 & 17 Vict., c. 97, their admission under an order (or rather request) and medical certificates, the records and returns relating to them, their treatment and ill-treatment, correspondence, escape, and recapture, transfer, removal, and discharge, and death, are similar to, though not precisely identical with, those already detailed with respect to private inmates in licensed houses and registered hospitals.

Under the head of *Restraint* it has been held in Anderdon *v.* Burrows, M.D. (4 C. & P. 210) that a medical man is not warranted, merely on statements made to him by the relations of a person supposed to be insane, in sending men to take him into custody and confine him, unless he is satisfied from those statements that such a step is necessary to prevent the lunatic from doing some immediate injury to himself or others. Lord Tenterden, C.J., observed:— "From the statement made by Dr. Burrows, when the parties were before the magistrate, it seems that it is usual, on the application of the family, to act in this manner. I confess I am sorry to hear it so said, for it certainly is not right; and although there may be difficulty in getting access to a party labouring under insanity, yet the proper course is, if access cannot be obtained, to apply to the high authority which has cognizance over such matters, to get the party taken up, in order that he may be examined." This case occurred in November, 1829.

Symm *v.* Fraser and another, 3 F. & F. 859, is on the other hand a very recent case. It was an action brought in the year 1863 against two medical men for having imprisoned the plaintiff and put her under bodily restraint under the false and unfounded pretence that she was a person of unsound mind, incapable of taking care of herself, and unfit to be at liberty, &c. The defence was that she was suffering from a fit of delirium tremens caused by drink.

Dr. Tunstall, a physician, in the course of his evidence stated, that he never knew an instance of an habitual drinker, with a habit of delirium tremens, ever getting over it, because the habit of drinking was so strong on such persons that they could not abstain from drink. The witness said, delusions of the senses were symptoms of delirium tremens. Both defendants were called as witnesses to sustain the defence, and one of them, Mr. Andrews, the usual medical attendant of the plaintiff, was permitted to give his general opinion, as a scientific witness, although he was one of the defendants.

At the close of the case, which lasted several days, *Chief Justice Cockburn* (to the jury).—" The case was one of great importance, involving as it did the question how far medical men, acting honestly and to the best of their judgment for the good of their patients, were responsible; and while on the one hand, the jury ought jealously to watch over and uphold the personal liberty of the individual, yet on the other hand they ought to be equally careful not to impair the efficacy of medical assistance by exposing medical practitioners to be harassed by vexatious actions. . . . . . . No doubt the power which the law gave to medical men in these cases might be, and in rare instances had been abused, for interested purposes; but these instances were so rare that he hoped the public might continue to repose that confidence in the members of this noble profession which had hitherto been reposed in them, and he believed was deserved. . . . . . . There had been some restraint, no doubt, upon natural liberty, but was it not necessary ? Let the jury put themselves in the position of these gentlemen, or the friends and relatives of the plaintiff, and let them ask whether, even supposing that the defendants were responsible for all that had been done to prevent her from getting out into the street, or throwing herself out of the window. the jury would not consider that these gentlemen, so far from being proper subjects of condemnation and of censure, were not rather fit objects of gratitude and regard ?"

In conclusion, the Lord Chief Justice desired them to consider the case, "not only with reference to the interests of the individuals committed to the care of medical men, but also with a view to the interests of the public, taking care not to impair or neutralize the energy and usefulness of medical assistance by exposing medical men unjustly to vexatious and harassing actions."

Verdict for the defendants.

This verdict gave universal satisfaction both to the legal profession and the public. Not long after, a somewhat similar case was tried in another court, and with a similar result. The Lord Chief Justice, in effect, directed the jury that if the restraint exercised, under the sanction, and even directions of the defendants, was no greater degree of restraint than was medically necessary, they were legally justified. The case is the exact converse of Hall v. Semple, vide ante.

CHAPTER XII.

CORONERS' INQUESTS.

As the office of Coroner is not only not necessarily a medical one, but not even a medico-legal one; indeed, as it is not necessary in point of law that a coroner should be either a lawyer or a doctor, his sole legal qualification for the office being that of a freeholder, it might at first sight appear that the duties, like the qualification of such an officer, were scarcely a subject within the province of this work. But when it is considered that although in theory the coroner is neither a medical nor a legal professional, yet that in practice he is required to be versed in both medical and legal science, it will no longer be thought that either he or his duties are not fitting topics to be discussed in a work on the Laws relating to the Medical Profession.

It is not proposed, however, to take otherwise than a practical view of this subject, and therefore it will not be desirable to trace the origin, or descant upon the antiquity of the office—to assert its dignity by reminding the reader that the Lord Chief Justice of the Court of Queen's Bench is by virtue of his office supreme Coroner over all England, and the Puisne Judges of that Court are also Sovereign Coroners, or by showing that originally the qualification for the office was that of knighthood, which qualification was defined by statute so early as 3 Edw. I., c. 10, thus, "of the most wise and discreet knights, which know, will, and may best attend upon such offices;" or to lament that in consequence of a deficiency of knights it was at length enacted by 28 Edw. III., c. 6, that coroners should be elected of "the most meet and lawful people that shall be found in the said counties to execute the said office." No precise amount of estate is defined by the statute 14 Edw. III., s. 1, c. 8, which enacts "that no coroner be chosen unless he have land in fee sufficient in the same county, whereof he may answer to all manner of people," but it was evidently intended that he should be a person of sufficient property to maintain the dignity of his office, and to answer any fine that might be set upon him for misbehaviour, although for the office of coroner of *a borough* no qualification by estate, residence, or otherwise is required; and perhaps, after all, the qualification by title or estate is the least important qualification for the office: for although in the present day we may look in vain among coroners for the individual who, in the language of Chaucer, was

> "Lord and sire,
> Full oftentime was knight of the shire,
> A shreve had been, and Coronour,"

yet we may hope to recognise him in the descriptive qualities which Sir Edward Coke considers should distinguish the office. A coroner, according to him, should be *probus homo; legalis homo;* of sufficient knowledge and understanding; of good ability and power to execute his office according to his knowledge; and, lastly, of diligence and attendance for the due execution of his office. And this for three purposes. I. The law presumes that he will do his duty, and not offend the law, at least for fear of punishment, whereunto his lands and goods are subject. II. That he be able to answer the king all such fines and duties as belong to him, and to discharge the county thereof. And, lastly, that he may execute his office without bribery. And these three properties are necessary to every officer.

As by the term *legalis homo*, Sir Edward Coke, speaking as a lawyer, could not have meant literally a member of the legal profession, such requirement forming no part of the legal qualification of a coroner, he probably intended to indicate a person of *a judicial mind*, and such is doubtless, at the present day, the fittest qualification for the office, and together with that of *probus homo* should be esteemed the choicest attribute of an officer of justice, especially of one who derives his very name, *de coronâ*, from the Crown itself, having cognizance of those pleas, emphatically styled *placita coronæ*. But whatever spell may attach to a name, for it is doubtful, speaking poetically, whether "a rose by any other name would smell as sweet," romance and imagination having much to do with the appreciation of the senses, it is certain that this office is one of great antiquity; for not only is the coroner spoken of in the reign of Henry II. as *Servicus regis* and *coronarius;* in the reign of Richard I. as *custos placitorum Coronæ*, and in Magna Charta and subsequent statutes as Coronator; but he is mentioned in the charter granted by King Athelstane to the Monastery of St. John of Beverley, A.D. 925. As, however, was said before, it is not intended in this treatise to deal with this functionary or his duties, otherwise than in a practical way as affecting the medical profession especially; but that course of dealing will not extend to an analysis of the judicial functions of a public officer, but will be confined to the duties of medical practitioners who may be summoned to this as to any other court of justice.

By the 6 & 7 Wm. IV., c. 89, s. 1, power is given to the coroner to direct the performance of a post-mortem examination, with or without an analysis of the contents of the stomach or intestines of a deceased, by the medical witness or witnesses summoned to attend the inquest, provided that if any person shall state upon oath before the coroner that in his or her belief the death of the deceased was caused partly or entirely by the improper or negligent treatment of any medical practitioner or other person, such medical practitioner or other person shall not be allowed to perform or assist at the post-mortem examination of the deceased. And it may here be remarked that by the 29 & 30 Vict., c. 90 (Sanitary Act, 1866), s. 28, any "nuisance authority" (therein defined) shall have power to provide places for the reception of dead bodies during the time required for post-mortem examinations, otherwise than at a mortuary house or workhouse.

In all cases of sudden or violent death a medical man should be examined as a witness, and by the 3rd section of the first-named Act every legally qualified medical practitioner is entitled to a fee of one guinea for attending to give evidence at any coroner's inquest whereat no post-mortem examination has been made by such practitioner. And for the making a post-mortem examination, either with or without an analysis of the contents of the stomach or intestines, and for attending to give evidence thereon, the fee or remuneration shall be two guineas, and the 1 Vict., c. 68, s. 2, directs that "the coroner shall, immediately after the termination of the proceedings at any inquest, advance and pay such remuneration or fee to every medical witness summoned under the provisions of the said Act, and the amount thereof shall be repaid to the said coroner in manner thereinafter mentioned."

By the 6th section—" where any order for the attendance of any medical practitioner shall have been personally served upon him, or where any such order not personally served shall have been received by any medical practitioner in sufficient time for him to have obeyed such order, or where any such order has been served at the residence of any medical practitioner, and in every case where any medical practitioner has not obeyed such order, he shall for such neglect or disobedience forfeit the sum of 5*l*. And the 5th section exempts from the operation of the Act inquests held on the bodies of persons dying in any public hospital or infirmary, or in any building or place belonging thereto, or dying in any county or other lunatic asylum, or in any public infirmary or other medical institution, whether supported by endowments or by voluntary subscriptions: and declares that in such case the medical officer whose duty it may have been to attend the deceased as such medical officer shall not be entitled to any fee or remuneration.

The evidence of medical witnesses upon coroners' inquests is frequently all-important. Not only in cases of palpable violence are they required to explain the nature of the injuries inflicted on the deceased, but sometimes upon their evidence depends the solution of the problem whether the deceased died by his own hand or by that of another. By the post-mortem examination the presence of poison is perhaps discovered by the aid of chemical tests, no matter with what subtlety the agent poison may have been disguised, or however evanescent may have been its operation. Also in a charge of child-murder almost everything depends upon the evidence of the medical witness, the fact of whether the child was born alive, and had an existence separate from its mother, being essential to constitute the crime of murder, otherwise it in all probability resolves itself into a case of concealment of birth only, which at once ousts the coroner of his jurisdiction. The inquest of the coroner is an inquest to ascertain the cause of death, with which the concealment of the birth has no connexion. The first question, therefore, which presents itself is, whether the child was born alive. As a test of this, it was formerly usual to immerse the lungs in water, it being supposed that if they floated the child must have respired. But this test is now quite exploded; for it is obvious that if the child made but one gasp and instantly died, the lungs would swim in the water as readily as if the child had breathed for a lengthened time; and it is not uncommon for an infant to breathe as soon as its head is protruded from the mother, although it may die the next moment. Air may pass into the lungs by inflation, or may be generated by putrefaction, and both will produce the same effect. The question of course is less difficult in cases of immature birth, especially under the fifth month, in which case no fœtus can be born alive; and even from the fifth to the seventh the presumption either that the child was not born alive, or almost immediately ceased, from natural causes, to maintain an existence, would probably be in favour of the suspected person. But yet such serious cases should not be determined by the rules of presumption, but by facts proved and explained.

The real question is the cause of death. Now, the child may die in the womb, during the labour, by pressure, or by strangulation from the umbilical cord, in which latter case especially the body presents appearances which to a common observer would seem to be the marks of a violent death. To kill a child in its mother's womb is no murder, because, as Sir E. Coke says, the person killed must be "a reasonable creature in being, and under the King's peace." (3 Iust.)

To constitute murder it must be proved that the entire child was actually born into the world in a living state (R. v. Poulton, 5 C. & P. 329); and the fact of its having breathed is not conclusive proof thereof (R. v. Sellis, 1 Mood. C. C. 850; R. v. Crutchley, 7 C. & P. 814). There must be an independent circulation in the child before it can be accounted alive (R. *v.* Enoch, 5 C. & P. 539 ; R. *v.* Wright, 9 C. & P. 754). But the fact of the child's being still connected with the mother by the umbilical cord will not prevent the killing from being murder (R. *v.* Crutchley, *supra;* R. *v.* Rewes, 9 C. & P. 25; R. *v.* Trilloe, 2 Mood. C. C. 260). But the child, though safely born, may still be so weak as to die, and yet without any criminal act on the part of the mother, or the child may be suffocated by being left upon its face.

These and a variety of other causes may contribute to the death of new born infants, particularly where the mother is delivered in secret by herself, and, either from exhaustion or unconsciousness, inflicts an unintentional injury upon the child.

But of all the forms of murder, that by poison is the most detestable and the most difficult to guard against, and the detection of this secret and subtle agent frequently requires much chemical knowledge and delicate manipulation, and the perfect cleanliness of the apparatus used in the analysis. The medical practitioner, perhaps the medical attendant of the deceased, supposing the case to be one of suspected poisoning, may be able to produce most valuable corroborative or confirmatory evidence through having noted many circumstances in the course of his attendance upon the deceased. Indeed, many valuable suggestions may proceed from him, as, for instance, even upon a post-mortem examination no poison may be found in the stomach, because it may have been vomited up ; the vomited matter may thus be procured. If none be procurable, other things impregnated with it, such as clothes, may be secured, or even a portion of deal flooring may be scraped or cut out and reserved for analysis. The vessel in which vomited matter has been contained will often furnish valuable evidence, since heavy mineral poisons fall to the bottom, adhere to the sides, or leave a sediment. Any suspected articles of food should be at once taken possession of and carefully sealed up in a clean glass vessel. In the event of the death of the patient (which event alone can bring the case before the coroner's court) the exact time of death should be noted down, the attitude and position of the body should be observed, any bottles, phials, or packets, should be collected together and preserved, and many other important but minute facts may thus be observed and brought to bear on the subsequent investigation. But this, of course, is in a case of suspected poisoning where the suspicion was aroused during the life of the deceased. In other cases, the evidence probably is that which alone can be collected after death and sometimes after a long period has elapsed, and even, perhaps, after the exhumation of the body. Sudden death from natural causes might sometimes be mistaken for death by poison ; but one of the means for distinguishing death by narcotic poisoning from that through apoplexy or disease of the heart is the difference in the rapidity with which death takes place. The only common poison likely to operate with equal rapidity is *prussic acid;* opium is much slower in its operation, but there is nothing to prevent arsenic destroying life in an hour. In the absence of finding the poison in the stomach, which, in the case of prussic acid, is not always possible, great care should be taken in such cases before an opinion is expressed as to the real cause of death; at the same time, it is not necessary in order to cause death that prussic acid should be received into the stomach. A drop placed on the tongue might be sufficient, and a case is known to the author of this work of simple inhalation of the fume having a fatal effect. The gentleman who so destroyed himself, was, however, enabled hurriedly to leave the shop (a chemist's) where the occurrence took place, and, after walking into a neighbouring shop, to request that he might be accommodated with a chair, seated in which he died. This would appear scarcely credible, but these facts were adduced in evidence at the coroner's inquest on the body. He had obtained possession of the bottle containing the acid by announcing himself as a medical man, and asking for hydrocyanic acid of Scheele's strength, when, suddenly snatching the bottle containing it out of the hands of the chemist—between whom and himself the counter opposed a temporary barrier—he thus effected his purpose. He was immediately followed, on his exit, by the chemist, who never lost sight of him; but, had this occurred in private, his death would doubtless (could he have made away with the bottle, for which he had ample time,) have been attributed either to apoplexy or disease of the heart. Even the very faint smell of such a poison might have been lost before discovery. In such an event even medical science might be baffled, and a murderer, through thus administering a subtle poison, might escape. Such a case, therefore, suggests that perhaps the state of the brain or the lungs might disclose the true cause of death when the state of the stomach had failed so to do.

Under the present law, regulating the sale of poisons, precisely similar cases, happily, are not likely to occur; yet the subject of Coroners' Inquests cannot be brought to a close without some further observations on the point.

The deceased in this last case was a gentleman of considerable position and eminence, and a very old friend of the author's family, but for obvious reasons the name cannot now be mentioned. The case seems to supply the hitherto lack of authenticated evidence of death caused by the vapour of prussic acid. Dr. Taylor in his work " On poisons in relation to medical jurisprudence," published in 1848, says, "The vapour of anhydrous prussic acid, if respired, would prove almost instantaneously mortal. Even the vapour of the diluted acid, accidentally respired, occasions very alarming symptoms." He then proceeds to give the particulars of a case as follows :—" A practitioner was showing to some friends the effects of Scheele's prussic acid on an animal, when, by accident, a quantity of the acid fell upon the dress of a lady who was standing before a fire. The poison was rapidly evaporated, and the lady was immediately seized with dizziness, stupor, inability to stand, and faintness. The pulse became feeble and irregular." (p. 645.) Remedies being promptly resorted to, the lady happily recovered. " Mr. Nunneley's experiments on animals prove that the action of the acid upon the lungs, when air impregnated with the vapour is breathed, is not only rapid but certain in its effects, and it forms one of the easiest methods of exhibiting the poison—one which it would be very easy to employ, but after a few hours most difficult to detect, because the vapour, from its great diffusibility, is very soon dissipated." (Prov. Trans. N. S. III., 84.) "There can be no doubt," says Dr. Taylor, "that, like arsenic and all poisons, prussic acid acts much more speedily and powerfully when taken into the lungs as a vapour than when introduced into the stomach as a liquid." . . . . . . " I am not aware," he continues, " that there is any well-authenticated case of death having been caused by the vapour. The celebrated Scheele died suddenly while making his researches on this poison, and it is alleged that he was killed by respiring the vapour. The anhydrous acid was not known to Scheele; this was prepared by Gay-Lussac in the early part of the present century." (p. 646.) The case of the lady, mentioned by Dr. Taylor, seems to confirm the author's impression of the facts he has narrated, the only difference in the two cases being *that* between voluntary and determined and involuntary and accidental respiration of the vapour. It is of great importance when the life of a deceased happens to be insured to ascertain whether the act of poisoning was the result of *accident* or *suicide*. In the event of litigation ensuing respecting a policy of insurance in such a case, it may be too late to discover not only any trace of poison in the body but even any appearances betokening death by such means. A coroner's inquest held shortly after death frequently affords the only opportunity of discovering the truth—and yet even upon such occasions the fact of the deceased having possessed the power of volition or locomotion up to the moment of death is too frequently taken as proof that his death resulted from natural causes, such as epilepsy, apoplexy, or disease of the heart; and yet this is one of the most important questions connected with death by prussic acid. No doubt a strong dose would take immediate effect and at once annihilate the sensorial functions. But what about a small dose, or the mere inhalation, or respiration of the vapour? Perhaps one of the most extraordinary cases of locomotion and volition is one related by Mr. Godfrey. A gentleman, after taking a sufficient quantity of prussic acid to destroy life, walked ten paces to the top of a flight of stairs, descended the stairs, seventeen in number, and went to a druggist's shop, at forty-five paces distance, where he had previously bought the poison, entered the shop, and said in his usual voice, "I want some more of that prussic acid." He lived from five to ten minutes after taking the poison. It has generally been supposed that in cases of *slow* death from a small dose of poison the body would be found convulsed, but this opinion is not borne out by facts. In the case just referred to, although it was a case of *slow* death, there were no *convulsions*, and, what is of further interest, there was no odour of prussic acid about the mouth, and the individual died in the presence of several medical men without any shriek or any symptom approaching to it. (Prov. Med. Jour., Sept. 25, 1844.)

Dr. Taylor mentions a case communicated to him in May, 1850, by Mr. Peake, of Newark. A man swallowed a large dose of prussic acid, and was afterwards observed walking and smoking his pipe. He was found dead in a privy very shortly afterwards; but although the body was warm, the smell of tobacco smoke from the mouth completely overpowered and concealed the odour of prussic acid. On opening the body the odour of the acid was at once perceptible. (Taylor's Med. Jurisprudence, p. 174.) In such a case, had there been no knowledge or suspicion of the man's having taken poison, he might readily have been considered to have died from natural causes—probably from apoplexy, under the circumstances. On the other hand, it is sometimes too hastily concluded that acts which, after all, are but only indicative of this consciousness, volition, and locomotion, can only be accounted for upon the supposition that the poison was administered by some other than the deceased.

In a case (Rex *v.* Freeman) tried at the Leicester Spring Assizes, 1829, reported in the *Medical Gazette* (vol. viii. p. 759), owing to the position of the body and other circumstances, it was inferred that the deceased could not have taken the poison herself. The opinion of four out of five medical witnesses was strongly against the possibility of the acts having been performed by the deceased. All the acts to which the opinion referred might be performed in from *five to eight seconds*. There are now numerous facts which show that the symptoms may be often protracted for *several minutes*. Fortunately the medical opinion was completely set aside by circumstances, and an innocent man, charged with murder, was thus saved from being the victim of medical opinion.

"*Strychnia.*—The symptoms produced by strychnia very much

resemble those of tetanus, but in that disease they are more slowly formed, and can only be coincidentally connected with the taking of some kind of solid or liquid. Death is a much more rapid effect of the poison than of the disease as it is produced by natural causes. Medical men may, however, be easily deceived respecting the origin of the symptoms when the dose is small and frequently repeated." (Taylor on Poisons, p. 776.) This poison has acquired of late years a sad notoriety. The rigidity which the body of a person poisoned by strychnia acquires soon after death is retained for a very considerable period. In the case of the unfortunate John Parsons Cook, who was poisoned by his medical friend William Palmer, this rigidity was observable after two months' interment. At this trial, emptiness of the heart was set down as an indication that strychnia was not the cause of death, and a theory of death from angina pectoris, or some "latent and undiscoverable disease of the nervous system," was set up. Even Dr. Letheby supported this view by swearing that he had destroyed "some dozens" of animals by strychnia, and the heart was always full. This, however, is not in accordance with experiments on the human body. "Out of ten inspections in poisoning by strychnia," says Dr. Taylor, "in *human* subjects, the heart has been found either empty or deficient of blood in *six*." To these two other cases may be added, in which the cavities of this organ were found empty (see Guy's Hospital Reports, October, 1856, Poisoning by Strychnia). So much for "animal" experience, or the evidence of witnesses who rely upon what they have seen in animals in preference to facts derived from an actual examination of human bodies. On an average, in poisoning by strychnia, the symptoms appear in from five to twenty minutes. In two cases at least an hour has elapsed (*Lancet*, August 31, 1850; "On Poisoning by Strychnia," 1856, p. 139). The longest interval recorded is that of *two hours and a half* before the appearance of symptoms. (Poisoning by Strychnia, p. 42.) In a case which occurred to Drs. Lawrie and Cowan in June, 1853, an hour and a half elapsed. In the case of Cook the symptoms commenced in an hour and a quarter, and he died in twenty minutes; and yet in spite of these facts an attempt was made by the medical witnesses in the defence of Palmer to prove that an interval of *an hour and a quarter* in the case of Cook rendered it impossible that the symptoms could have been caused by strychnia (Reg. v. Palmer, C. C. C., May, 1856). Such was the substance of the evidence given by Dr. Letheby and Mr. Nunnely, from their experiments upon animals. Whilst giving these gentlemen full credit for their evidence as regards animals, such analogies, it is clear, can no longer be depended on.

It is here as well to remark that, as has been before observed, as the symptoms produced by strychnia very much resemble those of tetanus, so as to require much judgment and discernment on the part of a medical practitioner in detecting the difference, so great caution is necessary before such a practitioner pronounces an opinion as to the cause of a fatal attack of tetanus itself, whether the result of violence, or accident, or natural causes—such as exposure to cold and wet (Cormack's Monthly Journal, Dec., 1845, p. 902), " it being scarcely possible to distinguish, by the symptoms, tetanus from wounds from that which occurs spontaneously as a result of natural causes" (Taylor's Med. Jur., p. 333). Where tetanus proves fatal, and there is any particular wound or personal injury on the body, before connecting it with the wound or injury, it will be proper, says Dr. Taylor (*ib.* p. 335), to make the following inquiries:— 1. Whether there were any symptoms indicative of it before the maltreatment. 2. Whether any probable cause could have intervened to produce it, between the time of its appearance and the time at which the violence was inflicted. 3. Whether the deceased ever rallied from the effects of the violence. The time at which tetanus sets in from the effect of a wound is from about the third to the sixth day. A case occurred in St. Bartholomew's Hospital, in September, 1853, which exemplified the necessity of making a rigorous inquiry into all such circumstances. A boy, aged fifteen, while quarrelling with another received a blow on the back from the other's fist, and this was followed by a kick, but not of a severe character. In about two hours he complained of stiffness about the jaw; the stiffness continued to increase, and the jaw became partially fixed. Spasms of the muscles of the back supervened, and, in one word, there was confirmed tetanus. He died on the fourth day after receiving the blow. It turned out that, six days previously to the first appearance of the tetanic symptoms, the boy had accidentally driven a rusty nail into his foot, and that the suppurating wound resulting from this injury had only closed on the day on which the blow was inflicted, and there could be no doubt, from the whole of the circumstances, that this, and not the blow on the back, had been the cause of the fatal attack of tetanus. (See *Lancet*, Dec. 10, 1853, p. 550.) It is often the case that the science and skill of the medical practitioner, afterwards called as a witness, can alone distinguish the innocent from the guilty. A man may have received a wound, and yet that wound may not have been the cause of death, or have in any way accelerated it; or a man may have received two wounds, one inflicted by himself and one by another person, and may certainly have died of one of them. It remains for the medical practitioner to determine which of the two was the cause of death.

Arsenic.—The following case of poisoning by arsenic shows how even circumstantial evidence of guilt may, in reality, vindicate the innocence of the accused, when submitted to the opinion of an intelligent medical witness. A woman, aged sixty-five, accused her husband, aged seventy, of having attempted to poison her. She handed to the authorities a vessel containing arsenic in coarse powder, and some food, which she stated had been prepared for her by her husband. On analysis the food was found to contain a large quantity of arsenic. The husband was immediately committed to prison. The wife remained apparently quite well for eight days afterwards, no symptoms of poisoning having manifested themselves about her. She was then seized with a fit of nausea, and was guilty of many extravagant acts. She died the following day—*i.e.*, nine days after she had accused her husband of having administered arsenic to her in her food. On a post-mortem examination it was evident that she had died from the effects of arsenic. The poison was found in large quantities in the alimentary canal, and there were the usual morbid changes in the stomach and intestines. The husband all along denied his guilt; but yet that guilt, to all appearance, rested upon the medical question, whether a large quantity of arsenic could be taken by a person and remain dormant in the system, without producing any of its usual effects, for the long period of *eight* days?

The witnesses answered the question in the negative. While the prisoner was with his wife, she did not suffer from the symptoms of poisoning, nor was there any proof that he had administered poison. When, however, he was in prison, she died from the effects of it. It was fortunate for the accused that during the whole eight days he could not even have been accessory to its administration, and that the medical witnesses were versed in the properties and effects of the poison. The woman had doubtless, in her endeavours to rid herself of her husband, tampered with herself, but overdid her part in her desire to get up a case against him. (Annales d'Hygiène, 1836, ii. 391.)

M. Raspail, who at least is a distinguished chemist, and a French *savant*, some years ago, when dealing with a famous poisoning case, in which the main evidence against the prisoner was that arsenic had been found in the stomach of the victim, startled all France by the doctrine that nature had spread arsenic through the human system, and that, unless found in considerable quantities, its presence in the stomach after death gave no proof of murder. Allowing for the grossest exaggeration in this opinion, it certainly has been a matter of controversy whether arsenic is a constituent part of human bones, and even of the muscular system; and that it is found in the soil of cemeteries is an acknowledged fact. These scientific differences of opinion, and the admission that zinc and sulphuric acid as well as copper, used in experiments for the detection of arsenic, are far from being free from arsenic themselves, at least lead to the deduction that the metallic deposit may be proved to be arsenical, and yet the arsenic may not be derived from the suspected liquid, the subject of analyzation.

Without venturing more than to counsel great caution in all cases in which the deposit obtained through experiments is but small, it may fairly be remarked that unless arsenious acid can also be obtained from the deposit there can be no certainty that the deposit itself is arsenical, and especially when the other symptoms are not unlike those of cholera. Antimony has been mistaken for arsenic, and as tartarized antimony and James's powder are frequently employed in medicine, and may exist in the stomach at the time of death, it is highly necessary to guard against such a serious fallacy. But if no arsenic be found in the stomach, and only a minute quantity in the tissues, without fully accepting even Orfila's opinion that arsenic exists as a natural constituent of bones, and, as he at one time alleged, of tissues, such conflicting opinions of scientific men of eminence, even if now exploded, might well cause less distinguished practitioners to hesitate before committing themselves to a decision as to the origin of the deposit involving the life or death of a fellow-creature.

This, perhaps, may not be an inappropriate place for stating, once for all, that whether in civil or criminal cases, a witness is not bound to disclose matters which have come to his knowledge as a professional *legal* adviser; but other professional persons, it is now considered, either *medical* or clerical, have no such privilege. It is now conceded that a Protestant clergyman not only may, but must, disclose what has been revealed to him as a matter of religious confession, and therefore, *a fortiori*, must a medical attendant reveal communications made to him.

The office of Coroner does not determine by the demise of the sovereign (1 Lev. 120; 3 Salk. 100; Dyer, 165). He is chosen for life, but he may be removed by the ancient writ *De Coronatore exonerando*, for a cause to be therein assigned (1 Bl. Com. 348; F. Nat. Brev. 163-4). It is unnecessary to particularize the causes which have from time to time been deemed sufficient to justify a coroner's removal under this writ, as now, by the 23 & 24 Vict., c. 116, s. 6, the Lord Chancellor is empowered, "if he shall think fit, to remove for inability or misbehaviour in his office," any existing or future coroner. The only offices that disqualify for the appointment of *Borough* coroner are those of alderman or councillor, the right of such appointment being vested in the council of every borough in England and Wales in which, under the provisions of the Municipal Corporation Act, (5 & 6 Wm. 4, c. 76,) a separate Court of Quarter Sessions is or shall be holden.

CHAPTER XIII.

REGISTRATION OF BIRTHS, DEATHS, AND MARRIAGES.

Although the Act 6 & 7 Wm. IV., c. 86, has no special reference to the medical profession, yet as, under it, medical practitioners are frequently appointed registrars, and as any one present at either the birth or the death of a person may be required under the provisions of the Act to perform certain duties which may be deemed of a public character, and as such, almost necessarily, devolve on medical attendants more frequently than upon others, a brief allusion to this Act will probably not be considered altogether out of place, although its provisions must be already familiar to almost every family throughout the country. The Poor Law Guardians are by the Act empowered to appoint superintendent registrars and registrars for the different districts into which unions or parishes may be divided, and the clerk to the guardians, if he be qualified to the satisfaction of the Registrar-General, may if he please accept such office of superintendent registrar. And in the event of his refusal or disqualification, the guardians may appoint any other person with such qualifications as the Registrar-General may declare to be necessary, one of which is residence within the district. The father or mother of any child born, or the occupier of any house or tenement in which any birth or death shall happen, may within forty-two days after the day of such birth, or within five days after the day of such death, give notice of such to the registrar of the district; and in case any new-born child, or any dead body shall be found exposed, the overseers of the poor in the former case, and the coroner in the latter case, shall give notice thereof, and of the place where such child or dead body was found, to the registrar. And the father or mother of every child born after the passing of the Act, or in case of the death, illness, absence, or inability of the father and mother, the occupier of the house or tenement in which such child is born shall, within forty-two days after its birth, give information to the registrar, upon being requested so to do, to the best of their knowledge and belief, of the particulars required. But after the expiration of such forty-two days it is made unlawful for the registrar to register such birth, unless within six months from the date of the birth any person present at the birth, or the father or guardian make a declaration of all the particulars required—in which case the registrar may, in the presence of the superintendent registrar, register such birth. The superintendent registrar, in such case, before whom the declaration is made, as well as the registrar, shall sign the entry of birth, and unless the entry shall be so signed no register of births shall be received as evidence of the birth of such child; and every person who shall knowingly register or cause to be registered the birth of any child, otherwise than as directed, after the expiration of forty-two days from the day of its birth, shall forfeit 50*l.* And after six months from the birth of any child, except one born at sea, no registrar shall register such, nor shall any register to prove the birth of any such child, with the exception before-named, be given in evidence wherein it shall appear that six calendar months have intervened between the day of birth and that of registration. Any one offending against this enactment shall also forfeit for such offence 50*l.*

If any child duly registered shall, within six months after such registration, have any name given it in Baptism, the person procuring such name to be given may, within seven days after such baptism, deliver to the registrar, or superintendent-registrar, a certificate in the form of Schedule (G) signed by the minister who has baptized it, and upon the receipt of such the registrar, or superintendent-registrar, shall register that the child was baptized in such name, and shall certify upon the certificate the additional entry, and send the said certificate to the Registrar-General.

In the case of a death, some person present at it, *or in attendance during the last illness of the deceased,* or in case of the death or default of such person, the occupier of the house or tenement, or if the occupier be the person deceased, some inmate of the house or tenement in which the death took place is required, within eight days after such death, to give information, upon being requested so to do, to the registrar, according to the best of his ability, of all particulars touching the death of such person, except in the case of inquests, when the coroner's jury shall make the requisite inquiry, and the coroner shall inform the registrar of their finding, and the registrar shall make the entry accordingly.

And every registrar, immediately upon registering any death, or as soon as required, shall deliver to the undertaker, or any one having charge of the funeral of the deceased, a certificate, according to form of Schedule (E), of such registration, and the undertaker, or other person, shall deliver the same to the officiating minister; and if any body be buried without such certificate, the person burying, or performing any such funeral or religious service, shall forthwith give notice thereof to the registrar.—Provided that a coroner in his discretion may order the burial before registration, and in such case shall give a certificate of his order to the undertaker or person having charge of the funeral. And any one who shall bury, or perform any funeral or religious service for the burial of a dead body without receiving one or other such certificate shall forfeit 10*l*. Also any one wilfully making, or causing to be made, for the purpose of being inserted in any register of birth, death, or marriage, any false statement touching any of the particulars required to be registered, shall be deemed guilty of perjury. And further, every registrar who shall refuse, or, without reasonable cause, omit to register any birth or death of which he shall have had due notice, and any person having the custody of the register-book, or certified copy of any part thereof, who shall carelessly injure, or allow to be injured the same, shall forfeit 50*l*. for every such offence.

And every person wilfully destroying or injuring or causing to be destroyed or injured any such register book or certified copy of any part thereof, or falsely counterfeiting or causing to be counterfeited any part of such, or who shall wilfully insert or cause to be inserted in any such book or certified copy of any part thereof, any false entry of any birth, death, or marriage, or shall wilfully give any false certificate, or shall certify any writing to be a copy or extract of any register book, knowing the same register to be false in any part, or shall forge or counterfeit the seal of the Register Office, shall be guilty of felony.

Penalties under this Act are recoverable before two justices of the peace of the locality, with power to levy them by distress and sale of goods; and for want of such distress such justices may commit the offender to prison for one month, an appeal being given in all cases above 5*l*. to the next General or Quarter Sessions, to be held not sooner than twelve days after such conviction, upon the usual written notice of appeal being served, together with the grounds of appeal, within three days after such conviction, and seven clear days before such Sessions, and the appellant entering into recognizances with two sureties to prosecute such appeal.

Such are the requirements of the Act, at least as far as they even indirectly bear upon the Medical Profession. It may be added, that by 1 & 2 Vict., c. 49, the production of the certificate of any superintendent registrar or registrar of births and deaths, attested by two credible witnesses, certifying that such is a true copy of the register book, or of the certificate of the Registrar-General under the seal of the Register Office, is sufficient evidence of the birth, death, or marriage of any nominee for enabling the Commissioners for the reduction of the National Debt to grant life annuities, accompanied with a declaration of the identity of such nominee;

and whenever any such certificate is produced of any other than the Registrar-General, a declaration must be annexed to it of the witnesses who attested its execution, or one of them, to be made before a magistrate of the place wherein the place of the birth, death, or marriage of the nominee *shall be situate*, setting forth that such witness or witnesses compared the copy of the register with the register itself, and that it is a true and literal copy thereof, and that he or they saw the registrar or superintendent registrar sign the certificate, and that the names of such witnesses are of their own handwriting.

Suggestions as to the duties of officers to public institutions, whether those institutions have their head-quarters, like those of the Registrar-General in the metropolis, and themselves are distributed in local districts throughout the country, or whether they present the more tangible and corporeal form of prisons or workhouses, should rather come from those thus brought into practical working with such institutions than from those who can do little more than call attention to the requirements which the law exacts of its officers. Whether as surgeon to a county or government prison, to a union workhouse, or in any official medical capacity at an hospital, or as holding the appointment of public vaccinator, or even superintendent registrar or assistant registrar of births, deaths, and marriages (which latter, as before mentioned, are not necessarily medical appointments), any suggestions for practically and efficiently executing the duties of the office would assuredly present themselves to any well-regulated mind, attuned by good common sense, in a far more effective form than they could be stereotyped on paper by those who have had no experience of that force of circumstances which creates the suggestions desired. There is no royal road to practical experience; and however a young practitioner, and perhaps a still younger public officer, might desire directions or instructions by which to regulate his conduct, a determination efficiently to discharge his duty, and a zealous ambition to do with his might whatever his hand has undertaken, no matter how humble or matter of routine may be the requirements, will readily supply the place of any leading-strings, in the shape of suggestions from others, which after all might turn out to be crude; and, even though they bore the impress of originality, would perhaps be found more to be admired in theory than to be implicitly followed as a rule in practice.

CHAPTER XIV.

POOR-LAW LEGISLATION AS AFFECTING MEDICAL PRACTITIONERS. PAUPER LUNATICS, INCLUDING CRIMINALS AND PRISONERS FOR DEBT.

Section I.

PAUPERS IN WORKHOUSES OR POOR LAW DISTRICTS.

The only medical practitioners liable to be affected by Poor Law legislation are the medical officers appointed by the Poor Law Guardians of their respective unions, either to districts into which such unions are for this purpose divided, or to workhouses, or as public vaccinators; and here it may be remarked that a medical officer of any such district or workhouse cannot, as a matter of right, claim to be appointed a public vaccinator, but, as a matter of practical convenience it has been found desirable, as the readiest mode of complying with the statutes, to confer the appointment on such medical officers. The salary, therefore, assigned as medical officer for the district or workhouse should have no reference to the remuneration which such officer is to receive as public vaccinator, if holding, or about to hold, that office; for, although appointed to both offices by the same authorities, the appointments and remuneration respectively are based upon two distinct contracts. It is desirable to bear this distinction in mind, as, in the event of having to sue for the recovery of either or both salaries, such distinction might become a matter of practical importance. The appointment of public vaccinator is made under a written contract between the guardians and a duly qualified medical practitioner, the form of such contract being furnished under a general order of the Poor Law Board, subject to such modifications as the guardians, with the approval of the Poor Law Board, may determine upon; whereas, the appointment of medical officer to a district or workhouse under the Poor Law Amendment Act may be either by agreement between a medical practitioner and the guardians, written and under seal, in which case the remedy would be by action on the contract;

or, if there is no such contract, but there has been an election, at such salary, or the Poor Law Board have fixed the same by order, the proper remedy is by *mandamus* from the Ct. Q. B. (Addison *v.* Mayor of Preston, 12 C. B. 108; and Smart *v.* West Ham Union, 10 Ech. R. 567; and 11 Ech. R. 867.) The salary is payable up to the date at which the medical officer ceases to hold office; but if he be suspended by the Board of Guardians from the discharge of his duties, and is dismissed by the Poor Law Board without such suspension having been previously removed, he is only entitled to his salary up to the date of his suspension. (See Art. 173 and 175 of the Consolidated Order of the Poor Law Commissioners (24 July, 1847,) relating to the medical relief of the poor.) Although no attempt will be made in this work to unravel the intricacy of official regulations, or to elucidate the complex character of duties—perhaps at all times "hard to be" thoroughly "understood," more especially as red-tapeism is an inexorable taskmaster —a hint or two may yet be dropped, or a suggestion offered, which may save many an honestly-intentioned medical officer from falling into the meshes with which verbose or ambiguously worded Acts of Parliament, and sometimes even the more carefully drawn general orders of the Poor Law Board seem to environ them. For instance, by 9 & 10 Vict., c. 66, s. 6, "If *any officer* of any parish or union do, contrary to law, with *intent* to cause any poor person to become chargeable to any parish to which such person was not then chargeable, convey any poor person out of the parish for which such officer acts, or cause or procure any poor person to be so conveyed, or give directly or indirectly any money, relief, or *assistance*, or afford or procure to be afforded any *facility for such conveyance*, or make any offer or promise, or use any threat to induce any poor person to depart from such parish, and if, in consequence of such conveyance or departure, any poor person become chargeable to any parish to which he was not then chargeable, such officer, on conviction thereof before any two justices, shall forfeit and pay for every such offence any sum not exceeding 5*l*., and not less than 2*l*."

Medical officers rendering *assistance*, though of a strictly professional character, as, for instance, administering stimulants or medical nutriment with a view to alleviate the fatigue of travelling, if done with the intent to assist in the conveyance or departure of such poor person out of the parish for which such officer acts, and if in consequence of such departure, so facilitated, such person becomes chargeable to another parish, the medical officer so rendering assistance, even though dictated by humanity, is amenable under the above section, provided he was aware of the object of the poor-law officers in effecting the removal, viz., to cause such person to become chargeable to another parish to which he was not then chargeable. See Walsh's "Poor-Law Medical Officers' Vade Mecum," where an instance of such liability on the part of a medical officer is given, as having resulted in a conviction of the officer. The case thus referred to is as follows :—The medical officer of the Tadcaster Union was summoned, at eight o'clock on the morning of the day on which the offence is charged as having been committed, to attend a woman in labour at the Tadcaster railway station. Upon his arrival there, finding that another surgeon was in attendance, he withdrew, and heard no more of the case until one o'clock in the day, when, at the request of the overseer, he accompanied him and the assistant overseer to the railway station, where they found the woman sitting in an arm-chair in the waiting-room. She was, under the circumstances, unusually strong and well. She had not been put to bed; she herself stating, in answer to inquiry, that she was very well as she was, and nothing had taken place since delivery to cause debility. She said she had walked from Leeds the day before, and expressed a wish to return there, as she had at that place a home and a mother. The medical officer thereupon gave it as his opinion that she might, without risk, be permitted to return, provided she were conveyed with all proper care, &c. The woman even offered to pay her own fare by train, and repeatedly expressed her wish to go home, thus disarming the medical officer of any suspicion that she would upon her arrival become chargeable to the township of Leeds. The magistrates gave their decision in the following terms :—" This person, at the time of her affliction, was in a state of pauperism, and in her condition she ought to have been taken care of, till she was sufficiently recovered, by the officers of the place in which she was found. Instead of doing so, they assist her, and cause her to be removed to Leeds, where she had not before been chargeable, but immediately afterwards was so. This seems to us to have been the natural and inevitable consequence of the removal; and we think the intent, as charged, has been proved against the defendants." The penalty inflicted was 5*l*., or in default seven days' imprisonment.

In concluding this section of the chapter, it may be remarked that by section 2 of the 4 & 5 Vict., c. 32, (one of the Vaccination Acts,) vaccination shall not be considered to be parochial relief, and that no persons availing themselves of it shall, by reason thereof, be deprived of any right or subjected to any disability.

Section II.

PAUPER LUNATICS, INCLUDING CRIMINALS AND PRISONERS FOR DEBT.

Medical certificate.—The form of the certificate is prescribed by the Act 16 & 17 Vict., c. 97, s. 67, and Schedule F, No. 3 (see Appendix). If given alone by the medical man who is called in, it is not conclusive, but if given by the *medical officer*, in addition to such medical man, the justice or officiating clergyman and relieving officer (or overseer) will have no alternative, and must make an order accordingly for the admission of the pauper into an asylum (s. 7, last proviso). The order, however, will not authorize the admission of the pauper after the lapse of seven clear days from the date of the certificate (s. 73). Where the certificate is signed by a medical man called in by a justice, the latter may order the guardians (or overseers) to pay a reasonable remuneration for the examination (s. 69). If any certificate be found incorrect, it may be amended within fourteen days after the pauper's admission, with the sanction of the Commissioners in Lunacy, who, if it be not amended, may discharge the pauper (16 & 17 Vict., c. 97, s. 87; 25 & 26 Vict., c. 111, s. 27). By the 66th section of the former Act, "every pauper lunatic not in an asylum or an hospital registered or a house licensed for the reception of lunatics" is to be personally visited once in every quarter of a year " by the medical officer of or for the parish or union, or district of a parish or union, in which such lunatic is resident," who is to make his report in the form prescribed, the object of such visitation being to show whether all pauper lunatics, not in confinement, "are or are not properly taken care of, and may or may not properly remain out of an asylum ;" and with respect to the lunatics in any workhouse, the medical officer as aforesaid is required, upon his quarterly visits of such inspection, to certify whether, in his opinion, the workhouse is or is not sufficient for the accommodation of the lunatics detained therein, and whether or not the lunatics so detained are proper persons to be kept in a workhouse (16 & 17 Vict., c. 97, s. 66; 25 & 26 Vict., c. 111, s. 21, and Schedule B). A special remuneration is assigned to the medical officer in respect of "each such quarterly visit to any pauper not being in a workhouse," and the forms are to be supplied to him by the guardians.

On becoming aware that any pauper resident in his district is or is deemed to be a lunatic, and a proper person to be sent to an asylum, the medical officer is to give written notice thereof, within three days, to the relieving officer ; or if there be no relieving officer, to the overseers, subject to a penalty not exceeding 10*l*. for neglect (16 & 17 Vict., c. 97, s. 67 & s. 70). Pauper lunatics wandering abroad and away from control may be dealt with in the same way as private lunatics, under 16 & 17 Vict., c. 97, s. 68. This enactment has been already adverted to in a former chapter. If an order be made for his admission into an asylum it must be in the form prescribed by the Act, (Schedule F, No. 1,) and will have the same effect as an order made under s. 67. (The provisions of ss. 75, 76, and 122 will apply to the medical certificate required under s. 68.)

Any two visitors, with the advice in writing of the medical officer of the asylum, may permit *any patient* to be absent from the asylum on trial for such period as they may think fit, making an allowance to him; the amount whereof is to be charged and paid as if he were in the asylum (16 & 17 Vict., c. 97, s. 79). If he do not return at the proper time he may be retaken within fourteen days afterwards, unless a medical certificate be sent to the visitors, certifying that his detention in an asylum is no longer necessary (s. 79). And by the same section any two of the visitors of a lunatic asylum, with the advice in writing of the medical officer of the asylum, may order the discharge of any pauper lunatic who has been received into such asylum.

In the case of the death of a pauper lunatic in an asylum, a notice and statement of the death, in the prescribed form (Schedule F, No. 5), and the cause of death, and the names of any persons present at the death, must be drawn up and signed by the clerk and medical officer, and copies transmitted to the following persons:—1. The Registrar of Deaths for the District; 2. The Commissioners in Lunacy, the relieving officer, or the overseers of the union or parish to which the deceased was chargeable. As to legal proceedings against the superintendent of a county asylum for furnishing a defective certificate of the cause of death, which omitted to mention the post-mortem examination, see Seventeenth Report of the Commissioners, 1863, pp. 32-34.

The Attorney-General *v.* Pearson (10 Jurist, 651), arose out of a case which occurred in the parish of Islington in 1842.

James Elliot, an alleged lunatic, brought an action for assault and false imprisonment against the overseers, the relieving officer, the master of the workhouse, *the medical officer*, and a person named Beavan, which was tried before Ch. Justice Tindal in the Court of Common Pleas, at Westminster, in May, 1844. The jury found a verdict for the plaintiff, with 400*l*. damages. A motion was afterwards made for a new trial (see 14 Law Jour. Rep. (N.S.) C. P., 136, and 1 C. B. R., 18), and by consent the Court reduced the damages to 200*l*., and a *stet processus* was entered in a second action which had been commenced. The medical officer (Semple) and Beavan suffered judgment to go by default. The damages and costs on both sides amounted to nearly 900*l*., and were paid by the trustees of the parish of Islington out of the rates ; and an information was filed by two ratepayers, seeking to charge the trustees personally with the payment of the amount. The Court (V.C. Knight Bruce) under all the circumstances of the case, and being of opinion that the facts were such as to justify the parish officers in interfering, dismissed the information with costs. This case is referred to principally as an instance of the amount of litigation that may ensue through a mistake, partly that of the medical officer of the parish. As the Lord Chief Justice put it to the jury, the case resolved itself into whether "at the time of the committing of the trespasses the plaintiff was a *dangerous lunatic and in a state of destitution*." On the occasion of the motion for a new trial his lordship, whilst suggesting the above compromise by reducing the damages, said :— " The defendants stand wholly absolved from the suspicion of having acted from sinister or malignant motives. The most that can be said is, that they mistook their course."

Up to the year 1808 the only enactments affecting pauper lunatics were certain provisions in successive Vagrant Acts, which authorized the detention of dangerous lunatics, in chains, if necessary (rather with a view to the protection of the public than for their own benefit), and which provided for their removal to the parishes of their settlement, and the application of their property towards their maintenance. In the year 1808, however, the 48 Geo. III., c. 96, was passed, and laid the foundation of that system of providing public lunatic asylums at the expense of the respective counties and boroughs in England and Wales, which has since been gradually, though not yet completely, carried into effect. The scheme of this Act, in its general features, has been in the main adhered to in the subsequent legislation on this subject ; though improvements have been introduced from time to time.

This Act, as also a subsequent Act (the 9 Geo. IV., c. 40), left it entirely at the option of the justices to provide asylums or not, as they might think proper; but the 8 & 9 Vict., c. 126, introduced an important change upon this point, by rendering it incumbent upon the justices of every county and borough, not already provided with an asylum, to obtain one, either separately, or jointly with other counties and boroughs, and by empowering the Secretary of State, after the lapse of three years, to require the justices to take the necessary steps for that purpose, in every case in which they might then have omitted to do so. Notwithstanding this enactment, however, it appeared from the Seventh Annual Report of the

Commissioners in Lunacy (1853), that in 1852 there were still several counties unprovided with asylums, and that there were only four boroughs which separately possessed one. Looking at this deficiency, and the importance of supplying it, the legislature, by the Act of 1853 (16 & 17 Vict., c. 97) made it compulsory upon the justices of every county not having an asylum to provide one forthwith, either by erecting or otherwise providing an asylum for the county alone, or by uniting for that purpose with any county or counties, borough or boroughs; and also (if deemed expedient) with the subscribers to any hospital for lunatics established, or in course of erection, or afterwards to be established; but so that no agreement for this purpose should interfere with the reception of as many lunatics not paupers as might otherwise have been received into such hospital. The Act likewise imposed a similar obligation upon the justices (or council) of every borough not having an asylum, excepting any borough in which at the passing of the 8 & 9 Vict., c. 126, there were not six justices besides a recorder, (which boroughs, as well as cities, towns, or other districts, not being boroughs within the meaning of the Act, were to be annexed for the purposes of the Act to the adjacent counties) though it allowed such boroughs as might prefer this alternative, to contract for the reception of their lunatics into the asylums of other boroughs or counties; and it further provided, that boroughs thus contributing to county asylums should be considered as possessing asylums.

It is not necessary here to describe further the provisions of this Act, or to do more than state, that all necessary powers are conferred upon the justices for carrying the Act into effect, and for providing visitors to manage the asylum when provided. Such visitors may also enter into temporary contracts for sending pauper lunatics into the asylums of other counties or boroughs, or into registered hospitals or licensed houses, as well as for receiving pauper and other lunatics of other counties or boroughs into their asylums, where there is accommodation to spare. The proceedings are to be subject, in certain respects, to the supervision and control of the Commissioners of Lunacy and the Secretary of State.

By 16 & 17 Vict., c. 97, s. 72, no pauper shall be received into any asylum, registered hospital, or licensed house (save under the provisions with respect to removal), without an order according to the form required in Schedule (E), No. 1, signed by one justice, or an officiating clergyman, and one of the overseers or the relieving officer of the parish or union from which such pauper is sent, together with a statement of particulars (see Schedule) and a medical certificate, according to the form in Schedule (E), No. 3, signed by *one* medical practitioner who shall have personally examined him not more than seven clear days previous to his reception: And every one receiving any pauper into any asylum without such order and certificate (save under any of the said provisions) shall be guilty of a misdemeanour.

As to the liabilities of persons signing the order, see Fletcher v. Fletcher, 1 E. & E. 420, L. J. R. (N. S.) Q. B. 134. But it is rather with persons signing the medical certificate that this treatise is concerned. The Commissioners in Lunacy, in a letter addressed to the Lord Chancellor on the 4th July, 1849, (which was printed by order of the House of Commons on 1st August, 1849; Sess. Papers, No. 620,) entered very fully into this important question. They said—"It is of vital importance that no mistake or misconception should exist, and that every medical man who may be applied to for advice on the subject of lunacy, and every relative and friend of any lunatic, as well as every magistrate and parish officer (each of whom may be called upon to act in cases of this sort) should know, and be well assured that, according to law, any person of unsound mind, whether he be pronounced dangerous or not, may legally and properly be placed in a county asylum, lunatic hospital, or licensed house, on the authority of the preliminary order and certificate prescribed by the Acts." With respect to resorting to a Commission of Lunacy, the Commissioners added— "It is obvious that the finding of a jury is in no case essential, in order legally to justify the confinement of a person of unsound mind." (See also the Fourth Annual Report of the Commissioners, 30th June, 1849, p. 16; and Fifth Report, 30th June, 1850, p. 14.) At common law the power of restraining and confining a lunatic is limited to cases in which it would be dangerous, either as regards others or himself, for the lunatic to be at large, and though the powers given under the Lunacy Acts are not so limited, it is important that the provisions of those Acts should be strictly observed. By the 5 & 6 Vict., c. 22, it was enacted, that if any prisoner confined in the Queen's Prison should become insane, it should be lawful for one of the principal Secretaries of State, upon the certificate of two physicians or surgeons to that effect, to remove him to Bethlehem Hospital, there to remain until it shall be duly certified by two physicians or surgeons that such prisoner had become of sound mind, whereupon the Secretary of State might order him to be re-delivered into the custody of the keeper of the Queen's Prison.

Under this enactment a prisoner for debt might, upon becoming insane, be thus removed. (Gore v. Sir George Grey and others, 9 Jurist (N. S.) 752.)—Such judgment affirmed in the Exch. Chamber, 33 L. J. R. (N. S.) C. P. 109, but the Queen's Prison no longer exists as a debtors' prison. Similar Acts have been passed having reference to convict prisons alone—as, for instance, 5 & 6 Vict., c. 29 (Pentonville), by which any convict may, upon being reported insane by the Commissioners for governing the prison, be sent by order of the Secretary of State to such lunatic asylum as the Secretary of State shall direct, there to remain until it be certified by two physicians or surgeons that he has become sane, whereupon, if his time of imprisonment shall not have expired, he shall be remanded to Pentonville. A similar Act was passed with respect to Millbank, (6 & 7 Vict., c. 26,) with the exception that the prisoner must not only be reported by the Inspectors of the prison as insane, but must be found to be so upon the certificates of two physicians or surgeons.

By 23 & 24 Vict., c. 75, entitled an Act to make better provision for the custody and care of Criminal Lunatics (6th August, 1860), under section 14—" Two or more of the Commissioners in Lunacy, one at least of whom shall be a physician or surgeon, and one at least a barrister, shall, once or oftener in each year, on such day or days, and at such hours of the day, and for such length of time as they think fit, and also at any time when directed by the Secretary of State, visit every asylum for criminal lunatics, and shall inquire as to the condition, as well mental as bodily, of the persons confined therein, or any of them, and shall also make such other inquiries as to such asylum as to them may seem proper, or as such Secretary of State may direct."

And by the 15th section the Commissioners are directed in the mouth of March in every year, to report to one of the principal Secretaries of State the visits made as aforesaid in the preceding year, with such particulars as they may think deserving of notice.

SCHEDULE (A) OF THE ACT.

Statement respecting Criminal Lunatics to be filled up and transmitted to the Medical Superintendent with every criminal lunatic.

Name.
Age.
Date of Admission.
Former Occupation.
From whence brought.
Married, Single, or Widowed.
How many Children.
Age of Youngest.
Whether first attack.
When previous attacks occurred.

Duration of existing attack.
State of bodily health.
Whether suicidal, or dangerous to others.
Supposed cause.
Chief delusions or indications of Insanity.
Whether subject to Epilepsy.
Whether of temperate habits.
Degree of Education.
Religious persuasion.
Crime.
Where and when tried.
Verdict of Jury.
Sentence.

Such are the particulars required to be embodied in every certificate, duly filled up and authenticated, to be transmitted with every person whom the Secretary of State may direct, under the provisions of this Act, to be removed to any asylum for criminal lunatics.

By the 24 & 25 Vict., c. 134 (entitled an Act to amend the Law relating to Bankruptcy and Insolvency in England, 1861), "if any person being or alleged to be of unsound mind shall be in prison for debt, the gaoler shall forthwith require a justice of the peace for the county or place to visit such debtor, and inquire into his state of mind; and such justice shall call to his assistance two duly qualified medical practitioners, each of whom shall be a physician, surgeon, or apothecary, and each of whom shall separately examine such debtor, and if such two medical practitioners shall each sign a certificate with respect to such debtor, according to the form in Schedule H, and the justice be also satisfied from his own view that such debtor is of unsound mind, he shall certify the same to the proper Court, and thereupon the Court may appoint some one to represent such debtor, and direct such proceedings to be taken for adjudication in Bankruptcy against him as the Court shall think fit. And such prisoner may thereupon be removed to the county asylum as a lunatic. Such prisoner to be dealt with in all respects as a pauper lunatic, and in the event of his recovery shall, if still liable to be detained in custody as a debtor, be remitted to the gaol from whence he was received."

SCHEDULE (H) REFERRED TO.

Form of Medical Certificate.

I, the undersigned,   being a (*here set forth the*

qualification entitling the person to practise as a physician, surgeon, or apothecary, ex. gra., Fellow of the Royal College of Physicians in London, Licentiate of the Apothecaries Company, *or as the case may be*), and being in actual practice as (Physician, Surgeon, or Apothecary, *as the case may be*), hereby certify that I, on the day of   , at the Gaol of   , at   , in the county of   , separately from any other medical practitioner, personally examined   , a prisoner for debt in the same gaol, and that the said   is a (lunatic, *or* an idiot, *or* a person of unsound mind), and a proper person to be taken charge of and detained under care and treatment, and that I have formed this opinion upon the following grounds, viz. :—

1. Facts indicating insanity observed by myself (*here state the facts*).
2. Other facts (if any) indicating insanity communicated to me by others (*here state the information and from whom*).

(Signed)   Name.
              Place of abode.
Dated this   day of   , one thousand eight hundred and   .

This Act is still referred to and the form of certificate given, because although imprisonment for debt is commonly said to have been abolished by the recent Bankruptcy Bill, debtors may still be committed to prison under certain circumstances, which it is not necessary to enumerate here, by order of the Court or of a Judge, and whether they should be said to be committed for contempt of Court in not obeying its judgment or order to pay the debt, or for the debt itself, they would probably still come under the definition of "prisoners for debt," and if insane would have to be removed to a lunatic asylum. But even should a short Act be required to facilitate that object, doubtless the same form of medical certificate would be annexed to it.

By the 27 & 28 Vict., c. 29 (entitled, an Act to amend the Act 3 & 4 Vict., c. 54, for making further provision for the confinement and maintenance of insane prisoners,) (23 June, 1864), which was passed in consequence of the proceedings which occurred in the case of Townley, mentioned in a previous part of this work, the first section of 3 & 4 Vict., c. 54, is repealed, and other provisions are made enabling the Secretary of State to remove insane prisoners, confined under any other than civil process, from prison to a lunatic asylum, under certain circumstances and conditions. This Act also transfers the charge of the maintenance of such prisoners from the parish of the settlement to the common fund of the union. (S. 5.)

The 3 & 4 Vict., c. 54, did, and the 27 & 28 Vict., c. 29, does, empower the Secretary of State to remove the insane prisoners to such county lunatic asylum or other proper receptacle as he may judge and appoint; and, with reference to 39 & 40 Geo. III., c. 94, and 3 & 4 Vict., c. 54, the Metropolitan Commissioners in Lunacy, in their general report of 1844 (p. 198), stated as follows:—" It has been assumed in practice, that the Acts above cited are compulsory upon the visitors of county asylums, and that they cannot refuse to receive persons committed under Royal warrant." By the second section, the justices therein described are empowered and required in the case of all prisoners under any sentence short *of death*, or under any charge, or for want of finding sureties for good behaviour, or in consequence of any summary conviction, to call to their assistance two physicians or surgeons, or one physician and one surgeon, duly registered as such respectively under the provisions of the Medical Act (1858), and to be selected by them for that purpose, and to inquire with their aid as to the insanity of such person; and, if it be duly certified by such justices and such medical practitioners as aforesaid that such person is insane, the Secretary of State may, upon receipt of such certificate, if he shall think fit, direct the removal of such person to a lunatic asylum: and in the case of a prisoner under sentence of death, if there is good reason to believe that he is insane, and such shall be made to appear to the Secretary of State either by means of a certificate in writing to that effect in the form given in Schedule A, transmitted to him by two or more of the visiting justices, or by any other means whatsoever, such Secretary of State shall appoint two or more physicians or surgeons duly registered as aforesaid to inquire into the insanity of such prisoner, and, upon receipt of a certificate in writing to that effect, signed by such persons, the Secretary of State may order the removal of the prisoner to a lunatic asylum, there to be detained until it be duly certified to the Secretary of State by two physicians or surgeons, or one physician and one surgeon duly registered as aforesaid, that such person is sane, in which case he shall be removed back to prison to undergo his sentence of death.

Schedule (A) referred to.

We   being visiting justices of   hereby certify under our hands that we believe a prisoner in the said prison under sentence of death to be now insane.

By 27 & 28 Vict., c. 119, s. 75, entitled, an Act to make provisions for the discipline of the Navy (29 July, 1864), if any person imprisoned by virtue of that Act shall become insane, and a certificate to that effect shall be given by two physicians or surgeons, the Admiralty shall direct the removal of such person to a lunatic asylum; and, if such person in the same manner be certified to be again of sound mind, the Admiralty may order him to be removed back to prison to undergo the remainder of his sentence, if unexpired.

CHAPTER XV.

CHEMISTS, DRUGGISTS, AND DENTISTS.

Section I.

CHEMISTS AND DRUGGISTS.

As the Pharmacy Acts (15 & 16 Vict., c. 56; 31 & 32 Vict., c. 121; and 33 Vict., c. 117,) have given a Parliamentary recognition to chemists and druggists, by requiring their registration as such, it is necessary here to refer to them. The first of these Acts (1852) enacts, in section 6, that "all such persons as shall at the time of the passing of this Act be members, associates, apprentices, or students of the Pharmaceutical Society of Great Britain (incorporated by charter, 18th Feb., 1843), according to the terms of the said charter, shall be registered as pharmaceutical chemists, assistants and apprentices, or students, respectively." And by section 8 the examiners appointed under the said charter are empowered to "examine all persons who shall present themselves for examination, in their knowledge of the Latin language, in botany, and in materia medica, and in pharmaceutical and general chemistry, and such other subjects as may from time to time be determined by any bye-law: provided that such examination shall not include the theory and practice of medicine, surgery, or midwifery." Such examiners may grant or refuse certificates of competent skill, knowledge, and qualifications to exercise the calling of pharmaceutical chemists, or to be engaged or employed as students, apprentices or assistants respectively. The amended Act of 1868 (31 & 32 Vict., c. 121,) enacts that from and after the 31st December, 1868, it shall be unlawful for any person to sell or keep open shop for retailing, dispensing, or compounding poisons, or to assume or use the title "chemist and druggist," or chemist, or druggist, or pharmacist, or dispensing chemist or druggist, in any part of Great Britain, unless such person shall be a pharmaceutical chemist, or a chemist and druggist, within the meaning of that Act, and be registered under the Act, and conform to such regulations as to the keeping, dispensing and selling of such poisons as the

Pharmaceutical Society may from time to time prescribe with the consent of the Privy Council. The 3rd section defines chemists and druggists to consist of all persons who at any time before the passing of the Act have carried on in Great Britain the business of a chemist and druggist, in the keeping of open shop for the compounding of the prescriptions of duly qualified medical practitioners, also of assistants and associates who before the passing of the Act shall have been duly registered under the existing Pharmacy Act (15 & 16 Vict., c. 56,) and all such as may be duly registered under this Act;" the qualifications for such latter registration to consist of being twenty-one years of age, and having been for the three years immediately preceding the passing of the Act of 1868 employed in the occupation of an assistant to a pharmaceutical chemist, or a chemist and druggist, and having passed a modified examination to the satisfaction of the Pharmaceutical Society established by the Act of 1852, confirming their charter dated February 18, 1843. Pharmaceutical chemists, under the Act of 1852, and existing chemists and druggists, as before defined, shall be entitled to be registered without fee, upon a declaration duly certified as therein mentioned. And any one contravening the provisions of this Act shall for every such offence forfeit the penalty of 5l. And by the 16th sec. nothing contained in this Act shall interfere with the business of a legally qualified apothecary. By 32 & 33 Vict., c. 117, nothing contained in the first 15 sections of the Act of 1868 shall affect any person registered as a legally qualified medical practitioner before the passing of the Act of 1869, and the said clauses shall not apply to any person who may hereafter be registered as a legally qualified practitioner, and who in order to obtain his diploma for registration shall have passed an examination in pharmacy. And it is further provided by the original Act (s. 11) that no member of the medical profession, or who is practising under the right of a degree of any University, or under a diploma, or license of a medical or surgical corporate body, is to be registered; and if any registered pharmaceutical chemist obtains such diploma, his name is not to be retained on the register. And by the Act (1868) it is further enacted, as in the Medical Act, that every registrar of deaths on receiving notice of the death of any pharmaceutical chemist shall give the requisite information of such death to the proper authorities mentioned in the Act, that the name of the deceased may be erased from the register. It is almost needless to add, although it is expressly provided, that registration under either of the Pharmacy Acts does not entitle any one so registered to practise medicine or surgery, or any branch of medicine or surgery (31 & 32 Vict., c. 121, s. 18).

Thus it is seen that henceforth the business of a chemist and druggist is clearly defined. He may make up prescriptions of duly qualified medical practitioners, compound and dispense medicines, but in no case prescribe or practise medicine, surgery, or any branch of medicine or surgery, under penalties. He is in no sense a member of the profession, and it would therefore be inconsistent with the character of the present work to occupy more of its pages with further remarks on this topic, beyond a few observations on the subject of "poisons" and the Pharmacopœia. The list of poisons is contained in Schedule A of the Act. By the 17th sec. it is enacted that " it shall be unlawful to sell any poison by wholesale or by retail unless the box, bottle, vessel, wrapper, or cover in which such poison is contained be distinctly labelled with the name of the article and the word Poison, and with the name and address of the seller of the poison, and it shall be unlawful to sell any poison of those which are in the first part of Schedule A to this Act, or may hereafter be added thereto, to any person unknown to the seller, unless introduced by some person known to the seller; and the seller before delivery shall cause to be made an entry in a book in the form set down in Schedule F, the date of the sale, the name and address of the purchaser, the name and quality of the article sold, and the purpose for which it is stated by the purchaser to be required, signed by the purchaser and by the person, if any, who introduced him." Any infringement of this enactment renders the seller " liable, upon summary conviction, to a penalty of 5l. for the first offence, and 10l. for the second or any subsequent offence, and the person on whose behalf any sale is made by an apprentice or servant shall be deemed to be the seller." The provisions of this section are not to apply to articles to be exported from Great Britain by wholesale dealers, nor to sales by wholesale to retail dealers in the ordinary course of business, nor shall any of the provisions of this Act apply to any medicine supplied by a legally qualified apothecary to his patient, nor to any article when forming part of the ingredients of any medicine dispensed by a person registered under this Act, provided the medicine be duly labelled with the name and address of the seller, and the ingredient thereof entered, with the name of the person to whom it is sold or delivered, in a book kept for that purpose." And by the 3rd section of the amended Act of 1868, " nothing contained in section 17 of the former Act shall apply to any legally qualified medical practitioner, provided any medicine supplied by him be distinctly labelled with the name and address of the seller, and the ingredients be entered, with the name of the person to whom it is sold or delivered, in a book to be kept for that purpose." Unfortunately the Act of 1868 contains an exception in favour of " the making or dealing in patent medicines." This is its one great blot.

L

By section 54 of 21 & 22 Vict., c. 90, the "Medical Council shall cause to be published under their direction a book containing a list of medicines and compounds, and the manner of preparing them, together with the true weights and measures, by which they are to be prepared and mixed, and containing such other matter and things relating thereto as the General Council shall think fit, to be called "British Pharmacopœia," and the General Council shall cause to be altered, amended, and republished such Pharmacopœia as often as they shall deem necessary." And by 25 & 26 Vict., c. 91, "the General Council of Medical Education and Registration of the United Kingdom" is created a body corporate, and the exclusive right of publishing, printing, and selling the Pharmacopœia is vested in the said Council, and the Treasury is empowered to fix the price of copies from time to time.

SECTION II.

DENTISTS.

By the 55th section of the Medical Act it is provided that " nothing in the Act contained shall be construed to prejudice or in any way affect the lawful occupation, trade, or business of dentists ;" and by a previous section (48) her Majesty is empowered by charter to grant to the Royal College of Surgeons of England power to institute and hold examinations for the purpose of testing the fitness of persons to practise as dentists who may be desirous of being so examined, and to grant certificates of such fitness.

By 32 Hen. VIII., c. 42, s. 3, "drawing of teeth" is excepted from among those "things belonging to surgery" which any one "using barbery or shaving within the city of London, suburbs, or one mile circuit of the said city," was thenceforth prohibited from practising, and hence this art fell into, or rather perhaps remained almost entirely in, the hands of the barbers, and until a recent period was not cultivated by the surgeons as even ancillary to their calling. Now, however, there seems a probability of giving a professional surgical status to those who make this specialty their almost exclusive study, to the great relief, comfort, and security of those whom fate compels to seek their aid.

CHAPTER XVI.

MEDICAL ETHICS.

MEDICAL ETHICS, by which may be understood the conventional rules that prevail for regulating the intercourse among medical practitioners, and in which also their patients are indirectly interested, do not, strictly speaking, form a subject that can be treated of in these pages, inasmuch as they are no part of the law of the land, but rather relate to the social amenities which, like those of the Bar, form a code of etiquette, upon the due observance of which the profession most properly relies for the cultivation of a high tone of character among its members, alike beneficial to the practitioner and those requiring the aid of his services. But as far as they bear upon the legal responsibility of those whose conduct should be guided by them, they have already been indirectly suggested under the various heads of the Duties, Rights, and Liabilities of Medical Practitioners, whether as witnesses in courts of justice, as authors, medical referees, or as holding the relation of partners, principals, or assistants, or as undertaking the delicate and responsible duty of drawing up and attesting the wills of their patients, or as confidants of communications made to them by those who so naturally look upon them in the light of their truest friends in the hour of need. Utterly deprecating any attempt to weaken the bonds of the latter sacred relationship, it is perhaps better, nevertheless, when legal or clerical assistance can be readily obtained, to leave each profession to do its own particular work — the lawyer to do his part in assisting the patient in the settlement of his worldly affairs, and the clergyman to perform the functions of his sacred office. A consultation with the latter as to the most fitting moment when, without distressing or fatiguing the patient, he may minister to him that consolation for which he sometimes appeals even to his earthly physician, and which, by inspiring hope and resignation, may not infrequently operate beneficially in aiding his bodily infirmity, would in many instances be invaluable. With these few remarks the subject of Medical Ethics must, for the reasons before given, be dismissed; as to other works, such as

"Percival's Medical Ethics," and "Langley's Via Medica,"* the practitioner or student must refer for the regulation of his conduct both in the hospital, and in private practice, in all matters relating to the amenities to be cultivated both morally and socially. At the same time it is as well to observe that the former work, having been originally published so long ago as 1803, although republished in 1849, would in some respects lead any one relying on its legal accuracy into most serious errors. The chapter on "Professional duties in certain cases which require a knowledge of the law," is, in respect to the law of wills, entirely revolutionized, and indeed the whole of that chapter, which forms so large a portion of the work, is more or less affected by modern legislation. Mr. Langley's "Via Medica," being of comparatively recent date, of course is free from any such objection, and is replete with many valuable suggestions on the subject of which it mainly treats, viz., the *mores medici*, which clearly cannot form a matter of discussion in the present treatise. Suffice it to say—

Ingenuas didicisse fideliter artes, emollit mores.

At the same time the following valuable rules of conduct, carefully revised, by an eminent member of the medical profession, will be read with interest :—

Of Professional Conduct, whether in Hospital, Private, or General Practice.

1. The same moral rules of conduct should prevail, whether towards hospital patients or in private and general practice. It is unnecessary to say that every case committed to the charge of a physician or surgeon should be treated with care, attention, and humanity; every allowance should be made for the whims and caprices of the sick, secrecy and delicacy should be strictly observed, and the familiarity and confidential intercourse to which members of the faculty are admitted should ever be used with the most scrupulous regard to fidelity and honour.

2. The strictest temperance should at all times be practised, as the exercise of the duties and functions of medical men requires a clear head, a vigorous mind, and oftentimes a steady hand, and, on emergencies, even the life of a fellow-creature may depend upon the possession by the professional man of all these essentials.

* See also "On the Duties and Qualifications of a Physician," by John Gregory, M.D., &c. London: 1820.
"The Moral Aspects of Medical Life, &c.," by James Mackness, M.D., &c. London: 1846.
"A Manual of Clinical Medicine and Physical Diagnosis." by T. H. Tanner, M.D. The Second Edition, revised by Dr. Tilbury Fox. London: 1869.

3. The medical practitioner should, wherever the case justifies it, encourage the patient with hopes of recovery; but, on the other hand, without indulging in gloomy prognostications, he should not fail on suitable occasions to warn the friends of the patient, or even the patient himself, of any real or imminent danger when it has actually arisen.

4. Officious intercourse in a case under the charge of another should be carefully avoided; no meddling inquiries should be made concerning the patient; no officious hints given relative to the nature or treatment of the disorder, nor any innuendoes indulged in that may tend either directly or indirectly to diminish the trust reposed in the medical attendant in charge of the case.

5. When a medical practitioner is called to a patient who has been under the care of another medical gentleman, a consultation with the latter should be proposed, even though he may have discontinued his visits. His treatment of the case had better not be criticised; but if it be noticed, it ought to be regarded with candour, and even justified, so far as probity and truth will permit; and, a good understanding thus established between a former and a subsequent practitioner, may serve to suggest further means of relief without at all impeaching the professional skill or knowledge exhibited in the primary treatment of the case, though such treatment may have been unsuccessful.

6. In large and wealthy communities, the distinction between physic and surgery should, as far as possible, be steadily maintained, for the division of skill and labour is no less advantageous in the liberal sciences than in the mechanical arts; but, in small communities or rural districts, such an attempt at distinction in the present day would virtually leave patients without the necessary combined assistance so much required, and perhaps lead them to fly for succour to any audacious empiric who would not be so scrupulous on the subject of "division of labour." As medical men can now, if they possess varied qualifications, be registered in all of them, cases of breach of this rule of etiquette are not in future likely to arise.

7. Consultations should be promoted in difficult or protracted cases, as they give rise to confidence and an interchange of ideas to the advantage of the sick person; and every respect should be shown towards the medical man first engaged, and too much stress should not be laid upon the strict precedency of seniority or rank.

8. A medical practitioner may sometimes be hastily summoned, through the anxiety or sudden alarm of the patient or his family, to visit a patient already under the care of another, to whom notice of this call has not been given. In such a case, no change in the treatment of the patient should be made until a consultation has

taken place, unless the lateness of the hour precludes such a meeting, or the symptoms of the case are too pressing to admit of delay.

9. Theoretical discussions at consultation should be avoided, as occasioning much perplexity and seldom leading to any practical result.

10. Punctuality, as far as possible, should be observed in meeting in consultation; visits to the sick, either singly or in consultation, should not be unreasonably repeated, because, when too frequent, they tend to diminish medical authority or influence, and might be objectionable to the patient himself.

11. When one medical man officiates for another who is sick or absent during any considerable length of time, he should receive a portion or the whole of the fees accruing during such interval, according to previous arrangement; but if this neighbourly act be of short duration, it had better be gratuitously performed, and the utmost delicacy should be exhibited towards the interests of the absent brother practitioner.

12. The legal claim to fees and their amount are subjects which have been dealt with in a previous part of this work; but in the consideration of fees it should be ever borne in mind that the characteristic benevolence of the profession is inconsistent with sordid views or avaricious greed; and although "the labourer is worthy of his reward," and duty to himself and his family demands a consideration of his material interests, yet knowledge, benevolence, and their practical application to the benefit of his fellow creatures, are of the highest importance, and in the cultivation of them he will not lose his reward; for his character will always in such case be his recommendation, even to those who do not desire to derive any undue advantage from his liberality, and who may be willing as well as able to repay his services in a substantial form.

13. All members of the profession, together with their wives and children, should be attended gratuitously by any other members of the Faculty, provided at least they reside within a reasonable distance. Great care should be taken not to obtrude visits officiously, as such over-civility, however well intended, may cause embarrassment, or interfere with the exercise of a free choice, on which so much of confidence depends.

14. Clergymen, or at least the unbeneficed clergy, are sometimes visited gratuitously. As a rule, however, even curates with large families will make an effort to maintain their self-respect and independence by offering some remuneration.

15. Should a medical man have to certify to the illness of any official, who through such temporary incapacity claims exemption from the duties of his situation, whether in the army, navy, or civil

service, or as taking part in the administration of justice as a juryman, truth and probity should be the paramount consideration. Such certificates are assurances to the public, and should never be given without a strict and faithful scrutiny, and even private friendship should not be permitted to sway the judgment in such cases.

16. A medical man should cautiously guard against whatever may injure the general respectability of the profession, and be careful to avoid any affectation of scepticism, even in a vein of jocularity, concerning the efficacy and utility of the healing art.

17. Should a diversity of opinion or controversy arise as to the professional conduct of any one of the Faculty, the point in dispute should be referred to the arbitration of two other medical men, one of whom to be selected by the party complained of; and in case of their differing, the decision to be referred to an umpire, previously named by the two arbitrators. The adjudication should not be made public, but submissively acquiesced in and accepted *con amore*.

18. When a case has terminated fatally, it may be useful and instructive to indulge in a retrospect of the treatment adopted. Regrets may possibly follow; but if the object is really honestly to review a course of treatment with a view to avoiding errors, even though only of judgment and not of science, for the future, the consciousness of good intentions will sufficiently justify the past, and the desire thus evinced to subserve a more matured experience for the future will be a sufficient atonement for those errors which the imperfection of human skill, and a subsequent knowledge of events, must constantly, though perhaps tardily, disclose.

19. One grown old in the useful and honourable exercise of the healing art may yet continue to enjoy, and justly to enjoy, the unabated confidence of the public; the quickness of the eye, the delicacy of the touch, and the steadiness of hand, which are so essential to the skilful performance of operations, may no longer be his, but he may still retain an unclouded judgment, and must have the advantage of great experience. If he can happily, under such circumstances, perceive wherein his deficiencies consist, and retire to the exercise of the revered and valuable duties of an adviser and a counsellor, rather than attempt the continuance of those more practical and physical functions involved in general practice, and requiring energy, activity, and nerve, he may yet apply his knowledge and experience to the benefit of mankind, and himself enjoy to the last the unspeakable satisfaction of a "mens conscia recti."

One remark may be made on the difference between the position of the medical man in private and in public practice (whether as medical officer of a hospital, a prison, or workhouse, or of a Poor-Law district, or even as a surgeon or assistant surgeon in the army or navy). The difference consists in this, that in the case of the public medical officer no choice is allowed to the patients who come under his care, and yet personal confidence is not less important to the comfort and relief of the sick poor, or even criminal class, than of the rich, when by sickness distinctions are thus levelled. More than ordinary pains should therefore be taken to inspire with confidence those who sometimes look with suspicion upon eleemosynary relief of any kind, and to treat scruples with respect, and even prejudices with consideration.

Conjoint Medical Examination Board for each of the three Divisions of the United Kingdom.

One uniform qualification for practice is of course the object here aimed at. Sound practical medical education is the general desideratum, and if this can be accomplished by the process of levelling, provided it be in the direction of levelling up and not levelling down, a great object would be effected. The General Medical Council have already invited co-operation. It, however, scarcely seems necessary that the general practitioner should be required to possess the amount of knowledge requisite for the higher medical examinations, especially considering the number of highly-educated medical men now scattered over the country whose assistance in difficult cases can generally be obtained at a comparatively small expense to the patient. Indeed, without underrating the scholastic attainments of the bulk of those intended for general practitioners, it may fairly be doubted whether the levelling up system could for some time to come be fairly responded to. If the examination, on the other hand, were simply of a practical character, and that the license to practise, founded upon it, should be for every member of the Profession one and the same, this might operate as a discouragement to academical training, and thus, by granting no privilege to the University graduates, but reducing them to one uniform level with all other practitioners, would tend to the depreciation of such distinctions, or at least to their acknowledgment merely as ornamental appendages to be enjoyed by the wealthy and dilettante classes alone, culminating in a dead level of ambition and laudable emulation.

It becomes a further matter of serious consideration with the Profession whether, whilst an improvement might be effected in general medical education, it is desirable to abolish all distinctions in the relative departments of medical and surgical science. Upon this question the author of this work, not being a medical man, cannot presume to offer an opinion, and apologizes for even venturing the above hints.

SUPPLEMENTARY CHAPTER

TO THE BODY OF THE WORK.

DISSERTATION ON THE TITLE OF DOCTOR OF MEDICINE, OR DOCTOR OF PHYSIC.

ACCORDING to the draft of an Act of Parliament, 9 Hen. V.—
"No one shall use the mysterie of fysyk, unless he hath studied it in some university, and is at least a bachelor in that science. The sheriff shall inquire whether any one practises in his county contrary to this regulation; and if any one so practise, he shall forfeit 40*l.*, and be imprisoned." So great a regard, not merely preference, was shown at this time to those medically educated in some university, but such regard was not confined to those who had pursued the study at Oxford or Cambridge; the phrase, "some university," being probably used advisedly, in order to embrace foreign universities where the science of physic was cultivated with much success. It appears, however, that this draft Act, or, as it would be called in the present day, "Bill," never had the effect of an Act of Parliament—why, does not appear, but possibly parliament declined so to limit the practitioners in physic, notwithstanding the wish of the king.

In re Dr. Bonham, 8 Co. 107-114, Mic. Ter. 6 Jac. I., which was an action for false imprisonment of the plaintiff by the President and censors of the College of Physicians and their servants, the question was raised whether the College had any authority over doctors of the University, the plaintiff being a doctor of physic of Cambridge, practising in London, but without the license or authority of the College. The following remarks of Lord Chief Justice Coke show the impression existing shortly after the incorporation of the College, of the status of its doctors, this being the first judgment concerning fine and imprisonment given since the granting of the Charter and the passing of the Acts of Parliament concerning it.

The Chief Justice, because much was said in commendation of the *doctors of physic of the College in London*, and somewhat, as he conceived, in derogation of the dignity of the doctors of the Universities, attributed first much to *the doctors* of the said College in London, and confessed that nothing was spoken in their recommendation which was not due to their merits; but yet that no comparison was to be made between that private College and either of the Universities of Cambridge and Oxford: no more than between father and his children, or than between the fountain and the small rivers that descend from it. The University is *alma mater*, from whose breasts those of that private college have sucked all their science and knowlege, which I acknowledge to be great and profound; but the law saith, "erubescit lex filios castigare parentes:" the university is the fountain, "et melius est petere fontes quam sectari rivulos." Briefly, "Academiæ Cantabrigiæ et Oxoniæ sunt Athenæ nostræ nobilissimæ, regni soles, occuli et animæ regni, unde religio, humanitas, et doctrina in omnes regni partes uberrime diffunduntur;" but it is true, "nunquam sufficiet copia lanctatoris, quia nunquam deficiet materia laudis," and therefore those universities exceed and excel all private colleges, "quantum inter viburna cupressus." According to the Act 14 Hen. VIII., five manner of persons were to be promoted to the office of physician, viz., those who were, I., profound; II., sad; III., discreet; IV., groundedly learned; V., profoundly studied. And it was well ordained that the professors of physic should be profound, sad, discreet, &c., and not youths who have no gravity and experience; for as one saith, "in juvene theologo conscientiæ detrimentum, in juvene legista bursæ detrimentum, in juvene medico cœmeterii incrementum." And it ought to be presumed every doctor of any of the universities to be within the statutes: *sc.*, to be profound, sad, discreet, groundedly learned, and profoundly studied; for none can there be master of arts, *who is a doctor of philosophy*, under the study of seven years, and cannot be doctor in physic under seven years more in the study of physic; and that is the reason that the plaintiff is named in the declaration doctor of philosophy and doctor of physic; "quia oportet medicum esse philosophum, ubi enim philosophus desinit medicus incipit." It will be observed in this judgment that Lord Coke speaks of the "*doctors of physic* of the college in London," and proceeds to contrast their status with that of the "doctors of the universities." The plaintiff was described in the declaration as a doctor in philosophy and physic, but yet there is no such academical degree as doctor in philosophy, conferred by diploma or otherwise by either University of Oxford or Cambridge, but that designation is included, as Lord Coke puts it, in the degree of Master of Arts. And so "oportet *medicum* esse philosophum, ubi enim philosophus desinit medicus incipit." In other words every duly recognised "Medicus," like every "Magister Artium," is a doctor of philosophy, and "ubi philosophus desinit medicus incipit." He is, if one of the College of Physicians, also a doctor of physic in all its branches, including surgery, though not a Doctor of Medicine *par excellence*, which special designation pertains alone to doctors of the Universities whose practice is limited to pure medicine. "Medicus" is a more ancient designation than that of either doctor or bachelor of medicine. We find from the Exon Domesday, which furnished a portion of the materials from which the great Domesday, or General Survey of William the Conqueror, was compiled, that Nigel, the physician, is there described as "Medicus," and not Doctor or Bachelor of Medicine. "Medicus" and "Doctor" are both used in the Act of Henry VIII., incorporating the College of Physicians. The term "Medicus" is there applied individually to all the medical members of the corporation who are mentioned by name, and the body itself is declared to be "collegium perpetuum *doctorum* et gravium virorum, qui medicinam in urbe nostra Londino et suburbiis, intraque septem millia passuum ab ea urbe quaqueversus publice exerceant." The three first-named members are designated, as it were emphatically, "gravium virorum doctorum," and also as "medicorum nostrorum," or the king's physicians, and the others as simply "medicorum," but "medicus" has at the same time always been considered as implying the style and title of "doctor of physic," and hence it is that the College, by an express bye-law, forbids those of its licentiates who are permitted to practise pharmacy among their own patients to assume the title of Doctor of Medicine, or use any other name, title, designation, or distinction implying that they are graduates in medicine of an University, unless they be graduates in medicine of an University. It would therefore seem that, but for this bye-law so prohibiting them, such licentiates, though neither fellows nor members of the College, might assume the title at least of Doctor, which might lead to the above implication, without incurring any penalty, and indeed as of right; and *a fortiori*, therefore, may the members and fellows, who are of a higher grade in the College, do so. If this were otherwise, even though such might not be liable to be summarily convicted under the 40th section of the Medical Act, for "wilfully and falsely pretending to, and taking or using the name or title of a Doctor of Medicine," the above bye-law would surely be wholly unnecessary, or at least supererogatory.

APPENDIX.

Cases decided in the County Courts upon the claim of the Medical Referee of the Assured to the payment of a fee by the Company.

COLCHESTER COUNTY COURT (October, 1850).

PHILBRICK *v.* WHETHAM.

DEFENDANT was sued as one of the directors of the National Provident Institution for the fee of one guinea for furnishing particulars as to the health of a party proposing to assure his life in that office. The circular requesting the favour of answers to questions contained in that document was from the Secretary. The plaintiff admitted that he had furnished certificates to the office on previous occasions without charge; life assurances not being then so prevalent as to make the practice a burden to the profession, but since these applications had become so frequent he had refused to supply the information without payment; but it appeared in the two instances, mentioned by him, in which he had received payment from the Company's agent in Colchester, he was aware that the money came from the parties effecting the insurance, and not from the office. For the defence, but in answer to the Judge, the agent of the office said that the plaintiff repeatedly told him he would not furnish any more certificates without a fee; but he had never said that in case of applications to him he should hold the Company liable: that he believed this was the first circular sent to the plaintiff direct from the secretary, those which had been furnished gratuitously having all come from himself as agent. Upon this it was submitted, on behalf of the plaintiff, that the last answer was a very material one to the plaintiff's case, as showing a new mode of application to him, after he had declined to furnish more certificates without payment. It was also urged that upon the face of the secretary's letter itself the agreement was between the office and himself; for the letter could not be construed otherwise than as a request from the office to perform certain services on their account. His Honour said, if the letter had stood alone, and this had been the first transaction of the kind between the plaintiff and the office, he thought they would have been liable and must have paid him; but now, as it appeared to him, the whole depended upon what had previously taken place. Of course the plaintiff was not bound to continue furnishing certificates gratuitously, but he was of opinion that he could not claim payment without previous notice.

Judgment for the defendant.

HOOPER v. THE GRAHAM LIFE ASSURANCE SOCIETY.

IN THE SHOREDITCH COUNTY COURT.

Before Mr. Serjeant STORKS (Judge), June 7th, 1851.

The claim was of a like nature to the last.
The letter from the secretary of the society commenced thus:—

"PROPOSER'S MEDICAL REFEREE.

"To Mr. Hooper, Queen's Road, Dalston.

"SIR,—Mr. Jacob J. Davies, who has been proposed to this office for an assurance on his life, has given a reference to you for private and confidential answers to the following queries."

Then follow the queries, and there is a note at the foot, thus:—

"The directors would be obliged by your answering each of the questions separately."

Answers were accordingly returned to the society, for which the plaintiff some time afterwards claimed the fee of one guinea. This the society refused to pay.

At the conclusion of the trial, his Honour said:—"The question in this case is whether there was an implied assumpsit. The argument that the medical referee is the selection of the assured is certainly a strong one: for, although the questions are put down by the society, the name of the referee is given by the assured. I don't think there is any evidence of an implied assumpsit beyond the sending of the letter. That is my present impression, but it is a question of too general importance to be decided off-hand. I shall take time, therefore, to consider my judgment."

The following judgment was afterwards delivered by the Court.—
"The question turns upon whether there was a contract expressed or implied—what the law calls *assumpsit*. It is perfectly clear in this case there was no contract expressed—that there was nothing approaching a precise and specific contract. Then the question as to whether there was an implied assumpsit arises. Assurance offices

have made certain rules and regulations upon which they hold out to the public that they are willing to enter into a contract. Therefore, the contract is from the beginning as between the assured and the assurer. The contract is that the parties shall give certain securities to them, and they assure his life upon certain conditions; and one of the conditions is that answers by a medical practitioner shall be forwarded to the society in reply to certain inquiries (as set forth in their printed letter) as to the state of health of the proposed assurer. They know not the medical practitioner to whom they are referred, and have never had any communication with him, but they take it for granted that he must know something of the assured. Now, let us advert to the letter sent by the secretary of the society at the request of the assured. A proposal is made which the medical referee knows nothing of, and they agree to assure the party's life on condition that the inquiries are satisfactorily answered, and from the letter adverted to it is clear that the society had no communication whatever with the referee before the proposal is made. It would be very true, if there was nothing in the letter sent by the secretary to Mr. Hooper which showed that it was the act of the party who had proposed to the office for an insurance on his life, and that the opinion of a medical gentleman as to his state of health was one of the conditions upon which the society agrees to assure a party's life, that the medical referee might say, 'You have no right to draw upon my skill and services.' And if it was put upon that principle abstractedly, it might be a question whether the medical practitioner would not be entitled to remuneration. But the contract here is between the assurer and the assured. The society say, 'The medical man is not of our choosing, nor are we dependent upon his skill.' In my opinion, it is one of the conditions made with the assured, that a medical man shall answer the questions proposed by the society. This is the basis of the contract, and the meaning of it is 'not that we shall trouble ourselves to select a referee,' we leave it to your honour to choose some one who shall supply us with the information we require. I can see no contract in this between the company and the third party—the medical man. The plaintiff might have said, I shall not send answers unless I am paid. Upon the whole, I am decidedly of opinion that the plaintiff in this case is not entitled to recover."

DUPLEX AND ANOTHER, *v.* THE ECONOMIC LIFE OFFICE.

IN THE SHERIFF'S COURT, LONDON (August 24, 1852).

Before Mr. RUSSELL GURNEY, Q.C.

The same question was involved in this action as in the two previous cases. The only difference between this and the last case was, that in the present case the plaintiff in returning the answers to the questions asked by the defendants also inclosed a note stating that his fee was a guinea, which he requested should be forwarded to him.

His Honour thought that no case had been made out against the defendants. The plaintiff's need not have answered the questions without the fee, and the letter enclosing the questions stated that they were referred by the lady proposing her life for assurance; he felt bound to refuse the claim of the plaintiffs.

Verdict for defendants.

Opinions of the Judges on questions propounded by the House of Lords.

The following questions of law were propounded by the House of Lords to the Judges, in relation to alleged crimes committed by persons afflicted with insane delusions, and, together with the answers, were ordered by the House to be printed, on the 19th of June, 1843:—

"1. What is the law respecting alleged crimes committed by persons afflicted with insane delusion in respect of one or more particular subjects or persons; as, for instance, where, at the time of the commission of the alleged crime, the accused knew he was acting contrary to law, but did the act complained of with a view, under the influence of insane delusion, of redressing or revenging some supposed grievance or injury, or of producing some supposed public benefit?

"2. What are the proper questions to be submitted to the jury when a person alleged to be afflicted with insane delusion respecting one or more particular subjects or persons is charged with the commission of a crime (murder, for example), and insanity is set up as a defence?

"3. In what terms ought the question to be left to the jury as to the prisoner's state of mind at the time when the act was committed?

"4. If a person under an insane delusion as to existing facts commits an offence in consequence thereof, is he hereby excused?

"5. Can a medical man conversant with the disease of insanity, who never saw the prisoner previous to the trial, but who was present during the whole trial, and the examination of all the witnesses, be asked his opinion as to the state of the prisoner's mind at the time of the commission of the alleged crime; or his opinion whether the prisoner was conscious at the time of doing the act that he was acting contrary to law, or whether he was labouring under any and what delusion at the time?"

Mr. Justice MAULE.—I feel great difficulty in answering the questions put by your lordships on this occasion: first, because they do not appear to arise out of, and are not put with reference to a particular case, or for a particular purpose, which might explain or limit the generality of their terms, so that full answers to them ought to be applicable to every possible state of facts, not inconsistent with those assumed in the questions; and this difficulty is the greater from the practical experience both of the Bar and the Court being confined to questions arising out of the facts of particular cases: secondly, because I have heard no argument at your lordships' Bar, or elsewhere, on the subject of these questions, the want of which I feel the more, the greater is the number and extent of questions which might be raised in argument: and, thirdly, from a fear, of which I cannot divest myself, that, as these questions relate to matters of criminal law of great importance and frequent occurrence, the answers to them by the judges may embarrass the administration of justice when they are cited in criminal trials. For these reasons, I should have been glad if my learned brethren would have joined me in praying your lordships to excuse us from answering these questions: but, as I do not think they ought to induce me to ask that indulgence for myself individually, I shall proceed to give such answers as I can, after the very short time which I have had to consider the questions, and under the difficulties I have mentioned, fearing that my answers may be as little satisfactory to others as they are to myself.

The first question, as I understand it, is, in effect, What is the law respecting alleged crime, when at the time of the commission of it the accused knew he was acting contrary to the law, but did the act with a view, under the influence of some delusion, of redressing or revenging some supposed grievance or injury, or of producing some supposed public benefit? If I were to understand this question according to the strict meaning of its terms, it would require, in order to answer it, a solution of all questions of law which could arise on the circumstances stated in the question, either by explicitly stating and answering such questions, or by stating some principles or rules which would suffice for the solution. I am quite unable to do so, and, indeed, doubt whether it be possible to be done; and therefore request to be permitted to answer the question only so far as it comprehends the question whether a person, circumstanced as stated in the question, is for that reason only to be found not guilty of a crime respecting which the question of his guilt has been duly raised in a criminal proceeding; and I am of opinion that he is not. There is no law that I am aware of that makes persons in the state described in the question not responsible for their criminal acts. To render a person irresponsible for crime on account of unsoundness of mind, the unsoundness should, according to the law as it has long been understood and held, be such as to render him incapable of knowing right from wrong. The terms used in the question cannot be said (with reference only to the usage of language) to be equivalent to a description of this kind and degree of unsoundness of mind. If the state described in the question be one which involves or is necessarily connected with such an unsoundness, this is not a matter of law, but of physiology, and not of that obvious and familiar kind as to be inferred without proof.

Secondly: The questions necessarily to be submitted to the jury are those questions of fact which are raised out of the record. In a criminal trial the question commonly is, whether the accused be guilty or not guilty; but, in order to assist the jury in coming to a right conclusion on this necessary and ultimate question, it is usual and proper to submit such subordinate or intermediate questions as the course which the trial has taken may have made it convenient to direct their attention to. What those questions are, and the manner of submitting them, is a matter of discretion for the Judge, a discretion to be guided by a consideration of all the circumstances attending the inquiry. In performing this duty it is sometimes necessary, or convenient, to inform the jury as to the law; and if, on a trial such as is suggested in the question, he should have occasion to state what kind and degree of insanity would amount to a defence, it should be stated conformably to what I have mentioned in my answer to the first question, as being, in my opinion, the law on this subject.

Thirdly: There are no terms which the Judge is by law required to use. They should not be inconsistent with the law as above stated, but should be such as, in the discretion of the Judges, are proper to assist the jury in coming to a right conclusion as to the guilt of the accused.

Fourthly: The answer which I have given to the first question is applicable to this.

Fifthly: Whether a question can be asked depends, not merely on the questions of fact raised on the record, but on the course of the cause at the time it is proposed to ask it, and the state of an inquiry as to the guilt of a person charged with a crime, and defended on the ground of insanity, may be such that such a question as either of those suggested is proper to be asked and answered, though the witness has never seen the person before the trial, and though he has been present and heard the witnesses; these circumstances, of his never having seen the person before, and of his having

M

been present at the trial, not being necessarily sufficient, as it seems to me, to exclude the lawfulness of a question which is otherwise lawful, though I will not say that an inquiry might not be in such a state as that these circumstances should have such an effect. Supposing there is nothing else in the state of the trial to make the questions suggested proper to be asked and answered, except that the witness has been present and heard the evidence, it is to be considered whether that is enough to sustain the question. In principle it is open to this objection, that, as the opinion of the witness is founded on those conclusions of fact which he forms from the evidence: and, as it does not appear what those conclusions are, it may be that the evidence he gives is on such an assumption of facts as makes it irrelevant to the inquiry. But such questions have been very frequently asked, and the evidence to which they have been directed has been given, and has never, that I am aware of, been successfully objected to. Evidence most clearly open to this objection, and on the admission of which the event of a most important trial probably turned, was raised in the case of *The Queen v. McNaughten,* tried at the Central Criminal Court in March last, before the Lord Chief Justice, Mr. Justice Williams, and Mr. Justice Coleridge, in which counsel of the highest eminence were engaged on both sides; and I think the course and practice of receiving such evidence, confirmed by the very high authority of those Judges, who not only received it, but left it, as I understand, to the jury without any remark derogating from its weight, ought to be held to warrant its reception, notwithstanding the objection in principle to which it may be open. In cases even where the course of practice in criminal law has been unfavourable to parties accused, and entirely contrary to the most obvious principles of justice and humanity, as well as those of law, it has been held that such practice constituted the law, and could not be altered without the authority of Parliament.

Chief Justice TINDAL.—My Lords, her Majesty's judges, with the exception of Mr. Justice Maule, who has stated his opinion to your Lordships' House, think it right, in the first place, to state that they have forborne entering into any particular discussion upon these questions, from the extreme and almost insuperable difficulty of applying those answers to cases in which the facts are not brought judicially before them. The facts of each particular case must of necessity present themselves with endless variety, and with every shade of difference in each case; and, as it is their duty to declare the law upon each particular case, on facts proved before them, and after hearing arguments of counsel thereon, they deem it at once impracticable, and at the same time dangerous to the administration of justice, if it were practicable, to attempt to make minute applications of the principles involved in the answers given by them to your Lordships' questions.

They have, therefore, confined their answers to the statement of that which they hold to be the law upon the abstract questions proposed by your Lordships : and as they deem it unnecessary, in this particular case, to deliver their opinions seriatim, and as all concur in the same opinion, they desire me to express such their unanimous opinion to your Lordships. The first question proposed by your Lordships is this :—" What is the law respecting alleged crimes committed by persons afflicted with insane delusion in respect of one or more particular subjects or persons ; as for instance, where at the time of the commission of the alleged crime, the accused knew he was acting contrary to law, but did the act complained of with a view, under the influence of insane delusion, of redressing or revenging some supposed grievance or injury, or of producing some supposed public benefit?"

In answer to which question, assuming that your Lordships' inquiries are confined to those persons who labour under such partial delusions only, and are not in other respects insane, we are of opinion that, notwithstanding the party accused did the act complained of with a view, under the influence of insane delusion, of redressing or revenging some supposed grievance or injury, or of producing some public benefit, he is nevertheless punishable if he knew at the time of committing such crime that he was acting contrary to law—by which expression we understand your Lordships to mean the law of the land.

Your Lordships are pleased to inquire of us, secondly, "What are the proper questions to be submitted to the jury, when a person alleged to be afflicted with insane delusions respecting one or more particular subjects or persons is charged with the commission of a crime (murder, for example), and insanity is set up as a defence?" And, thirdly, "In what terms ought the question to be left to the jury as to the prisoner's state of mind at the time when the act was committed?" And as these two questions appear to us to be more conveniently answered together, we have to submit our opinion to be, that the jury ought to be told, in all cases, that every man is presumed to be sane, and to possess a sufficient degree of reason to be responsible for his crimes, until the contrary be proved to their satisfaction; and that, to establish a defence on the ground of insanity, it must be clearly proved that, at the time of the committing of the act, the party accused was labouring under such a defect of reason, from disease of the mind, as not to know the nature and quality of the act he was doing, or, if he did know it, that he did not know he was doing what was wrong. The mode of putting the latter part of the question to the jury on these occasions has generally been, whether the accused, at the time of doing the act, knew the difference between right and wrong; which mode, though rarely, if ever, leading to any mistake with the jury, is not, as we conceive, so accurate when put generally and in the abstract, as when put with reference to the party's knowledge of right and wrong in respect to the very act with which he is charged.

If the question were to be put as to the knowledge of the accused, solely and exclusively with reference to the law of the land, it might tend to confound the jury, by inducing them to believe that an actual knowledge of the law of the land was essential in order to lead to a conviction; whereas, the law is administered upon the principle that every one must be taken conclusively to know it, without proof that he does know it. If the accused was conscious that the act was one which he ought not to do, and if that act was at the same time contrary to the law of the land, he is punishable; and the usual course, therefore, has been to leave the question to the jury, whether the party accused had a sufficient degree of reason to know that he was doing an act that was wrong; and this course, we think, is correct, accompanied with such observations and explanations as the circumstances of each particular case may require.

The fourth question which your Lordships have proposed to us is this:—"If a person under an insane delusion as to existing facts commits an offence, in consequence thereof is he thereby excused?" To which question the answer must, of course, depend on the nature of the delusion; but, making the same assumption as we did before, namely, that he labours under such partial delusion only, and is not in other respects insane, we think he must be considered in the same situation as to responsibility as if the facts with respect to which the delusion exists were real. For example, if under the influence of his delusion, he supposes another man to be in the act of attempting to take away his life, and he kills that man, as he supposes, in self-defence, he would be exempt from punishment. If his delusion was that the deceased had inflicted a serious injury to his character and fortune, and he killed him in revenge for such supposed injury, he would be liable to punishment.

The question lastly proposed by your Lordships is:—"Can a medical man, conversant with the disease of insanity, who never saw the prisoner previous to the trial, but who was present during the whole trial and the examination of all the witnesses, be asked his opinion as to the state of the prisoner's mind at the time of the commission of the alleged crime, or his opinion whether the prisoner was conscious at the time of doing the act that he was acting contrary to law, or whether he was labouring under any and what delusion at the time?"

In answer thereto we state to your Lordships that we think the medical man, under the circumstances supposed, cannot in strictness be asked his opinion in the terms above stated, because each of those questions involved the determination of the truth of the facts deposed to, which it is for the jury to decide; and the questions are not mere questions upon a matter of science, in which case such evidence is admissible. But, where the facts are admitted or not disputed, and the question becomes substantially one of science only, it may be convenient to allow the question to be put in that general form, though the same cannot be insisted on as a matter of right.

ACTS OF PARLIAMENT PRINTED *IN EXTENSO*.

6 & 7 William IV., cap. 89 1 Victoria, cap. 68	Medical witnesses at Coroners Inquests.
8 & 9 Victoria, cap. 100 16 & 17 Victoria, cap. 96 16 & 17 Victoria, cap. 97 19 & 20 Victoria, cap. 87 18 & 19 Victoria, cap. 105 25 & 26 Victoria, cap. 86 25 & 26 Victoria, cap. 111	Lunacy.
15 & 16 Victoria, cap. 56 31 & 32 Victoria, cap. 121	Pharmacy (Pharmaceutical chemists).
16 & 17 Victoria, cap. 100 30 & 31 Victoria, cap. 84	Vaccination.
29 & 30 Victoria, cap. 35 32 & 33 Victoria, cap. 96	Contagious diseases.
18 & 19 Victoria, cap. 116 18 & 19 Victoria, cap. 121 23 & 24 Victoria, cap. 77 29 & 30 Victoria, cap. 90	Health (public
21 & 22 Victoria, cap. 90 22 Victoria, cap. 21 23 Victoria, cap. 7 23 & 24 Victoria, cap. 66 31 & 32 Victoria, cap. 29.	Medical Acts.

ANNO SEXTO & SEPTIMO GULIELMI IV. REGIS.
CAP. LXXXIX.
An Act to provide for the Attendance and Remuneration of Medical Witnesses at Coroners Inquests. [17th August 1836.]

Coroner empowered to summon Medical Witnesses, and to Direct the Performance of a post-mortem Examination.

WHEREAS it is expedient to provide for the Attendance of Medical Witnesses at Coroners Inquests, also Remuneration for such Attendance, and for the Performance of post-mortem Examinations at such inquests; be it therefore enacted by the King's most Excellent Majesty, by and with the Advice and Consent of the Lords Spiritual and Temporal, and Commons, in this present Parliament assembled, and by the Authority of the same, That from and after the passing of this Act, whenever upon the summoning or holding of any Coroner's Inquest it shall appear to the Coroner that the deceased Person was attended at his Death or during his last Illness by any legally qualified Medical Practitioner, it shall be lawful for the Coroner to issue his Order, in the Form marked (A.) in the Schedule hereunto annexed, for the Attendance of such Practitioner as a Witness at such Inquest; and if it shall appear to the Coroner that the deceased Person was not attended at or immediately before his Death by any legally qualified Medical Practitioner, it shall be lawful for the Coroner to issue such Order for the Attendance of any legally qualified Medical Practitioner being at the Time in actual Practice in or near the Place where the Death has happened; and it shall be lawful for the Coroner, either in his Order for the Attendance of the Medical Witness, or at any Time between the issuing of such Order and the Termination of the Inquest, to direct the Performance of a post-mortem Examination, with or without an Analysis of the Contents of the Stomach or Intestines, by the Medical Witness or Witnesses who may be summoned to attend at any Inquest; provided that if any Person shall state upon Oath before the Coroner that in his or her Belief the Death of the deceased Individual was caused partly or entirely by the improper or negligent Treatment of any Medical Practitioner or other Person, such Medical Practitioner or other Person shall not be allowed to perform or assist at the post-mortem Examination of the Deceased.

A Majority of the Jury may require the Coroner to summon additional Medical Evidence if the first be not satisfactory.

II. And be it further enacted, That whenever it shall appear to the greater Number of the Jurymen sitting at any Coroner's Inquest, that the Cause of Death has not been satisfactorily explained by the Evidence of the Medical Practitioner or other Witness or Witnesses who may be examined in the first instance, such greater Number of the Jurymen are hereby authorized and empowered to name to the Coroner in Writing any other legally qualified Medical Practitioner or Practitioners, and to require the Coroner to issue his Order, in the Form herein-before mentioned, for the Attendance of such last-mentioned Medical Practitioner or Practitioners as a Witness or Witnesses, and for the performance of a post-mortem Examination, with or without an Analysis of the Contents of the Stomach or Intestines, whether such an Examination has been performed before or not; and if the Coroner, having been thereunto required, shall refuse to issue such Order, he shall be deemed guilty of a Misdemeanor, and shall be punishable in like Manner as if the same were a Misdemeanor at Common Law.

Fees to Medical Witnesses to be paid out of Funds collected for Relief of the Poor.

III. And be it further enacted, That when any legally qualified Medical Practitioner has attended upon any Coroner's Inquest in obedience to any such Order as aforesaid of the Coroner, the said Practitioner shall for such Attendance at any Inquest in *Great Britain* be entitled to receive such Remuneration or Fee as is mentioned in the Table marked (B.) in the Schedule hereunto annexed ; and for any Inquest held in *Ireland*, the said Practitioner shall be paid in the Manner provided by the Laws in force in that Part of the United Kingdom ; and the Coroner is hereby required and commanded to make, according to the Form marked (C.) in the Schedule hereunto annexed, his Order for the Payment of such Remuneration or Fee, when the Inquest shall be held in *Great Britain*, and such Order may be addressed and directed to the Churchwardens and Overseers of the Parish or Place in which the Death has happened ; and such Churchwardens and Overseers, or any One of them, is and are hereby required and commanded to pay the Sum of Money mentioned in such Order of the Coroner to the Medical Witness therein mentioned, out of the Funds collected for the Relief of the Poor of the said Place.

No fee for a post-mortem Examination instituted without Order from the Coroner.

IV. Provided nevertheless, and be it further enacted, That no Order of Payment shall be given, or Fee or Remuneration paid, to any Medical Practitioner for the Performance of any post-mortem Examination which may be instituted without the previous Direction of the Coroner.

Inquests on Bodies of Persons dying in public Institutions.

V. Provided also, and be it further enacted, That when any Inquest shall be holden on the Body of any Person who has died in any public Hospital or Infirmary, or in any Building or Place belonging thereto, or used for the Reception of the Patients thereof, or who has died in any County or other Lunatic Asylum, or in any public Infirmary or other public Medical Institution, whether the same be supported by Endowments or by voluntary Subscriptions, then and in such Case nothing herein contained shall be construed to entitle the Medical Officer whose Duty it may have been to attend the deceased Person as a Medical Officer of such Institution as aforesaid to the Fees or Remuneration herein provided.

Penalty on Medical Practitioner for neglecting to attend.

VI. And be it further enacted, That where any Order for the Attendance of any Medical Practitioner as aforesaid shall have been personally served upon such Practitioner, or where any such Order not personally served shall have been received by any Medical Practitioner in sufficient Time for him to have obeyed such Order, or where any such Order has been served at the Residence of any Medical Practitioner, and in every Case where any Medical Practitioner has not obeyed such Order, he shall for such Neglect or Disobedience forfeit the Sum of Five Pounds Sterling upon Complaint thereof made by the Coroner or any Two of the Jury before any Two Justices having Jurisdiction in the Parish or Place where the Inquest under which the Order issued was held, or in the Parish where such Medical Practitioner resides ; and such Two Justices are hereby required, upon such Complaint, to proceed to the Hearing and Adjudication of such Complaint, and, if such Medical Practitioner shall not show to the said Justices a good and sufficient Cause for not having obeyed such Order to enforce the said Penalty by Distress and Sale of the Offender's Goods, as they are empowered to proceed by any Act of Parliament for any other Penalty or Forfeiture.

Act not to extend to Scotland.

VII. And be it enacted, That nothing in this Act contained shall extend to *Scotland.*

SCHEDULE to which this Act refers.
(A.)
Form of Summons.

| Coroner's Inquest at this my Order as Coroner for | upon the body of you are required to appear before | By virtue of |

me and the Jury at   on the   Day of   One thousand eight hundred and   , at   of the Clock, to give Evidence touching the Cause of Death of   [and then add, when the Witness is required to make or assist at a post-mortem Examination, and make or assist in making a post-mortem Examination of the Body, with [or without] an Analysis, as the Case may be], and report thereon at the said Inquest.
                              (Signed)   Coroner.
To   Surgeon [or M.D., as the Case may be].

(B.)

Table of Fees.

1. To every legally qualified Medical Practitioner for attending to give Evidence under the Provisions of this Act at any Coroner's Inquest whereat no post-mortem Examination has been made by such Practitioner, the Fee or Remuneration shall be One Guinea.
2. For the making of a post-mortem Examination of the Body of the Deceased, either with or without an Analysis of the Contents of the Stomach or Intestines, and for attending to give Evidence thereon, the Fee or Remuneration shall be Two Guineas.

(C.)

Coroner's Order for the Payment of Medical Witnesses.

BY virtue of an Act of Parliament passed in   Session of holden in the   intituled   I,   the Coroner of and for   do order you, the Overseers of the Parish [or Township, as the Case may be], to pay to   the Sum of [One Guinea, or Two Guineas, as the Case may be], being the Fee [or Fees] due to him for having attended as a Medical Witness at an Inquest holden before me this   Day of   upon the Body of   about the Age of   who was found dead at   [or other Particulars or Description], and at which said Inquest the Jury returned a Verdict of
                              (Signed)   Coroner.
Witnessed by me   of
To the Overseers, et cœtera.

ANNO PRIMO VICTORIÆ REGINÆ.

CAP. LXVIII.

An Act to provide for Payment of the Expenses of holding Coroners Inquests.   [15th *July* 1837.]

Schedule to be made of Fees payable on holding Inquests. 6 & 7 W. 4. c. 89.

WHEREAS the holding of Coroners Inquests on dead Bodies is attended with divers necessary Expenses, for the Payment whereof no certain Provision is made by Law, and such Expenses have usually been discharged without any lawful Authority for that Purpose out of the Monies levied for the Relief of the Poor ; and it is expedient to make adequate legal Provision for the Payment of such Expenses: Be it therefore enacted by the Queen's most Excellent Majesty, by and with the Advice and Consent of the Lords Spiritual and Temporal, and Commons, in this present Parliament assembled, and by the Authority of the same, That the justices of the Peace for every County, Riding, Division, or District in *England* and *Wales*, in General or Quarter Sessions assembled, shall, at the General or Quarter Sessious of the Peace to be holden next after the passing of this Act, or at some subsequent General or Quarter Sessions, and the Town Council of every Borough having a Coroner shall at the Quarterly Meeting of such Council which shall be holden next after the passing of this Act, or at some subsequent Quarterly meeting thereof, make or cause to be made a Schedule of the several Fees, Allowances, and Disbursements which, on the holding of any Inquest on any dead Body within such County, Riding, Division, District, or Borough, may be lawfully paid and made by the Coroner holding such Inquest (other than the Fees payable to Medical Witnesses under and by virtue of an Act passed in the last Session of Parliament, intituled *An Act to provide for the Attendance and Remuneration of Medical Witnesses at Coroners Inquests*); and it shall be lawful for such Justices in General or Quarter Sessions assembled, and for such Town Council at any such Quarterly Meeting as aforesaid, from Time to Time to alter and vary such Schedule as to such Justices and Town Council respectively may seem fit; and the said Justices and Town Council respectively shall cause a Copy of every such Schedule to be deposited with the Clerk of the Peace of such County, Riding, Division, District, or B·rough, and one other Copy thereof to be delivered to every Coroner acting in and for such County, Riding, Division, District, or Borough as aforesaid; and whenever any Inquest shall be holden on any dead Body the Coroner holding the same shall immediately after the Termination of the Proceedings advance and pay all Expenses reasonably incurred in and about the holding thereof, not exceeding the sums set forth in the said Schedule, and which Sums so advanced and paid shall be repaid to the said Coroner in manner herein-after mentioned: Provided always, that until such Schedule as aforesaid shall have been made the Coroner shall advance and pay, at his Discretion, all reasonable Expenses of holding every Inquest within the Limits of his Jurisdiction, and shall be repaid the amount thereof, in the same Manner as if the Sums so paid had been included in a Schedule duly made according to the Provisions of this Act.

Coroners to pay Medical Witnesses.

II. And be it enacted, That so much of the said Act passed in the last Session of Parliament as directs the Coroner to make out an Order on the Churchwardens and Overseers of the Parish in which any Death shall have happened for Payment of the Remuneration or Fee payable under the Provisions of that Act to any Medical Practitioner, and as directs such Churchwardens and Overseers to pay the same out of the Funds collected for the Relief of the Poor of such Parish, shall be and the same is hereby repealed, and in lieu thereof the Coroner shall, immediately after the Termination of the Proceedings at any Inquest, advance and pay such Remuneration or Fee to every Medical Witness summoned under the Provisions of the said Act, and the Amount thereof shall be repaid to the said Coroner in manner herein-after mentioned.

Coroners of Counties to lay their Accounts before the Sessions, and Coroners of Boroughs to lay them before the Town Council. The Coroner to be repaid out of the County Rates or the Borough Fund.

III. And be it enacted, That every Coroner acting in and for any County, Riding, Division, or District shall, within Four Months after holding any Inquest, cause a full and true Account of all Sums paid by him under the Provisions of this Act, including all Sums paid to any Medical Witness as aforesaid, to be laid before the Justices of the Peace of such County, Riding, Division, or District in General or Quarter Sessions assembled, or at any Adjournment thereof; and every Coroner of any Borough shall, within Four Months after holding any Inquest, cause a full and true Account of all Sums paid by him under the Provisions of this Act, including as aforesaid, to be laid before the Town Council of such Borough; and all such Accounts shall be accompanied by such Vouchers as under the Circumstances may to such Justices or Council respectively seem reasonable; and such Justices or Council respectively may, if they shall think fit, examine the said Coroner on Oath as to such Account, and on being satisfied of the Correctness thereof such Justices or Council respectively shall make an Order on the Treasurer of the said County, Riding, Division, or District, or of the said Borough (as the Case may be), for Payment to the said Coroner not only of the Sum due to him on such Account, but also of a Sum of Six Shillings and Eight-pence for every Inquest holden by him as aforesaid, over and above all other Fees and Allowances to which he is now by

Law entitled; and the Treasurer of any County, Riding, Division, or District on whom any such Order shall be made shall, out of the Monies in his Hands arising from the County Rates, and the Treasurer of any Borough on whom any such Order shall be made shall, out of the Monies in his Hands on account of the Borough Fund, pay to the said Coroner the Sum mentioned in such Order, without any Abatement or Deduction whatever; and every such Treasurer shall, on passing his Accounts, be allowed all Sums which he shall pay in pursuance of any such Order as aforesaid.

Act applicable to London.

IV. And be it enacted, That this Act and the several Provisions herein contained shall extend and be applicable to the City of *London* and the Town and Borough of *Southwark*.

Act may be altered.

V. And be it enacted, That this Act may be altered or repealed by any Act in this present Session of Parliament.

ANNO OCTAVO & NONO VICTORIÆ REGINÆ.

CAP. C.

An Act for the Regulation of the Care and Treatment of Lunatics.
[4th *August* 1845.]

The following Acts repealed, except as they repeal other Acts: 2 & 3 W. 4. c. 107. 3 & 4 W. 4. c. 64. 5 & 6 W. 4. c. 22. 1 & 2 Vict. c. 73. 5 Vict. c. 4. 5 & 6 Vict. c. 87. Proviso that present Visitors and Clerk shall act under this Act till new ones are appointed; and that Licences heretofore granted shall remain in force, unless, &c.

BE it enacted by the Queen's most Excellent Majesty, by and with the Advice and Consent of the Lords Spiritual and Temporal, and Commons, in this present Parliament assembled, and by the Authority of the same, That from and after the passing of this Act an Act passed in the Session of Parliament holden in the Second and Third Years of the Reign of His late Majesty King *William* the Fourth, intituled *An Act for regulating for Three Years, and from thence until the End of the then next Session of Parliament, the Care and Treatment of Insane Persons in* England; and an Act passed in the Session of Parliament holden in the Third and Fourth Years of the Reign of His said late Majesty, intituled *An Act to amend an Act of the Second and Third Year of His present Majesty, for regulating the Care and Treatment of Insane Persons in* England; and an Act passed in the Session of Parliament holden in the Fifth and Sixth Years of the Reign of His said late Majesty, intituled *An Act to continue for Three Years, and from thence to the End of the then next Session of Parliament, Two Acts of the Second and Third Year and the Third and Fourth Year of His present Majesty, relating to the Care and Treatment of Insane Persons in* England; and an Act passed in the Session of Parliament holden in the First and Second Years of the Reign of Her present Majesty, intituled *An Act to continue for Three Years, and from thence to the End of the then next Session of Parliament, Two Acts relating to the Care and Treatment of Insane Persons in* England; and an Act passed in the Session of Parliament holden in the Fifth Year of the Reign of Her said present Majesty, intituled *An Act to continue for Three Years, and from thence to the End of the then next Session of Parliament, Two Acts relating to the Care and Treatment of Insane Persons in* England; and an Act passed in the Session of Parliament holden in the Fifth and Sixth Years of the Reign of Her said present Majesty, intituled *An Act to amend, and continue for Three Years, and from thence to the End of the next Session of Parliament, the Laws relating to Houses licensed by the*

APPENDIX. 171

Metropolitan Commissioners and Justices of the Peace for the Reception of Insane Persons, and for the Inspection of County Asylums and Public Hospitals for the Reception of Insane Persons, shall be and the same are hereby repealed, save and except so far as they or any of them repeal any other Act: Provided always, that until the Appointment for any Jurisdiction of Visitors and their Clerk under the Provisions of this Act the Visitors and Clerk appointed for such Jurisdiction under the said repealed Acts or any of them shall respectively have and perform the Powers, Authorities, and Duties which they would have respectively had or performed if appointed under this Act: Provided also, that all Licences heretofore granted shall remain in force for the Periods for which they were respectively granted, unless revoked as herein-after provided; and that all Orders, Matters, and Things which have been granted, made, done, or directed to be done in pursuance of the said repealed Acts or any of them shall be and remain as good, valid, and effectual to all Intents and Purposes as if the said repealed Acts had not been repealed, except so far as such Orders, Matters, or Things are expressly made void or affected by this Act; and that all Fees, Charges, and Expenses which have become payable under the said repealed Acts or any of them, shall be payable in the same Manner and from the same Funds as would have been applicable thereto in case such Acts had not been repealed.

Commissioners in Lunacy under 5 & 6 Vict. c. 84. to be henceforth called " The Masters in Lunacy."

II. And be it enacted, That the Persons already appointed and hereafter to be appointed under an Act passed in the Session of Parliament holden in the Fifth and Sixth Years of the Reign of Her present Majesty, intituled *An Act to alter and amend the Practice and Course of Proceeding under Commissions in the Nature of Writs De lunatico inquirendo,* whereby the Lord Chancellor is empowered to appoint Two Persons, to be called " The Commissioners in Lunacy," shall henceforth be and be called "The Masters in Lunacy," and shall take the same Rank and Precedence as the Masters in Ordinary of the High Court of Chancery.

Appointment of " The Commissioners in Lunacy."

III. And be it enacted, That the Right Honourable Lord *Ashley,* the Right Honourable Lord *Seymour,* the Right Honourable *Robert Vernon Smith, Robert Gordon* of *Lewiston,* in the County of *Dorset,* Esquire, *Francis Barlow* of *Montagu Square,* Esquire, *Thomas Turner* of *Curzon Street,* Esquire, *Henry Herbert Southey* of *Harley Street,* Esquire, *John Robert Hume* of *Curzon Street,* aforesaid, Esquire, *Bryan Waller Procter* of *Gray's Inn,* Esquire, *James William Mylne* of *Lincoln's Inn,* Esquire, and *John Hancock Hall* of the *Middle Temple,* Esquire, (which said *Thomas Turner, Henry Herbert Southey,* and *John Robert Hume,* and no other of the said Persons, are Physicians, and which said *Bryan Waller Procter, James William Mylne,* and *John Hancock Hall,* and no other of the said Persons, are practising Barristers at Law of Ten Years standing at the Bar and upwards,) and their respective Successors, to be appointed as herein-after provided, shall be Commissioners for the Purposes of this Act, to be called " The Commissioners in Lunacy;" and that such Commissioners for the Time being shall respectively hold their Offices during good Behaviour, and shall not, so long as they shall remain such Commissioners, and receive any Salary under this Act, accept, hold, or carry on any other Office or Situation, or any Profession or Employment, from which any Gain or Profit shall be derived; and that there shall be paid to each of the Six Commissioners for the Time being who shall be Physicians, Surgeons, or Barristers of Five Years standing and upwards, out of the Monies or Funds hereinafter mentioned, over and above their respective travelling and other Expenses whilst employed in visiting any Houses, Hospitals, Asylums, Gaols, Workhouses, or other Places, in pursuance of this Act, the yearly Salary of One thousand and five hundred Pounds, by Four equal quarterly Payments, on the Twenty-ninth Day of *September,* the Twenty-fifth Day of *December,* the Twenty-fifth Day of *March,* and the Twenty-fourth Day of *June* in every Year, the first of each such Payments (or a proportionate Part thereof, to be computed, in the Case of the

Commissioners appointed by this Act, from the passing of the Act, and in case of the Commissioners to be appointed as herein-after provided, from the Time of the respective Appointments of such Commissioners,) to be made to such Commissioners respectively on such of the same Days of Payment as shall first happen after the passing of this Act, or after the Dates of their respective Appointments, as the Case may be.

In case of Death, Disqualification, Refusal, or Inability of Commissioners, others to be appointed.

IV. And be it enacted, That as often as any Commissioner appointed by this Act or to be appointed under this present Provision shall die, or be removed for Ill-behaviour, or be disqualified, or resign, or refuse to act, or become unable by Illness or otherwise to perform the Duties or exercise the Powers of this Act, the Lord Chancellor shall appoint a Person to be a Commissioner in the Room of the Commissioner who shall die or be removed, or be disqualified, or resign, or refuse or become unable to act as aforesaid, but so that every Person so appointed in the Room of a Physician shall be a Physician or Surgeon, and every Person so appointed in the Room of a Barrister of Five Years standing at the Bar and upwards shall be a practising Barrister of not less than Five Years standing at the Bar, and every Person appointed in the Room of any other Commissioner shall be neither a Physician nor a Surgeon, nor a practising Barrister, and until such Appointment it shall be lawful for the continuing Commissioners or Commissioner to act as if there were no such Vacancy.

Provision for retiring Pension to incapacitated Commissioners. 4 & 5 W. 4. c. 24.

V. And be it enacted, That any Superannuation Allowance to be granted to any Commissioner appointed or to be appointed, under this Act shall be granted only in respect of Services performed under this Act, and shall be subject to the Provisions of an Act passed in the Fourth and Fifth Years of His late Majesty King *William* the Fourth. intituled *An Act to alter, amend, and consolidate the Laws for regulating the Pensions, Compensation, and Allowances to be made to Persons in respect of their having held Civil Offices in His Majesty's Service*, so far as such Provisions relate to Officers and Clerks who had entered or might enter the Public Service subsequent to the Fourth Day of *August* One thousand eight hundred and twenty-nine.

Commissioners to take the following Oath.

VI. And be it enacted, That every Person hereby or hereafter appointed a Commissioner under this Act shall, before he acts in the Execution of his Duty as a Commissioner, take an Oath to the following effect ; (that is to say,)

' I *A.B.* do swear, That I will discreetly, impartially, and faithfully execute all the
' Trusts and Powers committed unto me by virtue of an Act of Parliament made
' in the Ninth Year of the Reign of Her Majesty Queen Victoria, intituled [*here
' insert the Title of the Act*] ; and that I will keep secret all such Matters as shall
' come to my Knowledge in the Execution of my Office (except when required to
' divulge the same by legal Authority, or so far as I shall feel myself called upon
' to do so for the better Execution of the Duty imposed on me by the said Act.
So Help me GOD.'

Which Oath it shall be lawful for the Lord Chancellor to administer to every such Commissioner ; and any Three of the Commissioners who shall have previously taken the Oath are hereby authorized to administer such Oath to any other Commissioner.

Commissioners to have a Common Seal.

VII. And be it enacted, That the Commissioners shall cause to be made a Seal of the Commission, and shall cause to be sealed or stamped therewith all Licences, Orders, and Instruments granted or made, or issued, or authorized by the Commissioners, in pursuance of this Act, except such Orders or Instruments as are herein-

APPENDIX. 173

after required or directed to be given or signed and sealed by One Commissioner or Two Commissioners; and all such Licences, Orders, and Instruments, or Copies thereof, purporting to be sealed or stamped with the Seal of the Commission, shall be received as Evidence of the same respectively, and of the same respectively having been granted, made, issued, or authorized by the Commissioners, without any further Proof thereof; and no such Licence, Order, or Instrument, or Copy thereof, shall be valid, or have any Force or Effect, unless the same shall be so sealed or stamped as aforesaid.

Commissioners to elect a Permanent Chairman.

VIII. And be it enacted, That the Commissioners or any Five of them shall as soon as may be after the passing of this Act meet at the usual Office or Place of Business now occupied or used by the Metropolitan Commissioners in Lunacy, or at such other Place as the Lord Chancellor shall direct, and elect one of the same Commissioners (not being a Physician or a Barrister receiving any Salary by virtue of this Act) to be the permanent Chairman of the Commission; and in Case such permanent Chairman, or any other permanent Chairman who shall thereafter be elected in pursuance of this Provision, shall die, or decline or become incapable to act as Chairman, or shall cease to be a Commissioner, then and as often as the same shall happen the Commissioners for the Time being, or any Five of them, at any Meeting to be specially summoned for that Purpose, shall elect another Person to be the permanent Chairman of the Commission in the Place of the Chairman who shall so die or decline or become incapable to act, or cease to be a Commissioner as aforesaid; and in case the permanent Chairman for the Time being shall be absent from any Meeting it shall be lawful for the Majority of the Commissioners present at any such Meeting to elect a Chairman for that Meeting; and in all Cases every Question shall be decided by a Majority of Voters (the Chairman, whether permanent or temporary, having a Vote), and in the event of an Equality of Votes the Chairman for the Time being shall have an additional or casting Vote.

Appointment of Secretary.

IX. And be it enacted, That *Robert Wilfred Skeffington Lutwidge* of *Lincoln's Inn*, Esquire, shall be the Secretary to the Commissioners; and that the said *Robert Wilfred Skeffington Lutwidge*, and every Secretary to be hereafter appointed, shall be removable from his Office by the Lord Chancellor, on the Application of the Commissioners; and that as often as the said *Robert Wilfred Skeffington Lutwidge*, or any Secretary to be appointed under this present provision, shall die, or resign, or be removed from his Office, the Commissioners, with the Approbation of the Lord Chancellor, shall appoint a Person to be Secretary in the Room of the said *Robert Wilfred Skeffington Lutwidge*, or other the Secretary who shall die, or resign, or be removed as aforesaid: and that the Secretary for the Time being shall, in the Performance of all his Duties, and in all respects, be subject to the Inspection, Direction, and Control of the Commissioners; and that there shall be paid to the Secretary for the Time being, out of the Monies and Funds herein-after mentioned, the yearly Salary of Eight hundred Pounds, by Four equal quarterly Payments, on the Twenty-ninth Day of *September*, the Twenty-fifth Day of *December*, and the Twenty-fifth Day of *March*, and the Twenty-fourth Day of *June* in every Year, the first of such Payments (or a proportionate part thereof, to be computed, in the Case of the said *Robert Wilfred Skeffington Lutwidge*, from the passing of this Act, and in case of every other Secretary from the Time of his Appointment,) to be made to the said *Robert Wilfred Skeffington Lutwidge* on such of the same Days of Payment as shall first happen after the passing of this Act, and to every other Secretary for the Time being on such of the same Days of Payment as shall first happen after his Appointment.

Provision for retiring Pension to Secretary. 4 & 5 W. 4. c. 24.

X. And be it enacted, That any Superannuation Allowance to be granted to any Secretary appointed or to be appointed under this Act shall be granted only

174 APPENDIX.

in respect of Services performed under this Act, and shall be subject to the Provisions of an Act passed in the Fourth and Fifth Years of His late Majesty King *William* the Fourth, intituled *An Act to alter, amend, and consolidate the Laws for regulating the Pensions, Compensation, and Allowances to be made to Persons in respect of their having held Civil Offices in His Majesty's Service,* so far as such Provisions relate to Officers and Clerks who had entered or might enter the Public Service subsequent to the Fourth Day of *August* One thousand eight hundred and twenty-nine.

Power for the Commissioners to appoint Two Clerks.

XI. And be it enacted, That it shall be lawful for the Commissioners to appoint, during Pleasure, any Two Persons as Clerks to the Commissioners, and to allow to such Two Clerks any such yearly or other Salaries (not exceeding in the whole the yearly Sum of Two hundred Pounds for such Two Clerks) as the Commissioners shall think proper ; and further, that it shall be lawful for the Commissioners, at any Time hereafter, in case they shall find it expedient so to do, for the due Performance of the Business of the Commission, with the Consent of the Lord High Treasurer, or of the Commissioners of Her Majesty's Treasury, or of any Three or more of them, to appoint One or Two other Clerks (in addition to the Two Clerks firstly herein-before mentioned), and to allow to such One or Two additional Clerk or Clerks any such yearly or other Salaries as the Commissioners shall think fit (not exceeding in the whole the yearly Sum of Two Hundred Pounds); and such Salaries shall be paid out of the Monies or Funds herein-after mentioned.

Secretary and Clerks to take an Oath.

XII. And be it enacted, That every Person appointed to be Secretary or Clerk as aforesaid shall, before he shall act as such Secretary or Clerk, take the following Oath, to be administered by any One of the Commissioners :

' I A.B. do swear, That I will faithfully execute all such Trusts and Duties as shall be committed to my Charge as Secretary to the Commissioners in Lunacy [*or as* ' Clerk to the Commissioners in Lunacy, *as the Case may be*]; and that I will keep ' secret all such Matters as shall come to my Knowledge in the Execution of my ' Office (except when required to divulge the same by legal Authority).
So help me GOD.'

Clerk of the Metropolitan Commissioners to deliver all Documents to the Commissioners under this Act.

XIII. And be it enacted, That immediately after the passing of this Act the Clerk to the Metropolitan Commissioners in Lunacy appointed under the said Act of the Second and Third Years of the Reign of His late Majesty King *William* the Fourth, or under any of the other Acts hereby repealed, shall forthwith deliver up every Book, Paper, and Document, and all Goods, Property, and Effects which may be in his Possession by Virtue of his said Office, or in consequence thereof, or connected with the Business thereof, to the Commissioners in Lunacy hereby appointed ; and every Book, Paper, and Document, and all Goods, Property, and Effects respectively, which shall be so delivered unto or shall hereafter come into the Possession of the Commissioners in Lunacy by virtue of their Office, shall thereupon be vested in and shall be deemed to be the Property of the Commissioners in Lunacy for the Time being.

Jurisdiction within which Commissioners are to grant Licences, and termed their immediate Jurisdiction, defined.

XIV. And be it enacted, That it shall be lawful for the Commissioners (if and when they shall think fit) to grant a Licence to any Person to keep a House for the Reception of Lunatics, or of any Sex or Class of Lunatics, within the Places following ; (that is to say,) the Cities of *London* and *Westminster,* the County of *Middlesex,* the Borough of *Southwark,* and the several Parishes and Places hereinafter mentioned ; (that is to say,) *Brixton, Battersea, Barnes, Saint Mary Mag-*

dalen *Bermondsey, Christ Church Clapham, Saint Giles Camberwell, Dulwich, Saint Paul Deptford, Gravenay, Kew Green, Kennington, Saint Mary Lambeth, Mortlake, Merton, Mitcham, Saint Mary Newington, Norwood, Putney, Peckham, Saint Mary Rotherhithe, Roehampton, Streatham, Stockwell, Tooting, Wimbledon, Wandsworth*, and *Walworth*, in the County of *Surrey ; Blackheath, Charlton, Deptford, Greenwich, Lewisham, Lee, Southend*, and *Woolwich*, in the County of *Kent ;* and *East Ham, Layton, Laytonstone, Low Layton, Plaistow, West Ham*, and *Walthamstow*, in the County of *Essex ;* and also within every other Place (if any) within the Distance of Seven Miles from any Part of the said Cities of *London* or *Westminster*, or of the said Borough of *Southwark ;* all which Cities, County, Borough, Parishes, and Places aforesaid shall be and are hereafter referred to as the immediate Jurisdiction of the Commissioners.

Commissioners to hold quarterly and special Meetings for granting Licences.

XV. And be it enacted, That the Commissioners or some Five of them shall meet at the usual Office or Place of Business which shall for the Time being be occupied or used by the said Commissioners, or at such other Place as the Lord Chancellor may direct, on the First *Wednesday*, in the Months of *February*, *May*, *July*, and *November*, in every Year, in order to receive Applications from Persons requiring Houses to be licensed for the Reception of Lunatics within the immediate Jurisdiction of the Commissioners, and (if they shall think fit) to license the same ; and in case on any such Occasion Five Commissioners shall not be present the Meeting shall take place on the next succeeding *Wednesday*, and so on weekly until Five Commissioners shall be assembled ; and the Commissioners assembled at every such Meeting shall have Power to adjourn such Meeting from Time to Time and to such Place as they shall see fit : Provided always, nevertheless, that it shall be lawful for any Five of the Commissioners at any other Time, at any Meeting duly summoned under the Provisions in that Behalf herein-after contained, to receive Applications from Persons requiring Houses, to be licensed as aforesaid, and, if they shall think fit, to license the same.

Provision for summoning Special Meetings.

XVI. And be it enacted, That when and so often as any Commissioner shall by Writing under his Hand require the Secretary to convene a Meeting of the Commissioners for a Purpose or Purposes specified in such Writing, or for the general Despatch of Business, such Secretary is hereby required to convene such Meeting by summons to the other Commissioners, such of them as shall be then in *England* and shall have an Address known to the Secretary, and to give them, as far as Circumstances will admit, not less than Twenty-four Hours Notice of the Place, Day, and Hour where and on and at which such Meeting is intended to be held, and also to state in the Summons the Purpose or Purposes of such Meeting, as specified by the Commissioner requiring the same to be convened ; and then and in every such Case it shall be lawful for any Three of the Commissioners to assemble themselves to consider, and (if they shall think fit) to execute the Purpose or Purposes of such Meeting : Provided always, nevertheless, that nothing shall be done at any such Meeting, at which less than five Commissioners shall be present which by this Act is required to be done by Five Commissioners : Provided also, that every such Meeting shall, as far as circumstances will admit, be held at the usual Office or Place of Business of the Commissioners.

The Justices of the Peace in General or Quarter Sessions in all other Parts of England to license Houses for the Reception of Lunatics, and to appoint Visitors.

XVII. And be it enacted, That in all Places not being within the immediate Jurisdiction of the Commissioners the Justices for the County or Borough assembled in General or Quarter Sessions shall have the same Authority, within their respective Counties or Boroughs to license Houses for the Reception of Lunatics as the Commissioners within their immediate Jurisdiction ; and that the said Justices shall at the *Michaelmas* General or Quarter Sessions in every Year, appoint Three or more Justices, and also One Physician, Surgeon, or Apothecary, or more, to act as Visitors of every or any House or Houses licensed for the Reception of Lunatics within the said Counties or Boroughs respectively ; and such Visitors shall at their First Meeting take the Oath required by this Act to be taken by the Commissioners, *mutatis mutandis*, such Oath to be administered by a Justice.

For Appointment of a Visitor in the Place of one dying, being unable, disqualified, &c.

XVIII. And be it enacted, That in case at any Time of the Death, Inability, Disqualification, Resignation, or Refusal to act of any Person so appointed a Visitor as aforesaid, it shall be lawful for the Justices of the County or Borough, at any General or Quarter Sessions, to appoint a Visitor in the Room of the Person who shall die, or be unable or be disqualified, or resign or refuse to act as aforesaid.

Lists of Visitors to be published by the Clerk of the Peace, in a Newspaper, and to be sent to the Commissioners. Penalty for Default.

XIX. And be it enacted, That a List of the Names, Places of Abode, Occupations, or Professions of all Visitors appointed as herein-before is directed shall, within Fourteen Days from the Date of their respective Appointments, be published by the Clerk of the Peace of the County or Borough for which they shall be respectively appointed in some Newspaper commonly circulated within the same County or Borough, and shall, within Three Days from the Date of their respective Appointments, be sent by the Clerk of the Peace to the Commissioners ; and every Clerk of the Peace making default in either of the respects aforesaid shall for every such Default forfeit a Sum not exceeding Two Pounds.

Every Visitor, being a Physician, Surgeon, or Apothecary, to be remunerated.

XX. And be it enacted, That every such Visitor as aforesaid, being a Physician, Surgeon, or Apothecary, shall be paid out of the Monies or Funds herein-after mentioned for every Day during which he shall be employed in executing the Duties of this Act such Sum as the Justices of the County or Borough shall in General or Quarter Sessions direct.

Clerk of the Peace, or some other Person, to be appointed to be Clerk to Visitors. His Duties and Remuneration.

XXI. And be it enacted, That the Clerk of the Peace, or some other Person to be appointed by the Justices for the County or Borough in General or Quarter Sessions, shall act as Clerk to the Visitors so appointed as aforesaid, and such Clerk shall summon the Visitors to meet at such Time and Place, for the Purpose of executing the Duties of this Act, as the said Justices in General or Quarter Sessions shall appoint ; and every such Appointment, Summons, and Meeting shall be made and held as privately as may be, and in such Manner that no Proprietor, Superintendent, or Person interested in or employed about or connected with any House to be visited shall have notice of such intended Visitation ; and such Clerk to the Visitors shall, at their First Meeting, take the Oath required by this Act to be taken by the Secretary of the Commissioners, *mutatis mutandis*, such Oath to be administered by One of the Visitors, being a Justice : and the Name, Place of Abode, Occupation, and Profession of the Clerk to the Visitors (whether the same shall be the Clerk of the Peace or any other Person) shall within Fourteen Days after the Appointment be published by the Clerk of the Peace for the County or Borough in some Newspaper commonly circulated therein, and within Three Days from the Date of the Appointment be communicated by the said Clerk of the Peace to the Commissioners ; and every Clerk of the Peace making default in either of the respects aforesaid shall for every such Default forfeit a Sum not exceeding Two Pounds ; and every such Clerk to the Visitors shall be allowed such Salary or Remuneration for his Services (to be paid out of the Monies or Funds herein-after mentioned) as the Justices for the County or Borough shall in General or Quarter Sessions direct.

Provision for Assistants to the Clerk of the Visitors. Oath of Assistant.

XXII. And be it enacted, That if the Clerk of any Visitors shall at any Time desire to employ an Assistant in the Execution of the Duties of his Office, such Clerk shall certify such Desire and the Name of such Assistant to one of the Visitors, being a Justice ; and if such Visitor shall approve thereof he shall administer the following Oath to such Assistant :

'I *A.B.* do solemnly swear, That I will faithfully keep secret all such Matters and
' Things as shall come to my Knowledge in consequence of my Employment as
' Assistant to the Clerk of the Visitors appointed for the County [or Borough] of
' by virtue of an Act of Parliament passed in the Ninth Year of the
' Reign of Her Majesty Queen Victoria, intituled [*here insert the Title of the Act,*]
' unless required to divulge the same by legal Authority.
So help me GOD.'
And such Clerk may thereafter, at his own Cost, employ such Assistant.

Persons interested in any licensed House, or being Medical Attendant on any Patient therein, disqualified to act as Commissioner, Visitor, Secretary, Clerk, or Assistant. Disqualified Persons acting a Misdemeanor. Physicians, &c. contravening, Penalty 10l.

XXIII. And be it enacted, That no Person shall be or act as a Commissioner, or Visitor, or Secretary, or Clerk to the Commissioners, or Clerk or Assistant Clerk to any Visitors, or act in granting any Licence, who shall then be, or shall within One Year then next preceding have been, directly or indirectly interested in any House licensed for the Reception of Lunatics, or the Profits of such Reception ; and no Physician or Surgeon (being a Commissioner), and no Physician, Surgeon, or Apothecary, (being a Visitor,) shall sign any Certificate for the Admission of any Patient into any licensed House or Hospital, or shall professionally attend upon any Patient in any licensed House or Hospital, unless he be directed to visit such Patient by the Person upon whose Order such Patient has been received into such licensed House or Hospital, or by the Lord Chancellor, or Her Majesty's Principal Secretary of State for the Time being for the Home Department, or by a Committee appointed by the Lord Chancellor ; and if any such Commissioner, or Visitor, or Secretary, or Clerk to the Commissioners, or Clerk or Assistant Clerk to any Visitors, shall after his Appointment be or become so interested in any House licensed for the Reception of Lunatics or the Profits of such Reception, such Commissioner, Visitor, Secretary, or Clerk, or Assistant Clerk, as the Case may be, shall immediately thereupon be disqualified from acting and shall cease to act in such Capacity ; and if any Person, being disqualified as aforesaid, shall take the Office of Commissioner, Visitor, Secretary, Clerk, or Assistant Clerk, or, being a Commissioner, Visitor, Secretary, Clerk, or Assistant Clerk, shall become disqualified as aforesaid, and shall afterwards continue to act in such Capacity, such Person shall be guilty of a Misdemeanor ; and if any Physician or Surgeon (being a Commissioner), or any Physician, Surgeon, or Apothecary, (being a Visitor,) shall sign any Certificate for the Admission of any Patient into any licensed House or Hospital, or shall professionally attend any Patient in any licensed House or Hospital (except as aforesaid), such Physician, Surgeon, or Apothecary (as the Case may be) shall for each Offence against this Provision forfeit the Sum of Ten Pounds.

Fourteen Days previous Notice of intended Application for and Plan of licensed House to be given to the Commissioners or Clerk of the Peace.

XXIV. And be it enacted, That every Person who shall desire to have a House licensed for the Reception of Lunatics shall give a Notice, if such House be situate within the immediate Jurisdiction of the Commissioners, to the Commissioners, and if elsewhere to the Clerk of the Peace for the County or Borough in which such House is situate, Fourteen clear Days at the least prior to some

N

quarterly or other Meeting of the Commissioners, or to some General or Quarter Sessions for such County or Borough, as the Case may be; and such Notice shall contain the true Christian and Surname, Place of Abode, and Occupation of the Person to whom the Licence is desired to be granted, and a true and full Description of his Estate or Interest in such House ; and in case the Person to whom the Licence is desired to be granted does not propose to reside himself in the licensed House, the true Christian and Surname and Occupation of the Superintendent who is to reside therein ; and such Notice, when given for any House which shall not have been previously licensed, shall be accompanied by a Plan of such House, to be drawn upon a Scale of not less than One Eighth of an Inch to a Foot, with a Description of the Situation thereof, and the Length, Breadth, and Height of and a Reference by a Figure or Letter to every Room and Apartment therein, and a Statement of the Quantity of Land, not covered by any Building, annexed to such House, and appropriated to the exclusive Use, Exercise, and Recreation of the Patients proposed to be received therein, and also a Statement of the Number of Patients proposed to be received into such House, and whether the Licence so applied for is for the Reception of Male or Female Patients, or of both, and if for the Reception of both, of the Number of each Sex proposed to be received into such House, and of the Means by which the one Sex may be kept distinct and apart from the other ; and such Notice, Plan, and Statement, when sent to the Clerk of the Peace, shall be laid by him before the Justices of the County or Borough at such Time as they shall take into their Consideration the Application for such Licence : Provided always, that it shall be lawful for any Person to whom a Licence shall be granted to remove the Superintendent named in the Notice, and at any Time or Times to appoint another Superintendent, upon giving a Notice containing the true Christian and Surname and Occupation of the new Superintendent to the Commissioners or the Visitors of the House, as the Case may require: Provided always, that all Plans heretofore delivered shall be deemed sufficient for the Purposes of this Act, if the Commissioners or Justices, as the Case may be, shall so think fit.

No Licence to include more than One House ; but detached Buildings in certain Cases, to be considered Part of the House.

XXV. And be it enacted, That no One Licence shall include or extend to more than One House ; but if there be any Place or Building detached from a House to be licensed, but not separated therefrom by Ground belonging to any other Person, and if such Place or Building be specified, delineated, and described in the Notice, Plan, and Statement herein-before required to be given, in the same Manner in all Particulars as if the same had formed Part of such House, then such detached Place or Building may be included in the Licence for the House, if the Commissioners or Justices, as the Case may be, shall think fit, and if so included shall be considered Part of such House for the Purposes of this Act: Provided always, that no Person hereafter receiving a Licence for the first Time shall receive any Licence for the Reception of Lunatics in any Lunatic Asylum who shall not reside on the Premises for which he is licensed.

Notice of all Additions and Alterations to be given to the Commissioners or Clerk of the Peace.

XXVI. And be it enacted, That no Addition or Alteration shall be made to, in, or about, any licensed House, or the Appurtenances, unless previous Notice in Writing of such proposed Addition or Alteration, accompanied with a Plan of such Addition or Alteration, to be drawn upon the Scale aforesaid, and to be accompanied by such Description as aforesaid, shall have been given by the Person to whom the Licence shall have been granted to the Commissioners or to the Clerk of the Peace, as the Case may be, and the Consent in Writing of the Commissioners, or of Two of the Visitors, as the Case may be, shall have been previously given.

Untrue Statement a Misdemeanor.

XXVII. And be it enacted, That if any Person shall wilfully give an untrue or incorrect Notice, Plan, Statement, or Description of any of the Things herein-before required to be included in any Notice, Plan, or Statement, he shall be guilty of a Misdemeanor.

A Copy of every Licence granted by Justices to be sent to the Commissioners.

XXVIII. And be it enacted, That in every Case in which a Licence for the Reception of Lunatics shall after the passing of this Act be granted by any Justices, the Clerk of the Peace for the County or Borough shall, within Fourteen Days after such Licence shall have been granted, send a Copy thereof to the Commissioners; and any Clerk of the Peace omitting to send such Copy within such Time shall for every such Omission forfeit a Sum not exceeding Two Pounds.

Every Person applying for the Renewal of a Licence to furnish a Statement of the Number and Class of Patients then detained.

XXIX. And be it enacted, That in every Case in which any Person shall apply for the Renewal of a Licence already granted or hereafter to be granted, such Person, if applying to the Commissioners, shall with such Application transmit to the Commissioners, and if applying to any Justices shall with such Application transmit to the Clerk of the Peace for the County or Borough, and also at the same Time to the Commissioners, a Statement signed by the Person so applying, containing the Names and Number of the Patients of each or either Sex then detained in such House, and distinguishing whether such Patients respectively are private or pauper Patients; and any Person who shall hereafter obtain the Renewal of a Licence without making such Return or Returns shall for every such Offence forfeit the Sum of Ten Pounds; and any Person who shall make any such Return untruly shall be guilty of a Misdemeanor.

Licences to be made out in a given Form, &c., and to be for not more than Thirteen Months.

XXX. And be it enacted, That every Licence shall, as nearly as conveniently may be, be according to the Form in the Schedule (A.) annexed to this Act, and shall be stamped with a Ten Shilling Stamp, and shall be under the Seal of the Commissioners, if granted by them, and if by any Justices under the Hands and Seals of Three or more such Justices in General or Quarter Sessions assembled, and shall be granted for such Period, not exceeding Thirteen Calendar Months, as the Commissioners or Justices, as the Case may be, shall think fit.

No Licence, &c. in any Borough without Consent of Recorder.

XXXI. And be it enacted, That no Licence shall be granted or Visitor or Clerk appointed by the Justices for any Borough without the Consent in Writing of the Recorder of such Borough to such Grant or Appointment.

Charge for Licences to be granted in pursuance of this Act. Power to reduce the Charge for the Licence in certain Cases.

XXXII. And be it enacted, That for every Licence to be hereafter granted there shall be paid to the Secretary of the Commissioners, or to the Clerk of the Peace, according as the Licence shall be granted by the Commissioners or Justices (exclusive of the Sum to be paid for the Stamp), the Sum of Ten Shillings, and no more, for every Patient not being a Pauper, and the Sum of Two Shillings and Sixpence and no more for every Patient being a Pauper, proposed to be received into such House, and if the total Amount of such Sums of Ten Shillings and Two Shillings and Sixpence shall not amount to Fifteen Pounds, then so much more as shall make up the Sum of Fifteen Pounds; and no such Licence shall be delivered until the Sum payable for the same shall be paid: Provided always, that if the Period for which a Licence shall be granted be less than Thirteen Calendar Months, it shall be lawful for the Commissioners or the Justices, as the Case may be, to reduce the Payment to be made on such Licence to any Sum not less than Five Pounds.

N 2

Application of Monies received for Licences by the Secretary of the Commissioners.

XXXIII. And be it enacted, That all Monies received for Licences granted by the Commissioners, and for Searches made in pursuance of the Provision for that Purpose herein-after contained, shall be retained by the Secretary of the Commissioners, and be applied by him in or towards the Payment of the Salaries and travelling and other Expenses of the Commissioners and of their Secretary and Clerks, and in or towards the Payment or Discharge of all or any Costs, Charges, and Expenses incurred by or under the Authority of the Commissioners in the Execution of or under or by virtue of this Act.

Secretary of the Commissioners to make out an annual Account, to be laid before the Lords Commissioners of the Treasury, of all Receipts and Payments by him under this Act.

XXXIV. And be it enacted, That the Secretary of the Commissioners shall make out an Account of all Monies received and paid by him as afore-aid, and of all Monies otherwise received and paid by him, and of all Charges and Expenses incurred under or by virtue of or in the Execution of this Act; and such Account shall be made up to the First Day of *August* in each Year, and shall be signed by Five at least of the Commissioners; and such Account shall specify the several Heads of Charge and Expenditure, and shall be transmitted to the Lord High Treasurer, or to the Commissioners of Her Majesty's Treasury, who shall thereupon audit such Account, and, if he or they shall deem it expedient, direct the Balance (if any) remaining in the Hands of the said Secretary to be paid into the Exchequer to the Account of the Consolidated Fund; and such Accounts shall be laid before Parliament on or before the Twenty-fifth Day of *March* in each Year, if Parliament be then sitting, or if Parliament be not then sitting then within One Month after the then next Sitting of Parliament.

Balance of Payments over Receipts may be paid out of the Consolidated Fund.

XXXV. And be it enacted, That it shall be lawful for the Lord High Treasurer, or the Commissioners of Her Majesty's Treasury, or any Three or more of them, and they are hereby directed and empowered, from Time to Time (on an Application to them; agreed to at some quarterly or other Meeting of the Commissioners, attended by Five at least of the Commissioners, and certified under their Hands,) to cause to be issued and paid out of the Consolidated Fund to the Secretary of the Commissioners such a Sum of Money as the Commissioners shall in such Application have certified to be requisite to pay and discharge so much of the Salaries, Costs, Charges, and Expenses herein-before directed to be paid out of the Monies received by the said Secretary for Licences and otherwise as aforesaid as such Monies shall be inadequate to pay, and the said Secretary shall thereupon apply such Money in or towards the Payment or Discharge of such Salaries, Costs, Charges, and Expenses respectively; and that it shall be lawful for the Lord High Treasurer, or the Commissioners of Her Majesty's Treasury, or any Three or more of them, from Time to Time to advance by way of Imprest to the said Secretary such Sum or Sums of Money as to such Lord High Treasurer or Commissioners of Her Majesty's Treasury may appear requisite and reasonable, for or towards the Payment or Discharge of all or any such Salaries, Costs, Charges, or Expenses as aforesaid, such Sum or Sums to be accounted for by the said Secretary in his then next Account.

Application of Monies received for Licences by Clerks of the Peace.

XXXVI. And be it enacted, That all Monies to be received for Licences granted by any Justices shall be applied by the Clerk of the Peace for the County or Borough in or towards the Payment of the Salary or Remuneration of the Clerk to the Visitors for such County or Borough, and in or towards the Remuneration of such of the same Visitors as are herein-before directed to be remunerated, and in or towards the Payment or Discharge of all Costs, Charges, and Expenses incurred by or under the Authority of the same Justice or Visitors in the Execution of or under or by virtue of this Act.

APPENDIX.

Clerks of the Peace to make out annual Accounts, to be laid before the Justices in Sessions, of all Receipts and Payments made under this Act.

XXXVII. And be it enacted, That the Clerk of the Peace for every County or Borough shall keep an Account of all Monies received and paid by him as aforesaid, and of all Monies otherwise received or paid by him under or by virtue of or in the execution of this Act; and such Account shall respectively be made up to the First Day of *August* in each Year, and shall be signed by Two at least of the Visitors for the County or Borough; and every such Account shall be laid by the Clerk of the Peace before the Justices at the *Michaelmas* General or Quarter Sessions, who shall thereupon direct the Balance (if any) remaining in the Hands of the Clerk of the Peace to be paid into the Hands of the Treasurer for such County or Borough, in aid and as Part of the County or Borough Rate.

Balance of Payments over Receipts may be paid out of the Fund of the County or Borough.

XXXVIII. And be it enacted, That it shall be lawful for the Justices for any County or Borough, in General or Quarter Sessions assembled, if they shall think fit, to order to be paid to the Clerk of the Peace of such County or Borough, out of the Rates or Funds thereof, such Sum or Sums of Money as they shall on Examination deem to be necessary to pay and discharge so much of the Salary, Remuneration, Costs, Charges, and Expenses herein-before directed to be paid out of the Monies received by such Clerk of the Peace for Licences and otherwise as aforesaid as such Monies shall be inadequate to pay; and also that it shall be lawful for the Justices in General or Quarter Sessions assembled, if they shall think fit, from Time to Time to order to be advanced out of the Rates or Funds of such County or Borough, to the Clerk of the Peace, such Sum or Sums of Money as to such Justices may appear requisite and reasonable for or towards the Payment or Discharge of any such Salary, Remuneration, Costs, Charges, or Expenses as last aforesaid; and every such Sum of Money as aforesaid shall be paid and advanced out of the Rates or Funds of such County or Borough by the Treasurer thereof, and shall be allowed in his Accounts, on the Authority of the aforesaid Order by the Justices for the Payment or Advance thereof.

Provision in case of the Incapacity or Death of the Person licensed.

XXXIX. And be it enacted, That if any Person to whom a Licence shall have been granted under this Act or under any of the Acts herein-before repealed shall by Sickness or other sufficient Reason become incapable of keeping the licensed House, or shall die before the Expiration of the Licence, it shall be lawful for the Commissioners or for any Three Justices for the County or Borough, as the Case may be, if they shall respectively think fit, by Writing endorsed on such Licence under the Seal of the Commissioners or under the Hands of such Three Justices, to transfer the said Licence, with all the Privileges and Obligations annexed thereto, for the Term then unexpired, to such Person as shall at the Time of such Incapacity or Death be the Superintendent of such House, or have the Care of the Patients therein, or to such other Person as the Commissioners or such Justices respectively shall approve, and in the meantime such Licence shall remain in force and have the same Effect as if granted to the Superintendent of the House; and in case a Licence has been or shall be granted to Two or more Persons, and before the Expiration thereof any or either of such Persons shall die, leaving the other or others surviving, such Licence shall remain in force and have the same Effect as if granted to such Survivors or Survivor.

In case of a licensed House being taken for public Purposes, or accidentally rendered unfit, or of the Keeper wishing to transfer his Patients to a new House.

XL. And be it enacted, That if any licensed House shall be pulled down or occupied under the Provisions of any Act of Parliament, or shall by Fire, Tempest, or other Accident be rendered unfit for the Accommodation of Lunatics, or if the Person keeping such House shall desire to transfer the Patients to another House, it shall be lawful for the Commissioners (if the new House shall be within their immediate Jurisdiction), at any quarterly or other Meeting, or for any Two or more of the Visiting Justices for the County or Borough within which the new House is situate, as the Case may be, upon the Payment to the Secretary of the Commissioners or the Clerk of the Peace, as the Case may be, of not less than One Pound for the Licence (exclusive of the Sum to be paid for the Stamp), to grant to the Person whose House has been so pulled down, occupied, or so rendered unfit, or who shall desire to transfer his Patients as aforesaid, a Licence to keep such other House for the Reception of Lunatics, for such Time as the Commissioners or the said Justices, as the Case may be, shall think fit : Provided always, that the same Notice of such intended Change of House, and the same Plans and Statements and Descriptions of and as to such intended new House, shall be given as are required when Application is first made for a Licence for any House, and shall be accompanied by a Statement in Writing of the Cause of such Change of House ; and that, except in Cases in which the Change of House is occasioned by Fire or Tempest, Seven clear Days previous Notice of the intended Removal shall be sent by the Person to whom the Licence for keeping the original House shall have been granted, to the Person who signed the Order for the Reception of each Patient, not being a Pauper, or the Person by whom the last Payment on account of such Patient shall have been made, and to the Relieving Officer or Overseer of the Union or Parish to which each Patient being a Pauper is chargeable, or the Person by whom the last Payment on account of such Patient shall have been made.

Power of Revocation of Licences granted by Justices.

XLI. And be it enacted, That if a Majority of the Justices of any County or Borough in General or Quarter Sessions assembled, shall recommend to the Lord Chancellor, that any Licence granted by the Justices for such County or Borough, either before or after the passing of this Act, shall be revoked, it shall be lawful for the Lord Chancellor to revoke the same by an Instrument under his Hand and Seal, such Revocation to take effect at a Period to be named in such Instrument, not exceeding Two Calendar Months from the Time a Copy or Notice thereof shall have been published in the "*London Gazette ;*" and a Copy or Notice of such Instrument of Revocation shall be published in the "*London Gazette,*" and shall before such Publication be transmitted to the Person to whom such Licence shall have been granted, or to the resident Superintendent of the licensed House, or be left at the licensed House : Provided always, that in case of any such Revocation being recommended to the Lord Chancellor, Notice thereof in Writing shall, Seven clear Days previously to the Transmission of such Recommendation to the Lord Chancellor, be given to the Person the Revocation of whose Licence shall be recommended, or to the resident Superintendent of the licensed House, or shall be left at the licensed House.

Power of Revocation and of Prohibition of Renewal of Licences granted by the Commissioners or by Justices.

XLII. And be it enacted, That if the Commissioners shall recommend to the Lord Chancellor that any Licence granted either by the Commissioners or by any Justices, either before or after the passing of this Act shall be revoked or shall not be renewed, it shall be lawful for the Lord Chancellor, by an Instrument under his Hand and Seal to revoke or prohibit the Renewal of such Licence ; and in the Case of a Revocation the same shall take effect at a Period to be named in such Instrument, not exceeding Two Calendar Months, from the Time a Copy or Notice thereof shall have been published in the "*London Gazette ;*" and a Copy or Notice of such Instrument of Revocation shall be published in the "*London Gazette,*" and shall before such Publication be transmitted to the Person to whom such Licence shall have been granted, or to the resident Superintendent of the licensed House, or shall be left at the licensed House : Provided always, that in case of any such Revocation or Prohibition to renew being recommended to the Lord Chancellor, Notice thereof in Writing shall, Seven clear Days previously to the Transmission of such Recommendation to the Lord Chancellor, be given to the Person the Revocation or Prohibition of Renewal of whose Licence shall be recommended, or to the resident Superintendent of the licensed House, or shall be left at the licensed House.

Hospitals receiving Lunatics to have their Regulations printed and a resident Medical Attendant, and to be registered.

XLIII. And be it enacted, That the Regulations as to Lunatics of every Hospital in which Lunatics are or shall be received shall be printed, and complete Copies thereof shall be sent to the Commissioners, and also kept hung up in the Visitors Room of such Hospital; and that every such Hospital shall have a Physician, Surgeon, or Apothecary resident therein, as the Superintendent and Medical Attendant thereof; and such Superintendent shall immediately after the passing of this Act (or immediately after the Establishment of such Hospital, as the Case may be,) apply to the Commissioners to have such Hospital registered, and thereupon such Hospital shall be registered in a Book to be kept for that Purpose by the Commissioners; and in case the Superintendent of any such Hospital shall at any time omit to have Copies of such Regulations sent or hung up as aforesaid, or to apply to have such Hospital registered as aforesaid, he shall for every such Omission forfeit a Sum not exceeding Twenty Pounds.

No House to be kept for the Reception of Two or More Lunatics without a Licence.

XLIV. And be it enacted, That after the passing of this Act it shall not be lawful for any Person to receive Two or more Lunatics into any House, unless such House shall be an Asylum or an Hospital registered under this Act, or a House for the Time being duly licensed under this Act, or One of the Acts herein-before repealed; and any Person who shall receive Two or more Lunatics into any House other than a House for the Time being duly licensed as aforesaid, or an Asylum or an Hospital duly registered under this Act, shall be guilty of a Misdemeanor.

No Person (not a Pauper) to be received without an Order and Medical Certificate.

XLV. And be it enacted, That no Person (not a Pauper), whether being or represented to be a Lunatic, or only a Boarder or Lodger, in respect of whom any Money shall be received or agreed to be received for Board, Lodging, or any other Accommodation, shall be received into or detained in any licensed House, and no Person (not a Pauper) shall be received into or detained as a Lunatic in any Hospital, without an Order under the Hand of some Person according to the Form and stating the Particulars required in Schedule (B.) annexed to this Act, nor without the Medical Certificates, according to the Form in Schedule (C.) annexed to this Act, of Two Physicians, Surgeons, or Apothecaries who shall not be in Partnership, and each of whom shall separately from the other have personally examined the Person to whom it relates, not more than Seven clear Days previously to the Reception of such Person into such House or Hospital, and shall have signed and dated the same on the Day on which such Person shall have been so examined; and every Person who shall receive or detain any such Person as aforesaid in any such House or Hospital as aforesaid, without such Order and Medical Certificates as aforesaid, and any Physician, Surgeon, or Apothecary who shall knowingly sign any such Medical Certificate as aforesaid which shall untruly state any of the Particulars required by this Act, shall be guilty of a Misdemeanor.

Medical Practitioner signing such Certificate to specify Facts upon which Opinion formed.

XLVI. Provided always, and be it enacted, That every Physician, Surgeon, or Apothecary signing such Certificate shall specify therein any Fact or Facts (whether arising from his own Observation or from the Information of any other Person) upon which he has formed his Opinion that the Person to whom such Certificate relates is a Lunatic or an Insane Person, or an Idiot, or a Person of unsound Mind.

Proviso that in certain Cases a Person may be received on a Certificate signed by One Medical Practitioner only.

XLVII. Provided always, nevertheless, and be it enacted, That any Person (not a Pauper) may, under special Circumstances, be received into any such House or Hospital as aforesaid, upon such Order as aforesaid, with the Certificate of One Physician, Surgeon, or Apothecary alone, provided that such Order state the special Circumstances which have prevented the Person from being examined by Two Medical Practitioners; but in every such Case another such Certificate shall be signed by some other Physician, Surgeon, or Apothecary not being connected with any such House or Hospital, who shall have especially examined such Person within Three Days after his Reception into such House or Hospital; and every Person who, having received any Person into any House or Hospital as aforesaid upon the Certificate of One Medical Practitioner alone as aforesaid, shall keep or permit such Person to remain in such House or Hospital beyond the said Period of Three Days without such further Certificate as aforesaid, shall be guilty of a Misdemeanor.

No Pauper to be received into any House or Hospital for Lunatics without a certain Order and Certificate.

XLVIII. And be it enacted, That no Pauper shall be received into or detained in any licensed House, or any Hospital, without an Order and Statement according to the Form and stating the Particulars required in Schedule (D.) annexed to this Act, under the Hands of One Justice or an Officiating Clergyman, with the Relieving Officer or One of the Overseers of the Union or Parish from which such Pauper shall be sent, (which said Justice or which said Clergyman and Relieving Officer or Overseer, as the Case may be, shall have personally examined such Pauper previously to signing such Order,) nor without a Medical Certificate according to the Form in the said Schedule (D.) annexed to this Act, and dated not more than Seven clear Days previously to the Reception of such Pauper into such House or Hospital; and every such Certificate shall be signed by a Physician, Surgeon, or Apothecary (not being the Medical Officer of such Parish or Union) on the Day whereon he shall examine such Pauper; and every Person who shall receive any Pauper into any such House or Hospital as aforesaid without such Order and Medical Certificate as last aforesaid shall be guilty of a Misdemeanor.

No Medical Practitioner who is interested in or attends a licensed House or Hospital to sign a Certificate for Admission of a Patient into such Place.

XLIX. And be it enacted, That no Physician, Surgeon, or Apothecary who, or whose Father, Brother, Son, or Partner, is wholly or partly the Proprietor of or a regular professional Attendant in a licensed House or an Hospital, shall sign any Certificate for the Reception of a Patient into such House or Hospital; and no Physician, Surgeon, or Apothecary who, or whose Father, Brother, Son, or Partner, shall sign the Order herein-before required for the Reception of a Patient, shall sign any Certificate for the Reception of the same Patient; and any Physician, Surgeon, or Apothecary who shall sign any Certificate contrary to any of the Provisions herein-before contained, or without having complied with all the Provisions hereby required in the Case of the Patient to whom the same shall relate, or who shall in such Certificate describe his Medical Qualification untruly, or shall untruly state anything therein, shall be guilty of a Misdemeanor.

Every Person receiving a Person as a Lunatic into any House or Hospital to make an Entry thereof in a certain Form.

L. And be it enacted, That every Proprietor or Superintendent who shall receive any Patient into any licensed House or any Hospital shall, within Two Days after the Reception of such Patient, make an Entry with respect to such Patient in a Book to be kept for that Purpose to be called "The Book of Admissions," according to the Form and containing the Particulars required in Schedule (E.) annexed to this Act, so far as he can ascertain the same, except as to the Form of the mental Disorder, and except also as to the Discharge or Death of the Patient, which shall be made when the same shall happen; and every Person who shall so receive any such Patient, and shall not within Two Days thereafter make such Entry as aforesaid (except as aforesaid), shall forfeit a Sum not exceeding Two Pounds; and every Person who shall knowingly and willingly in any such Entry untruly set forth any of the Particulars shall be guilty of a Misdemeanor.

Form of Patient's Disorder to be entered in " The Book of Admissions" by the Medical Attendant.

LI. And be it enacted, That the Form of the mental Disorder of every Patient received into any licensed House or any Hospital shall within Seven Days after his Reception be entered in the said Book of Admissions by the Medical Attendant of such House or Hospital; and every such Medical Attendant who shall omit to make any such Entry within the Time aforesaid shall for every such Offence forfeit a Sum not exceeding Two Pounds.

Every Person receiving a Patient into any House or Hospital to transmit a Notice thereof to the Commissioners, and if within the Jurisdiction of any Visitors, then also to the Clerk of such Visitors.

LII. And be it enacted, That the Proprietor or resident Superintendent of every licensed House (whether licensed by the Commissioners or by any Justices), and the Superintendent of every Hospital, shall, after Two clear Days, and before the Expiration of Seven clear Days from the Day on which any Patient shall have been received into such House or Hospital, transmit a Copy of the Order and Medical Certificates or Certificate on which such Person shall have been received, and also a Notice and Statement according to the Form in Schedule (F.) annexed to this Act, to the Commissioners; and the Proprietor or resident Superintendent of every House licensed within the Jurisdiction of any Visitors shall also within the same Period transmit another Copy of such Order and Certificates or Certificate, and a Duplicate of such Notice and Statement, to the Clerk of the Visitors; and every Proprietor or Superintendent of any such House or Hospital who shall neglect to transmit such Copy, Notice, or Statement to the Commissioners, or (where the same is required) to the Clerk of the Visitors, shall be guilty of a Misdemeanor.

Notices to be given in case of the Escape of any Patient, and of his being brought back.

LIII. And be it enacted, That whenever any Patient shall escape from any licensed House or any registered Hospital the Proprietor or Superintendent of such House or Hospital shall within Two clear Days next after such Escape transmit a written Notice thereof to the Commissioners, and if such House be within the Jurisdiction of any Visitors then also to the Clerk of such Visitors; and such Notice shall state the Christian and Surname of the Patient who has so escaped, and his then State of Mind, and also the Circumstances connected with such Escape; and if such Patient shall be brought back to such House or Hospital such Proprietor or resident Superintendent shall, within Two clear Days next after such Person shall be so brought back, transmit a written Notice thereof to the Commissioners, and also, if such House be within the Jurisdiction of any Visitors, to the Clerk of such Visitors; and such Notice shall state when such Person was so brought back, and the Circumstances connected therewith, and whether with or without a fresh Order and Certificates or Certificate; and every Proprietor or resident Superintendent omitting to transmit such Notice, whether of Escape or of Return, shall for every such Omission forfeit a Sum not exceeding Ten Pounds.

Entry to be made, and Notice given, in case of the Death, Discharge, or Removal of any Patient.

LIV. And be it enacted, That whenever any Patient shall be removed or discharged from any licensed House or any Hospital, or shall die therein, the Proprietor or Superintendent of such House or Hospital shall, within Two clear Days next after such Removal, Discharge, or Death, make an Entry thereof in a Book to be kept for that Purpose according to the Form and stating the Particulars in Schedule (G. 1.) annexed to this Act, and shall also within the same Two Days transmit a written Notice thereof, and also of the Cause of his Death, to the Commissioners, and also, if such House shall be within the Jurisdiction of any Visitors, to the Clerk of such Visitors, according to the Form and containing the Particulars in Schedule (G. 2.) annexed to this Act; and every Proprietor or Superintendent of any such House or Hospital who shall neglect to make such Entry or transmit such Notice or Notices, or shall therein set forth anything untruly, shall be guilty of a Misdemeanor.

In case of the Death of a Patient, a Statement of the Cause of Death to be transmitted to the Commissioners, and, if within the Jurisdiction of any Visitors, to the Clerk of the Visitors also.

LV. And be it enacted, That in case of the Death of any Patient in any licensed House or any Hospital, a Statement of the Cause of the Death of such Patient, with the Name of any Person present at the Death, shall be drawn up and signed by the Medical Attendant of such House or Hospital, and a Copy thereof, duly certified by the Proprietor or Superintendent of such House or Hospital, shall by him be transmitted to the Commissioners, and also to the Person signing the Order for such Patient's Confinement, and to the Registrar of Deaths for the District, and if such House be within the Jurisdiction of any Visitors, then also to the Clerk of such Visitors, within Forty-eight Hours after the Death of such Patient; and every Medical Attendant, Proprietor, or Superintendent who shall neglect or omit to draw up, sign, certify, or transmit such Statement as aforesaid shall for every such Neglect or Omission forfeit and pay a Sum not exceeding Fifty Pounds.

Abuse or Ill-treatment or (in certain Cases) Neglect of a Patient to be a Misdemeanor.

LVI. And be it enacted, That if any Superintendent, Officer, Nurse, Attendant, Servant, or other Person employed in any licensed House or registered Hospital shall in any way abuse or ill-treat any Patient confined therein, or shall wilfully neglect any such Patient, he shall be deemed guilty of a Misdemeanor ; and that in the event of the Release of any Person from Confinement in any Asylum or private House who shall consider himself to have been unjustly confined, a Copy of the Certificates and Order upon which he has been confined shall at his Request be furnished to him or to his Attorney by the Clerk to the Commissioners, without any Fee or Reward for the same ; and it shall be lawful for the Home Secretary, on the Report of the Commissioners, or Visitors of any Asylums, to direct Her Majesty's Attorney General to prosecute on the part of the Crown any Person who shall have been concerned in the unlawful taking or Confinement of any of Her Majesty's Subjects as an Insane Patient, and likewise any Person who shall have been concerned in the Neglect or Ill-treatment of any Patient or Person so confined.

Houses having 100 Patients to have a resident Medical Attendant, and Houses having less to be visited by a Medical Attendant.

LVII. And be it enacted, That in every House licensed for One hundred Patients or more there shall be a Physician, Surgeon, or Apothecary resident as the Superintendent or Medical Attendant thereof; and that every House licensed for less than One hundred and more than Fifty Patients (in case such House shall not be kept by or have a resident Physician, Surgeon, or Apothecary,) shall be visited daily by a Physician, Surgeon, or Apothecary ; and that every House licensed for less than Fifty Patients (in case such House shall not be kept by or have a resident Physician, Surgeon, or Apothecary,) shall be visited twice in every Week by a Physician, Surgeon, or Apothecary: Provided always, that it shall be lawful for the Visitors of any licensed House to direct that such House, and for the Commissioners to direct that any licensed House shall be visited by a Physician, Surgeon, or Apothecary at any other Time or Times, not being oftener than once in every Day.

The Commissioners and Visitors, in Houses licensed for less than 11 Persons, may lessen the Number of Medical Visits.

LVIII. Provided always, and be it enacted, That when any House is licensed to receive less than Eleven Lunatics, it shall be lawful for any Two of the Commissioners or any Two of the Visitors of such House, if they shall respectively so think fit, by any Writing under their Hands, to permit that such House shall be visited by a Physician, Surgeon, or Apothecary at such Intervals more distant than twice in every Week as such Commissioners or Visitors shall appoint, but not at a greater Interval than once in every Two Weeks.

A Book to be kept, to be called " The Medical Visitation Book," in which a weekly Entry is to be made, showing the Condition of the House and of the Patients.

LIX. And be it enacted, That every Physician, Surgeon, or Apothecary, where there shall be only One, keeping or residing in or visiting any licensed House or any Hospital, and where there shall be Two or more Physicians, Surgeons, or Apothecaries keeping or residing in or visiting any licensed House or any Hospital, then One at least of such Physicians, Surgeons, or Apothecaries, shall once in every Week (or, in the Case of any House at which Visits at more distant Intervals than once a Week are permitted, on every Visit), enter and sign in a Book to be kept at such House or Hospital for that Purpose, to be called " The Medical Visitation Book," a Report, showing the Date thereof, and also the Number, Sex, and State of Health of all the Patients then in such House or Hospital, the Christian and Surname of every Patient who shall have been under Restraint, or in Seclusion, or under Medical Treatment, since the Date of the last preceding Report, the Condition of the House or Hospital, and every Death, Injury, and Act of Violence which shall have happened to or affected any Patient since the then last preceding Report, according to the Form in Schedule (H.) annexed to this Act ; and every such Physician, Surgeon, or Apothecary who shall omit to enter or sign such Report as aforesaid shall for every such Omission forfeit and pay the Sum of Twenty Pounds; and every such Physician, Surgeon, or Apothecary who shall in any such Report as aforesaid enter anything untruly shall be guilty of a Misdemeanor.

A Medical Case Book to be kept.

LX. And be it enacted, That there shall be kept in every licensed House and in every Hospital a Book to be called "The Case Book," in which the Physician, Surgeon, or Apothecary keeping or residing in or visiting such House or Hospital shall from Time to Time make Entries of the mental State and bodily Condition of each Patient, together with a correct Description of the Medicine and other Remedies prescribed for the Treatment of his Disorder ; and that it shall be lawful for the Commissioners from Time to Time, by any Order under their Common Seal, to direct the Form in which such Case Book shall be kept by such Physician, Surgeon, or Apothecary ; and immediately after a Copy of such Order shall have been transmitted by the Secretary of the Commissioners to such Physician, Surgeon, or Apothecary ; such Physician, Surgeon, or Apothecary shall thereupon keep such Case Book in the Form which shall be directed by such Order ; and that it shall be lawful for the Commissioners (whenever they shall see fit) to require, by an Order in Writing under their Common Seal, such Physician, Surgeon, or Apothecary to transmit to the Commissioners a correct Copy of the Entries or Entry in any Case Book kept under the Provisions of this Act, relative to the Case of any Lunatic who is or may have been confined in any such licensed House or Hospital ; and every such Physician, Surgeon, or Apothecary who shall neglect to keep the said Case Book, or to keep the same according to the Form directed by the Commissioners, or to transmit a Copy of the said Entry or Entries, pursuant to such Order or Orders as aforesaid, shall for every such Neglect forfeit any Sum not exceeding Ten Pounds.

APPENDIX.

All licensed Houses and Hospitals to be visited by the Commissioners.

LXI. And be it enacted, That every licensed House shall, without any previous Notice, be visited by Two at least of the Commissioners (one of whom shall be a Physician or Surgeon, and the other a Barrister), Four Times at the least in every Year, if such House shall be within the immediate Jurisdiction of the Commissioners, and if not, twice at least in every Year; and every Hospital in which Lunatics shall be received shall, without any previous Notice, be visited by Two at least of the said Commissioners, (one of whom shall be a Physician or Surgeon, and the other a Barrister), once at least in every Year; and every such Visit shall be made on such Day or Days, and at such Hours of the Day, and for such Length of Time as the Visiting Commissioners shall think fit, and also at such other Times (if any) as the said Commissioners in Lunacy shall direct; and such Visiting Commissioners, when visiting such House or Hospital, may and shall inspect every Part of such House or Hospital, and every Outhouse, Place, and Building communicating with such House or Hospital, or detached therefrom, but not separated by Ground belonging to any other Person, and every Part of the Ground or Appurtenances held, used, or occupied therewith, and see every Patient then confined in such House or Hospital, and inquire whether any Patient is under Restraint, and why, and inspect the Order and Certificates or Certificate for the Reception of every Patient who shall have been received into such House or Hospital since the last Visit of the Commissioners, and in the Case of any House licensed by Justices shall consider the Observations made in the Visitors Book for such House by the Visitors appointed by the Justices, and enter in the Visitors Book of such House or Hospital a Minute of the then Condition of the House or Hospital, and of the Patients therein, and the Number of Patients under Restraint, with the Reasons thereof, as stated, and such Irregularity (if any) as may exist in any such Order or Certificates as aforesaid, and also whether the previous Suggestions (if any) of the Visiting Commissioners or Visitors have or have not been attended to, and any Observations which they may deem proper as to any of the Matters aforesaid or otherwise, and also, if such Visit be the first after the granting a Licence to the House, shall examine such Licence, and if the same be in conformity with the Provisions of this Act, sign the same, but if it be informal enter in such Visitors Book in what respect such Licence is informal: Provided also, that it shall be lawful for the Lord Chancellor, on a Representation by the Commissioners setting forth the Expediency of such Alteration, by any Writing under his Hand, to direct that any House licensed by Justices shall (during such Period as he shall therein specify, or until such his Direction shall be revoked), be visited by the Commissioners once only in the Year, and also to direct that any House licensed by the Commissioners, and not receiving any Pauper Patients therein, shall (during such Period as he shall therein specify, or until such his Direction shall be revoked), be visited by the Commissioners twice only in the Year.

Licensed Houses not within the immediate Jurisdiction of the Commissioners to be inspected Four Times a Year at least by the Visitors.

LXII. And be it enacted, That every licensed House within the Jurisdiction of any Visitors appointed by Justices shall be visited by Two at least of the said Visitors (one of whom shall be a Physician, Surgeon, or Apothecary,) Four Times at the least in every Year, on such Days, and at such Hours in the Day, and for such Length of Time as the said Visitors shall think fit, and also at such other Times (if any) as the Justices by whom such House shall have been licensed shall direct; and such Visitors when visiting any such House may and shall inspect every Part of such House, and every House, Outhouse, Place, and Building communicating therewith or detached therefrom, but not separated by Ground belonging to any other Person, and every Part of the Ground or Appurtenances held, used, or occupied therewith, and see every Patient then confined therein, and inquire whether any Patient is under Restraint, and why, and inspect the Order and Certificates or Certificate for the Reception of every Patient who shall have been received into such House since the last Visit of the Visitors, and enter in the

Visitors Book a Minute of the then Condition of the House, of the Patients therein, and the Number of Patients under Restraint, with the Reasons thereof as stated, and such Irregularity (if any) as may exist in any such Order or Certificates as aforesaid, and also whether the previous Suggestions (if any) of the Visitors or Visiting Commissioners have or have not been attended to, and any Observations which they may deem proper as to any of the Matters aforesaid or otherwise.

The Proprietor or Superintendent of every House and Hospital to show every Part and every Patient to the Visiting Commissioners and Visitors.

LXIII. And be it enacted, That the Proprietor or Superintendent of every licensed House or Hospital shall show to the Commissioners and Visitors respectively visiting the same every Part thereof respectively, and every Person detained therein as a Lunatic; and every Proprietor or Superintendent of any licensed House or any Hospital who shall conceal or attempt to conceal, or shall refuse or wilfully neglect to show, any Part of such House or Hospital, or any House, Outhouse, Place, or Building communicating therewith or detached therefrom, but not separated as aforesaid, or any Part of the Ground or Appurtenances held, used, or occupied therewith, or any Person detained or being therein, from any Visiting Commissioners or Visitors, or from any Person authorized under any Power or Provision of this Act to visit and inspect such House or Hospital, or the Patients confined therein or any of them, shall be guilty of a Misdemeanor.

Inquiries to be made by the Commissioners and Visitors on their several Visitations.

LXIV. And be it enacted, That the Visiting Commissioners and Visitors respectively, upon their several Visitations to every licensed House and to every Hospital, shall inquire when Divine Service is performed, and to what Number of the Patients, and the Effect thereof; and also what Occupations or Amusements are provided for the Patients, and the Result thereof; and whether there has been adopted any System of Non-coercion, and, if so, the Result thereof; and also as to the Classification of Patients; and also as to the Condition of the Pauper Patients (if any) when first received; and also as to the Dietary of the Pauper Patients (if any); and shall also make such other Inquiries as to such Visiting Commissioners or Visitors shall seem expedient; and every Proprietor or Superintendent of a licensed House or an Hospital who shall not give full and true Answers to the best of his Knowledge to all Questions which the Visiting Commissioners and Visitors respectively shall ask in reference to the Matters aforesaid shall be guilty of a Misdemeanor.

Books and Documents to be produced to Visiting Commissioners and Visitors.

LXV. And be it enacted, That upon every Visit of the Visiting Commissioners to any licensed House or to any Hospital, and upon every Visit of the Visitors to any licensed House, there shall be laid before such Visiting Commissioners or Visitors (as the Case may be), by the Proprietor or Superintendent of such licensed House or of such Hospital, a List of all the Patients then in such House or Hospital (distinguishing Pauper Patients from other Patients, and Males from Females, and specifying such as are deemed curable), and also the several Books by this Act required to be kept by the Proprietor or Superintendent and by the Medical Attendant of a licensed House or an Hospital, and also all Orders and Certificates relating to Patients admitted since the last Visitation of the Commissioners or Visitors (as the Case may be), and also, in the Case of a licensed House, the Licence then in force for such House, and also all such other Orders, Certificates, Documents, and Papers relating to any of the Patients at any Time received into such licensed House or Hospital as the Visiting Commissioners or Visitors shall from Time to Time require to be produced to them; and the said Visiting Commissioners or Visitors, as the Case may be, shall sign the said Books as having been produced to them.

A Book to be kept, called "The Visitors Book," for the Result of Inspection and Inquiries; and a Book called "The Patients Book," for Observations as to State of Patients.

LXVI. And be it enacted, That there shall be hung up in some conspicuous Part of every licensed House a Copy of the Plan given to the Commissioners or Justices on applying for the Licence for such House; and that there shall be kept in every licensed House and in every Hospital in which Lunatics shall be received a Queen's Printer's Copy of this Act, bound up in a Book to be called "The Visitors Book," and that the said Visiting Commissioners and Visitors respectively shall at the Time of their respective Visitations enter therein the Result of the Inspections and Inquiries herein-before directed or authorized to be made by them respectively, with such Observations (if any) as they shall think proper; and that there shall also be kept in every such House and Hospital a Book to be called "The Patients Book," and that the said Visiting Commissioners and Visitors respectively shall at the Times of their respective Visitations enter therein such Observations as they may think fit respecting the State of Mind or Body of any Patient in such House or Hospital.

Proprietor or resident Superintendent to transmit all Entries by Visitors and Visiting Commissioners to the Clerk of the Visitors and to the Commissioners.

LXVII. And be it enacted, That the Proprietor or resident Superintendent of every licensed House and of every Hospital shall, within Three Days after every such Visit by the Visiting Commissioners as aforesaid, transmit a true and perfect Copy of the Entries made by them in "The Visitors Book," "The Patients Book," and "The Medical Visitation Book," respectively (distinguishing the Entries in the several Books) to the Commissioners, and shall within Three Days after every such Visitation by the Visitors, transmit a true and perfect Copy of the Entries made by them as aforesaid (distinguished as aforesaid) to the Commissioners and also to the Clerk of the Visitors; and the Copies so transmitted to the Clerk of the Visitors of all such Entries relating to any licensed House, and made since the Grant or last Renewal of the Licence thereof, shall be laid before the Justices on taking into consideration the Renewal of the Licence to the House to which such Entries shall relate; and every such Proprietor or Superintendent as aforesaid who shall omit to transmit, as herein-before directed, a true and perfect Copy of every or any such Entry as aforesaid, shall for every such Omission forfeit a Sum not exceeding Ten Pounds.

Commissioners visiting a House licensed by Justices to make an Entry in the Patients Book as to the State of Mind of any doubtful Patient, and the same to be sent to the Clerk of the Visitors, who are thereupon to visit such Patient.

LXVIII. And be it enacted, That the Commissioners visiting any House licensed by Justices shall carefully consider and give special Attention to the State of Mind of any Patient therein confined, as to the Propriety of whose Detention they shall doubt (or as to whose Sanity their Attention shall be specially called), and shall, if they shall think that the State of Mind of such Patient is doubtful, and that the Propriety of his Detention requires further Consideration, make and sign a Minute thereof in the Patients Book of such House; and a true and perfect Copy of every such Minute shall, within Two clear Days after the same shall have been made, be sent by the Proprietor or Superintendent of such House to the Clerk of the Visitors of such House, and such Clerk shall forthwith communicate the same to the said Visitors, or some Two of them, (of whom a Physician, Surgeon, or Apothecary shall be One), and such Visitors shall thereupon immediately visit such Patient and act as they shall see fit; and every such Proprietor or Superintendent who shall omit to send a true and perfect Copy, as herein-before directed, of every or any such last-mentioned Minute, and every Clerk who shall neglect to communicate the same to Two of the Visitors as aforesaid, shall be guilty of a Misdemeanor.

APPENDIX.

Visiting Commissioners to report on every House and Hospital not within their immediate Jurisdiction.

LXIX. And be it enacted, That the Visiting Commissioners shall after every Visitation by them to every licensed House not being within their immediate Jurisdiction, and to every Hospital, report in Writing the general Result of their Inspection thereof (together with such special Circumstances, if any, as they may deem proper to notice), to the Commissioners, and the Secretary of the Commissioners shall thereupon enter the same in a Book to be kept for that Purpose.

Power for the Commissioners or any Five of them to make Rules.

LXX. And be it enacted, That it shall be lawful for the Commissioners or any Five of them, at any Quarterly or Special Meeting, by any Resolution or Resolutions under their Common Seal, or to be entered in a Book to be kept for that Purpose, and signed by Five at least of the Commissioners present at such Meeting, from Time to Time to make such Orders and Rules as they shall think fit for regulating the Duties of the Commissioners or any of them, or of their Secretary, Clerks, and Servants, or for the due or better Performance of the Business of the Commission : Provided nevertheless, that the Secretary of the Commissioners shall give to every Commissioner, so far as Circumstances will admit, not less than Seven Days Notice of every such Special Meeting, and shall in the Summons for such Special Meeting state the Purposes for which the same is intended to be held.

Power in certain Cases to visit by Night.

LXXI. And be it enacted, That it shall be lawful for any Two or more of the Commissioners, or any Two Visitors, to visit and to inspect any licensed House or Hospital at such Hour of the Night as they shall think fit : Provided nevertheless, that no such Visitor shall make any such Visitation or Inspection except of a licensed House within their Jurisdiction.

The Person who signed the Order for the Reception of a private Patient may order his Discharge or Removal.

LXXII. And be it enacted, That if and when any Person who signed the Order on which any Patient (not being a Pauper) was received into any licensed House or into any Hospital shall by Writing under his Hand direct that such Patient shall be discharged or removed, then and in such Case such Patient shall forthwith be discharged or removed, as the Person who signed the Order for his Reception shall direct.

Provision for the Discharge of a private Patient when the Person who signed the Order for his Reception is incapable.

LXXIII. And be it enacted, That if the Person who signed the Order on which any Patient (not being a Pauper) was received into any licensed House or into any Hospital be incapable by reason of Insanity or Absence from *England*, or otherwise, of giving an Order for the Discharge or Removal of such Patient, or if such Person be dead, then and in any of such Cases the Husband or Wife of such Patient, or if there be no such Husband or Wife, the Father of such Patient, or if there be no Father, the Mother of such Patient, or if there be no Mother, then any one of the nearest of Kin for the Time being of such Patient, or the Person who made the last Payment on account of such Patient, may by any Writing under his or her Hand give such Direction as aforesaid for the Discharge or Removal of such Patient, and thereupon such Patient shall be forthwith discharged or removed as the Person giving such Direction shall direct.

Mode of Removal or Discharge of Pauper Patients.

LXXIV. And be it enacted, That the Guardians of any Parish or Union may by a Minute of their Board, or an Officiating Clergyman of any Parish not under a Board of Guardians, and One of the Overseers thereof, or any Two Justices of the County or Borough in which such last-mentioned Parish is situate, may by Writing under the Hands respectively of such Clergyman and Overseer or of such Justices direct that any Pauper Patient belonging to such Parish or Union, and detained in any licensed House or any Hospital, shall be discharged or removed therefrom, and may direct the Mode of such Discharge or Removal; and if a Copy of such Minute or such Writing be produced to the Proprietor or Superintendent of such licensed House or such Hospital, he shall forthwith discharge or remove such Patient, or cause or suffer such Patient to be discharged or removed accordingly.

No Patient to be removed under any of the preceding Powers if certified to be dangerous, unless the Commissioners or Visitors consent, or for the Purpose of Transfer to some other Asylum.

LXXV. Provided always, nevertheless, and be it enacted, That no Patient shall be discharged or removed, under any of the Powers herein-before contained, from any licensed House or any Hospital, if the Physician, Surgeon, or Apothecary by whom the same shall be kept, or who shall be the regular Medical Attendant thereof, shall by Writing under his Hand certify that in his Opinion such Patient is dangerous and unfit to be at large, together with the Grounds on which such Opinion is founded, unless the Commissioners visiting such House or the Visitors of such House shall, after such Certificate shall have been produced to them, give their Consent in Writing that such Patient shall be discharged or removed; provided that nothing herein contained shall prevent any Patient from being transferred from any licensed House or any Hospital to any other licensed House or any other Hospital, or to any Asylum, but in such Case every such Patient shall be placed under the Control of an Attendant belonging to the licensed House, Hospital, or Asylum, to or from which he shall be about to be removed for the Purpose of such Removal, and shall remain under such Control until such Time as such Removal shall be duly effected.

Commissioners may discharge any Patient confined in a House licensed by themselves.

LXXVI. And be it enacted, That it shall be lawful for any Two or more of the Commissioners to make Visits to any Patient detained in any House licensed by the Commissioners, on such Days and at such Hours as they shall think fit; and if after Two distinct and separate Visits so made (Seven Days at least to intervene between such Visits) it shall appear to such Visiting Commissioners that such Patient is detained without sufficient Cause, it shall be lawful for the Commissioners if they shall think fit, to make such Order as to the Commissioners shall seem meet for the Discharge of such Patient, and such Patient shall be discharged accordingly.

Two Commissioners may make special Visits to Discharge any Patient confined in a House licensed by Justices or in an Hospital.

LXXVII. And be it enacted, That it shall be lawful for any Two or more of the Commissioners, of whom One shall be a Physician and One a Barrister, to make special Visits to any Patient detained in any Hours licensed by the Justices or in any Hospital, on such Days and at such Hours as they shall think fit; and if after Two distinct and separate Visits so made it shall appear to such Visiting Commissioners that such Patient is detained without sufficient Cause, they may make such Order as to them shall seem meet for the Discharge of such Patient, and such Patient shall be discharged accordingly.

Similar Powers for Two Visitors as to Houses within their Jurisdiction.

LXXVIII. And be it enacted, That it shall be lawful for any Two or more of the Visitors of any licensed House, of whom One shall be a Physician, Surgeon, or Apothecary, to make special Visits to any Patient detained in such House, on such Days and at such Hours as they shall think fit; and if after Two distinct and separate Visits so made it shall appear to such Visitors that such Patient is detained without sufficient Cause, they may make such Order as to them shall seem meet for the Discharge of such Patient, and such Patient shall be discharged accordingly.

Every Order for the Discharge of a Patient under the last preceding Powers to be signed by the Persons exercising them, and to be subject to certain Restrictions.

LXXIX. Provided always, and be it enacted, That every such Order by any Commissioners or Visitors for the Discharge of a Patient from any House licensed by Justices, or from any Hospital, shall be signed by them, and that each of such special Visits shall be by the same Commissioners or Visitors; and that it shall not be lawful for such Commissioners or Visitors to Order the Discharge of any Patient from any such last-mentioned House or Hospital without having previously, if the Medical Attendant of such House or Hospital shall have tendered himself for that Purpose, examined him as to his Opinion respecting the Fitness of such Patient to be discharged ; and if such Commissioners or Visitors shall, after so examining such Medical Attendant discharge such Patient, and such Medical Attendant shall furnish them with any Statement in Writing containing his Reasons against the Discharge of such Patient, they shall forthwith transmit such Statement to the Commissioners or to the Clerk of the Visitors, as the Case may require, to be kept and registered in a Book for that Purpose.

The last preceding Powers to be exercised under certain other Restrictions.

LXXX. Provided also, and be it enacted, That not less than Seven Days shall intervene between the First and Second of such special Visits ; and that such Commissioners or Visitors shall, Seven Days previously to the Second of such special Visits, give Notice thereof, either by Post or by an Entry in the Patients Book, to the Proprietor or Superintendent of the House licensed by Justices or of the Hospital in which the Patient intended to be visited is detained ; and that such Proprietor or Superintendent shall forthwith, if possible, transmit by Post a Copy of such Notice, in the Case of a Patient not being a Pauper, to the Person by whose Authority such Patient was received into such House or by whom the last Payment on account of such Patient was made, and in the Case of a Pauper to the Guardians of his Parish or Union, or if there be no such Guardians, to One of the Overseers for the Time being of his Parish, and also in the Case of any Patient detained in a House licensed by Justices, to the Clerk of the Visitors of such House.

Preceding Powers not to extend to Persons found lunatic by Inquisition, or confined under Authority of Secretary of State.

LXXXI. Provided always, nevertheless, and be it enacted, That none of the Powers of Discharge herein-before contained shall extend to any Person who shall have been found lunatic by Inquisition or under any Inquiry directed by the Lord Chancellor, in pursuance of the Powers in that Behalf herein-after given to him, nor to any Lunatic confined under any Order or Authority of Her Majesty's Principal Secretary of State for the Home Department, or under the Order of any Court of Criminal Jurisdiction.

Power for Visitors and Visiting Commissioners to regulate the Dietary of Pauper Patients.

LXXXII. And be it enacted, That it shall be lawful for the Visitors of any licensed House at any Time to determine and regulate the Dietary of the Pauper Patients therein ; and that it shall be lawful for the Visiting Commissioners at any Time to determine and regulate the Dietary of the Pauper Patients in any licensed House or in any Hospital ; and that if such Determination and Regulation of any Visitors and of the Visiting Commissioners shall not agree with each other, then the Determination and Regulation of the Visiting Commissioners shall be followed : Provided always, nevertheless, that every such Regulation shall be made to take effect only from such Time as not to affect any Contract existing on the First Day of June last for the Maintenance of Pauper Patients before the First Day of June One thousand eight hundred and forty-six, or the Expiration of such Contract, whichever shall first happen.

Power for any Visitor to give an Order to the Clerk of the Visitors to search and give Information.

LXXXIII. And be it enacted, That if any Person shall apply to any Visitor in order to be informed whether any particular Person is confined in any licensed House within the Jurisdiction of such Visitor, the said Visitor, if he shall think it reasonable to permit such Inquiry to be made, shall sign an Order to the Clerk of the Visitors, and the said Clerk shall, on Receipt of such Order, and on Payment to him of a Sum not exceeding Seven Shillings for his Trouble, make search amongst the Returns made to him in pursuance of this Act whether the Person inquired after is or has been within the then last Twelve Calendar Months confined in any licensed House within the Jurisdiction of such Visitor; and if it shall appear that such Person is or has been so confined the said Clerk shall deliver to the Person so applying a Statement in Writing, specifying the Situation of the House in which the Person so inquired after appears to be or to have been confined, and of the Name of the Proprietor or resident Superintendent thereof, and also the Date of the Admission of such Person into such licensed House, and (in case of his having been removed or discharged) the Date of his Removal or Discharge therefrom.

Power for any Commissioner to give an Order to the Secretary of the Commissioners to search and give Information whether any particular Person is or has been within Twelve Months confined in any House or Hospital.

LXXXIV. And be it enacted, That if any Person shall apply to any Commissioner in order to be informed whether any particular Person is confined in any licensed House, or in any Hospital, Asylum, or other Place by this Act made subject to the Visitation of the Commissioners, such Commissioner, if he shall think it reasonable to permit such Inquiry to be made, shall sign an Order to the Secretary of the Commissioners, and the Secretary shall, on the Receipt of such Order, and on Payment to him of a Sum not exceeding Seven Shillings (to be applied as herein-before provided), make search amongst the Returns made in pursuance of this Act, or of any of the Acts hereby repealed, whether the Person inquired after is or has been within the last Twelve Calendar Months confined in any House, Hospital, Asylum, or Place by this Act made subject to the Visitation of the Commissioners; and if it shall appear that such Person is or has been so confined the Secretary shall deliver to the Person so applying a Statement in Writing, specifying the Situation of the House, Hospital, Asylum, or Place in which the Person so inquired after appears to be or to have been confined, and also (so far as the said Secretary can ascertain the same from any Register or Return in his Possession) the Name of the Proprietor, Superintendent, or principal Officer of such House, Hospital, Asylum, or Place, and also the Date of the Admission of such Person into such licensed House, Hospital, Asylum, or other Place. and (in case of his having been removed or discharged) the Date of his Removal or Discharge therefrom.

Any One Commissioner or Visitor may give an Order for the Admission to any Patient of any Friend or Relation, or any Person named by a Friend or Relation.

LXXXV. And be it enacted, That it shall be lawful for any One of the Commissioners, as to Patients confined in any House, Hospital, or other Place (not being a Gaol) hereby authorized to be visited by the Commissioners, and also for any One of the Visitors of any licensed House as to Patients confined in such House, at any Time to give an Order in Writing under the Hand of such One Commissioner or Visitor for the Admission to any Patient of any Relation or

Friend of such Patient (or of any Medical or other Person whom any Relation or Friend of such Patient shall desire to be admitted to him), and such Order of Admission may be either for a single Admission, or for an Admission for any limited Number of Times, or for Admission generally at all reasonable Times, and either with or without any Restriction as to such Admission or Admissions being in the Presence of a Keeper or not, or otherwise ; and if the Proprietor or Superintendent of any such House, Hospital, or Place shall refuse Admission to, or shall prevent or obstruct the Admission to any Patient of, any Relation, Friend, or other Person who shall produce such Order of Admission as aforesaid, he shall for every such Refusal, Prevention, or Obstruction forfeit a Sum not exceeding Twenty Pounds.

Proprietor or Superintendent, with consent of Two Commissioners or Visitors, may take or send a Patient to any Place for his Health.

LXXXVI. And be it enacted, That it shall be lawful for the Proprietor or Superintendent of any licensed House or of any Hospital, with the consent in Writing of any Two of the Commissioners, or in the Case of a House licensed by Justices of any Two of the Visitors of such House, to send or take, under proper Control, any Patient to any specified Place for any definite Time for the Benefit of his Health : Provided always, nevertheless, that before any such Consent as aforesaid shall be given by any Commissioners or Visitors the Approval in Writing of the Person who signed the Order for the Reception of such Patient, or by whom the past Payment on account of such Patient was made, shall be produced to such Commissioners or Visitors, unless they shall, on Cause being shown, dispense with the same.

In case of the Removal of a Patient, or of his Escape and Recapture, within Fourteen Days, the original Order for his Reception to remain in Force.

LXXXVII. And be it enacted, That in every Case in which any Patient shall, under any of the Powers or Provisions of this Act, be removed temporarily from the House or Hospital into which the Order for his Reception was given, or be transferred from such House or Hospital into any new House, and also in every case in which any Patient shall escape from any House or Hospital and shall be retaken within Fourteen Days next after such Escape, the Certificate or Certificates relating to and the original Order for the Reception of such Patient shall respectively remain in force, in the same Manner as the same would had done if such Patient had not been so removed or transferred, or had not so escaped and been retaken.

Commissioners to report to the Lord Chancellor periodically.

LXXXVIII. And be it enacted, That the Commissioners shall, at the expiration of every Six Calendar Months, report to the Lord Chancellor the Number of Visits which they shall have made, the Number of Patients whom they shall have seen, and the Number of Miles which they shall have travelled during such Months, and shall on the First Day of *January* in each Year make a Return to the Lord Chancellor of all Sums received by them for travelling Expenses, or upon any other and what Account, and shall also in the Month of *June* in every Year make to the Lord Chancellor a report of the State and Condition of the several Houses, Hospitals, Asylums, and other Places visited by them under this Act, and of the Care of the Patients therein, and of such other Particulars as they shall think deserving of Notice ; and a true Copy of such Reports, showing the Number of Visits made, the Number of Patients seen, and the Number of Miles travelled, and also a Copy of such Return of Sums received for travelling Expenses, or on any other and what Account shall be laid before Parliament within Twenty-One Days next after the Commencement of every Session of Parliament.

Constitution of the Private Committee.

LXXXIX. And be it enacted, That the permanent Chairman for the Time being of the Commissioners, and Two other of the Commissioners to be appointed by the Lord Chancellor from Time to Time as Occasion may require (One of whom shall be a Physician, or Surgeon, and the other a Barrister), shall be a Committee, to be called "The Private Committee," for the Purposes herein-after mentioned.

No Person (except a Person deriving no Profit, or a Committee,) to take charge of a single Lunatic, except upon such Order and Medical Certificates as aforesaid, and under certain Obligations.

XC. And be it enacted, That no Person (unless he be a Person who derives no Profit from the Charge, or a Committee appointed by the Lord Chancellor,) shall receive to board or lodge in any House, other than an Hospital registered under this Act, or an Asylum, or a House licensed under this Act, or under One of the Acts herein-before repealed, or take the Care or Charge of any One Patient as a Lunatic or alleged Lunatic, without the like Order and Medical Certificates in respect of such Patient as are herein-before required on the Reception of a Patient (not being a Pauper) into a licensed House ; and that every Person (except a Person deriving no Profit from the Charge, or a Committee appointed by the Lord Chancellor,) who shall receive to board or lodge in any unlicensed House, not being a registered Hospital, or an Asylum, or take the Care or Charge of any One Patient as a Lunatic or alleged Lunatic, shall, within Seven clear Days after so receiving or taking such Patient transmit to the Secretary of the Commissioners a true and perfect Copy of the Order and Medical Certificates on which such Patient has been so received, and a Statement of the Date of such Reception, and of the Situation of the House into which such Patient has been received, and of the Christian and Surname and Occupation of the Occupier thereof and of the Person by whom the Care and Charge of such Patient has been taken ; and every such Patient shall at least once in every Two Weeks be visited by a Physician, Surgeon, or Apothecary not deriving and not having a Partner, Father, Son or Brother who derives, any Profit from the Care or Charge of such Patient ; and such Physician, Surgeon, or Apothecary shall enter in a Book, to be kept at the House or Hospital for that Purpose, to be called "The Medical Visitation Book," the Date of each of his Visits, and a Statement of the Condition of the Patient's Health, both mental and bodily, and of the Condition of the House in which such Patient is, and such Book shall be produced to the Visiting Commissioner on every Visit, and shall be signed by him as having been so produced ; and the Person by whom the Care or Charge of such Patient has been taken, or into whose House he has been received as aforesaid, shall transmit to the Secretary of the Commissioners the same Notices and Statements of the Death, Removal, Escape, and Re-capture of such Lunatic, and within the same Periods as are herein-before required in the case of the Death, Removal, Escape, and Re-capture of a Patient (not being a Pauper) received into a licensed House ; and that every Person who shall receive into an unlicensed House, not being a registered Hospital nor an Asylum, or take the Care or Charge of any Person therein as a Lunatic, without first having such Order and Medical Certificates as aforesaid or who, having received any such Patient, shall not within the several Periods aforesaid transmit to the Secretary of the Commissioners such Copy, Statement, and Notices as aforesaid, or shall fail to cause such Patient to be so visited by a Medical Attendant as aforesaid, and every such Medical Attendant who shall make an untrue Entry in the said Medical Visitation Book, shall be guilty of a Misdemeanor.

Copy of the Order and Certificates, &c., with respect to Lunatics received into an unlicensed House to be entered in a Private Register.

XCI. And be it enacted, That the Secretary to the Commissioners shall preserve every Copy transmitted as aforesaid of the Order and Certificates for the Reception of any Patient as a Lunatic into an unlicensed House, and every Statement and Notice which may be transmitted to such Secretary with respect to any such Patient as aforesaid, and shall enter the same (in such form as the Private Committee shall direct) in a Book to be kept for that Purpose, to be called "The Private Register ;" and such Private Register shall be kept by such Secretary in his own Custody, and shall be inspected only by the Members for the Time being of the said Private Committee, and by such other Persons as the Lord Chancellor shall by Writing under his Hand appoint.

APPENDIX. 197

Members of the Private Committee to visit unlicensed Houses receiving a single Patient, and report.

XCII. And be it enacted, That it shall be lawful for any One Member of the said Private Committee, on the Direction of such Committee, or of any Two Members thereof (of whom the One Member aforesaid may be One,) at all reasonable Times to visit every or any unlicensed House in which One Patient only is received as a Lunatic (unless such Patient be so received by a Person deriving no Profit from the Charge, or by a Committee appointed by the Lord Chancellor), and to inquire and report to the said Private Committee on the Treatment and State of Health, both bodily and mental, of such Patient; and a Copy of every or any such Report shall be entered in a Private Register, to be kept for that Purpose, by the Secretary of the Commissioners, and another Copy thereof shall, if such Private Committee think it expedient, be laid before the Lord Chancellor.

The Lord Chancellor on such Report, and the Representation of the Private Committee, may order a Lunatic to be removed.

XCIII. And be it enacted, That it shall be lawful for the Lord Chancellor, on the Representation of the said Private Committee, accompanied with a Copy of a Report made as last aforesaid as to any Patient received or detained as a Lunatic in an unlicensed House as aforesaid, to make an Order that such Patient shall be removed from such House, and from the Care and Charge of the Person under whose Care and Charge such Lunatic may be; and any Person detaining such Lunatic in such House, or in such Care or Charge, for the Space of Three Days after a Copy of such Order shall have been left at such House or served on such Person, shall be guilty of a Misdemeanor.

Commissioners to report if Property of Lunatics be not duly protected or applied.

XCIV. And be it enacted, That whenever the Commissioners shall have reason to suppose that the Property of any Person detained or taken charge of as a Lunatic is not duly protected, or that the Income thereof is not duly applied for his Maintenance, such Commissioners shall make such Inquiries relative thereto as they shall think proper, and report thereon to the Lord Chancellor.

The Lord Chancellor to direct the Master in Lunacy to report as to the Lunacy of any Person detained as a Lunatic, and to appoint Guardians of his Person and Estate, and direct the Application of his Income.

XCV. And be it enacted, That when any person shall have been received or taken charge of as a Lunatic upon an order and Certificates, or an Order and Certificate, in pursuance of the Provisions of this Act, or of any Act herein-before repealed, and shall either have been detained as a Lunatic for the Twelve Months then last past, or shall have been the Subject of a Report by the Commissioners in pursuance of the Provision lastly herein-before contained, it shall be lawful for the Lord Chancellor to direct that One of the said Masters in Lunacy shall, and thereupon One of the said Masters shall personally examine such Person, and shall take such Evidence and call for such Information as to such Master shall seem necessary to satisfy him whether such Person is a Lunatic, and shall report thereon to the Lord Chancellor, and such Report shall be filed with the Secretary of Lunatics; and it shall be lawful for the Lord Chancellor from Time to Time to make Orders for the Appointment of a Guardian or otherwise for the Protection, Care, and Management of the Person of any Person who shall by any such Report as last aforesaid be found to be a Lunatic, and such Guardian shall have the same Powers and Authorities as a Committee of the Person of a Lunatic found such by Inquisition now has, and also to make Orders for the Appointment of a Receiver or otherwise for the Protection, Care, and Management of the Estate of such Lunatic, and such Receiver shall have the same Powers and Authorities as a Receiver of the Estate of a Lunatic found such by Inquisition now has, and also to make Orders for the Application of the Income of such Lunatic, or a sufficient Part thereof, for his Maintenance and Support, and in Payment of the Costs,

Charges, and Expenses attending the Protection, Care, and Management of the Person and Estate of such Lunatic, and also as to the Investment or other Application for the Purpose of Accumulation of the Overplus, if any, of such Income, for the Use of such Lunatic, as to the Lord Chancellor shall from Time to Time in each Case seem fit: Provided always that such Protection, Care, and Management shall continue only during such Time as such Lunatic shall continue to be detained as a Lunatic upon an Order and Certificates or Certificate as aforesaid, and for such further Time, not exceeding Six Months, as the Lord Chancellor may fix: Provided also, that it shall be lawful for the Lord Chancellor in any such Case, either before or after directing such Inquiry by such Master as aforesaid, and whether such Master shall have made a Report as aforesaid or not, to direct a Commission in the Nature of a Writ De lunatico inquirendo to issue, to inquire of the Lunacy of such Person.

Masters in Lunacy to have all necessary Powers of Inquiry, and to make Inquiries referred to them.

XCVI. And be it enacted, That such Masters shall have Power, in the Prosecution of all Inquiries and Matters which may be referred to them as aforesaid or otherwise under this Act, to summon Persons before them, and to administer Oaths, and take Evidence, either *vivâ voce* or on Affidavit, and to require the Production of Books, Papers, Accounts, and Documents; and that the Lord Chancellor may by any Order (either general or particular) refer to such Masters any Inquiries under the Provisions of this Act relating to the Person and Estate of any Lunatic as to whom a Report shall be made by a Master as aforesaid, in like Manner as Inquiries relating to the Persons and Estates of Lunatics found such by Inquisition are now referred to them.

Lord Chancellor to make Orders and Regulations, and fix Fees.   5 & 6 Vict. c. 84.

XCVII. And be it enacted, That it shall be lawful for the Lord Chancellor from Time to Time to make such Orders as shall to him seem fit for regulating the Form and Mode of Proceeding before the Lord Chancellor and before the said Masters, and of any other Proceedings pursuant to the Provisions of this Act, for the due Protection, Care, and Management of the Persons and Estates of Lunatics as to whom such Reports shall be made by the said Masters as aforesaid, and also for fixing, altering, and discontinuing the Fees to be received and taken in respect of such Proceedings, as to the Lord Chancellor shall from Time to Time seem fit: Provided nevertheless, that all Fees to be so received and taken shall be paid into the Bank of *England*, and placed to the Credit of the Accountant General of the Court of Chancery, to the Account intituled "The Suitors Fee Fund Account," in like Manner as and together with the Fees payable under the Act passed in the Fifth and Sixth Years of Her present Majesty, intituled *An Act to alter and amend the Practice and Course of Proceeding under Commissions in the Nature of Writs De lunatico inquirendo*, and be applied in like Manner as such last-mentioned Fees.

Masters Expenses how to be paid.

XCVIII. And be it enacted, That the travelling and other Expenses of the said Masters and their Clerks shall be paid to them, by virtue of any Order or Orders of the Court of Chancery, out of the said Fund, intituled "The Suitors Fee Fund Account," in the same Manner as their Expenses under the said last-mentioned Act.

Proprietors, Superintendents, and other authorized Persons may plead the Order and Certificate for receiving any Lunatic in bar of all Proceedings at Law.

XCIX. And be it enacted, That every Proprietor and Superintendent of a licensed House or registered Hospital, and every other Person hereby or by any of the Acts herein-before repealed authorized to receive or take charge of a Lunatic upon an Order, and who shall receive or has received a proper Order, in pursuance of this Act or any of the said repealed Acts, accompanied with the required

Medical Certificates or Certificate for the Reception or taking charge of any Person as a Lunatic, and the Assistants and Servants of such Proprietor, Superintendent, or other Person, shall have Power and Authority to take charge of, receive, and detain such Patient until he shall die or be removed or discharged by due Authority, and in case of the Escape at any Time or Times of such Patient to retake him at any Time within Fourteen Days after such Escape, and again to detain him as aforesaid; and in every Writ, Indictment, Information, Action, and other Proceeding which shall be preferred or brought against any such Proprietor, Superintendent, or other Person authorized as aforesaid, or against any Assistant or Servant of any such Proprietor, Superintendent, or authorized Person, for taking, confining, detaining, or retaking any Person as a Lunatic, the Party complained of may plead such Order and Certificates or Certificate in defence to any such Writ, Indictment, Information, Action, or other Proceeding as aforesaid, and such Order and Certificates or Certificate shall, as respects such Party, be a Justification for taking, confining, detaining, or retaking such Lunatic or alleged Lunatic.

Commissioners and Visitors may summon Witnesses to give Evidence, with a Penalty for Noncompliance.

C. And be it enacted, That it shall be lawful for the Commissioners, or any Two of them, and also for the Visitors of any licensed House, or any Two of such Visitors, from Time to Time, as they shall see Occasion, to require, by Summons under the Common Seal of the Commission, if by the Commissioners, and if by Two only of the Commissioners or by Two Visitors, then under the Hands and Seals of such Two Commissioners or Two Visitors, as the Case may be, (according to the Form in Schedule (I.) annexed to this Act, or as near thereto as the Case will permit,) any Person to appear before them to testify on Oath the Truth touching any Matters respecting which such Commissioners and Visitors respectively are by this Act authorized to inquire (which Oath such Commissioners or Visitors are hereby empowered to administer); and every Person who shall not appear before such Commissioners or Visitors pursuant to such Summons, or shall not assign some reasonable Excuse for not so appearing, or shall appear and refuse to be sworn or examined, shall, on being convicted thereof before One of Her Majesty's Justices for the County or Borough within which the Place at which such Person shall have been by such Summons required to appear and give Evidence is situate, shall for every such Neglect or Refusal forfeit a Sum not exceeding Fifty Pounds.

Provision for the Payment of Witnesses Expenses.

CI. And be it enacted, That it shall be lawful for any Commissioners or Visitors who shall summon any Person to appear and give Evidence as aforesaid to direct the Secretary of the Commissioners or the Clerk of such Visitors, as the Case may be, to pay to such Person all reasonable Expenses of his Appearance and Attendance in pursuance of such Summons, the same to be considered as Expenses incurred by such Commissioners and Visitors respectively in the Execution of this Act, and to be taken into account and paid accordingly.

Upon Complaint made of any Offence against this Act, Justices to require the Attendance of the Person charged, and adjudicate thereon. Recovery of Penalties, and Application thereof.

CII. And be it enacted, That every Complaint or Information of or for any Offence against this Act, where any pecuniary Penalty is hereby imposed, (except when hereby otherwise provided for,) may be made before One Justice; and when any Person shall be charged upon Oath before a Justice for any such Offence against this Act, such Justice may summon the Person charged to appear at a Time and Place to be named in such Summons, and if he shall not appear accordingly, and upon Proof of the due Service of the Summons (either personally or by leaving the same at his last or usual Place of Abode) any Two Justices may either proceed to hear and determine the Case, or may issue their Warrant for apprehending such Person, and bringing him before any Two Justices; and any Two Justices shall and may, upon the appearing of such Person pursuant to such Summons, or upon such Person being apprehended with such Warrant, or upon the Nonappearance of such Person, hear the Matter of every such Complaint or Information, and make any such Determination thereon as such Justices shall think proper; and upon Conviction of any Person such Justices may, if they shall think fit, reduce the Amount of the Penalty by this Act imposed for such Offence to any Sum not less than One Fourth of the Amount thereof, and shall and may issue a Warrant under their Hands and Seals for levying such Penalty or reduced Penalty, and all Costs and Charges of such Summons, Warrant, and Hearing, and all incidental Costs and Charges, by Distress and Sale of the Goods and Chattels of the Person so convicted; and it shall be lawful for any such Two Justices to order any Person so convicted to be detained and kept in the Custody of any Constable or other Peace Officer until Return can be conveniently made to such Warrant of Distress, unless the said Offender shall give Security, to the Satisfaction of such Justices, by way of Recognizance or otherwise, for his Appearance before such Justices on such Day as shall be appointed for the Return of such Warrant of Distress, such Day not being more than Seven Days from the Time of taking any such Security; but if upon the Return of such Warrant of Distress it shall appear that no sufficient Distress can be had whereupon to levy the said Penalty, and such Costs and Charges as aforesaid, and the same shall not be forthwith paid, or in case it shall appear to the Satisfaction of such Justices, either by the Confession of the Offender or otherwise, that the Offender hath not sufficient Goods and Chattels whereupon the said Penalty, Costs, and Charges may be levied, such Justices shall and may, by Warrant under their Hands and Seals, commit such Offender to the Common Gaol or House of Correction for any Term not exceeding Three Calendar Months, unless such Penalty, and all such Costs and Charges as aforesaid, shall be sooner paid; and all such Penalties, when recovered, shall be paid, when the Complaint or Information shall be laid or brought by or by the Direction of the Commissioners, to the Secretary of the Commissioners, to be by him applied and accounted for as herein-before directed with respect to Monies received for Licences granted by the Commissioners, and when the Complaint or Information shall be laid or brought by the Direction of any Visitors, to the Clerk of the Peace for the County or Borough, to be by him applied and accounted for as herein-before directed with respect to Monies received for Licences granted by the Justices of such County or Borough; and the Overplus (if any) arising from such Distress and Sale, after Payment of the Penalty and all Costs and Charges as aforesaid, shall be paid, upon Demand, to the Owner of the Goods and Chattels so distrained.

Form of Conviction before Justices.

CIII. And be it enacted, That the Justices before whom any Person shall be convicted of any Offence against this Act for which a pecuniary Penalty is imposed may cause the Conviction to be drawn up in the following Form, or in any other Form to the same Effect, as the case may require; and that no Conviction under this Act shall be void through Want of Form:

'BE it remembered, That on the   Day of   in
' the Year of our Lord   at   in the
' County [or Borough] of   .   A. B. was convicted before
' us   of Her Majesty's Justices of the Peace
' for the said County [or Borough], for that he the said
' did   and we the said   adjudge the
' said   for his Offence to pay the Sum of   .'

Appeal to Quarter Sessions.

CIV. Provided always, and be it enacted, That any Person who shall think himself aggrieved by any Order or Determination of any Justices under this Act may, within Four Calendar Months after such Order made or given, appeal to the Justices at General or Quarter Sessions, the Person appealing having first given at least Fourteen clear Days Notice in Writing of such Appeal, and the Nature and Matter thereof, to the Person appealed against, and forthwith after such Notice entering into a Recognizance before some Justice, with Two sufficient Sureties, conditioned to try such Appeal, and to abide the Order and Award of the said Court thereupon; and the said Justices at General or Quarter Sessions, upon the Proof of such Notice and Recognizance having been given and entered into, shall in a summary Way hear and determine such Appeal, or, if they think proper, adjourn the Hearing thereof until the next General or Quarter Sessions, and, if they see Cause, may mitigate any Penalty to not less than One Fourth of the Amount imposed by this Act, and may order any Money to be returned which shall have been levied in pursuance of such Order or Determination, and shall and may also award such further Satisfaction to be made to the Party injured, or such Costs to either of the Parties, as they shall judge reasonable and proper; and all such Determinations of the said Justices at General or Quarter Sessions shall be final, binding, and conclusive upon all Parties to all Intents and Purposes whatsoever.

Actions to be commenced within Twelve Calendar Months. Act may be given in Evidence.

CV. And be it enacted, That if any Action or Suit shall be brought against any Person for anything done in pursuance of this Act or of any of the Acts hereby repealed, the same shall be commenced within Twelve Calendar Months next after the Release of the Party bringing the Action, and shall be laid or brought in the County or Borough where the Cause of Action shall have arisen, and not elsewhere; and the Defendant in every such Action or Suit may, at his Election, plead specially or the General Issue Not Guilty, and give this Act and the special Matter in Evidence at any Trial to be had thereupon, and that the same was done in pursuance and by the Authority of this Act; and if the same shall appear to be so done, or that such Action or Suit shall be brought in any other County or Borough than as aforesaid, or shall not have been commenced within the Time before limited for bringing the same, then the Jury shall find a Verdict for the Defendant; and upon a Verdict being so found, or if the Plaintiff shall be nonsuited, or discontinue his Action or Suit after the Defendant shall have appeared, or if upon Demurrer Judgment shall be given against the Plaintiff, then the Defendant shall recover Double Costs, and have such Remedy for recovering the same as any Defendant hath or may have in any other Cases by Law.

Offenders to be prosecuted, and Penalties sued for by the Secretary of the Commissioners and the Clerk of any Visitors, and by no Person without the Authority of the Commissioners or Visitors.

CVI. And be it enacted, That it shall be lawful for the Secretary of the Commissioners, on their Order, to prosecute any Person for any Offence against the Provisions of this Act, and to sue for and recover any Penalty to which any Person is made liable by this Act; and all Penalties sued for and recovered by such Secretary shall be paid to him, and be by him applied and accounted for as hereinbefore directed with respect to Monies received for Licences granted by the Commissioners; and that it shall be lawful for the Clerk of any Visitors, on their Order, to prosecute any Person for any Offence against the Provisions of this Act committed within the Jurisdiction of such Visitors, and to sue for and recover any Penalty to which any Person within the Jurisdiction of such Visitors is made liable by this Act; and all Penalties sued for and recovered by any such Clerk shall be paid to him, and be by him paid to the Clerk of the Peace for such County or Borough, and be by such Clerk of the Peace applied and accounted for as herein-before directed with respect to Monies received for Licences by such Clerk of the Peace; and it shall not be lawful for any one to prosecute any Person for any Offence against the Provisions of this Act, or to sue for any Penalty to which any Person is made liable by this Act, except by Order of the Commissioners or of Visitors having Jurisdiction in the Place where the Cause of Prosecution has arisen or the Penalty being incurred, or with the Consent of Her Majesty's Attorney General or Solicitor General for *England* for the Time being.

Offenders against the Provisions of any of the repealed Acts may be prosecuted under this Act.

CVII. And be it enacted, That notwithstanding the Repeal of the several Acts herein-before repealed, every Offence heretofore committed against any of the Provisions of any of the same Acts may be prosecuted, and every Penalty heretofore incurred by any Person for any Offence against the Provisions of any of the same Acts may be sued for and recovered, by the Secretary of the Commissioners, in the same Manner and with all the same Powers and Rights as if such Offence had been committed or such Penalty incurred for an Offence against the Provisions of this Act; and every Penalty so recovered shall be applied in the same Manner as a Penalty recovered for an Offence against the Provisions of this Act.

No Person to be punishable for omitting to send any Copy, &c., if proved to have been put in the Post, or left at the proper Office.

CVIII. And be it enacted, That when any Person shall be proceeded against, under the Provisions of this Act, for omitting to transmit or send any Copy, List, Notice, Statement, or other Document herein-before required to be transmitted or sent by such Person, and such Person shall prove by the Testimony of One Witness upon Oath that the Copy, List, Notice, Statement, or Document in respect of which such Proceeding is taken was put into the Post in due Time, or (in case of Documents required to be transmitted or sent to the Commissioners or a Clerk of the Peace) left at the Office of the Commissioners or of the Clerk of the Peace, and shall have been properly addressed, such Proof shall be a Bar to all further Proceeding in respect of such Omission.

Costs incurred by the Commissioners to be paid by their Secretary, and Costs incurred by Visitors by the Clerk of the Peace.

CIX. And be it enacted, That the Costs, Charges, and Expenses incurred by or under the Authority or Order of the Commissioners in Proceedings under this Act shall be paid by the Secretary of the Commissioners, and included by him in the Account of Receipts and Payments herein-before directed to be kept by him ; and that the Costs, Charges, and Expenses incurred by or under the Order of any Visitors in Proceedings under this Act shall be paid by the Clerk of the Peace of their County or Borough, and included by him in the Account of Receipts and Payments herein-before directed to be kept by him.

Commissioners to visit Asylums and Gaols.

CX. And be it enacted, That Two or more of the Commissioners, One at least of whom shall be a Physician or Surgeon, and One at least a Barrister, shall and may, once or oftener in each Year, on such Day or Days, and at such Hours of the Day, and for such Length of Time as they shall think fit, visit every Asylum for Lunatics, and every Gaol in which there shall be or alleged to be any Lunatic, and shall inquire whether the Provisions of the Law have been carried out as to the construction of each Asylum visited, and as to its Visitation and Management, and also as to the Regularity of the Admissions and Discharges of Patients therein and therefrom ; and whether Divine Service is performed therein ; and whether any System of Coercion is in practice therein, and the Result thereof ; and as to the Classification or Nonclassification of Patients therein, and the Number of Attendants on each Class ; and as to the Occupations and Amusements of the Patients, and the Effects thereof ; and as to the Condition as well mental as bodily, of the Pauper Patients when first received ; and also as to the Dietary of the Pauper Patients ; and shall also make such other Inquiries as to every or any such Asylum, and all such Inquiries as to the Lunatics in any Gaol, as to such Visiting Commissioners shall seem meet.

Commissioners to visit Workhouses.

CXI. And be it enacted, That Two or more of the Commissioners, One at least of whom shall be a Physician or Surgeon, and One at least a Barrister, shall and may, once or oftener in each Year, on such Day or Days, and such Hours in the Day, and for such length of Time as they shall think fit, visit every Parish and Union Workhouse in which there shall be or alleged to be any Lunatic, and shall inquire whether the Provisions of the Law as to Lunatics have been carried out as to the Arrangements, Visitation, and Management of such Workhouse, and as to the Dietary, Accommodation, and Treatment of the Lunatics in such Workhouse, and shall report in Writing thereon to the Poor Law Commissioners for *England and Wales.*

Provision for the Visitation of Lunatics under the Care of Committees, and also of State and Criminal Lunatics, and other Lunatics not comprised in the preceding Provisions.

CXII. And be it enacted, That it shall be lawful for the Lord Chancellor, in the Case of any Lunatic under the Care of a Committee appointed by the Lord Chancellor, and for the Lord Chancellor or Her Majesty's Principal Secretary of State for the Home Department, in the Case of any Lunatic under the Care of any Person receiving or taking the Charge of such One Lunatic only, and deriving no Profit from the Charge, and in the Case of any Person confined as a State Lunatic or as a Lunatic under the Order of any Criminal Court of Justice, and in the Case of every other Person detained or taken Charge of as a Lunatic, or represented to be a Lunatic, or to be under any Restraint as a Lunatic, at any Time, by an Order in Writing under the Hand of the Lord Chancellor or the said Secretary of State, as the Case may be, directed to the Commissioners or any of them, or to any other Person, to require the Persons or Person to whom such Order shall be directed, or any of them, to visit and examine such Lunatic or supposed Lunatic, and to make a Report to the Lord Chancellor, or to Her Majesty's Principal Secretary of State for the Home Department, of such Matters as in such Order shall be directed to be inquired into.

Power for the Lord Chancellor and Secretary of State for the Home Department, to authorize a special Visitation of any Place where a Lunatic is represented to be confined.

CXIII. And be it enacted, That it shall be lawful for the Lord Chancellor or Her Majesty's Principal Secretary of State for the Home Department to employ any Commissioner appointed under this Act, or other Person, to inspect or inquire into the State of any Asylum, Hospital, Gaol, House, or Place wherein any Lunatic, or Person represented to be lunatic, shall be confined or alleged to be confined, and to report to him the Result of such Inspection and Inquiry ; and every such Person so employed, and not being a Commissioner, may be paid such Sum of Money for his Attendance and Trouble as to the Lord Chancellor or Her Majesty's Principal Secretary of State for the Home Department shall seem reasonable ; and every such Person so employed, whether a Commissioner or not, shall be allowed his reasonable travelling or other Expenses while so employed ; and such Sum of Money for Attendance and Trouble, and such Expenses, shall be charged on and shall be paid out of the Contingency Fund of the Home Office.

Interpretation Clause. 48 G. 3. c. 96. 9 G. 4. c. 40.

CXIV. And be it enacted, That in this Act and the Schedules thereto the Word and Expressions following shall have the several Meanings hereby assigned to them, unless there shall be something in the Subject or Context repugnant to such Construction ; (that is to say,)

- "Borough" shall mean every Borough, Town, and City Corporate having a separate Quarter Sessions, Recorder, and Clerk of the Peace :
- "County" shall mean every County, Riding, Division of a County, County of a City, County of a Town, Liberty, and other Place having a separate Commission of the Peace, and not being a " Borough" within the Meaning aforesaid :

APPENDIX.

"The Lord Chancellor" shall mean the Lord High Chancellor, the Lord Keeper or Commissioners of the Great Seal of *Great Britain*, and other the Person or Persons for the Time being intrusted, by virtue of the Queen's Sign Manual, with the Care and Commitment of the Custody of the Persons and Estates of Persons found idiot, lunatic, or of unsound Mind:

"Barrister" shall mean a Barrister and a Serjeant at Law; and a Serjeant at Law who shall have been called to the Bar Five Years or more before his Appointment to be a Commissioner shall be considered as a Barrister of Five Years standing:

"Lunatic" shall mean every Insane Person, and every Person being an Idiot or Lunatic or of unsound Mind:

"Parish" shall mean any Parish, Township, Hamlet, Vill, Tithing, Extra-parochial Place, or Place maintaining its own Poor:

"Officiating Clergyman of a [or the] Parish" shall mean a Clergyman regularly officiating and acting as the Minister or One of the Ministers of a Parish, Chapelry, or Ecclesiastical District:

"Borough Rate" shall mean a Borough Rate, and any Funds assessed upon or raised in or belonging to any Borough in the Nature of a Borough Rate, and applicable to the Purposes to which Borough Rates are applicable:

"County Rate" shall mean a County Rate, and any Funds assessed upon or raised in or belonging to any County in the Nature of a County Rate, and applicable to the Purposes to which County Rates are applicable:

"Pauper" shall mean every Person maintained wholly or in part at the Expense of any Parish, Union, County, or Borough:

"Patient" shall mean every Person received or detained a a Lunatic, or taken care or charge of as a Lunatic:

"Private Patient" shall mean every Patient who is not a Pauper:

"Proprietor" shall mean every Person to whom any Licence has been granted under the Provisions of any Act hereby repealed, or shall be granted under the Provisions of this Act, and every Person keeping, owning, having any Interest or exercising any Duties or Powers of a Proprietor in any licensed House:

"Clerk of the Peace" shall mean every Clerk of the Peace and Person acting as such, and every Deputy duly appointed:

"Medical Attendant" shall mean every Physician, Surgeon, and Apothecary who shall keep any licensed House, or shall in his Medical Capacity attend any licensed House, or any Asylum, Hospital, or other Place where any Lunatic shall be confined:

"Justice" shall mean a Justice of the Peace:

"Asylum" shall mean any Lunatic Asylum already erected and established under an Act passed in the Forty-eighth Year of the Reign of His late Majesty King *George* the Third, intituled *An Act for the better Care and Maintenance of Lunatics, being Paupers or Criminals, in* England, or erected and established, or hereafter to be erected and established, under or which have been made subject or liable to any of the Provisions of an Act passed in the Ninth Year of the Reign of His late Majesty King *George* the Fourth, intituled *An Act to amend the Laws for the Erection and Regulation of County Lunatic Asylums, and more effectually to provide for the Care and Maintenance of Pauper and Criminal Lunatics, in* England, or hereafter to be erected and established under the Provisions of any Act for the Erection or Regulation of County or Borough Lunatic Asylums:

"Hospital" shall mean any Hospital or Part of an Hospital or other House or Institution (not being an Asylum) wherein Lunatics are received, and supported wholly or partly by voluntary Contributions, or by any Charitable Bequest or Gift, or by applying the Excess of Payments of some Patients for or towards the Support, Provision, or Benefit of other Patients:

"Licensed House" shall mean a House licensed under the Provisions of this Act, or of some Act hereby repealed, for the Reception of Lunatics:

"Oath" shall mean an Oath, and every Affirmation or other Declaration or Solemnity lawfully substituted for an "Oath" in the Case of Quakers

APPENDIX.

or other Persons exempted by Law from the Necessity of taking an Oath:

Words importing the Singular Number shall include the Plural Number, and Words importing the Plural Number shall include the Singular Number, and Words importing the Masculine Gender shall include Females.

Boroughs and Counties to comprise all Places therein not having separate Commission of the Peace.

CXV. And be it enacted, That for the Purposes of this Act every Borough and County shall include every Place situate within the Limits of such Borough or County, and not having a separate Commission of the Peace; and for the Purposes of this Act every Place situate within the Limits of any Borough or County, and not having a separate Commission of the Peace, shall be within the Jurisdiction of the Justices of such Borough or County; and that the Justices of every Borough shall, for the Purposes of this Act, assemble in Special Sessions at such Times as the Quarter Sessions for such Borough shall be holden; and that all Acts herein-before required to be done by the Justices of Counties in General or Quarter Sessions assembled may be done by the Justices of Boroughs at such Special Sessions.

Act not to extend to Bethlehem Hospital.

CXVI. And be it enacted, That nothing in this Act contained shall extend to the Royal Hospital of *Bethlehem*, or any Building adjacent thereto and used therewith: Provided always, that it shall be lawful for any Commissioner or other Person whom the Lord Chancellor or any One of Her Majesty's Principal Secretaries of State shall at any Time, by an Order in Writing under the Hand of the said Lord Chancellor or Secretary of State, direct, to visit and examine the Royal Hospital of *Bethlehem*, and every or any Building adjacent thereto as aforesaid, and every or any Person confined therein.

Act to be confined to England and Wales.

CXVII. And be it enacted, That this Act shall extend only to *England* and *Wales*.

Alteration of Act.

CXVIII. And be it enacted, That this Act may be amended or repealed by any Act to be passed in this present Session of Parliament.

SCHEDULES referred to by the foregoing Act.

SCHEDULE (A.) Section 30.

FORM OF LICENCE.

KNOW ALL MEN, that we, the Commissioners in Lunacy, [or we the undersigned Justices of the Peace, acting in and for   in General [or Quarter or Special] Sessions assembled,] do hereby certify, That *A.B.* of   in the Parish of   in the County of   hath delivered to us [or the Clerk of the Peace] a Plan and Description of a House and Premises proposed to be licensed for the Reception of Lunatics, situate at   in the County of   [or, *in the Case of a renewed Licence*, hath delivered to us [or the Clerk of the Peace] a List of the Number of Patients now detained in a House and Premises licensed on the   Day of   last, for the Reception of Lunatics, situate at   in the County of   ], and we, having considered and approved the same, do hereby authorize and empower the said *A.B.*   (he intending [or not intending] to reside therein) to use and employ the said House and Premises for the Reception of   Male [or   Female, or

Male and   Female] Lunatics, of whom not more than
shall be private Patients, for the Space of   Calendar Months from
this Date.
Sealed with our Common Seal [or given under our Hands and Seals], this
      Day of   in the Year of our Lord 18 .
      Witness,
Y.Z., Secretary to the Commissioners of Lunacy,
      [or Clerk of the Peace].

SCHEDULE (B.) Section 45.

ORDER FOR THE RECEPTION OF A PRIVATE PATIENT.

I, the undersigned, hereby request you to receive *A.B.* a Lunatic [or an insane Person, *or* an Idiot, *or* a Person of unsound Mind], as a Patient into your House [*or* Hospital]. Subjoined is a Statement respecting the said *A.B.*
      (Signed)   *Name.*
      Occupation (if any).
      Place of Abode.
      Degree of Relationship (if any).
      or other Circumstances of
      Connexion with the Patient.
Name of Patient, with Christian Name at length.
Sex and Age.
Married, single, or widowed.
Condition of Life, and Previous Occupation (if any).
Previous Place of Abode.
Religious Persuasion, so far as known.
Duration of existing Attack.
Whether First Attack.
Age (if known) on First Attack.
Whether subject to Epilepsy.
Whether suicidal or dangerous to others.
Previous Place of Confinement (if any).
Whether found lunatic by Inquisition, and Date of Commission.
Special Circumstances (if any) preventing the Patient being examined, before Admission, separately, by Two Medical Practitioners.
Special Circumstances (if any) preventing the Insertion of any of above Particulars.
      (Signed)   *Name.*
   Dated this   Day of   One thousand eight hundred and
   To   Proprietor [*or* Superintendent] of
[*describing the House or Hospital by Situation and Name, if any*].

SCHEDULE (C.) Section 45.

FORM OF MEDICAL CERTIFICATE in the Case of PRIVATE PATIENTS.

I   being a Physician, *or* Surgeon, *or* an Apothecary, duly authorized to practise as such, hereby certify, that I have this Day, separately from any other Medical Practitioner, visited and personally examined *A.B.*, the Person named in the accompanying Statement and Order, and that the said *A.B.* is a Lunatic [*or* an insane Person, *or* an Idiot, *or* a Person of unsound Mind], and a proper Person to be confined, and that I have formed this Opinion from the following Fact or Facts; viz.
      (Signed)   *Name.*
      Place of Abode.
   Dated this   Day of   One thousand eight hundred and

SCHEDULE (D.) Section 48.

Order for the Reception of a Pauper Patient.

WE, the undersigned, having called to our Assistance a Physician [*or* Surgeon *or* Apothecary, *as the Case may be*], not being the Medical Officer of the Parish or Union to which the said *A.B.* belongs, and having personally examined *A.B.* a Pauper, and being satisfied that the said *A.B.* is a Lunatic [*or* an insane Person, *or* an Idiot, *or* a Person of unsound Mind], and a proper Person to be confined, hereby request you to receive the said *A.B.* as a Patient into your House or Hospital.

Subjoined is a Statement respecting the said *A.B.*

     (Signed) *Name.*
    A Justice of the Peace for the City or Borough
    of   [*or* an *or* the Officiating
    Clergyman of the Parish of   ].
         Name.
    With the Relieving Officer of the Union or Parish
    of   [*or* with an Overseer of the
    Parish of   ].

Statement.

Name of Patient, and Christian Name at length.
Sex and Age.
Married, single, or widowed.
Condition of Life, and previous Occupation (if any).
Previous Place of Abode.
Religious Persuasion, so far as known.
Length of Time insane.
Whether First Attack.
Age (if known) on First Attack.
Whether subject to Epilepsy.
Whether suicidal or dangerous to others.
Previous Places of Confinement (if any).

  I certify that to the best of my Knowledge the above Particulars are correctly stated.
    (Signed)
  [*To be signed by the Relieving Officer or Overseer signing the Order.*]

Dated this   Day of   One thousand eight hundred

To   Proprietor [*or* Superintendent] of
  [*describing the House or Hospital by Situation and Name, if any.*]

Medical Certificate.

  I,   being a Physician, *or* Surgeon, *or* an Apothecary, duly authorized to practise as such, hereby certify, That I have this Day personally examined *A.B.*, the Person named in the Statement and Order, and that the said *A.B.* is a Lunatic [*or* an insane Person, *or* an Idiot, *or* a Person of an unsound Mind], and a proper Person to be confined.

    (Signed) *Name.*
        Place of Abode.

  Dated this   Day of   One thousand eight hundred and

APPENDIX.

SCHEDULE (E.) Section 50.

REGISTRY OF ADMISSIONS.

REGISTER OF PATIENTS.

No. in Order of Admission	Date of Admission	Christian and Surname at length	Sex and Class - Private M	Private F	Pauper M	Pauper F	Age	Married	Single	Widowed	Condition of Life, and previous Occupation	Previous Place of Abode	County, Union, or Parish to which chargeable	By whose Authority sent	D.tes of Medical Certificates, and by whom signed	Bodily Condition	Name of Disorder (if any)	Form of Mental Disorder	Supposed Cause of Insanity	Epileptics	Congenital Idiots	Duration of existing Attacks - Years	Months	Weeks	Number of previous Attacks	Age on first Attack	Date of Discharge or Death	Discharged - Recovered	Relieved	Not improved	Died	Observations
1	1846: Jan. 3	William Johnson			1		23		1		Carpenter							Melancholia					4		2	17	1846: Sept. 1	1				
2																																
3																																
4	1848: June 9	William Johnson			1		25		1													7			3		1848: Dec. 2	1				
5																																
6	1852: May 6	William Johnson			1		29	1														3			4		1853: June 8	1				
7																																
8																																

SCHEDULE (F.) Section 52.

NOTICE of ADMISSION.

I HEREBY give you Notice, That *A.B.* was received into this House [*or* Hospital] as a Private [*or* Pauper] Patient on the   Day of
and I hereby transmit a Copy of the Order and Medical Certificates [*or* Certificate] on which he was received.
Subjoined is a Statement with respect to the mental and bodily Condition of the above-named Patient.

(Signed)
Superintendent
[*or* Proprietor] of

Dated this   Day of   One thousand eight hundred and

STATEMENT.

I HAVE this Day seen and personally examined   the Patient named in the above Notice, and hereby certify that with respect to mental State he [*or* she]   and that with respect to bodily Health and Condition he [*or* she]
(Signed)
Medical Proprietor [*or* Superintendent, *or* Attendant].

Dated this   Day of   One thousand eight hundred and

SCHEDULE (G. 1.) Section 54.

REGISTER of DISCHARGES and DEATHS.

Date of Discharge or Death.	Date of last Admission.	No. in Register of Patients.	Christian and Surname at length.	Sex and Class.				Discharged.								Died.		Assigned Cause of Death.	Age at Death.		Observations.
				Private.		Pauper.		Recovered.		Relieved.		Not improved.									
				M.	F.	M.	F.	M.	F.	M.	F.	M.	F.			M.	F.		M.	F.	
1846: Sept. 1	1846: Jan. 3	1	William Johnson	-	-	1	-	1													
1849: Dec. 2	1848: June 9	4	William Johnson	-	-	1	-	1													
1853: June 8	1852: May 6	7	William Johnson	-	1	-	-	-	-	-	-	-	-			1	-	Phthisis	27		

SCHEDULE (G. 2.) Section 54.

FORM of NOTICE of DISCHARGE or DEATH.

I HEREBY give you Notice, That   a Private [*or* Pauper] Patient, received into this House [*or* Hospital] on the   Day of   was

APPENDIX.

discharged therefrom recovered [or relieved, or not improved,] by the Authority of [or died therein, on the   Day of   ].
(Signed)
Superintendent [or Proprietor] of   House [or Hospital] at
Dated this   Day of   One thousand eight hundred and   .
In case of Death, add "and I further certify that A.B. was present at the Death of the said ; and that the apparent Cause of Death of the said [ascertained by post mortem examination (if so)] was   ."

SCHEDULE (H.) Section 59.

FORM of MEDICAL JOURNAL and WEEKLY REPORT.

Date.	Number of Patients.				Names of Patients under Restraint (and by what Means) or in Seclusion.		Names of Patients under Medical Treatment.		Report on State of Health of Patients and Condition of House or Hospital.	Deaths, Injuries, and Violences to Patients.
	Private.		Pauper.							
	M.	F.	M.	F.	Males.	Females.	Males.	Females.		

SCHEDULE (I.) Section 100.

FORM of SUMMONS.

WE, the Commissioners in Lunacy [or we whose Names are hereunto set and Seals affixed, being Two of the Commissioners in Lunacy, or Visitors] appointed under or by virtue of an Act passed in the   Year of the Reign of Her present Majesty, intituled [here insert the Title of the Act], do hereby summon and require you personally to appear before us at   in the Parish of   in the County of   on next the   Day of   at the Hour of   in the   noon of the same Day, and then and there to be examined, and to testify the Truth touching certain Matters relating to the Execution of the said Act.

Sealed with the Common Seal of "The Commssioners in Lunacy" [or given under our Hands and Seals], this   Day of   in the Year of our Lord One thousand eight hundred and

APPENDIX. 211

ANNO DECIMO SEXTO & DECIMO SEPTIMO
VICTORIÆ REGINÆ.

CAP. XCVI.

An Act to amend an Act passed in the Ninth Year of Her Majesty, "for the Regulation of the Care and Treatment of Lunatics."

[20th *August* 1853.]

8 & 9 *Vict.* c. 100.

WHEREAS an Act was passed in the Ninth Year of Her Majesty, "for the Regulation of the Care and Treatment of Lunatics:" And whereas it is expedient to amend the said Act as herein-after mentioned: Be it therefore enacted by the Queen's most Excellent Majesty, by and with the Advice and Consent of the Lords Spiritual and Temporal, and Commons, in this present Parliament assembled, and by the Authority of the same, as follows:

Section 25 of recited Act repealed, and Provision as to what may be included in One Licence.

I. Section Twenty-five of the said recited Act shall be repealed, and any One Licence to be granted for the Reception of Lunatics may, in the Discretion of the Commissioners or Justices granting such Licence, include Two or more Houses belonging to One Proprietor or to Two or more joint Proprietors, provided that no One of such Houses be separated from the other or others of them otherwise than by Land in the same Occupation, and by a Road, or by either of such Modes; and all Houses, Buildings, and Lands intended to be included in any Licence shall be specified, delineated, and described in the Plan required by Section Twenty-four of the said recited Act.

The Person or One of the Persons receiving a Licence to reside on the Premises.

II. No Person having, after the passing of the said recited Act, received for the First Time a Licence for the Reception of Lunatics, or hereafter receiving for the First Time such Licence, shall receive a Licence unless he shall reside on the Premises licensed; and no Two or more Persons having after the passing of the said recited Act received for the First Time a joint Licence for the Reception of Lunatics, or hereafter receiving for the First Time such Licence, shall receive such Licence unless they or One of them shall reside on the Premises licensed.

Sections 45, 46, 47, 48, and 49 of 8 & 9 Vict. c. 100, repealed.

III. Sections Forty-five, Forty-six, Forty-seven, Forty-eight, and Forty-nine of the said recited Act shall be repealed; but such Repeal shall not prevent or defeat any Prosecution for any Offence committed before the commencement of this Act, and every such Offence shall and may be prosecuted, and every pending Prosecution continued, as if this Act had not been passed.

No Person not a Pauper to be received into a Hospital or licensed House without a certain Order and Certificates.

IV. Save as herein-after otherwise provided, no Person (not being a Lunatic) for or in respect of whom any Money shall be paid or agreed to be paid shall be boarded or lodged in any licensed House; and, save where otherwise provided or authorized under this or any other Act, no Person (not being a Pauper) shall be received as a Lunatic into any licensed House or Hospital without an Order under the Hand of some Person according to the Form in Schedule (A.) No. 1. annexed to this Act, together with such Statement of Particulars as is contained in the same Schedule, nor without the Medical Certificates, according to the Form in Schedule

P 2

APPENDIX.

(A.) No. 2. annexed to this Act, of two Persons, each of whom shall be a Physician, Surgeon, or Apothecary, and shall not be in Partnership with or an Assistant to the other, and each of whom shall separately from the other have personally examined the Person to whom the Certificate signed by him relates not more than Seven clear Days previously to the Reception of such Person into such House or Hospital; and such Order as aforesaid may be signed before or after the Medical Certificates or either of them ; and every Person who shall receive any such Person as aforesaid into any such House or Hospital as aforesaid (save where otherwise provided or authorized under this or any other Act) without such Order and Medical Certificates as aforesaid shall be guilty of a Misdemeanor.

Proviso that in certain Cases any Person may be received on a Certificate signed by One Medical Practitioner only.

V. Provided always, That any Person (not a Pauper) may, under special Circumstances preventing the Examination of such Person by Two Medical Practitioners as aforesaid, be received as a Lunatic into any licensed House or any Hospital upon such Order as aforesaid, and with the Certificate of one Physician, Surgeon, or Apothecary alone, provided that the Statement accompanying such Order set forth the special Circumstances which prevent the Examination of such Person by Two Medical Practitioners; but in every such Case Two other such Certificates shall, within Three clear Days after his Reception into such House or Hospital, be signed by Two other Persons, each of whom shall be a Physician, Surgeon, or Apothecary, not in Partnership with or an Assistant to the other or the Physician, Surgeon, or Apothecary who signed the Certificate on which the Patient was received, and not connected with such House or Hospital, and shall within such Time and separately from the other of them have personally examined the Person so received as a Lunatic ; and every Person who, having received any Person as a Lunatic into any House or Hospital as aforesaid upon the Certificate of One Medical Practitioner alone as aforesaid, shall keep or permit such Person to remain in such House or Hospital beyond the said Period of Three clear Days without such further Certificates as aforesaid, shall be guilty of a Misdemeanor.

Any Person discharged may, with Assent of Visitors or Commissioners, be retained in licensed House, and a Relative or Friend may, with like Assent, be received therein.

VI. Provided also, That it shall be lawful for the Proprietor or Superintendent of any licensed House, with the previous Assent in Writing of Two of the Commissioners, such Assent not to be given until after such Commissioners have, by personal Examination of the Patient, satisfied themselves of his Desire to remain, to entertain and keep in such House as a Boarder any Person who may have been discharged as a Patient from such House for such Time after such Discharge as he may desire to remain, not exceeding the Time specified in such Assent, and also, for the Benefit of any Patient in such House, and with the previous Assent in Writing of Two of the Commissioners, to receive and accommodate as a Boarder therein, for a Time to be specified in the Assent, any Relative or Friend of such Patient, and any Two of the Commissioners may from Time to Time, by any Writing under their Hands, extend or revoke any such Assent as aforesaid ; and every such Patient so retained after Discharge, and every such Relative or Friend so accommodated, shall, if required, be produced to the Commissioners and Visitors respectively at their respective Visits.

Paupers not to be received without a certain Order and Certificate.

VII. Save where otherwise provided or authorized under any Act, no Pauper shall be received into any licensed House or any Hospital without an Order according to the Form in Schedule (B.) No. 1. annexed to this Act, under the Hand of One Justice, or under the Hands of an Officiating Clergyman, and the Relieving Officer or One of the Overseers of the Union or Parish from which such Pauper shall be sent, together with such Statement of Particulars as is contained in the

APPENDIX.

same Schedule, nor without the Medical Certificate, according to the Form in Schedule (B.) No. 2. annexed to this Act, of a Physician, Surgeon, or Apothecary, who shall have personally examined the Pauper to whom it relates not more than Seven clear Days previously to his Reception ; and every Person who shall receive any Pauper into any such House or Hospital as aforesaid (save where otherwise provided or authorized under any Act) without such Order and Medical Certificate as last aforesaid shall be guilty of a Misdemeanor : Provided always, that this Enactment shall not by Implication or otherwise give any Power or Authority to make such Order, or extend, alter, or affect any Power or Authority expressly given by any Act to any Justice, Officiating Clergyman, Relieving Officer or Overseer to make or join in making any such Order, or any Provisions giving or relating to such Power or Authority.

The like Order and Certificates for Reception of a single Patient as for Reception of a private Patient into a licensed House.

VIII. Where, under Section Ninety of the said recited Act, the like Order and Medical Certificates are required on the Reception or taking the Charge or Care of any One Person as a Lunatic or alleged Lunatic as are therein-before required on the Reception of a Patient (not being a Pauper) into a licensed House, the like Order and Medical Certificates (in lieu of those required as first aforesaid) shall hereafter be required on the Reception or taking the Charge or Care of any such Person as are by this Act required on the Reception of a Patient (not being a Pauper) into a licensed House.

Penalty on Officers, &c., ill-treating Lunatics.

IX. If any Superintendent, Officer, Nurse, Attendant, Servant, or other Person employed in any registered Hospital, or licensed House, or any Person having the Care or Charge of any single Patient, or any Attendant of any single Patient, in any way abuse, or ill-treat, or wilfully neglect any Patient in such Hospital or House, or such single Patient, or if any Person detaining, or taking or having the Care or Charge, or concerned or taking Part in the Custody, Care, or Treatment, of any Lunatic or Person alleged to be a Lunatic, in any way abuse, ill-treat, or wilfully neglect such Lunatic or alleged Lunatic, he shall be guilty of a Misdemeanor, and shall be subject to Indictment for every such Offence, or to forfeit for every such Offence, on a summary Conviction thereof before Two Justices, any Sum not exceeding Twenty Pounds.

Medical Certificate to specify Facts upon which Opinion of Insanity has been formed.

X. Every Physician, Surgeon, and Apothecary signing any Certificate under or for the Purposes of this Act shall specify therein the Facts upon which he has formed his Opinion that the Person to whom such Certificate relates is a Lunatic, an Idiot, or a Person of unsound Mind, and distinguish in such Certificate Facts observed by himself from Facts communicated to him by others ; and no Person shall be received into any registered Hospital or licensed House, or as a single Patient, under any Certificate which purports to be founded only upon Facts communicated by others.

Orders and Medical Certificates may be amended.

XI. If after the Reception of any Lunatic it appear that the Order or the Medical Certificate, or (if more than One) both or either of the Medical Certificates, upon which he was received, is or are in any respect incorrect or defective, such Order and Medical Certificate or Certificates may be amended by the Person signing the same at any Time within Fourteen Days next after the Reception of such Lunatic : provided nevertheless, that no such Amendment shall have any Force or Effect unless the same shall receive the Sanction of One or more of the Commissioners.

APPENDIX.

Who not to sign Certificates, &c.

XII. No Physician, Surgeon, or Apothecary who, or whose Father, Brother, Son, Partner, or Assistant, is wholly or partly the Proprietor of, or a regular Professional Attendant in, a licensed House or a Hospital, shall sign any Certificate for the Reception of a Patient into such House or Hospital ; and no Physician, Surgeon, or Apothecary shall himself, or by his Servants or Agents, receive to board or lodge in any unlicensed House, or take the Charge or Care of any Person upon or under any Medical Certificate signed by himself or his Father, Brother, Son, Partner, or Assistant, and no Physician, Surgeon, or Apothecary having (either before or after the passing of this Act) signed any Certificate for the Reception of any Person shall be the regular professional Attendant of such Person while under Care or Charge under such Certificate ; and no Physician, Surgeon, or Apothecary who, or whose Father, Brother, Son, Partner, or Assistant, shall sign the Order herein-before required for the Reception of a Patient, shall sign any Certificate for the Reception of the same Patient.

A Medical Man giving false Certificates, &c., and a Person not being a Medical Man giving Certificates as such, guilty of a Misdemeanor.

XIII. Any Physician, Surgeon, or Apothecary who shall sign any Certificate, or do any other Act (not declared to be a Misdemeanor) contrary to any of the Provisions herein contained, shall for every such Offence forfeit any Sum not exceeding Twenty Pounds; and any Physician, Surgeon, or Apothecary who shall falsely state or certify anything in any Certificate under this Act, and any Person who shall sign any Certificate under this Act in which he shall be described as a Physician, Surgeon, or Apothecary, not being a Physician, Surgeon, or Apothecary respectively within the Meaning of this Act, shall be guilty of a Misdemeanor.

Commissioners may permit Medical Visitation of any single Patient less frequently than once a Fortnight, but if Patient be in the Care of a Medical Man he is to make an Entry once a Fortnight as to Patient's Health.

XIV. It shall be lawful for the Commissioners, by an Order under their Common Seal, where they see fit so to do, to permit the Visitation of any single Patient by a Physician, Surgeon, or Apothecary less frequently than once in every Two Weeks, as required by Section Ninety of the said recited Act, and to prescribe from Time to Time how often any single Patient shall be visited by such a Physician, Surgeon, or Apothecary as therein mentioned ; but where such Visitation of any single Patient so often as once in every Two Weeks is so dispensed with, and such Patient is in the Care or Charge of a Physician, Surgeon, or Apothecary, such Physician, Surgeon, or Apothecary shall once at the least in every Two Weeks make an Entry in a Book to be kept for that Purpose, to be called " The Medical Journal," of the Condition of the Patient's Health, both mental and bodily, together with the Date of such Entry, and such Book shall be produced to the Visiting Commissioner on every Visit, and shall be signed by him as having been so produced, and every such Physician, Surgeon, or Apothecary who shall make an untrue Entry in the said Book shall be guilty of a Misdemeanor.

Visitors of licensed Houses may visit single Patients on Request of Commissioners.

XV. It shall be lawful for One or more of the Visitors appointed in or for any County or Borough under the said recited Act, upon the Request in Writing of the Commissioners, or any Two of them, under their Hands so to do, to visit any Person detained in any unlicensed House in such County or Borough as a single Patient, and to inquire into and report to the Commissioners on the Treatment and State of Health, bodily and mental, of such Patient, and to inspect the Order and Certificates on which such Person was received ; and the Provisions of the said recited Act for and concerning the Remuneration or Payment of any such Visitor, being a Physician, Surgeon, or Apothecary, in respect of the Execution of the Duties of that Act, and for the Payment of the Costs, Charges, and Expenses incurred by any Visitor, in Proceedings under that Act, shall extend and be applicable to and for the Remuneration or Payment of any Visitor, being a Physician, Surgeon, or Apothecary, visiting as aforesaid any single Patient, and to and for the Payment of the Costs, Charges, and Expenses incurred by any Visitor in or about such Visit as aforesaid.

Annual Report to be made to the Commissioners by every Medical Man visiting or having Charge of a single Patient.

XVI. Every Physician, Surgeon, and Apothecary who visits any single Patient, or under whose Care or Charge any single Patient shall be, shall on the Tenth Day of *January*, or within Seven Days from that Time, in every Year, report in Writing to the Commissioners the State of Health, bodily and mental, of such Patient, with such other Circumstances as he may deem necessary to be communicated to the Commissioners; and it shall be lawful for the Commissioners, at any other Time and from Time to Time as they see occasion, to call for and require from any such Physician, Surgeon, or Apothecary a Report in Writing relative to any single Patient visited by him or under his Care or Charge in such Form and specifying such Particulars as the Commissioners may direct.

Provisions concerning Discharge of Patients from licensed Houses by Relatives extended to single Patients.

XVII. The Provisions contained in Sections Seventy-two and Seventy-three of the said recited Act for the Discharge of Patients (not being Paupers) from licensed Houses shall extend and be applicable to and for the Discharge of any single Patient : Provided always, that this Enactment shall not extend to authorize the Discharge of any single Patient, if the Physician, Surgeon, or Apothecary who has the Care or Charge of or visits such Patient certify in Writing under his Hand that in his Opinion such Patient is dangerous, and unfit to be at large, together with the Grounds on which such Opinion is founded, unless One of the Commissioners shall consent in Writing to the Discharge of such Patient.

Lord Chancellor, upon Report of Commissioners, may order Discharge, &c. of any single Patient.

XVIII. It shall be lawful for the Lord Chancellor, upon the Report of the Commissioners in Lunacy, to order the Discharge of any Person received or detained as a single Patient, or to give such Orders and Directions in reference to such Patient as the Lord Chancellor shall think fit ; and any Person detaining any such Patient for the Space of Three Days after a Copy of such Order for his Discharge shall have been served on him, or left at the House in which such Person so ordered to be discharged is detained, shall be guilty of a Misdemeanor.

On Recovery of a Patient Notice to be given to Friends, and in the Case of a Pauper to Guardians, &c., and in default of Discharge or Removal, to Commissioners and Visitors. Provision in Case of Death of Patient in any Hospital or licensed House.

XIX. The Superintendent or Proprietor of every registered Hospital and licensed House, and every Person having the Care or Charge of any single Patient, shall forthwith, upon the Recovery of any Patient in such Hospital or House, or of such single Patient, transmit Notice of such Recovery in the Case of a Patient not a Pauper to the Person who signed the Order for his Reception, or by whom the last Payment on account of such Patient was made, and in the Case of a Pauper to the Guardians of his Union or Parish, or if there be no such Guardians to One of the Overseers of the Poor of his Parish, or if such Pauper be chargeable to any County to the Clerk of the Peace thereof, and in case such Patient be not discharged or removed within Fourteen Days from the giving of such Notice, such Superintendent, Proprietor, or Person as aforesaid shall immediately after the

Expiration of such Period transmit Notice of the Recovery of such Patient to the Commissioners, and also, in the Case of a licensed House within the Jurisdiction of any Visitors, to the Clerk of such Visitors, with the Date of the Notice firstly in this Enactment mentioned, and where Notice is so given to the Clerk of any Visitors he shall forthwith communicate the same to the Visitors, or Two of them. One of whom shall be a Physician, Surgeon, or Apothecary ; and in case of the Death of any Patient in any Hospital or licensed House, a Statement setting forth the Time and Cause of the Death, and the Duration of the Disease of which such Patient died, shall be prepared and signed by the Medical Person or Persons who attended the Patient during the Illness which terminated in Death, and such Statement shall be entered in the "Case Book," and a Copy of such Statement, certified by the Superintendent or Proprietor, shall, within Two Days of the Date of the Death, be transmitted to the Coroner for the County or Borough, and in case such Coroner, after receiving such Statement, shall think that any reasonable Suspicion attends the Cause and Circumstances of the Death of such Patient, he shall summon a Jury to inquire into the Cause of such Death.

Provision authorizing Transfer of private and single Patients.

XX. Any Person, having Authority to order the Discharge of any Patient (not being a Pauper) from any Asylum, registered Hospital, or licensed House, or of any single Patient, may, with the previous Consent in Writing of Two of the Commissioners, direct, by an Order in Writing under his Hand, the Removal of such Patient to any Asylum, registered Hospital, or licensed House, or to the Care or Charge of any Person mentioned or named in such Order ; and every such Order and Consent shall be made and given respectively in duplicate, and One of the Duplicates shall be delivered to and left with the Superintendent or Proprietor of the Asylum, Hospital, or House from which or the Person from whose Care or Charge the Patient is ordered to be removed, and the other Duplicate shall be delivered to and left with the Superintendent or Proprietor of the Asylum, Hospital, or House into which or the Person into whose Care or Charge the Patient is ordered to be removed; and such Order for Removal, together with such Consent in Writing, shall be a sufficient Authority for the Removal of such Patient, and also for his Reception into the Asylum, registered Hospital, or licensed House into which or by the Person into whose Care or Charge he is ordered to be removed : Provided always, that a Copy of the Order and Certificates upon which such Patient was received into the Asylum, Hospital, or House from which he is removed, or as a single Patient, by the Person from whose Care he is removed, certified under the Hand of the Superintendent or Proprietor of such Asylum, Hospital, or House, or of such Person as last aforesaid, to be a true Copy, shall be furnished by him free of Expense, and shall be delivered, with one Duplicate of the said Order of Removal and Consent, to the Superintendent or Proprietor of the Asylum, Hospital, or House, to which or to the Person to whose Care or Charge such Patient is removed.

Notice of Discharge of single Patients to be sent to the Commissioners.

XXI. Every Person from whose Care or Charge any single Patient shall be discharged shall transmit to the Commissioners a written Notice of such Discharge within the like Period, and under the like Penalty for Default, as by the said recited Act is required and provided in the Case of the Discharge of a Patient from a licensed House.

Provisions as to Change of Residence of Persons having Charge of single Patients, and temporary Removal of Patients for Benefit of Health.

XXII. It shall be lawful or any Person having the Care or Charge of a single Patient to change his Residence, and remove such Patient to any new Residence of such Person, in *England*, provided that Seven clear Days before such Change of

Residence he give Notice in Writing thereof, and of the Place of such new Residence, to the Commissioners and to the Person who signed the Order for the Reception of such Patient, or by whom the last Payment on account of such Patient was made; and it shall be lawful for any Person having the Care or Charge of any single Patient, having first obtained the Consent of Two of the Commissioners, to take or send such Patient, under proper Control, to any specified Place or Places, for any definite Time, for the Benefit of his Health: Provided always, that before any such Consent shall be given, the Approval in Writing of the Person who signed the Order for the Reception of such Patient, or by whom the last Payment on account of such Patient was made, shall be produced to such Commissioners, unless they shall, on Cause being shown, dispense with the same.

On Representation of Commissioners Lord Chancellor may require Statement of Property of Lunatic detained for One Year.

XXIII. Where any Person has already been received as a Lunatic under Order and Certificates, and shall be detained thereunder, and where any Person shall hereafter be in like Manner received and detained, and the Commissioners represent to the Lord Chancellor that it is desirable that the Extent and Nature of his Income should be ascertained, and the Application thereof, the Lord Chancellor may, if he think fit, through the Registrar in Lunacy, require that the Person signing the Order, or other the Person paying for the Care and Maintenance of the Lunatic or having the Management of the Property, shall transmit to the Lord Chancellor a Statement in Writing, to the best of his Knowledge, of the Particulars of the Property and Income of the Lunatic and of the Application of the Income.

Form of Notice of Admission.

XXIV. The Notice of Admission and Statement mentioned or referred to in Section Fifty-two of the said recited Act shall hereafter be according to the Form mentioned in Schedule (C.) annexed to this Act, in lieu of the Form set forth in Schedule (F.) to the said recited Act; and such Statement shall be signed by the Medical Superintendent, Proprietor, or Attendant of the Hospital or licensed House from which the same is sent, and the said Notice and Statement shall be accompanied by a Copy of the several Documents mentioned in the said Notice.

Form of Medical Visitation Book.

XXV. The Medical Visitation Book mentioned in Section Fifty-nine of the said recited Act shall henceforth be kept in the Form set forth in Schedule (D.) annexed to this Act, in lieu of the Form set forth in Schedule (H.) to the said recited Act; and the said Section shall be construed as if the Particulars mentioned in the several Heads of the said Form in the said Schedule (D.) had by the said Section been required to be entered in the said Book in lieu of the Particulars mentioned in the said Section.

Notice of Dismissal for Misconduct of Attendants to be sent to Commissioners.

XXVI. The Superintendent or Proprietor of every registered Hospital or licensed House shall, within One Week after the Dismissal for Misconduct of any Nurse or Attendant employed in such Hospital or House, transmit to the Commissioners, by the Post, Information in Writing under his Hand of such Dismissal, and of the Cause thereof; and every Superintendent or Proprietor neglecting to transmit such Information to the Commissioners within the Period aforesaid shall for every such Offence forfeit any Sum not exceeding Ten Pounds.

Powers vested in Private Committee to be vested in the Commissioners.

XXVII. Section Eighty-nine of the said recited Act, constituting from among the Commissioners a Private Committee for the Purposes in the said Act mentioned shall be repealed, and all the Powers vested in, and all the Provisions of the said Act applicable to, the said Private Committee, or One or Two Members thereof, shall be vested in and be applicable to the Commissioners, or One Commissioner, or Two Commissioners, (as the Case may require,) as if, where in the said Act the said Private Committee, or One Member or Two Members thereof, (as the Case may be,) is or are mentioned or referred to, the Commissioners, or One Commissioner, or Two Commissioners, (as the Case may require,) had been mentioned or referred to, instead thereof.

Repeal of Section 111. of recited Act, and Provision as to Visitation of Workhouses.

XXVIII. Section One hundred and eleven of the said recited Act shall be repealed, and any One or more of the Commissioners shall and may on such Day or Days, and at such Hours in the Day, and for such Length of Time as he or they shall think fit, visit all such Parish and Union Workhouses in which there shall be or be alleged to be any Lunatic, as the Commissioners shall by any Resolution or Resolutions of the Board direct, and shall inquire whether the Provisions of the Law as to Lunatics in such Parish or Union have been carried out, and also as to the Dietary, Accommodation, and Treatment of the Lunatics in such Workhouses, and shall report in Writing thereon to the Poor Law Board.

Commissioners may in any special Case employ Persons to make the necessary Inquiries, and report to them thereon.

XXIX. It shall be lawful for the Commissioners, where, for any Reasons to be entered upon the Minutes of the Board, any Case appears to them specially to call for immediate Investigation, to authorize and direct, by an Order under their Common Seal, any competent Person or Persons to visit and examine and report to them upon the mental and bodily State and Condition of any Lunatic or alleged Lunatic in any Asylum, Hospital, or licensed House, or of any Pauper Lunatic in a Workhouse or elsewhere, or of any Lunatic or alleged Lunatic under the Care or Charge of any Person as a single Patient, and to inquire into and report upon any Matters into which the Commissioners are authorized to inquire; and every such Person shall, for the special Purposes mentioned in such Order, have all the Powers of a Commissioner; and the Commissioners may allow to every such Person a reasonable Sum for his Services and Expenses, such Sum to be paid in Manner provided by the said recited Act with regard to Expenses incurred by or under the Authority of the Commissioners in Proceedings thereunder; but this Enactment shall not be taken to exonerate the Commissioners from the Performance of any Duty by Law imposed on them.

Regulations for Hospitals to be submitted to Secretary of State.

XXX. The Committee having the Management or Government of every registered Hospital shall, within Three Months after the passing of this Act in the Case of every Hospital now registered, and within Three Months after the Registration of every Hospital hereafter to be registered under the said recited Act, submit the existing Regulations, or Regulations to be framed by such Committee, to One of Her Majesty's Principal Secretaries of State, for his Approval, and any such Committee may, with the like Approbation, alter and vary such Regulations as they think necessary; and all such Regulations so approved shall be printed, abided by, and observed, and a Copy thereof shall be sent to the Commissioners, and another Copy thereof kept hung up in the Visitors' Room of the Hospital.

Commissioners may make Regulations for the Government of licensed Houses.

XXXI. It shall be lawful for the Commissioners, with the Sanction and Approbation of One of Her Majesty's Principal Secretaries of State, from Time to Time to make Regulations for the Government of any House licensed for the Reception

APPENDIX.

of Lunatics ; and such Regulations of the Commissioners, or a Copy thereof, shall be transmitted by their Secretary to the Proprietor or resident Superintendent of every licensed House to which the same relate, and shall be abided by and observed therein.

Time at which Reports of Commissioners to the Lord Chancellor as to State of Asylums, &c., are to be made.

XXXII. The Report required by Section Eighty-eight of the said recited Act to be made by the Commissioners to the Lord Chancellor in the Month of *June* in every Year of the State and Condition of the several Houses, Hospitals, Asylums, and other Places visited by them under that Act, and of the Care of the Patients therein, and of such other Particulars as they think deserving of Notice, shall be made in or before the Month of *March* in every Year, and shall be made up to the End of the preceding Year.

Provision for Payment of Persons employed to inspect Places where Lunatics are confined extended to Persons visiting under s. 112. of 8 & 9 Vict. c. 100.

XXXIII. The Provision in Section One hundred and thirteen of the said recited Act, for and concerning the Payment for Attendance and Trouble of any Person (not being a Commissioner) employed under that Enactment, and of the travelling or other Expenses of any Person so employed, and as to the Fund out of which such Payment is to be made, shall extend and be applicable to and in the Case of any Person (not being a Commissioner) required to visit and examine any Lunatic or supposed Lunatic under Section One hundred and twelve of the said recited Act.

Penalty on Persons obstructing Execution of Orders of Lord Chancellor or Secretary of State made under ss. 112. or 113. of recited Act, or of Commissioners made under this Act.

XXXIV. Any person who wilfully obstructs the Commissioners or any of them, or any other Person authorized by an Order in Writing under the Hand of the Lord Chancellor or Her Majesty's Principal Secretary of State for the Home Department, pursuant to the Provisions of Section One hundred and twelve or One hundred and thirteen of the said recited Act, to visit and examine any Lunatic or supposed Lunatic, or to inspect or inquire into the State of any Asylum, Hospital, Gaol, House, or Place wherein any Lunatic or Person represented to be lunatic is confined or alleged to be confined, in the Execution of such Order, and any Person who wilfully obstructs any Person authorized under this Act by any Order of the Commissioners to make any Visit and Examination or Inquiry in the Execution of such Order, shall (without Prejudice to any Proceedings, and in Addition to any Punishment to which such Person obstructing the Execution of such Order would otherwise be liable,) forfeit for every such Offence any Sum not exceeding Twenty Pounds.

Sect. 116 of recited Act repealed, and Bethlehem Hospital to be subject to this Act.

XXXV. Section One hundred and sixteen of the said recited Act shall be repealed, and the Royal Hospital of *Bethlehem* shall henceforth be subject to the Provisions of the said recited Act and of this Act, in the same Manner as if the same had not been exempted from the said recited Act, and shall be forthwith registered as an Hospital accordingly, in pursuance of Section Forty-three of the said recited Act.

Interpretation of Terms.

XXXVI. In the Construction of the said recited Act and of this Act the Words "Physician," "Surgeon," and "Apothecary" shall respectively mean a Physician, Surgeon, and Apothecary duly authorized or licensed to practise as such by or as a Member of some College, University, Company, or Institution legally constituted and qualified to grant such Authority or Licence in some Part of the United Kingdom, or having been in Practice as an Apothecary in *England* or *Wales* on or before the First Day of *August*, One thousand eight hundred and fifteen, and being in actual Practice as such Physician, Surgeon, or Apothecary; the Expression " Officiating Clergyman of the Parish" shall include the Chaplain of the Workhouse of the same Parish, or of the Workhouse of the Union to which such Parish belongs; the Expression "single Patient" shall mean any Person received or taken charge of as a Lunatic under Section Ninety of the said recited Act, or under such Section as amended by this Act; and the Expression "Attendant" shall mean any Person, whether Male or Female, who shall be employed either wholly or partially in the personal Care, Control, or Management of any Lunatic in any registered Hospital, or licensed House, or of any single Patient; and in the Construction of this Act the Word " Board," as used in relation to the Commissioners in Lunacy, shall mean any Three or more of the Commissioners assembled at a Meeting convened in pursuance of Section Sixteen of the said recited Act, or holden under any Order or Rule for the Time being in force made under Section Seventy of the said recited Act for regulating the Duties of the Commissioners.

Recited Act and this Act to be construed as One Act, &c.

XXXVII. The said recited Act and this Act shall be construed together as One Act, and a Queen's Printer's Copy of this Act shall be bound up in the " Visitors Book" of every Hospital and licensed House together with the said recited Act.

Act not to affect Provisions relating to Criminal Lunatics, 39 & 40 G. 3. c. 94., 1 & 2 Vict. c. 14., *and* 3 & 4 Vict. c. 54., *save as herein provided.*

XXXVIII. Nothing in this Act shall affect the Provisions of any of the following Acts; (that is to say,) an Act of the Session holden in the Thirty-ninth and Fortieth Years of King *George* the Third, Chapter Ninety-four; an Act of the Session holden in the First and Second Years of Her Majesty, Chapter Fourteen; and an Act of the Session holden in the Third and Fourth Years of Her Majesty, Chapter Fifty-four, or any other Provisions concerning Criminal Lunatics, save as hereinafter provided; that is to say, it shall be lawful for One of Her Majesty's Principal Secretaries of State to issue his Warrant to remove or discharge any insane Person who shall be in Custody under the Provisions of the said Act of the Third and Fourth Years of Her Majesty, Chapter Fifty-four, provided it shall be duly certified to such Secretary of State, by two Physicians or Surgeons, that such insane Person was harmless and might be discharged from Restraint as an insane Person without Danger to himself or to others, in like Manner as if it had been certified to such Secretary of State that such Person had become of sound Mind, anything in the said Act or any other Act to the contrary thereof in anywise notwithstanding.

Secretary to the Commissioners, if at the Time of his Appointment a practising Barrister of Five Years standing, eligible to be appointed a Commissioner.

XXXIX. And whereas by the said recited Act it is provided that every Person to be appointed in the Room of any Commissioner, being a Barrister of Five Years standing at the Bar and upwards, shall be a practising Barrister of not less than Five Years standing at the Bar: And whereas it is expedient to amend the said Provisions as herein-after mentioned; the present or any future Secretary to the Commissioners, if at the Time of his Appointment to be such Secretary he was or shall have been a practising Barrister of not less than Five Years standing at the Bar, shall be eligible to be appointed a Commissioner in the Room of any such Commissioner as aforesaid.

Commencement of Act.

XL. This Act shall commence and come into operation on the First Day of *November,* One thousand eight hundred and fifty-three.

APPENDIX.

SCHEDULES to the foregoing Act.

SCHEDULE (A.) No. 1. Sections 4, 8.

ORDER for the RECEPTION of a PRIVATE PATIENT.

I, the Undersigned, hereby request you to receive *A.B.*, a Lunatic, [*or* an Idiot, *or* a Person of unsound Mind,] as a Patient into your House [*or* Hospital]. Subjoined is a Statement respecting the said *A.B.*

      (Signed)   Name.
              Occupation (if any).
              Place of Abode.
              Degree of Relationship (if any), or other Circumstance of Connexion with the Patient.

Dated this   Day of   One thousand eight hundred and   .
To   Proprietor [*or* Superintendent] of [*describing the House or Hospital by Situation and Name, if any*].

STATEMENT.

[*If any Particulars in this Statement be not known, the Fact to be so stated.*]
Name of Patient, with Christian Name at Length.
Sex and Age.
Married, single, or widowed.
Condition of Life, and previous Occupation (if any).
The Religious Persuasion, as far as known.
Previous Place of Abode.
Whether First Attack.
Age (if known) on First Attack.
When and where previously under Care and Treatment.
Duration of existing Attack.
Supposed Cause.
Whether subject to Epilepsy.
Whether suicidal.
Whether dangerous to others.
Whether found lunatic by Inquisition, and Date of Commission or Order for Inquisition.
Special Circumstances (if any) preventing the Patient being examined, before Admission, separatly by Two Medical Practitioners.
    (Signed)   Name.

[*Where the Person signing the Statement is not the Person who signs the Order, the following Particulars concerning the Person signing the Statement are to be added; vizt.,*]

Occupation (if any).
Place of Abode.
Degree of Relationship (if any) or other Circumstance of Connexion with the Patient.]

SCHEDULE (A.) No. 2. Sections 4, 5, 8, 10, 11, 12, 13.

FORM OF MEDICAL CERTIFICATE.

I, the Undersigned [*here set forth the Qualification entitling the Person certifying to practise as a Physician, Surgeon, or Apothecary*, ex. gra., being a Fellow of the Royal College of Physicians in London], and being in actual Practice as a

[Physician, Surgeon, or Apothecary, *as the Case may be*], hereby certify, That I, on the   Day of   at
[*here insert the Street and Number of the House (if any) or other like Particulars*], in the County of   , separately from any other Medical Practitioner, personally examined *A.B.* of
[*insert Residence and Profession or Occupation, if any*], and that the said *A.B.* is a [Lunatic, *or* an Idiot, *or* a Person of unsound Mind], and a proper Person to be taken charge of and detained under Care and Treatment, and that I have formed this Opinion upon the following Grounds ; viz.
    1. Facts indicating Insanity observed by myself [*here state the Facts*].
    2. Other Facts (if any) indicating Insanity communicated to me by others [*here state the Information, and from whom*].
                                (Signed)
                                                              Place of Abode.
    Dated this   Day of   One thousand eight hundred and   .

SCHEDULE (B.) No. 1. Section 7.

ORDER for the RECEPTION of a PAUPER PATIENT.

    I, *C.D.* [*or, in the Case of a Clergyman and Relieving Officer, &c.*, we, *C.D.* and *E.F.*], the Undersigned, having called to my [*or* our] Assistance a Physician, [*or* Surgeon, *or* Apothecary, *as the Case may be*,] and having personally examined *A.B.*, a Pauper, and being satisfied that the said *A.B.* is a Lunatic [*or* an Idiot, *or* a Person of unsound Mind,] and a proper Person to be taken charge of and detained under Care and Treatment, hereby direct you to receive the said *A.B.* as a Patient into your House [*or* Hospital]. Subjoined is a Statement respecting the said *A.B.*
                                (Signed)   *C.D.*
                                A Justice of the Peace for the County, City, or Borough of
                                [*or* an *or* the Officiating Clergyman of the Parish of   ].
                                (Signed)   *E.F.*
                                The Relieving Officer of the Union or Parish of   [*or* an Overseer of the Parish of   ].
    Dated the   Day of   One thousand eight hundred   .
To   Proprietor [*or* Superintendent] of [*describing the House or Hospital*].

STATEMENT.

[*If any Particulars in this Statement be not known, to be so stated.*]
Name of Patient, and Christian Name at Length.
Sex and Age.
Married, single, or widowed.
Condition of Life, and previous Occupation (if any).
The Religious Persuasion, as far as known.
Previous Place of Abode.
Whether First Attack.
Age (if known) on First Attack.
When and where previously under Care and Treatment.
Duration of existing Attack.
Supposed Cause.
Whether subject to Epilepsy.
Whether suicidal.

Whether dangerous to others.
Parish or Union to which the Lunatic is chargeable.
Name and Christian Name and Place of Abode of nearest known Relative of the Patient, and Degree of Relationship (if known).
I certify that, to the best of my Knowledge, the above Particulars are correctly stated.

(Signed)
Relieving Officer [or Overseer].

SCHEDULE (B.) No. 2. Sections 7, 10, 11, 12, 13.

Form of Medical Certificate.

I, the Undersigned [here set forth the Qualification entitling the Person certifying to practise as a Physician, Surgeon, or Apothecary, ex. gra., being a Fellow of the Royal College of Physicians in London], and being in actual Practice as a [Physician, Surgeon, or Apothecary, as the Case may be], hereby certify, That I, on the   Day of   at   [here insert the Street and Number of the House (if any) or other like Particulars], in the County of   , personally examined A. B. of   [insert Residence and Profession or Occupation (if any)], and that the said A. B. is a [Lunatic, or an Idiot, or a Person of unsound Mind], and a proper Person to be taken charge of and detained under Care and Treatment, and that I have formed this Opinion upon the following Grounds ; viz.

1. Facts indicating Insanity observed by myself [here state the Facts].
2. Other Facts (if any) indicating Insanity communicated to me by others [here state the Information, and from whom].

(Signed)

Place of Abode.

Dated this   Day of   One thousand eight hundred and   .

SCHEDULE (C.) Section 24.

Notice of Admission.

I HEREBY give you Notice, That A.B. was admitted into this House [or Hospital] as a Private [or Pauper] Patient on the   Day of   and I hereby transmit a Copy of the Order and Medical Certificates [or Certificate] on which he was received. [*If a Private Patient be received upon One Certificate only, the special Circumstances which have prevented the Patient from being examined by Two Medical Practitioners to be here stated, as in the Statement accompanying the Order for Admission.*]

Subjoined is a Statement with respect to the mental and bodily Condition of the above-named Patient.

(Signed)
Superintendent [or Proprietor] of
Dated the   Day of   One thousand eight hundred and   .

Statement.

I have this Day [*some Day not less than Two clear Days after the Admission of the Patient*] seen and examined   the Patient mentioned in the above Notice, and hereby certify that with respect to mental State he [or she], and that with respect to bodily Health and Condition, he [or she]

(Signed)
Medical Proprietor [or Superintendent, or Attendant] of
Dated the   Day of   One thousand eight hundred and   .

SCHEDULE (D.) Section 25.

Form of Medical Visitation Book.

Date.	Number and Class of Patients.				Patients who are, or since the last Entry have been, under Restraint, or in Seclusion, when, and for what Period, and Reasons, and, in Cases of Restraint, by what Means.				Patients under Medical Treatment, and for what (if any) bodily Disorder.		Deaths, Injuries, and Violence to Patients since the last Entry.
	Private.		Pauper.		Restraint.		Seclusion.				
	M.	F.	M.	F.	Males.	Females.	Males.	Females.	Males.	Females.	

ANNO DECIMO SEXTO & DECIMO SEPTIMO VICTORIÆ REGINÆ.

CAP. XCVII.

An Act to consolidate and amend the Laws for the Provision and Regulation of Lunatic Asylums for Counties and Boroughs, and for the Maintenance and Care of Pauper Lunatics, in *England*.

[20th *August* 1853.]

8 & 9 *Vict. c.* 126., 9 & 10 *Vict. c.* 84., *and* 10 & 11 *Vict. c.* 43. *repealed, but not to affect Appointments, &c.*

BE it enacted by the Queen's most Excellent Majesty, by and with the Advice and Consent of the Lords Spiritual and Temporal, and Commons, in this present Parliament assembled, and by the Authority of the same, as follows:

I. The following Acts relating to Lunatic Asylums for Counties and Boroughs, and the Maintenance and Care of Pauper Lunatics, in *England*, (that is to say,) an Act of the Session holden in the Eighth and Ninth Years of Her Majesty, Chapter One hundred and twenty-six, an Act of the Session holden in the Ninth and Tenth Years of Her Majesty, Chapter Eighty-four, and an Act of the Session holden in the Tenth and Eleventh Years of Her Majesty, Chapter Forty-three, shall be repealed; but such Repeal shall not interfere with or affect any Appointment, Salary, or Annuity made or granted, or Act done, or Agreement or Contract entered into or made, or prevent or defeat any Prosecution or Proceeding for any Offence committed or any Penalty or Forfeiture incurred before the Commencement of this Act, but every such Agreement or Contract shall and may (subject to the Provisions herein-after contained in relation thereto) be carried into effect and enforced, and every such Offence prosecuted, and every such Penalty and Forfeiture sued for, recovered, and applied, and every pending Prosecution or Proceeding continued, in like Manner as if this Act had not been passed.

AS TO PROVIDING ASYLUMS AND APPOINTMENT OF COMMITTEES OF VISITORS.

Justices of County and Borough not having a Lunatic Asylum to provide one, and Justices of the County or Recorder of the Borough at or before a certain Time to direct Notice to be given of the Intention to appoint a Committee for that Purpose.

II. The Justices of every County and (save as herein-after otherwise provided) of every Borough not having an Asylum for the Pauper Lunatics thereof, shall provide an Asylum in manner herein directed, (that is to say,) the Justices of every such County and the Recorder of every such Borough shall at or before the General or Quarter Sessions for such County or Borough next after the Twentieth Day of *December* One thousand eight hundred and fifty-three direct public Notice to be given by the Clerk of the Peace of such County or Borough, in some Newspaper or Newspapers commonly circulated in such County or Borough, of the Intention of the Justices of such County or Borough to appoint at the then next General or Quarter Sessions for such County, or (in the Case of a Borough) at a Special Meeting of the Justices of such Borough to be fixed in such Notice, and to be holden within Three Months from the Date thereof, a Committee of Justices to provide an Asylum for the Pauper Lunatics of such County or Borough, under the Provisions of this Act ; and the Clerk of the Peace of such County or Borough shall, within Ten Days after being so directed as aforesaid, cause such Notice to be given accordingly.

Justices to appoint a Committee to superintend the providing of an Asylum, or to treat for uniting with some County, &c., or to effect one or other of such Purposes.

III. The Justices of every such County and Borough respectively (such Notice having been given as aforesaid) shall at the then next General or Quarter Sessions for such County, or at such Special Meeting as aforesaid of the Justices of such Borough, either themselves determine in which of the Modes herein-after mentioned an Asylum shall be provided for such County or Borough, or shall refer the Selection to the Committee to be appointed as herein-after mentioned, and shall elect some Justices of such County or Borough to be a Committee to provide such Asylum, and may authorize such Committee to provide such Asylum, in such of the Modes herein-after mentioned as the said Justices shall have determined, (that is to say,) to superintend the erecting or providing of an Asylum for the Pauper Lunatics of such County or Borough for such County or Borough alone, or to treat and enter into an Agreement for uniting with any County or Counties, Borough or Boroughs, alone or together with the Subscribers to any Hospital for the Reception of Lunatics, established or in course of Erection, or afterwards to be established, or for uniting with any County or Counties and Borough or Boroughs jointly, or jointly and also together with the Subscribers to any such Hospital as aforesaid, or otherwise providing an Asylum under or for the Purposes of this Act, as the Justices appointing such Committee may have determined, or in case the said Justices appointing such Committee think fit to refer the Selection of the Mode in which such Asylum shall be provided to the Committee, they may authorize such Committee to provide such Asylum in such of the Modes aforesaid as to the Committee may seem best ; and any Committee so authorized to treat and enter into an Agreement may treat and enter into such Agreement with any Committee or Committees having due Authority in that Behalf under this Act, or any former Act, for any County or Counties, Borough or Boroughs, or on behalf of any such Subscribers as aforesaid, and with any Committee of Visitors of any existing Asylum, and whether or not any previous

Q

Agreement for uniting may have been already entered into between some of the Parties under this Act or any former Act; and by any such Agreement to be entered into as aforesaid the several Committees, Parties thereto, may, to the Extent of their Authority, in lieu of agreeing to erect or provide an Asylum, or in addition thereto, and in consideration of any Payment in gross or of the Payment of any Sum in the Nature of Rent or otherwise, agree for the joint Use of any existing Asylum or Hospital, and, where they think fit, for enlarging the same.

Subscribers to any Hospital empowered to appoint a Committee to treat for uniting with any County or Borough, &c.

IV. It shall be lawful for the major Part of such of the Subscribers to any such Hospital as aforesaid as shall be present at any Meeting of such Subscribers called together expressly for this Purpose by Advertisement in a Newspaper commonly circulated in the Place where such Hospital is or is intended to be situate, to elect any Number of such Subscribers not exceeding Five to be a Committee to treat and enter into an Agreement for uniting with any County or Counties or Borough or Boroughs alone, or any County or Counties and Borough or Boroughs jointly, under and for the Purposes of this Act; and where any such Agreement has been or shall be entered into under any former Act or this Act, nothing in this Act shall prevent the Reception into the Asylum provided under such Agreement, or the Discharge therefrom, of so many of any Lunatics other than Pauper Lunatics as might have been received into such Hospital or Asylum if this Act had not been passed.

Committees of Visitors of existing Asylums may enter into Agreements to unite.

V. It shall be lawful for the Committee of Visitors of any Asylum already provided for any County or Borough, alone or otherwise, to enter into an Agreement for uniting for the Purposes of this Act with any County or Counties, Borough or Boroughs, alone or together with the Subscribers to any such Hospital as aforesaid, or for uniting with any County or Counties and Borough or Boroughs jointly, or jointly and also together with the Subscribers to any such Hospital.

Saving where a Committee is already appointed, or Proceedings for the Appointment of a Committee have been commenced.

VI. Provided always, That where a Committee has been appointed before the Commencement of this Act for any County or Borough for any of the Purposes aforesaid, or Proceedings have been taken for or towards the Appointment of a Committee for any of the said Purposes, nothing herein contained shall render it necessary to proceed afresh to the Appointment of a Committee for any of such Purposes; and any Proceedings already taken as aforesaid shall remain in force and be continued; and all the Provisions of this Act shall be applicable to any such Committee already appointed, or to be appointed under such Proceedings, in like Manner as if such Committee had been appointed under the Provisions of this Act.

Justices of Boroughs may Contract with Committees of Visitors, &c., for Reception of the Pauper Lunatics of the Borough.

VII. Provided also, That it shall be lawful for the Justices of any such Borough as aforesaid, at such Special Meeting, if they think fit, in lieu of electing a Committee to superintend the erecting or providing of an Asylum, or to treat for uniting, as herein-before mentioned, or to effect either of such Purposes, to elect a Committee of Justices of such Borough to contract with any Committee of Visitors of any existing Asylum, or any Committee providing or about to provide an Asylum, whether for any County or Borough, alone or otherwise, for the Reception of the Pauper Lunatics of such first-mentioned Borough into such Asylum, in consideration of such Payment in gross, or such annual or periodical Payment, and upon and subject to such Terms, Stipulations, and Conditions as to the Duration and Determination of the Contract, and otherwise, as may be agreed upon; and it shall be lawful for any Committee of Visitors of any existing Asylum, or any other such Committee as last aforesaid, to contract with the Committee for any such Borough accordingly; and during the Continuance of such Contract the Justices of such Borough shall, at a Special Meeting of such Justices to be holden within Twenty Days after the Twentieth Day of *December* in every Year, appoint a Committee of such Justices to visit the Pauper Lunatics sent from such Borough to such Asylum, and Two at least of the Members of such Committee shall together once at the least in every Six Months visit such Asylum, and see and examine as far as Circumstances will permit every Lunatic received into such Asylum under such Contract, and shall after each such Visit report the Result thereof, with such Remarks as they think fit, to the Justices of such Borough at a Special Meeting of such Justices; and the Justices making any such Visit may, if they see fit, be accompanied by some Physician, Surgeon, or Apothecary, other than a Medical Officer of the Asylum; and such Justices may by Writing under their Hands order the Payment to such Physician, Surgeon, or Apothecary of such reasonable Sum for his Services on any such Visit as they may think fit, and such Sum shall, upon the Production of such Order, be paid to such Physician, Surgeon, or Apothecary by the Treasurer of such Borough; and every Report of such Justices so visiting shall be entered among the Records of the Court of Quarter Sessions of such Borough, and shall be open to the Inspection of any of the Commissioners in Lunacy; and such Commissioners may, if they think fit, require a Copy of every or any such Report to be transmitted to them by the Clerk of the Peace of such Borough; and while any such Contract making adequate Provision for the Pauper Lunatics of such Borough is in force such Borough shall not be required to provide an Asylum for itself alone, or in Union, as hereinbefore mentioned.

Boroughs now contributing to a County Asylum deemed to have an Asylum, but upon Notice may separate from the County.

VIII. Provided also, That every Borough situate within a County having an Asylum for Pauper Lunatics, and which at the Time of the passing of the said Act of the Eighth and Ninth Years of Her Majesty contributed and still contributes to such Asylum, shall be considered as having an Asylum for the Pauper Lunatics of such Borough; but it shall be lawful for any such Borough, at any Time hereafter, upon giving Six Months Notice in Writing under the Hand of the Town Clerk, in pursuance of a Resolution of the Council of such Borough, to the Clerk of the Peace of the County, to separate itself, so far as relates to the Establishment of a Lunatic Asylum for such County, and the Maintenance of Lunatics therein, from such County, and from and after the Expiration of such Notice such Borough shall for the Purposes of this Act be deemed a Borough not having an Asylum for the Pauper Lunatics thereof; and from and after the Expiration of such Notice, and the Withdrawal from such County Asylum of all Lunatics from or belonging to such Borough, such Borough shall not be liable to pay or contribute towards the Expense of the Establishment of such Asylum, or the Maintenance of Lunatics therein, but until the Withdrawal from such County Asylum of all Lunatics from or belonging to such Borough such Borough shall be liable to contribute towards the Expenses of such Asylum, in the same Manner and to the same Extent as if such Notice had not been given.

Every Borough not having Six Justices, besides the Recorder, to be annexed to the County or One of the Counties in which it is situate, for the Purposes of this Act. Recorder to appoint Two Justices to be Members of Committee of Visitors. 5 & 6 W. 4. c. 76.

IX. Provided also, That every Borough in which at the passing of the said Act of the Eighth and Ninth Years of Her Majesty hereby repealed, there were not Six Justices besides a Recorder shall, for the Purposes of this Act, be annexed to and be Part of the County in which it is wholly situate, or in case it be not wholly situate in any One County shall for the Purposes of this Act be annexed to and be Part of such One of the Counties in which it is situate as such Borough may have been annexed to under the said Act of the Eighth and Ninth Years of Her Majesty, or if not already so annexed then the same shall be annexed to and be

Part of such One of the said Counties as One of Her Majesty's Principal Secretaries of State shall by Writing under his Hand direct; and the Recorder of every such Borough shall, at the General or Quarter Sessions next after the Twentieth Day of *December* in every Year, appoint Two Justices of such Borough to be Members of the Committee of Visitors of the Asylum of the County to which such Borough is or shall be annexed; and the Justices of every County to which any Borough is or shall be annexed as aforesaid shall, at their General or Quarter Sessions, from Time to Time fix the Sum to be contributed by such Borough towards the Expenses of and incident to erecting, providing, and maintaining the Asylum of such County, according to the comparative Population of such Borough and County as stated in the then last Returns made of the same under the Authority of Parliament, and cause Notice thereof in Writing to be given to the Treasurer of such Borough, and such Sum shall be raised by a Borough Rate to be made by the Council of the Borough in manner directed by the Act of the Session holden in the Fifth and Sixth Years of King *William* the Fourth, "to provide for the Regulation of Municipal Corporations in *England* and *Wales*," or out of the Borough Fund, if the Council think fit, and shall be paid by the Treasurer of the Borough to the Treasurer of the Asylum.

Boroughs neglecting to provide an Asylum or to contract for the Care of their Pauper Lunatics may be annexed by Secretary of State to the County. Justices of Borough so annexed shall appoint Two Justices to be Members of Committee of Visitors.

X. If at any Time after the Expiration of One Year after the passing of this Act it appear to One of Her Majesty's Principal Secretaries of State, upon the Report of the Commissioners in Lunacy, that the Justices of any Borough by this Act required to provide an Asylum, or contract for the Care of the Pauper Lunatics thereof, have not provided an Asylum, or entered into an Agreement for that Purpose, or into a subsisting Contract making adequate Provision for the Care of the Pauper Lunatics thereof in some Asylum, and that any Asylum belonging wholly or in part to the County or any of the Counties (if more than One) in which such Borough is locally situate, either wholly or in part, is capable of affording Accommodation for the Pauper Lunatics of such Borough, or may be conveniently enlarged so as to afford such Accommodation, it shall be lawful for such Secretary of State, with the Consent of the Committee of Visitors of such Asylum, by Writing under his Hand, to annex such Borough for the Purposes of this Act to such County; and the Justices of every Borough so annexed under this Provision shall, at a Special Meeting of such Justices to be holden within Twenty Days after the Twentieth Day of *December* in every Year, appoint Two Justices of such Borough to be Members of the Committee of Visitors of the Asylum of the County to which such Borough shall be annexed; and the Provision in the Enactment lastly herein-before contained in relation to the Contribution by a Borough annexed to a County under such Enactment to the Expenses of the Asylum of such County, shall extend to any Borough so annexed under this Provision.

Powers of Committees may be enlarged.

XI. Where any Committee has been appointed for any County or Borough (whether before or after the passing of this Act) for any of the Purposes herein-before mentioned, it shall be lawful for the Justices of such County or Borough, if they think fit, at any General or Quarter Sessions for such County, or (in the Case of a Borough) at any Special Meeting of the Justices of such Borough, after like public Notice as is required in the Case of the First Appointment of the Committee, to enlarge or alter the Powers of the Committee so as to vest in the Committee any such Powers as might be vested in any Committee on the original Appointment thereof under this Act, and, if the Justices see fit so to do, to appoint additional Members of the said Committee, and every such Committee shall have the like Powers, and the Provisions of this Act shall be applicable to such Committee in like Manner, as if such Committee had been originally appointed with the Powers so vested in them under such Enlargement or Alteration of their Powers.

APPENDIX. 229

New Committees to be appointed in lieu of Committees which have ceased or shall hereafter cease to exist, &c.

XII. Where any Committee appointed for any County or Borough (either before or after the passing of this Act) for any of the Purposes hereinbefore mentioned has ceased or shall hereafter cease to exist, without carrying into effect the Purposes for which it was appointed, or, if appointed for the Purpose only of treating for uniting or of contracting as aforesaid, has reported or shall hereafter report that it is not practicable or expedient to enter into an Agreement for uniting or into the proposed Contract, or to that Effect, the Justices of such County or the Recorder of such Borough shall, at or before the General or Quarter Sessions next after the passing of this Act, or next after the Occasion has arisen, cause public Notice to be given, in manner herein directed in the Case of the original Appointment of a Committee under this Act for any of the said Purposes, of the Intention of the Justices of such County or Borough to appoint at the then next General or Quarter Sessions for such County, or (in the Case of a Borough) at some Special Meeting of the Justices of such Borough to be fixed in the Notice and to be holden within Three Months from the Date thereof, a Committee in lieu of the Committee previously appointed as aforesaid; and such Notice having been so given, the Justices of such County or Borough shall, at the then next General or Quarter Sessions for such County, or at such Special Meeting as aforesaid of the Justices of such Borough, appoint a Committee accordingly, and shall have the like Discretion and Authority for determining the Purposes for which such Committee shall be appointed as in the Case of an original Appointment of a Committee under the Provisions hereinbefore contained; or such Justices may, if they think fit, in lieu of appointing a new Committee in the Place of any such Committee appointed only for the Purpose of treating for uniting or of contracting as aforesaid, and which may have reported that it is not practicable or expedient to enter into an Agreement for uniting or into the proposed Contract, or to that Effect, enlarge or alter the Powers of such Committee as hereinbefore provided, and, if such Justices think fit, appoint additional Members of such Committee.

Notice for Appointment of a Committee given at a Time subsequent to that required by this Act, and the Appointment of such Committee, to be valid.

XIII. Provided always, That where the Justices of any County or the Recorder of any Borough have or has not, in pursuance of any of the Provisions hereinbefore contained, at or before such General or Quarter Sessions as in that Behalf required, caused Notice to be given of the Intention of the Justices of such County or Borough to appoint a Committee under this Act, it shall be lawful for the Justices of such County or the Recorder of such Borough, at or before any subsequent General or Quarter Sessions, to cause such Notice to be given in manner required by this Act; and the Appointment of a Committee in pursuance of such Notice, or the Enlargement or Alteration of the Powers of any existing Committee, and the Appointment of any additional Members of such Committee, at the Sessions or Meeting for which such Notice has been given, shall be valid.

Committees uniting to enter into Agreement in the Form in Schedule (A.)

XIV. When Two or more Committees agree to unite for the Purposes of this Act, an Agreement shall be entered into and signed by the several Committees uniting, or the major Part of such Committees respectively, in the Form or to the Effect set forth in Schedule (A.) to this Act; and such Agreement, when signed by the major Part of each such Committee, and not before, shall be binding upon every County and Borough, and the Subscribers (if any) for or on behalf of which or whom such Agreement has been entered into; and every such Agreement shall specify the Proportion in which the Expenses necessary for carrying into execution the Purposes of this Act shall be charged upon each County and Borough, and the Subscribers (if any) so uniting; and the Proportions of the Counties and Boroughs uniting shall be calculated and fixed with reference to their respective Populations as stated in the then last Return made of the same under the Authority of Parliament; and where under any such Agreement a Right to the joint Use of any existing Asylum or Hospital is required by any County or Borough, or the Subscribers to any Hospital, such Agreement shall fix the Sum to be paid by such County, Borough, or Subscribers towards the Expenses already incurred in erecting or providing such Asylum or Hospital.

Additional Stipulations or Conditions may be inserted in Agreement, but not so as to subject Acts of Visitors to Control of General or Quarter Sessions.

XV. Provided always, That it shall be lawful for such Committees to insert in the Agreement to be entered into by them any Stipulations or Conditions, in addition to the Matters by this Act required to be specified in such Agreement, so that such additional Stipulations or Conditions do not in any way subject the Acts of the Committee of Visitors to the Approval or Control of any Court of General or Quarter Sessions, or of any Justices, in any Case not provided for by this Act, and the additional Stipulations and Conditions so inserted in the said Agreement shall be of the same Force and Effect as the Matters so required to be specified, notwithstanding that such additional Stipulations or Conditions may control in any other Manner than as herein-before specified and excepted the Discretion and Acts of the Committee of Visitors as regulated by this Act, or may require the Consent or Approval of, or may subject the Acts or Orders of the Visitors to be disallowed, modified, or controlled by, One of Her Majesty's Principal Secretaries of State, in Cases not provided for by this Act ; but any Stipulations or Conditions subjecting the Acts of the Committee of Visitors to the Approval or Control of any Court of General or Quarter Sessions, or of any Justices, in any Case not provided for by this Act, shall be void and of none effect.

With Consent of Visitors, Stipulations or Conditions may be repealed.

XVI. Provided also, That with the Consent in Writing under the Hands of the greater Number of Visitors of each County and Borough, and of the greater Number of Visitors of any Body of Subscribers united under any Agreement entered into under this Act or any former Act, and with the previous Consent in Writing under the Hand of One of Her Majesty's Principal Secretaries of State, the Committee of Visitors may from Time to Time repeal or alter any of the Stipulations or Conditions of such Agreement, but not so as to subject the Acts of the Committee of Visitors to the Approval or Control of any Court of General or Quarter Sessions, or of any Justices, in any Case not provided for by this Act.

Proportions of Expenses and of Visitors may be varied on any further Union being effected.

XVII. Where any Agreement for uniting has been entered into under this Act or any former Act, and the Union effected thereunder is added to by an Agreement for further Union, the Proportions in which any Expenses are under any former Agreement for Union to be charged on the Counties or Boroughs, or Counties and Boroughs, and the Subscribers, if any, uniting, and the Proportions in which Visitors are to be elected for and on behalf of such Counties or Boroughs, or Counties and Boroughs, and Subscribers (if any), may be altered as may be agreed upon.

As to Payment and Application of Money paid towards prior Expenses, or becoming repayable under Agreement for further Union.

XVIII. Where under an Agreement for Union any Money is to be paid towards the Expenses already incurred by any County or Borough in erecting or providing any Asylum, the same shall be paid to the Treasurer of such County or Borough, and shall be applied in Liquidation and Payment, *pro tanto*, of the Monies, if any, which shall have been raised by such County or Borough for the Purposes of this Act or the Acts hereby repealed, or any of them, in such Manner as the Justices of such County at any General or Quarter Sessions for the same, or the Council of such Borough, shall respectively order and direct, or if all such Monies shall have been paid, then the same shall be applied in diminution of any Rate to be made in pursuance of this Act.

Committee of Justices to report Agreement to Quarter Sessions, and the Original to be delivered to Clerk of the Peace of the County or Borough in which the Asylum is situate, and a Copy to Clerk of the Peace of each other County and Borough.

XIX. When any Agreement has been entered into and signed as aforesaid, the Committee for each County and Borough on behalf of which the same has been entered into shall report the same to the Justices of such County or the Recorder of such Borough at the then next General or Quarter Sessions; and the original Agreement shall, at such Sessions for the County or Borough in which the Asylum to which the same relates is situate or is intended to be situate, be delivered to the Clerk of the Peace of such County or Borough, to be by him entered among the Records thereof; and a Copy of such Agreement shall at such Sessions for each other County or Borough on behalf of which such Agreement has been entered into be delivered to the Clerk of the Peace of such County or Borough, to be by him entered among the Records thereof; and a Copy of every such Agreement shall be sent by the Clerk of the Peace to whom the original Agreement is delivered, within Twenty Days after the Delivery thereof to him, to the Commissioners in Lunacy; and any of the Justices of any County or Borough on behalf of which such Agreement has been entered into, and any Commissioner in Lunacy, shall be entitled, without Payment, to inspect the original Agreement so delivered to the Clerk of the Peace as aforesaid; and any Clerk of the Peace hereby required to send to the said Commissioners a Copy of any Agreement, who shall neglect so to do within the Time aforesaid, and any Clerk of the Peace who shall refuse to permit such Inspection as aforesaid, shall for every such Offence be liable to a Penalty not exceeding Five Pounds, and this Enactment shall extend and be applicable to and in respect of every Agreement by which any of the Stipulations or Conditions in any Agreement entered into under this Act or any former Act shall be repealed or altered.

After Agreement for uniting is reported, Visitors to be elected for carrying same into effect.

XX. When any Agreement for uniting has been entered into, signed, and reported as aforesaid, the Justices of every County to which the same relates shall, at the General or Quarter Sessions to which such Agreement is reported, elect from among the Justices of such County the Number of Visitors allotted to such County in the Agreement; and the Justices of every Borough to which such Agreement relates shall, at a Special Meeting of such Justices to be holden within Twenty Days after such Agreement has been reported to the General or Quarter Sessions for such Borough, elect from among the Justices of such Borough the Number of Visitors allotted to such Borough in the Agreement; and the Majority of such of the Subscribers to any Hospital to which such Agreement relates as shall be present at a Meeting of such Subscribers to be holden within Twenty-eight Days after the signing of such Agreement, and of which Meeting public Notice shall have been given by Advertisement in some Newspaper circulated in the Place in which such Hospital is situate or is intended to be situate, shall elect from among such Subscribers the Number of Visitors allotted to the Subscribers to such Hospital in such Agreement; and the Visitors so elected as aforesaid shall together form and be the Committee of Visitors for carrying such Agreement into effect.

Committee authorized to superintend the Erection of Asylums to be deemed Committee of Visitors.

XXI. Every Committee elected for any County or Borough as herein-before provided, and authorized to superintend the erecting or providing of an Asylum for such County or Borough, shall until the election of Visitors or a Committee of Visitors for such County or Borough or the Asylum thereof, under any of the Provisions herein contained, be deemed the Committee of Visitors for such County or Borough.

APPENDIX.

Visitors to be elected annually for Asylums.

XXII. At the General or Quarter Sessions to be held next after the Twentieth Day of *December* in every Year the Justices of every County, and at a Special Meeting to be held within Twenty Days after the Twentieth Day of *December* in every Year the Justices of every Borough, having for the Time being an Asylum (whether provided before or after the passing of this Act) either for the sole Use of such County or Borough or under any Agreement for uniting as aforesaid, shall elect some Justices of such County or Borough to be Visitors on behalf of such County or Borough for the said Asylum during the Year next ensuing the Election; and where such Asylum has been provided under any Agreement for uniting entered into with any such Subscribers as aforesaid, the Majority of such of the Subscribers as shall be present at a Meeting to be holden in the Month of *January* in every Year, of which Notice shall have been given by public Advertisement in some Newspaper circulated within the Place in which such Asylum is situate, shall elect some of such Subscribers to be Visitors for such Asylum during the Year then next ensuing ; and where such Asylum is for the sole Use of any One County or Borough, the Visitors elected for such County or Borough as aforesaid shall be " the Committee of Visitors" of such Asylum ; and where such Asylum has been provided under any Agreement for uniting, the Visitors elected as aforesaid on behalf of every County and Borough, and the Subscribers (if any) to which the Asylum belongs, shall together form and be "the Committee of Visitors" of such Asylum : Provided always, that the Number of the Committee of Visitors of any County or Borough having an Asylum for its sole Use shall not be less than Seven; and that in all other Cases the Number of Visitors to be elected on behalf of every County and Borough, and of any Body of Subscribers, to form and be the Committee of Visitors, shall be the Number provided for in the Agreement.

A separate Committee of Visitors to be appointed for every Asylum. Proviso.

XXIII. Where any County or Borough has more than one Asylum a separate Committee of Visitors shall be appointed as aforesaid for every such Asylum, each of which Committees shall have all the Powers and be subject to all the Provisions of this Act with regard to the Asylum for which it is appointed, as if it were the only Asylum for that County or Borough : Provided always, that it shall be lawful for the Justices of the County and Borough, if they think fit, with the Approval of One of Her Majesty's Principal Secretaries of State, to appoint the same Committee for Two or more such Asylums.

Meetings of Visitors. Every Committee to elect a Chairman. Number of Members to constitute a Meeting. Questions how to be decided.

XXIV. The several Persons elected Members of any Committee of Visitors shall within One Month after their Election assemble at some convenient Place to be named in a Notice in Writing given by Two or More of such Visitors, or by the Clerk to the outgoing Committee by the Direction of Two or more of the said Visitors, to the several Members so elected, such Notice to be given to each Member personally, or left at his Place of Abode, or transmitted to him through the Post Office, Seven Days at least before the Time appointed for such Meeting ; and the said Visitors may adjourn the said Meeting from Time to Time or from Place to Place, and meet where and as often as they think necessary ; and the said Visitors shall at their first Meeting after their Election elect One of their Members to be their Chairman, who shall preside at all Meetings at which he is present ; and in case of the Absence of the Chairman from any Meeting the Members of the Committee then present shall elect One of such Members to be Chairman for the Meeting, who shall preside at the Meeting; and to constitute a Meeting of a Committee there shall be present not less than Three Members thereof, except for Adjournment which may be made by less than Three ; and every Question shall be decided by a Majority of Votes (the Chairman, whether permanent or temporary, having a Vote), and in the event of an Equality of Votes on any Question the Chairman for the Time being shall have an additional or casting Vote.

APPENDIX.

Clerk, on Requisition of Chairman or Two Visitors or of Superintendent, to call Meetings of Visitors. Chairman may convene Meetings.

XXV. The Clerk of any Committee of Visitors shall, whenever required in Writing by the Chairman or Two of the Visitors or by the Superintendent of the Asylum, and the Chairman of any such Committee may, whenever he shall see fit, convene a Meeting of such Committee by a Notice in Writing to each Visitor of the Time and Place of such Meeting, such Notice to be delivered, left, or transmitted as aforesaid by such Clerk or Chairman Seven Days at least before the Time appointed for the Meeting.

Visitors to appoint a Clerk.

XXVI. Every Committee of Visitors shall appoint a Clerk to such Visitors for the Purposes of this Act, at such Salary or Remuneration as such Visitors think fit, and may, if and when they think fit, remove any Clerk appointed by them, and in any such Case, or in case of the Death or Resignation of any such Clerk, shall appoint a new Clerk; and the Clerk to any Committee of Visitors of any Asylum may also be the Clerk of such Asylum; and any Clerk to any Committee of Visitors shall, unless he sooner die, resign, or be removed, continue in Office so long as such Committee continue in Office.

Committee of Visitors to continue until First Meeting of new Committee, and in Default of Election of new Committee to continue as if re-elected.

XXVII. The Powers of any Committee of Visitors and of the Members of such Committee, whether appointed or elected before or after the Commencement of this Act, shall continue until the First Meeting of the Committee by which such first-mentioned Committee is to be succeeded, anything herein contained to the contrary notwithstanding; and if the Justices of any County, or the Justices or Recorders of any Borough, or any Body of Subscribers, neglect in any Year to make such Election or Appointment as required by this Act, then the Committee of Visitors lastly before elected, or the Members of such Committee elected or appointed for such County or Borough, or on behalf of such Body of Subscribers, or such of them as shall continue to act, shall be deemed and taken to be the Committee of Visitors, or to form Part of the Committee of Visitors, as if such Committee or Members had been re-elected or reappointed in such Year, and so from Time to Time so often as the said Justices, Recorder, or Subscribers so neglect.

Provision for supplying Vacancies in Committees. Continuing Members may act.

XXVIII. In case any Member of any Committee or any Visitor elected or appointed under this Act or any Act hereby repealed, die, resign, or become incapable to act, the Justices for the County or Borough for which such Member or Visitor was elected or appointed at any General or Quarter Sessions for such County, or at a Special Meeting of the Justices of such Borough or where such Visitor was appointed by the Recorder of a Borough, then the Recorder of such Borough, shall elect or appoint some other Justice in his Place; and where any such Member or Visitor has been elected on behalf of any Body of Subscribers, the Majority of such of the said Subscribers as shall be present at some Meeting called in manner provided with respect to the annual Election of Visitors shall elect some other Subscriber in his Place; but, notwithstanding any Vacancy in any Committee, the continuing Members or Visitors may act as if no such Vacancy had occurred.

Secretary of State may require any County or Borough not having an Asylum to provide one.

XXIX. In case at any Time after the Expiration of One Year from the Commencement of this Act it appear to One of Her Majesty's Principal Secretaries of State, upon the Report of the Commissioners in Lunacy, that any County or Borough has not an Asylum for the Pauper Lunatics thereof, it shall be lawful for such Secretary of State, by Writing under his Hand, to require the Justices of such County or Borough forthwith to provide a fit and sufficient Asylum for so many Pauper Lunatics as upon the Report of the said Commissioners such Secretary of State may think fit and direct, and such Justices shall forthwith proceed as herein-before mentioned to cause such Asylum to be provided : Provided always, that no Borough annexed to any County by virtue of this Act or any former Act, or on behalf of which a subsisting Contract making adequate Provision for the Care of the Pauper Lunatics thereof shall have been entered into under this Act, or which now contributes to any Asylum for the County in which it is situate, and shall not have been separated from such County, shall be required to provide an Asylum under any such Order.

Where Accommodation of existing Asylum is inadequate, additional Asylum to be provided, or existing Asylum enlarged.

XXX. It shall be lawful for the Justices of every County and Borough having an Asylum or Asylums for the Pauper Lunatics thereof, where it appears to such Justices at any General or Quarter Sessions, or (in the Case of a Borough) at any Special Meeting of such Justices, that the Asylum or Asylums of such County or Borough is or are inadequate or unfit for the proper Accommodation of the Pauper Lunatics of such County or Borough, to cause an additional Asylum, or a new Asylum in lieu of any existing Asylum of such County or Borough, to be provided for such County or Borough, in like Manner as herein-before directed in the Case of a County or Borough not having an Asylum, or to direct the Committee of Visitors of any existing Asylum to cause the same to be enlarged or improved, or, in any other Case where the said Justices deem it necessary or expedient, to direct the Committee of Visitors of any existing Asylum to improve the same ; but it shall not be incumbent on any such Committee under any such Direction as aforesaid to enlarge or improve such Asylum where the same does not belong to One County or Borough alone, without a like Direction from the Justices of every County or Borough to which the same belongs ; and in case at any Time it appear to One of Her Majesty's Principal Secretaries of State, upon the Report of the Commissioners in Lunacy, that any existing Asylum or Asylums for any County or Borough is or are inadequate or unfit for the proper Accommodation of the Pauper Lunatics thereof, it shall be lawful for such Secretary of State, by Writing under his Hand, to require the Justices of such County or Borough forthwith to cause an additional Asylum or a new Asylum in lieu of any existing Asylum, to be provided as aforesaid for such County or Borough, or the Committee or Committees of Visitors of any existing Asylum or Asylums forthwith to enlarge or improve the same, in such Manner as the said Secretary of State may see fit and direct, and the said Secretary of State may require Accommodation to be provided in and by such additional or new Asylum, or by means of the Enlargement of such existing Asylum or Asylums, for so many Pauper Lunatics as upon the Report of the said Commissioners such Secretary of State may think fit and direct ; and the said Justices or Committee or Committees shall forthwith carry such Requisition of the said Secretary of State into effect ; and the Powers and Provisions in this Enactment contained with respect to the Enlargement and Improvement of Asylums shall extend and be applicable to and for the Enlargement and Improvement of the Offices, Outbuildings, Yards, Courts, Outlets, Ground, Land, and Appurtenances belonging thereto.

When an Asylum or additional Asylum or Accommodation is required, the Visitors to procure and determine on Plans and Estimates, and to contract for the Purchase of Land and Buildings, and for erecting, &c., the necessary Buildings. Contractors to give Security. Contracts and Orders to be entered in a Book, to be deposited and to be open to Inspection. Visitors to report.

XXXI. It shall be lawful for any Committee of Visitors having Authority to provide an Asylum for Pauper Lunatics (but subject as herein-after mentioned) to procure, examine, and determine on Plans for the same, and Estimates, and contract for the Purchase of Lands and Buildings (and in the Case of Buildings, either with or without any Fittings-up and Furniture belonging thereto), and for building, erecting, altering, improving, restoring, furnishing, and completing, or otherwise providing such Asylum, and rendering the same in all respects fit and ready for the Reception of Lunatics, and for making, laying out, and completing the Offices, Outbuildings, Yards, Courts, Outlets, Grounds, Land, and Appurtenances of or for such Asylum, and for providing Clothing for Patients, and everything necessary for the opening of any such Asylum; and any Committee of Visitors having Authority to enlarge, alter, or improve any Asylum shall have like Powers for the Purpose of enlarging, altering, or improving such Asylum, or the Offices, Outbuildings, Yards, Courts, Outlets, Grounds, Land, and Appurtenances thereto belonging; and every Person contracting for building or doing any other such Work as aforesaid shall give to the Clerk of such Visitors sufficient Security for the due Performance of the Contract; and every such Contract, either for Purchase of Lands or Buildings, or for doing any such Work as aforesaid, and all Orders relating thereto, shall be entered in a Book to be kept by the Clerk to such Visitors; and when such Asylum and Appurtenances, or (as the Case may be) the Additions to or Alterations or Improvements thereof, are completed, such Book shall be deposited and kept among the Records of the County or Borough, or where more than one County or Borough is interested in such Contract by reason of an Agreement for Union, then among the Records of the County or Borough which has contributed the largest Proportion of the Expenses of such Contract; and every such Book may be inspected at all reasonable Times by any Person contributing to the Rates of the County or Borough, or in the Case of a Union, to the Rates of any of the Counties or Boroughs, and also, if any Part of such Expenses has been paid by voluntary Subscription, by any of such voluntary Subscribers; and a Copy of every such Book shall be kept at the Asylum to which the Contract relates: Provided always, that the said Visitors shall from Time to Time make their Report to the General or Quarter Sessions of the County or Borough, Counties or Boroughs, for which they, or such of them as have not been elected by Subscribers as aforesaid, have been elected, of the several Plans, Estimates, and Contracts which have been agreed upon, and of the Sum or Sums of Money necessary to be raised and levied for defraying the Purchase Monies and Expenses thereof on the County or Borough, or, in the Case of such Union as aforesaid, on each or every of the Counties or Boroughs; which Plans, Estimates, and Contracts shall be subject to the Approbation of the Court or Courts of General or Quarter Sessions of such County or Counties, and of the Justices of such Borough or Boroughs, before the same are completed or carried into execution, save where the Amount to be expended does not exceed an Amount previously fixed by the Court or Courts of General or Quarter Sessions of such County or Counties or by the Justices of such Borough or Boroughs.

Power to Visitors to purchase in consideration of a Rent reserved.

XXXII. It shall be lawful for any Committee of Visitors to purchase and take a Conveyance for the Purposes of this Act from any Person having absolute Power to sell and convey, independently of this Act, any Lands or Buildings, in consideration of a yearly Rentcharge or annual Sum to be limited to such Person, his Heirs and Assigns, or as he or they shall direct, out of the Lands or Buildings to be purchased, and the same shall accordingly be conveyed as aforesaid, subject thereto, and to Powers of Distress and Entry for securing the same.

Power for Visitors to take a Lease for Rent.

XXXIII. It shall be lawful for any Committee of Visitors, instead of purchasing any Land or Buildings which they are hereby authorized to purchase, to take a Lease thereof for any absolute Term of not less than Sixty Years, at such annual Rent and under such Covenants as the said Committee of Visitors think fit: and it shall also be lawful for such Committee to rent any Land by the Year for the Purpose of employing such of the Inmates of the Asylum as may be fit for such Employment, or otherwise for the Occupation and Use of the Patients.

Asylum may be erected beyond the Limits of any County or Borough, and Justices of such County or Borough may notwithstanding act therein.

XXXIV. The Asylum to be provided for any County or Borough either solely or jointly, may be without the Limits of such County or Borough, and when any Asylum provided or to be provided solely or in part for any County or Borough, or any Part of such Asylum, is situate within the Limits of any other County or Borough, then and in every such Case the Justices of the County or Borough to which such Asylum wholly or partly belongs shall have full Power and Authority to act in such other County or Borough, so far as concerns the Regulation of such Asylum, and the Powers conferred by this Act, in the like Manner as if such Asylum and every Part thereof were situate within such first-mentioned County or Borough.

Assessment to local Rates not to be increased after Purchases for the Purposes of this or any former Act.

XXXV. No Lands or Buildings already or to be hereafter purchased or acquired, under the Provisions of any former Act or this Act, for the Purposes of any Asylum, (with or without any additional Building erected or to be erected thereon,) shall while used for such Purposes be assessed to any County, Parochial, or other local Rates at a higher Value or more improved Rent than the Value or Rent at which the same were assessed at the Time of such Purchase or Acquisition.

Certain Provisions of 8 & 9 Vict. c. 18. incorporated, and extended authorize Exchanges.

XXXVI. The Provisions of "The Lands Clauses Consolidation Act, 1845," "with respect to the Purchase of Lands by Agreement," "with respect to the Purchase Money or Compensation coming to Parties having limited Interests, or prevented from treating, or not making Title," and all other Provisions of the said Act applicable to and in the Case of the Purchase of Lands by Agreement, shall be incorporated with this Act; and all Parties by the said Provisions empowered to sell any Lands may give Lands in exchange for the Purposes of this Act for other Lands, and enter into all necessary Agreements for that Purpose, and on any such Exchange Money may be paid by either Party by way of Equality of Exchange, and the said Provisions "with respect to Purchase Money or Compensation coming to Parties having limited Interests, or prevented from treating, or not making Title," shall apply to any Money coming to any such Parties on any such Exchange; and any Lands to be purchased or taken in exchange for the Purposes of this Act shall be conveyed to such Persons, being not less than Five in Number, and in such Manner as the Committee of Visitors purchasing the same or taking the same in exchange may direct, in trust for the Purposes of this Act; and any Conveyance to be so made shall have the like Force and Effect as a Conveyance made under Section Eighty-one of the said Lands Clauses Consolidation Act.

Provision for the Appointment of new Trustees of Land purchased or acquired for Asylum.

XXXVII. When and so often as any Land purchased or acquired under this Act or any former Act, for the Purposes of an Asylum, shall be vested in less than Three Trustees, or there shall not be any Trustee thereof living, it shall be lawful for the Committee of Visitors of such Asylum, or any Three or more of them, by an Instrument in Writing under the Hands of such Visitors or any Three or more of them, to appoint such Number of new Trustees of such Land as such Visitors may think fit; and such Appointment shall be deposited and kept among the Records of the County or Borough, or, where more than One County or Borough is interested in such Land, then among the Records of the County or Borough having the largest Interest therein; and all the Estate and Interest in such Land which at the Time of such Appointment may be vested in any Trustee or Trustees, in trust for the Purposes aforesaid, or in any other Person, as Heir, or Devisee, or otherwise, subject to such Trust, shall by virtue of such Appointment vest in the Trustees so appointed, either alone, or if there be any continuing Trustees or Trustee jointly with such continuing Trustees or Trustee, as the Case may require, without any Conveyance or Assignment for that Purpose.

Visitors to order all ordinary Repairs of Asylums, provided they do not exceed 400l. per Annum. As to Payment of Expenses of Repairs, &c. No Order for Payment of Money exceeding 100l. to be made unless Notice has been given of the Meeting at which the same shall be ordered.

XXXVIII. The Committee of Visitors of every Asylum may of their own Authority from Time to Time order all such ordinary Repairs as may be necessary for such Asylum, and any Additions, Alterations, or Improvements to or in such Asylum, or the Offices, Outbuildings, Yards, Courts, Outlets, Grounds, Land, and Appurtenances thereto belonging, which to them may seem necessary or proper for the further or better Accommodation of the Pauper Lunatics who may be received or taken care of therein, provided that the Expense of all such Additions, Alterations, and Improvements shall not exceed Four hundred Pounds in any One Year; and if such Asylum belong to One County or Borough only, they shall cause the Expense of such Repairs, Additions, Alterations, or Improvements to be paid by making an Order upon the Treasurer of such County or Borough for the Payment thereof, but if otherwise they shall apportion such Expense in the Proportion in which each County or Borough has contributed to the Erection thereof, or where any other Proportion is fixed by any Agreement for the Time being in force, then in such other Proportion, and where any such Agreement only provides in what Proportion the Expense of Repairs shall be defrayed, the said Committee shall apportion the Expense of such Additions, Alterations, and Improvements in the same Proportion unless it be otherwise provided by such Agreement, and the said Committee shall make an Order on the Treasurer of each County or Borough for the Payment of the Proportion to be paid by such County or Borough, and such Treasurer shall pay the same accordingly out of any Money of such County or Borough then in his Hands, or which may thereafter come to his Hands, not specifically appropriated to any other Purpose, and the same may be recovered from him, for the Benefit of such Asylum, by the Treasurer or Clerk thereof, together with all Costs and Expenses, in any of Her Majesty's Courts at *Westminster*, or in any other Court of competent Jurisdiction : Provided always, nevertheless, that no Order for any such Repairs, Additions, Alterations, or Improvements as aforesaid, or for the Payment of any Money for the Expenses thereof, where such Expenses exceed the Sum of One hundred Pounds, shall be made, unless Notice of the Meeting at which the same shall be ordered, and of the Intention to determine thereat the Question of such Expenditure, have been given in such Manner and so long before the Time appointed for the Meeting as is herein-before provided with respect to Notices of Meetings of Committees of Visitors, nor unless Three Visitors concur in and sign such Order : Provided also, that where any such Expenditure as aforesaid is incurred otherwise than for ordinary Repairs, the Visitors shall report the same to the next General or Quarter Sessions of the County or Borough, or each County and Borough, on behalf of which such Expenditure has been incurred.

Power of Visitors, with Consent of Secretary of State, to dissolve Unions.

XXXIX. It shall be lawful for every Committee of Visitors, with the Consent of One of Her Majesty's Principal Secretaries of State under his Hand, to determine and dissolve any Union, whether such Union have been formed under this Act or under any former Act, and upon such Dissolution to divide and allot the Lands, Buildings, Hereditaments, Chattels, Monies, and Effects of or belonging to such Union between or among every such County and Borough, and the Subscribers (if any) between which and whom such Union existed, in the Proportions in which they respectively have contributed thereto or are interested therein, or in such other Proportions and Manner as the said Visitors, with the Approbation of the said Secretary of State, think fit; and if on any such Division or Allotment there cannot be conveniently allotted to any County or Borough or Subscribers the proper Proportion of such County, Borough, or Subscribers in the Lands, Buildings, Hereditaments, Chattels, Monies, and Effects of such Union, there shall be paid to such County, Borough, or Subscribers such Sum of Money as the said Visitors, with the Approbation of the said Secretary of State, may direct, in full or in part Satisfaction, as the Case may require, of the aforesaid Proportion of such County, Borough, or Subscribers; and every such Sum of Money shall be raised by the County or Counties, Borough or Boroughs, to or between or among which the Lands, Buildings, Hereditaments, Monies, Chattels, and Effects of the said Union shall be allotted (if more than One) in such Shares as the said Visitors, with the Approbation of the said Secretary of State, think fit, in the same Manner and by the same Means as other Monies are appointed to be raised by Counties or Boroughs for the Purposes of this Act: Provided always, that no Union shall be so dissolved by any Committee of Visitors except under a Resolution of such Committee at a Meeting specially convened for the Purpose of determining the Question of such Dissolution by a Notice given in such Manner and so long before the Time appointed for such Meeting as is herein-before provided with respect to Notices of Meetings of Committees of Visitors, nor unless the Majority of the whole Number of the Committee of Visitors shall at such Meeting have concurred in such Resolution: Provided always, that in the Case of a Dissolution of Union, where any County or Borough having an Asylum shall be united with any County or Counties, Borough or Boroughs, not having an Asylum, and have erected additional Buildings and incurred any other Expense for their Benefit, and be in the Receipt of an annual fixed Sum or Rent as a Remuneration for the Expenses so incurred in lieu of the Payment of a Sum in gross, it shall be lawful for the said County or Counties, Borough or Boroughs, so paying such Rent, if they shall think fit, to raise, in the same Manner as is provided in the Act for the Purpose of erecting County Asylums, such a Sum of Money for the Purpose of compensating the County or Borough receiving such Rent for the Cessation of such Rent as may be agreed upon and approved of by the Committee of Visitors of such County or Counties, Borough or Boroughs, as may have been so united as aforesaid.

Power for Visitors, with Consent of Secretary of State, to sell or exchange Lands and Buildings. Application of Purchase Monies.

XL. It shall be lawful for every Committee of Visitors, with the previous Consent of One of Her Majesty's Principal Secretaries of Sate under his Hand, to sell, either by Public Auction or Private Contract, and subject to any Conditions, any Lands or Buildings or Parts of Lands or Buildings which may have belonged to and been used as or together with an Asylum, or which may have been purchased or otherwise acquired under any former Act or this Act, for the Purposes of an Asylum, and found unsuitable or otherwise not required for such Purposes, or to give the same in exchange for other Lands or Buildings, and to pay or receive through the Treasurer of such Asylum any Money by way of Equality of Exchange; and every Conveyance of Lands or Buildings so sold or given in exchange which shall be executed by the Persons in whom the same may then be vested as Trustees, or by any Three of the Members of the Committee of Visitors who sell the same, shall be effectual to convey the same for all the Estate or Interest then vested in such Trustees, in trust for the Purposes of such Asylum, and the Receipt of any Three of the Committee of Visitors shall be a sufficient Discharge for the Purchase Monies or for any Monies to be received for Equality of Exchange; and such Monies, in case the Sale or Exchange be made by a Committee of Visitors of any One County or Borough alone, shall be applied in carrying into execution the Powers and Purposes of this Act, or shall be paid to the Treasurer of such County or Borough and be applied for the general Purposes thereof, or otherwise, as the Justices of such County or Borough shall, at some General or Quarter Sessions for such County, or at some Special Meeting of the Justices of such Borough, direct; and in every other Case the Monies received shall be paid to the Treasurer of the

APPENDIX. 239

County, Borough, or Subscribers to which or to whom the Property sold or exchanged belonged, in case it belonged to any One of them, or if the same was joint Property then to the respective Treasurers of every County and Borough, and of the Subscribers, if any, in the Proportion in which such County, Borough, and Subscribers were respectively interested therein ; and such Monies shall be held and applied by every such Treasurer, in the Case of a County or Borough, as Part of the general Rates or Funds of such County or Borough, and in the Case of any Subscribers, as the Majority of such of the Subscribers as shall be present at any Meeting convened for that Purpose shall direct.

Visitors may, with Consent of Secretary of State, get released from Contracts.

XLI. Where any Committee of Visitors have (either before or after the passing of this Act) contracted for the Purchase of any Lands for the Purposes of an Asylum, or for any Exchange of any Lands for other Lands for such Purposes, and the Lands so contracted to be purchased or taken in exchange are found to be unsuitable or are not required for such Purposes, such Committee, or any other Committee appointed in their Place, may, with the Consent in Writing of One of Her Majesty's Principal Secretaries of State, (notwithstanding such Contract may have been approved as required by the said Acts hereby repealed, or this Act,) procure a Release from the said Contract, and in consideration of such Sum of Money (if any) as the said Committee, with such Consent as aforesaid, may agree to pay ; and the said Committee or any Three of such Committee may, in consideration of such Release, execute a Release to the other Party, to such Contract or other the Persons bound thereby ; and the Consideration Money (if any) by the said Committee agreed to be paid as aforesaid, and all Expenses in relation to the said Contract and Releases, shall be paid, defrayed, and raised in like Manner as if the same were payable in respect of the Purchase of Lands for the Purposes aforesaid.

Visitors empowered to contract for the Reception of Pauper Lunatics into Asylums of other Counties or Hospitals or licensed Houses. Period of such Contract limited. As to Money payable under Contract for Reception of Lunatics into any Asylum.

XLII. It shall be lawful for every Committee of Visitors to contract with the Committee of Visitors of any Asylum, or with the Subscribers to any Hospital registered or the Proprietor of any House licensed for the Reception of Lunatics, for the Reception into such Asylum, Hospital, or House of the whole or of a Portion of the Pauper Lunatics of the County or Counties, Borough or Boroughs, or Counties and Boroughs, or any of them respectively, for which such first-mentioned Committee is acting, or for the Use and Occupation of all or any Part of such registered Hospital or licensed House, at such Sum, either in gross or by way of annual or other periodical Payment or Rent, and under and subject to such Terms, Stipulations, and Conditions, as such Visitors shall think fit ; and it shall be lawful for the Committee of Visitors of any Asylum, or the Subscribers to any registered Hospital, or the Proprietor of any licensed House, to contract with any Committee of Visitors accordingly : Provided always, that no such Contract shall be made for any longer Period than for the Term of Five Years, and that any such Contract may be determined by Notice in Writing under the Hand of One of Her Majesty's Principal Secretaries of State, and that every such Contract with the Proprietor of a licensed House shall determine on such House ceasing to be duly licensed for the Reception of Lunatics ; provided also, that no such Contract shall exempt the Justices of any County or Borough or any Committee from the immediate Duty and Obligation of erecting or providing, or uniting in erecting or providing, an Asylum or additional Asylum, or of enlarging or improving any Asylum, as required by this Act, where One of Her Majesty's Principal Secretaries of State has caused Notice to be given as aforesaid for the Determination of such Contract, although the Term for which such Contract was entered into has not expired by Effluxion of Time : Provided also, that any Money which may be payable under such Contract for the Reception of the Lunatics of any County or Borough into any Asylum beyond the weekly Sums which may be charged under this Act for the Lodging, Maintenance, Medicine, Clothing, and Care of Lunatics in the Asylum belonging to the County or Borough to which such Lunatics shall belong, shall be paid, defrayed, and raised by such County or Borough out of any Monies in the Hands of the Treasurer for the County which shall be applicable for the Repairs or other ordinary Expenses of such Asylum; provided also, that any Hospital or licensed House with the Subscribers or Proprietor of which any such Committee so contract as aforesaid shall be subject to the Visitation of any of the Members of such Committee for the Time being.

When any Asylum can accommodate more than the Lunatics of the County or Borough, Visitors may order the Admission of other Lunatics.

XLIII. Whenever it appears to the Committee of Visitors of any Asylum that such Asylum is more than sufficient for the Accommodation of all the Pauper Lunatics of the County or Borough or each County and Borough to which the same wholly or in part belongs, and of any County or Counties, Borough or Boroughs with which any existing Contract for the Reception of all or any of the Pauper Lunatics thereof in such Asylum has been entered into, or which shall otherwise contribute to such Asylum, it shall be lawful for the Committee of Visitors, if they think fit, to give Notice thereof by Advertisement in some Newspaper commonly circulated in such County or Borough, or every such County or Borough as aforesaid, and (subject nevertheless and without Prejudice to any Agreement with any voluntary Subscribers,) by a Resolution of the said Committee, to permit the Admission of so many Pauper Lunatics of any other County or Borough, and (if such Committee think fit) Lunatics not Paupers, but who, in the Opinion of such Committee, may be proper Objects to be admitted into a public Asylum, as to such Committee may seem expedient, and at any Time to rescind or vary any such Resolution; and such Committee may, if they think fit, by such Resolution require that no Pauper Lunatic shall be admitted into such Asylum thereunder without an Undertaking by the Minute of the Guardians of the Union or Parish, or signed by Two of the Overseers of the Parish, to which such Lunatic is chargeable, or in the Case of a Lunatic not a Pauper by the Person signing the Order for the Admission of such Lunatic, for the due Payment of the weekly Charge for the Lodging, Maintenance, Medicine, Clothing, and Care of such Lunatic during his Continuance in such Asylum, and of the Expenses of his Burial in case he die therein, as well as for the Removal of such Lunatic from such Asylum within Six Days after Due Notice given in Writing by the Superintendent of such Asylum; and such Lunatic not being a Pauper shall have the same Accommodation in all respects as the Pauper Lunatics.

No Visitor to have any Interest in any Contract or Agreement.

XLIV. No Visitor of any Asylum shall have or take, or be capable of having or taking, any Interest or Concern whatsoever, either in his own Name or in the Name of any other Person, in any Contract or Agreement to be made under the Authority of this Act or in anywise relating to or connected with such Asylum, or shall for any Design or Plan he may deliver or produce, receive any Benefit or Emolument whatever, or otherwise have or take any Benefit or Emolument whatsoever, from or out of the Funds of the Asylum : Provided always, that this Enactment shall not extend to any such Interest, Benefit, or Emolument which any Visitor may have or derive by reason of his being a Shareholder of any Joint Stock Company established by Act of Parliament or by Charter, with which any Contract may be entered into on behalf of such Asylum, or which may otherwise receive any Benefit or Emolument out of the Funds of the Asylum ; provided that no Contract or Dealing between such Company and the Visitors of such Asylum be at or upon Rates or Terms more advantageous to such Company than in the Case of Contracts or Dealings by such Company with other Parties.

APPENDIX.

Plans, &c., to be submitted to Commissioners in Lunacy, and approved by Secretary of State.

XLV. Every Committee of Justices or Visitors shall submit all Agreements for uniting for the Purposes of this Act, and all Contracts under this Act, for the Reception of the Pauper Lunatics of any County or Borough, or any of them, into any Asylum, registered Hospital, or licensed House, or for the Use and Occupation of all or any Part of any such Hospital or licensed House, and all Plans for building or providing or enlarging or improving any Asylum for Pauper Lunatics, and all Contracts for Purchases of Lands or Buildings for any such Purpose, to the Commissioners in Lunacy, who shall make such Inquiries in reference thereto, and to the Amount of the Accommodation requiring to be provided, as they may deem proper, and shall report thereon in Writing to One of Her Majesty's Principal Secretaries of State, and such Committee shall submit to One of such Secretaries of State Estimates of the Cost and Expense of carrying into execution such Plans, and no such Agreement, Contract, or Plan shall be carried into effect until the same has been approved by such Secretary of State in Writing under his Hand.

HOW MONIES TO BE RAISED FOR PROVIDING ASYLUMS.

Provisions for raising Monies required for the Purposes of this Act by County and Borough Rates.

XLVI. In order to pay and defray the Monies, Costs, and Expenses payable for any of the Purposes of this Act, or the said Acts hereby repealed, by any County, the Justices of such County at any General or Quarter Sessions, for the same, may and shall assess and tax a General County Rate or Rates upon such County, and may and shall fix a Sum or Rate to be contributed by all Places whatsoever within such County, (other than any Borough being within such County or by this Act for the Purposes thereof annexed thereto,) and whether such Places be or be not liable to contribute to an ordinary County Rate; and in order to pay and defray the Monies, Costs, and Expenses payable as aforesaid by any Borough, the Council of such Borough may and shall assess a General Borough Rate in the Nature of a County Rate upon such Borough, and the said Rates shall be collected, levied, and recovered in the same Manner, and by the same Powers, Authorities, Ways, and Means, and under the same Penalties, as any ordinary Rate for such County or Borough respectively may by Law be collected, levied, and recovered; and the Monies, Costs, and Expenses to be paid and contributed by any County or Borough for the Purposes of this Act shall be paid by the Treasurer of such County or Borough, out of the Rates aforesaid, to the Treasurer of the Asylum to which such County or Borough shall either alone or jointly pay or contribute: Provided always, that it shall be lawful for the Council of any Borough, if they think fit, to direct that any Monies payable for the Purposes of this Act, or any Part thereof, shall be paid out of the Borough Fund of such Borough, and such Monies shall be paid by the Treasurer of such Borough out of such Fund accordingly.

Power for Justices of Counties and Councils of Boroughs to raise Money by Mortgage of the Rates.

XLVII. It shall be lawful for the Justices of every County in General or Quarter Sessions assembled, or the major Part of them, such major Part not being less than Five, and for the Council of every Borough, from Time to Time, to borrow and take up on Mortgage of the Rates to be made under this Act for such County or Borough, or on Mortgage of such Rates together with all other Rates or Funds, or any of them, of the same County or Borough, all or any of the Monies required for paying and defraying any such Monies, Costs, and Expenses, as aforesaid, payable by such County or Borough; and such Money may be so raised at any Rate of Interest not exceeding Five Pounds *per Centum per Annum*, and every such Mortgage may be made by an Instrument in the Form contained in the Schedule B. hereunto annexed, or to that or the like Effect, and shall be executed in the Case of a County by the Chairman and Two or more other Justices present

R

at the Time of making such Mortgage, and in the Case of a Borough by affixing the Common Seal of the Borough thereto ; and every such Mortgage shall be effectual for securing to the Person advancing the Sum of Money in such Mortgage expressed to be advanced, his Executors, Administrators, and Assigns, the Repayment thereof, with Interest for the same, after such Rate and at such Time and in such Manner as in such Mortgage provided ; and the said Mortgages shall be numbered in the Order of Succession in which they are granted ; and Copies or Extracts of all such Mortgages shall be kept by the Clerk of the Peace, or other proper Officer having the Custody of the Records of the Quarter Sessions of such County, or of the Records of such Borough, as the Case may be ; and every Person to whom any such Mortgage has been made under the Act hereby repealed, or any former Act, or is made under this Act, his Executors or Administrators, is hereby empowered, by endorsing his or their Name or Names on such Mortgage, to transfer the same, and his and their Right to the Principal Money and Interest thereby secured, unto any Person, and every Assignee under this Act or any former Act of any such Mortgage, his Executors and Administrators, may in like Manner transfer the same again, and so *toties quoties* ; and the Persons to whom such Mortgages or such Transfer thereof are made, their Executors and Administrators, shall be Creditors upon the Rates and Funds thereby expressed to be mortgaged in an equal Degree one with another, and shall not have any Preference or Priority other than is provided under the Powers of this Act.

Power to Public Works Loan Commissioners to lend Money for Purposes of this Act.

XLVIII. It shall be lawful for the Justices and Council of any County and Borough respectively to make Application for any Advance of any Sum necessary for the Purposes of this Act, or the said Acts hereby repealed, to the Commissioners acting in the Execution of an Act of the Session holden in the Fourteenth and Fifteenth Years of Her Majesty, Chapter Twenty-three, " to authorize for a further Period the Advance of Money out of the Consolidated Fund to a limited Amount for carrying on Public Works and Fisheries, and Employment of the Poor," and any Act or Acts amending or continuing the same, and the said Commissioners are hereby empowered, if they think fit, to make such Advance upon the Security of such Mortgage as aforesaid.

Provision for the Payment of the Interest on the Mortgages, and of a Portion of the Principal in each Year.

XLIX. The said Justices or Council, as the Case may be, shall in every Year charge the Rates or Funds of such County or Borough with the Sum for the Time being required to pay the Interest of the Money borrowed or any Mortgages under this Act or any former Act, or such of them as for the Time being remain unpaid, and also with the Payment of a further Sum, not less than One Thirtieth Part of the whole of such Mortgages at the Time of the same being first made, and such Sums shall be applied under the Direction of the said Justices or Council in discharge of the Interest on the said Mortgages or such of them as for the Time being remain unpaid, and of so many of the Principal Sums owing on the said Mortgages for the Time being remaining unpaid, as such Sums after Payment of the Interest as aforesaid will extend to discharge, until the whole of the Principal Monies for which such Mortgages shall have been made, and the Interest thereof shall be fully paid and discharged ; and the said Justices and Council, as the Case may be, are and is hereby required to fix One or more Days in each Year on which such Payment shall be made, and shall make Orders for Assessments in due Time, so as to provide for such Payments being regularly made ; and the said Justices or Council, as the Case may be, shall, by Agreement with the Parties, or others advancing any Money for the Purposes of this Act, determine the Order or Priority in which the several Sums advanced shall be respectively discharged; and the Justices of every County and the Council of every Borough so borrowing Money on Mortgage as aforesaid are and is hereby required to appoint a proper Person to keep an exact and regular Account of all Receipts and Payments in respect of Principal Monies borrowed or taken up as aforesaid under this Act or any former Act, and the Interest thereof, in a Book or Books separate and apart from all other Accounts, and the said Book and Books, duly adjusted and settled up to the Time being, to deliver annually, in the Case of a County into Court at some General or Quarter Sessions for such County, and in the Case of a Borough to the Council of the Borough, at such Time as such Council shall appoint; and the Justices for every such County at such Sessions, and the Council for every such Borough, are and is hereby required carefully to inspect all such Accounts, and to make such Orders for carrying the several Purposes aforesaid into execution as to them shall seem meet.

Provision to be made for paying Money borrowed within a limited Time not exceeding Thirty Years.

L. Provided always, That the Justices of every County and the Council of every Borough borrowing Money as aforesaid shall make Provision by means of the Rates which they are hereby respectively authorized to make, and by the Orders and Directions which they are hereby authorized to give, that the whole Principal Money to be borrowed under the Authority of this Act by such County or Borough, and all Interest for the same, shall be fully paid and discharged within a Time to be limited by such Justices or Council, not exceeding Thirty Years from the Time of borrowing the same.

Persons lending Money on Mortgage of Rates, &c., not bound to give Proof that Notices have been given, &c.

LI. No Person lending Money to any Justices of any County or the Council of any Borough, and taking a Mortgage for securing Repayment of the same, executed in manner directed by this Act, and purporting to be made under the Authority of this Act, shall be bound to require Proof that the several Provisions of this Act or of any former Act or Acts have been duly complied with; and if there be an Order of the Justices of any County in General or Quarter Sessions, or of the Council of any Borough making Application for the Loan, and any Mortgage have been thereupon duly executed, either before or after the passing of this Act, as by any Act then in force or this Act is provided, the Justices or Council (as the Case may be) shall have and be deemed to have had full Power to levy the Rates so mortgaged for Repayment of the Money so borrowed, with Interest, notwithstanding that the Provisions of this Act or any former Act or Acts may not have been complied with; and it shall not be competent to any Ratepayer or other Person to question the Validity of any such Rate or Mortgage on the Ground that such Provisions had not been complied with.

Power to raise Money to pay off Sums already borrowed.

LII. Provided also, That in every Case in which any Monies have been borrowed under the Powers of any former Act or this Act, it shall be lawful for the Justices of the County or Council of the Borough for which such Monies shall have been borrowed, (with the Consent of the Parties to whom the same shall be owing,) to pay off the Monies so borrowed, and to raise and borrow the Monies necessary for that Purpose, and also to repay the said last-mentioned Monies and the Interest thereof, under the Powers of this Act, as if such Monies were borrowed under the Powers herein-before contained: but so, nevertheless, that all Monies borrowed shall be discharged within Thirty Years from the Time of first borrowing the same.

REGULATION AND MANAGEMENT OF ASYLUMS, AND APPOINTMENT OF OFFICERS.

Visitors to submit General Rules to the Secretary of State, and, subject to such General Rules, to make Regulations and determine Diet of Lunatics.

LIII. Every Committee of Visitors shall, within Twelve Months after the passing of this Act, in the Case of every Asylum already established and General

Rules for the Government whereof have not been already submitted to One of Her Majesty's Principal Secretaries of State, and within Twelve Months after the Completion of every Asylum hereafter established, submit the existing General Rules, or General Rules to be prepared by such Committee, for the Government of the Asylum under their Superintendence to One of Her Majesty's Principal Secretaries of State for his Approval; and such Rules, when approved by him, shall be printed, abided by, and observed; and every such Committee shall have Power, with the like Approbation, to alter and vary such Rules from Time to Time as they think necessary; and every such Committee shall make from Time to Time such Regulations and Orders as they think fit, not inconsistent with the General Rules for the Time being in force for the Management and Conduct of the Asylum, and in such Regulations there shall be set forth the Number and Description of Officers and Servants to be kept, the Duties to be required of them, and the Salaries to be paid to them respectively; and every such Committee shall from Time to Time determine the Diet of the Patients; and in and by such Regulations such Committee may direct that any Number of Beds in such Asylum, and in such respective Parts thereof as such Committee may think fit, shall be always reserved for such Cases as in and by such Regulations shall be in this Behalf mentioned; and in such Case such Asylum shall, for the Purposes of this Act, as respects the Admission of all Cases not within the Description or Class for which such Beds are reserved, be deemed full when there are no vacant Beds in such Asylum except those so reserved, but nevertheless it shall be in the Power of the Committee of Visitors of such Asylum for the Time being to fill the Beds so reserved as they may deem expedient; and any such Committee may, if they see fit, by any such Regulations or Order, exclude from Admission into the Asylum Persons afflicted with any Disease or Malady which such Committee may deem contagious or infectious, and Persons coming from any District or Place in which any such Disease or Malady may be prevalent.

Visitors to fix weekly Rate to be paid for Maintenance of each Lunatic, not to exceed 14s. per Week. If the Rate be found insufficient, Justices in Quarter Sessions may increase it.

LIV. Every Committee of Visitors shall fix a weekly Sum to be charged for the Lodging, Maintenance, Medicine, Clothing, and Care of each Pauper Lunatic confined in such Asylum, of such Amount that the same may be sufficient to defray the whole Expense of the Lodging, Maintenance, Care, Medicine, and Clothing, and other Expenses requisite for each Pauper Lunatic, and that the total Amount of such weekly Sums, after defraying such Expenses, may also be sufficient to pay the Salaries of the Officers and Attendants, and such Committee may from Time to Time alter the Amount of such weekly Sum as Occasion may require; provided always, that any such Committee may, if they think fit, fix a greater weekly Sum to be charged as aforesaid in respect of Pauper Lunatics other than those sent to such Asylum from or settled in some Parish or Place situate in any County or Borough to which such Asylum belongs; provided also, that such Sum shall in no Case exceed the Rate of Fourteen Shillings *per* Week; but if the aforesaid Rate of Fourteen Shillings be found insufficient for the Purposes aforesaid, it shall be lawful for the major Part of the Justices of the County or Borough, or of each County or Borough to which such Asylum may belong, present at any General or Quarter Sessions for such County, or at a Special Meeting of the Justices of such Borough, or each such County or Borough respectively, to make such Addition to such Rate as to them respectively shall seem fit and necessary, and to make an Order or Orders accordingly, which Order or Orders shall be signed by the Clerk of the Peace for the County, or Clerk to the Justices for the Borough, and forthwith published in some Newspaper commonly circulated within such County or Borough.

Visitors to appoint a Chaplain. Patients allowed the Visits of any Minister of their own persuasion. Visitors to appoint Medical Officer, Clerk, and Teasurer, and such other Officers and Servants as they think fit.

LV. The Committee of Visitors of every Asylum shall appoint a Chaplain for the same, who shall be in Priest's Orders, and shall be licensed by the Bishop of the

Diocese, and the Licence of any such Chaplain as aforesaid shall be revocable by the Bishop whenever he shall think fit; and such Chaplain, or his Substitute approved by the Visitors, shall perform and celebrate, in the Chapel of or in some convenient Place within or belonging to such Asylum, Divine Service according to the Rites of the Church of *England* as established by Law, on every *Sunday*, *Christmas Day*, and *Good Friday*, and shall also perform and celebrate such Service within the said Asylum at such other Times, and also such other Services according to the Rites of the Church of *England* as established by Law at such Times, as the Visitors shall direct; and if any Patient be of a religious Persuasion differing from that of the Established Church, a Minister of such Persuasion, at the special Request of such Patient, or his Friends, shall, with the Consent of the Medical Officer of such Asylum, and under such Regulations as he shall direct, be allowed to visit such Patient at proper and reasonable Times; and the Committee of Visitors of every Asylum shall appoint a Medical Officer, who shall be resident in such Asylum, and who shall not be Clerk or Treasurer of such Asylum, and a Clerk and Treasurer, and such other Officers and Servants for the Asylum, as the Committee may think fit; and the Committee shall have Power to remove the Chaplain, Medical Officer, Clerk, and Treasurer, or any other Officer or Servant, and shall from Time to Time, upon every Vacancy, by Death, Removal, or otherwise, in the Office of the Chaplain, Medical Officer, Clerk, or Treasurer of the Asylum, appoint some other Person to such Office, subject to the Conditions and Restrictions affecting the original Appointment to such Office, and may from Time to Time fill up or not, as in their Discretion they may think fit, Vacancies among other Officers and Servants of the Asylum; and the Committee shall, if they think fit, have Power to appoint a Visiting Physician or Surgeon to every such Asylum, and shall from Time to Time appoint the Medical Officer or one of the Medical Officers (if more than One) of the Asylum, or where there is a separate Medical Officer of each Division, then the Medical Officer or One of the Medical Officers (if more than One) of each Division, to be the Superintendent of the Asylum, or of such respective Division thereof, and may remove any such Officer from being such Superintendent, and such Superintendent shall be resident in the Asylum: and the Committee shall from Time to Time fix the Salaries and Wages to be paid to the Officers and Servants of the Asylum: Provided always, that it shall be lawful for the said Committee, with the Sanction and Approbation of One of Her Majesty's Principal Secretaries of State, to appoint any Person other than such Medical Officer to be such Superintendent: Provided also, that where, on the Tenth Day of *February* One thousand eight hundred and fifty-three, any Person, other than a resident Medical Officer, was the Superintendent of any Asylum, such Person may continue to be such Superintendent as if this Act had not been passed, unless and until the Committee otherwise direct.

Clerk of Asylum to transmit to Commissioners in Lunacy Information of Dismissal of Attendants.

LVI. The Clerk of every Asylum shall, within One Week after the Dismissal for Misconduct of any Nurse or Attendant employed in such Asylum, transmit to the Commissioners in Lunacy, by the Post, Information in Writing under his Hand of such Dismissal, and of the Cause thereof; and every such Clerk neglecting to transmit such Information to the said Commissioners within One Week after the Dismissal of any such Nurse or Attendant shall for every such Offence forfeit any Sum not exceeding Ten Pounds.

Visitors may grant Superannuations to the Superintendent, &c., not exceeding Two Thirds of their Salaries.

LVII. In case any Superintendent, Chaplain, Matron, or any Officer or Servant of any Asylum, become, from confirmed Sickness, Age, or Infirmity, incapable of executing the Office in Person, or have been an Officer or Servant in the Asylum for not less than Twenty Years, and be not less than Fifty Years of Age, it shall be lawful for the Committee of Visitors of such Asylum, if in their

Discretion they think fit so to do, but not otherwise, to grant to such Superintendent, Chaplain, Matron, or other Officer or Servant such Annuity by way of Superannuation as they in their Discretion think proportionate to the Merits and Time of Service of such Superintendent, Chaplain, Matron, or other Officer or Servant (whether incapable from Sickness, Age, or Infirmity, or retiring from long Service and Age), and every such Annuity shall be payable out of the Rates lawfully applicable to the building or repairing of such Asylum: Provided always, that the annual Amount paid by way of Superannuation to any retired Superintendent, Chaplain, Matron, or other Officer or Servant of any Asylum shall not exceed the Amount of Two Thirds of the Salary payable at the Time of his or her Retirement, and that no such Superannuation shall be granted unless Notice of the Meeting at which the same shall be granted, and of the Intention to determine thereat the Question of such Superannuation, have been given, in such Manner and so long before the Time appointed for such Meeting as is herein-before provided with respect to Notices of Meetings of Committees of Visitors, nor unless Three Visitors concur in and sign the Order granting the same.

Clerk of the Asylum to keep Account of Monies paid and received, and send Abstract thereof annually to Secretary of State and Commissioners in Lunacy.

LVIII. The Clerk of every Asylum shall keep all Books, Documents, and Instruments which the Visitors of the Asylum are required to keep or direct to be kept, and shall also keep an Account of all Monies received or paid on account of the Asylum, either to or by the Treasurer of the Asylum or otherwise, and shall in the Month of *March* in every Year send an Abstract of such Account for the Year previous ending on the Thirty-first Day of *December* to One of Her Majesty's Principal Secretaries of State, and to the Clerk or Clerks of the Peace of the County or Borough, or of each County or Borough, to which the Asylum shall belong, and also to the Commissioners in Lunacy, such Abstract to contain such Particulars and be in such Form as the Commissioners in Lunacy may direct; and such Commissioners shall, within One Month from the Receipt of such Abstract, cause a Copy thereof to be laid before both Houses of Parliament.

Treasurer to keep Accounts.

LIX. The Treasurer of every Asylum shall keep Accounts of all Monies received and paid by him.

Visitors to audit Accounts.

LX. The Committee of Visitors of every Asylum shall, previously to the Month of *March* in every Year, audit the Accounts of the Treasurer and Clerk of such Asylum, and shall report the same to the next General or Quarter Sessions of the County or each of the Counties, and to the Council of the Borough or each of the Boroughs, to which the Asylum wholly or in part belongs.

Two Visitors at least to visit once in every Two Months every Asylum.

LXI. Not less than Two Members of every Committee of Visitors shall together, once at the least in every Two Months, inspect every Part of the Asylum of which they are Visitors, and see and examine, as far as Circumstances will permit, every Lunatic therein, and the Order and Certificate for the Admission of every Lunatic admitted since the last Visitation of the Visitors, and the General Books kept in such Asylum, and shall enter in a Book to be kept for that Purpose any Remarks which they may deem proper in regard to the Condition and Management of such Asylum and the Lunatics therein, and shall sign such Book upon every such Visit.

Annual Reports to be made by Committees of Visitors to Justices at Quarter Sessions, &c., and Copies sent to Commissioners in Lunacy.

LXII. The Committee of Visitors of every Asylum shall in every Year lay before the Justices of every County and Borough to which such Asylum wholly or in part belongs, at the Court of General or Quarter Sessions to be holden next after the Twentieth Day of *December* in every Year for such County, or at a Special Meeting of the Justices of such Borough to be holden within Twenty Days after the Twentieth Day of *December* in every Year, a Report in Writing of the State and Condition of such Asylum, and as to its Sufficiency for the proper Accommodation of the Number of Lunatics for whom it may be requisite to provide Accommodation, and as to the Management of such Asylum, and the Conduct of the Officers and Servants thereof, and the Care of the Patients therein, and such Committee may in such Report make such Remarks or Observations in relation to any Matters connected with such Asylum as they may think fit ; and the Clerk to such Committee shall transmit a Copy of such Report to the Commissioners in Lunacy, and if any such Clerk neglect so to do for Twenty-one Days after the laying of such Report before the Justices of any County or Borough, he shall for such Offence forfeit any Sum not exceeding Ten Pounds.

Lists of Pauper Patients in Asylums to be made half-yearly and laid before Visitors, and Copies transmitted to Clerks of the Peace and Commissioners in Lunacy. Lists of private Patients to be sent half-yearly to the Commissioners.

LXIII. The Clerk of every Asylum shall, on the First Day of *January* and the First Day of *July* in every Year, prepare a List of all Pauper Lunatics then in such Asylum, according to the Form in Schedule (C.) No. 1. to this Act annexed, and within Fifteen Days after such List shall have been prepared One Copy thereof shall be laid by such Clerk before the Visitors of the Asylum, and another shall be transmitted by him to the Clerk of the Peace of every or any County and to the Clerk to the Justices of every or any Borough to which such Asylum solely or jointly belongs, to be by him laid before the Justices of such County or Borough, and another Copy of such List shall within the same Time be transmitted by such Clerk to the Commissioners in Lunacy ; and the Clerk of every Asylum receiving private Patients shall also on the First Day of *January* and First Day of *July* in every Year prepare a List containing the Christian Names and Surnames of all the private Patients in such Asylum in the Form in Schedule (C.) No. 2. to this Act annexed, and shall within Fifteen Days after such List shall have been prepared transmit the same to the Commissioners in Lunacy ; and shall also within the same Time transmit to such Clerk of the Peace and Clerk to the Justices as aforesaid, for the Purposes aforesaid, a Certificate under his Hand of the Number of such private Patients of each Sex.

Clerks of Boards of Guardians, and Overseers where no Guardians, to make annual Returns of Pauper Lunatics.

LXIV. The Clerk of the Board of Guardians of every Union, and of every Parish under a Board of Guardians, and the Overseers of every Parish not in a Union nor under a Board of Guardians, shall, on the First Day of *January* in every Year, or as soon after as may be, make out and sign a true and faithful List of all Lunatics chargeable to the Union or Parish in the Form in Schedule (D.) hereunto annexed, and shall, on or before the First Day of *February* next succeeding, lay One Copy of such List before the Visitors of the Asylum or before the Visitors of each Asylum (if more than One) of the County or Borough in which such Union or Parish is situate, and shall transmit One Copy of such List to the Clerk of the Peace of the County, or the Clerk to the Justices of the Borough within which the Union or Parish to which each such Lunatic is chargeable is situate, to be by him laid before the Justices acting for such County at their next General or Quarter Sessions, or before the Justices of such Borough, and another Copy of such List to the Commissioners in Lunacy, and another Copy thereof to the Poor Law Board ; and any such Clerk or Overseer neglecting to make out and sign such List, or to transmit Copies thereof, as herein directed, shall for every such Offence forfeit any Sum not exceeding Twenty Pounds.

Power for Medical Persons, Guardians, and Overseers of Unions and Parishes, to visit Pauper Patients of such Unions and Parishes confined in any Asylum.

LXV. Any Physician, Surgeon, or Apothecary to be appointed by the Guardians of any Union or Parish, or the Overseers of any Parish, and also the Guardians of any Union or Parish, and the Overseers of any Parish, shall be permitted, whenever they see fit, between the Hours of Eight in the Morning and Six in the Evening, to visit and examine any or every Pauper Lunatic chargeable to such Union or Parish confined in any Asylum, registered Hospital, or licensed House: Provided always, that if the Medical Officer of any Asylum be of opinion that it will be injurious to any Lunatic to permit such Visit and Examination, and such Medical Officer state in Writing the Reasons why such Lunatic should not be visited and examined, and sign such Statement, and deliver the same to the Person or Persons so requiring to visit and examine such Lunatic, then and in such Case it shall be lawful for such Medical Officer to refuse such Visit and Examination; and in every such Case such Medical Officer shall forthwith enter in the Medical Journal the Reasons set forth in such Statement for such Refusal, and shall sign such Entry.

PROVISIONS CONCERNING VISITATION, CONFINEMENT, REMOVAL, AND DISCHARGE OF LUNATICS.

Every Pauper Lunatic not in an Asylum, registered Hospital, or licensed House, to be visited once a Quarter by the Medical Officer of the Parish or Union and List of such Lunatics to be sent to Commissioners in Lunacy.

LXVI. Every Pauper Lunatic not in an Asylum, or a Hospital registered or a House licensed for the Reception of Lunatics, shall be visited once in every Quarter of a Year (reckoning the several Quarters of the Year as ending on the Thirty-first Day of *March*, the Thirtieth Day of *June*, the Thirtieth Day of *September*, and the Thirty-first Day of *December*,) by the Medical Officer of or for the Parish or Union or District of a Parish or Union in which such Lunatic is resident; and such Medical Officer shall be paid the Sum of Two Shillings and Sixpence for each such quarterly Visit to any Pauper not being in a Workhouse, which Sum shall be paid by the same Persons, and be charged to the same Account as the Relief of such Pauper; and within Seven Days after the End of every such Quarter such Medical Officer shall prepare and sign a List according to the Form in the Schedule (E.) to this Act of all such Lunatics, and shall state therein whether in the Opinion of such Medical Officer all or any of such Lunatics are or are not properly taken care of, and may or may not properly remain out of an Asylum, and such Medical Officer shall within the Time aforesaid deliver or send such List to the Clerk to the Guardians of such Parish or Union, or if such Parish be not under a Board of Guardians to One of the Overseers thereof; and the Forms for such Lists shall be from Time to Time furnished to the Medical Officer of every Parish under a Board of Guardians, and to the Medical Officers of every Union, by the Guardians of such Parish or Union; but nothing in this Enactment shall be taken or construed to relieve any Medical Officer from any Obligation by this Act imposed upon him to give Notice to a Relieving Officer or Overseer where it appears to such Medical Officer that any Pauper Lunatic ought to be sent to an Asylum; and such Clerk or Overseer receiving any such List as aforesaid shall, within Three Days after the Receipt thereof, transmit the same to the Commissioners in Lunacy, and a Copy thereof to the Clerk to the Visitors of the Asylum for the County or Borough in which the Parish or Union for which he is Clerk or Overseer is situate; and every such Medical Officer, Clerk, or Overseer failing to comply with this Enactment shall for every such Offence forfeit any Sum not exceeding Twenty Pounds nor under Two Pounds.

Provision for sending Pauper Lunatics to Asylums.

LXVII. Every Medical Officer of a Parish or Union who shall have Knowledge that any Pauper resident in such Parish, or in any Parish within the District of such Medical Officer, is or is deemed to be a Lunatic, and a proper Person to be sent to an Asylum, shall within Three Days after obtaining such Knowledge give Notice thereof in Writing to a Relieving Officer of such Parish, or if there is no Relieving Officer then to One of the Overseers of such Parish, and every Relieving Officer of any Parish within a Union or under a Board of Guardians, and every Overseer of a Parish of which there is no Relieving Officer, who shall have Knowledge, either by such Notice or otherwise, that any Pauper resident in such Parish is or is deemed to be a Lunatic, and a proper Person to be sent to an Asylum, shall within Three Days after obtaining such Knowledge give Notice thereof to some Justice of the County or Borough within which such Parish is situate ; and thereupon the said Justice shall, by an Order under his Hand and Seal, require such Relieving Officer or Overseer to bring such Pauper before him, or some other Justice of the said County or Borough, at such Time and Place within Three Days from the Time of such Notice being given to such Justice as shall be appointed by the said Order ; and the said Justice before whom such Pauper shall be brought shall call to his Assistance a Physician, Surgeon, or Apothecary, and examine such Person ; and if such Physician, Surgeon, or Apothecary shall sign a Certificate with respect to such Pauper, according to the Form in Schedule (F.) No. 3. to this Act annexed, and such Justice be satisfied, upon View or personal Examination of such Pauper or other Proof, that such Pauper is a Lunatic, and a proper Person to be taken charge of and detained under Care and Treatment, he shall, by an Order under his Hand according to the Form in the said Schedule (F.) No. 1. to this Act annexed, direct such Pauper to be received into such Asylum as herein-after mentioned, or, where herein-after authorized in this Behalf, into some Hospital registered or some House duly licensed for the Reception of Lunatics ; and such Relieving Officer or Overseer shall immediately convey or cause the said Lunatic to be conveyed to such Asylum, Hospital, or House, and such Lunatic shall be received and detained therein : Provided always, that it shall be lawful for any Justice, upon Notice being given to him as aforesaid, or upon his own Knowledge, without any such Notice as aforesaid, to examine any Pauper deemed to be lunatic at his own Abode or elsewhere, and to proceed in all respects as if such Pauper were brought before him in pursuance of an Order for that Purpose ; provided also, that in case any Pauper deemed to be lunatic cannot, on account of his Health or other Cause, be conveniently taken before any Justice, such Pauper may be examined at his own Abode or elsewhere by an Officiating Clergyman of the Parish in which he is resident, together with a Relieving Officer, or if there be no Relieving Officer an Overseer of such Parish, and such Officiating Clergyman, together with such Relieving Officer or Overseer, shall call to their Assistance a Physician, Surgeon, or Apothecary ; and if such Physician, Surgeon, or Apothecary shall sign a Certificate with respect to such Pauper according to the said Form in the said Schedule (F.) No. 3., and if upon View or Examination of such Pauper such Officiating Clergyman and such Relieving Officer or Overseer be satisfied that such Pauper is a Lunatic, and a proper Person to be taken charge of and detained under Care and Treatment, such Officiating Clergyman, together with such Overseer or Relieving Officer, shall, by an Order under their Hands according to the said Form in the said Schedule (F.) No. 1., direct such Pauper to be received into such Asylum as herein-after mentioned, or, where herein-after authorized in this Behalf, into some such registered Hospital or licensed House as aforesaid, and such Relieving Officer or Overseer shall immediately convey, or cause such Pauper to be conveyed to such Asylum, Hospital, or House, and such Pauper shall be received and detained therein ; provided also, that if the Physician, Surgeon, or Apothecary by whom any such Pauper shall be examined shall certify in Writing that he is not in a fit State to be removed, his Removal shall be suspended until the same or some other Physician, Surgeon, or Apothecary shall certify in Writing that he is fit to be removed ; and every such Physician, Surgeon, and Apothecary is required to give such last-mentioned Certificate as soon as in his Judgment it ought to be given ; provided also, that where a Certificate in the Form in the said Schedule (F.) No. 3. is signed by the Medical Officer of the Parish or Union in which the Pauper named therein is resident, as well as by some other Person being a Physician, Surgeon, or Apothecary called to the Assistance of the Justice or

Clergyman and Overseer or Relieving Officer, as herein-before mentioned, such joint Certificate, or such Two Certificates, (as the Case may be,) shall be received by the Justice or Clergyman and Overseer or Relieving Officer by whom such Person is examined as herein-before mentioned as conclusive Evidence that the Person named therein is a Lunatic, and a proper Person to be taken charge of and detained under Care and Treatment, and he or they shall make an Order in the Form in the said Schedule (F.) No. 1. accordingly.

Provision as to Lunatics wandering at large, not being properly taken care of, or being cruelly treated, &c.

LXVIII. Every Constable of any Parish or Place, and every Relieving Officer and Overseer of any Parish, who shall have Knowledge that any Person wandering at large within such Parish or Place (whether or not such Person be a Pauper) is deemed to be a Lunatic, shall immediately apprehend and take or cause such Person to be apprehended and taken before a Justice; and it shall also be lawful for any Justice, upon its being made to appear to him by the Information upon Oath of any Person whomsoever that any Person wandering at large within the Limits of his Jurisdiction is deemed to be a Lunatic, by an Order under the Hand and Seal of such Justice, to require any Constable of the Parish or Place, or Relieving Officer or Overseer of the Parish where such Person may be found, to apprehend him and bring him before such Justice, or some other Justice having Jurisdiction where such Person may be found; and every Constable of any Parish or Place, and every Relieving Officer and Overseer of any Parish, who shall have Knowledge that any Person in such Parish or Place not a Pauper and not wandering at large as aforesaid is deemed to be a Lunatic, and is not under proper Care and Control, or is cruelly treated or neglected by any Relative or other Person having the Care or Charge of him, shall, within Three Days after obtaining such Knowledge, give Information thereof upon Oath to a Justice, and in case it be made to appear to any Justice, upon such Information or upon the Information upon Oath of any Person whomsoever, that any Person within the Limits of his Jurisdiction not a Pauper, and not wandering at large, is deemed to be a Lunatic, and is not under proper Care and Control, or is cruelly treated or neglected by any Relative or other Person having the Care or Charge of him, such Justice shall, either himself visit and examine such Person and make Inquiry into the Matters so appearing upon such Information, or by an Order under his Hand and Seal direct and authorize some Physician, Surgeon, or Apothecary to visit and examine such Person, and make such Inquiry, and to report in Writing to such Justice his Opinion thereupon; and in case upon such personal Visit, Examination, and Inquiry by such Justice, or upon the Report of such Physician, Surgeon, or Apothecary, it appear to such Justice that such Person is a Lunatic, and is not under proper Care and Control, or is cruelly treated or neglected by any Relative or other Person having the Care or Charge of him, it shall be lawful for such Justice, by an Order under his Hand and Seal, to require any Constable of the Parish or Place, or any Relieving Officer or Overseer of the Parish, where such Person is alleged to be, to bring him before any two Justices of the same County or Borough; and the Justice or Justices (as the Case may be) before whom any such Person as aforesaid in the respective Cases aforesaid is brought, under this Enactment, shall call to his or their Assistance a Physician, Surgeon, or Apothecary, and shall examine such Person, and make such Inquiry relative to such Person as he or they shall deem necessary; and if upon Examination of such Person or other Proof such Justice be satisfied that such Person so brought before him is a Lunatic, and was wandering at large, and is a proper Person to be taken charge of and detained under Care and Treatment, or such Two Justices be satisfied that such Person so brought before them is a Lunatic, and is not under proper Care and Control, or is cruelly treated or neglected by any Person having the Care or Charge of him, and that he is a proper Person to be taken Charge of and detained under Care and Treatment, and if such Physician, Surgeon, or Apothecary sign a Certificate with respect to every such Person so brought either before One Justice or Two Justices according to the Form in the Schedule (F.) No. 3. to this Act, it shall be lawful for the said Justice or Justices, by an Order under his or their Hand and Seal or Hands and

Seals, according to the Form in the Schedule (F.) No. 1, to this Act, to direct such Person to be received into such Asylum as herein-after mentioned, or, where herein-after authorized in this Behalf, into some Hospital registered or House licensed for the Reception of Lunatics, and the said Constable, Relieving Officer, or Overseer who may have brought such Person before the said Justice or Justices, or any Constable whom such Justice or Justices may require so to do, shall forthwith convey such Person to such Asylum, Hospital, or House accordingly: Provided always, that it shall be lawful for any Justice upon such Information on Oath as aforesaid, or upon his own Knowledge, and alone, in the Case of any such Person as aforesaid wandering at large and deemed to be a Lunatic, or with some other Justice, in any other of the Cases aforesaid, to examine the Person deemed to be a Lunatic, at his own Abode or elsewhere, and to proceed in all respects as if such Person were brought before him or them as herein-before mentioned; provided also, that it shall be lawful for the said Justice or Justices to suspend the Execution of any such Order for removing any such Person as aforesaid to any Asylum, Hospital, or House for such Period not exceeding Fourteen Days as he or they may deem meet, and in the meantime to give such Directions or make such Arrangements for the proper Care and Control of such Person as he or they shall consider necessary; provided also, that if the Physician, Surgeon, or Apothecary by whom such Person is examined certify in Writing that he is not in a fit State to be removed, the Removal of such Person shall be suspended until the same or some other Physician, Surgeon, or Apothecary certify in Writing that such Person is fit to be removed; and every such Physician, Surgeon, and Apothecary is hereby required to give such last-mentioned Certificate as soon as in his Judgment it ought to be given; provided also, that nothing herein contained shall be construed to extend to restrain or prevent any Relation or Friend from retaining or taking such Lunatic under his own Care, if such Relation or Friend shall satisfy the Justice or Justices before whom such Lunatic shall be brought, or the Visitors of the Asylum in which such Lunatic is or is intended to be placed, that such Lunatic will be properly taken care of.

Power to Justices to order Payment of a Fee to any Physician, &c. called in to examine any Person.

LXIX. It shall be lawful for any Justice or Justices causing any Person to be examined by any Physician, Surgeon, or Apothecary, under the Provisions herein-before contained, if he or they think fit so to do, to make an Order under his or their Hand and Seal or Hands and Seals upon the Guardians of the Union or Parish or the Overseers of the Parish to which such Person is chargeable, under the Provisions herein contained, for the Payment of such reasonable Remuneration to any such Physician, Surgeon, or Apothecary, for the Examination of such Person, and of all other reasonable Expenses in or about the Examination of such Person, and the bringing him before such Justice or Justices, and in case he be ordered to be conveyed to any Asylum, registered Hospital, or licensed House, of conveying him thereto, as to such Justice or Justices may seem proper.

Penalties on Medical Officers, Overseers, &c., omitting to give Notice as aforesaid.

LXX. If any Medical Officer of any Parish or Union omit for more than Three Days after obtaining Knowledge of any Pauper resident in such Parish, or in any Parish within his District, being or being deemed to be lunatic, and a proper Person to be sent to an Asylum, to give such Notice thereof as is herein-before required, or if any Relieving Officer of any Parish, or any Overseer of any Parish of which there is no Relieving Officer, omit for more than Three Days after obtaining Knowledge of any Pauper resident in such Parish, being deemed to be a Lunatic, and a proper Person to be sent to an Asylum, to give Notice thereof to a Justice as herein-before required, or if any Constable, Relieving Officer, or Overseer omit to apprehend and take before a Justice, as herein-before required, any Person wandering at large and deemed to be a Lunatic, or omit for Three Days after obtaining Knowledge that any Person deemed to be a Lunatic (not a Pauper and not wandering at large) is not under proper Care and Control, or is cruelly treated or neglected by any Person having the Care or Charge of him, to give Information thereof to a Justice as herein-before required, such Medical Officer, Relieving Officer, Overseer, or Constable, as the Case may be, shall for every such Offence forfeit any Sum not exceeding Ten Pounds.

Penalty on Relieving Officers, Overseers, and Constables, delaying to execute Orders.

LXXI. If any Relieving Officer, Overseer, or Constable by this Act required to convey any Person to any Asylum, registered Hospital, or licensed House, in pursuance of any Order under this Act, refuse or wilfully neglect to execute such Order with all reasonable Expedition, he shall for every such Offence forfeit any Sum not exceeding Ten Pounds.

Orders of Justices, &c. may extend to authorize Reception into Hospitals or licensed Houses, but Lunatics to be always sent to Asylum, if Circumstances permit.

LXXII. Every such Order by a Justice or Justices, or by a Clergyman and Overseer or Relieving Officer as aforesaid, for the Reception of a Lunatic into an Asylum, may authorize his Admission, not only into any Lunatic Asylum of the County or Borough in which the Parish or Place from which the Lunatic is sent is situate, but also into any other Asylum for the Reception of Pauper Lunatics of such County or Borough, and also into any Asylum for any other County or Borough, or any Hospital registered or House licensed for the Reception of Lunatics ; but every Lunatic shall under every such Order be sent to an Asylum of the County or Borough in which the Parish or Place from which he is sent is situate, unless there be no such Asylum, or there be a Deficiency of Room, or unless there be some special Circumstances by reason whereof such Lunatic cannot conveniently be taken to such Asylum, which Deficiency of Room or special Circumstances shall be stated in the Order for the Reception of such Lunatic into any Asylum other than such Asylum as aforesaid, or into any registered Hospital or licensed House ; and no Lunatic shall be sent to any registered Hospital or House licensed for the Reception of Lunatics, by virtue of such Order, except there be no such Asylum, or no such Asylum in which he can be received, or there be some special Circumstances by reason whereof he cannot be taken thereto, which shall be stated in like Manner as aforesaid.

No Pauper to be received into any Asylum without a certain Order and Certificate.

LXXIII. No Pauper shall be received into any Asylum, registered Hospital, or licensed House (save under the Provisions herein contained with respect to Removal of Lunatics) without an Order according to the Form required in the said Schedule (F.) No. 1., under the Hands of One Justice, or under the Hands of an Officiating Clergyman, and of One of the Overseers or the Relieving Officer of the Parish or Union from which such Pauper is sent as aforesaid, together with such Statement of Particulars as is contained in the same Schedule, nor without a Medical Certificate according to the Form in the said Schedule (F.) No. 3., signed by (ne Physician, Surgeon, or Apothecary, who shall have personally examined him not more than Seven clear Days previously to his Reception ; and every Person who receives any Pauper into any Asylum without such Order and Medical Certificate (save under any of the said Provisions) shall be guilty of a Misdemeanor.

No Person to be received into an Asylum, except under the Provisions of this Act, without an Order and Two Medical Certificates.

LXXIV. No Person, not a Pauper, shall be received into any Asylum (save under the Provisions herein contained) without an Order under the Hand of some Person according to the Form in Schedule (F.) No. 2. to this Act annexed, together with such Statement of Particulars as is contained in the same Schedule, nor without the Medical Certificate, according to the Form and containing the Particulars required in Schedule (F.) No. 3. annexed to this Act, of Two Persons, each of whom shall be a Physician, Surgeon, or Apothecary, and shall not be in Partnership with or an Assistant to the other, and each of whom shall separately from the other have personally examined the Person to whom it relates, not more than Seven clear Days previously to the Reception of such Person into such Asylum, and such Order as aforesaid may be signed before or after the Medical Certificates or either of them ; and every Person who receives any Person, not a Pauper, into any Asylum, save under the Provisions herein contained, without such Order and Medical Certificates as aforesaid, shall be guilty of a Misdemeanor : Provided always, nevertheless, that any Person may, under special Circumstances preventing the Examination of such Person by Two Medical Practitioners as aforesaid, be received into any Asylum upon the Certificate of One Physician, Surgeon, or Apothecary alone, provided that the Statement accompanying such Order set forth the special Circumstances which prevent the Examination of such Person by Two Medical Practitioners ; but in every such Case Two other such Certificates shall, within Three clear Days after the Reception of such Patient into such Asylum, be signed by Two other Persons, each of whom shall be a Physician, Surgeon, or Apothecary, not in Partnership with or an Assistant to the other, or the Physician, Surgeon, or Apothecary who signed the Certificate on which the Patient was received, and shall within such Time, and separately from the other of them, have personally examined the Person so received as a Lunatic ; and any Person who, having received any Person into any Asylum as aforesaid upon the Certificate of One Medical Practitioner alone as aforesaid, shall keep or permit such Person to remain in such Asylum beyond the said Period of Three clear Days, without such further Certificates as aforesaid, shall be guilty of a Misdemeanor.

Medical Certificate to specify Facts upon which Opinion of Insanity has been formed.

LXXV. Every Physician, Surgeon, and Apothecary signing any Certificate under or for the Purposes of this Act, shall specify therein the Facts upon which he has formed his Opinion that the Person to whom such Certificate relates is a Lunatic, an Idiot, or a Person of unsound Mind, distinguishing in such Certificate Facts observed by himself from Facts communicated to him by others; and no Person shall be received into any Asylum under any Certificate which purports to be founded only upon Facts communicated by others.

Who not to sign Certificate for Reception of a Patient.

LXXVI. No Physician, Surgeon, or Apothecary who, or whose Father, Brother, Son, Partner, or Assistant, shall sign the Order for the Reception of a Patient, shall sign any Certificate for the Reception of the same Patient, and no Patient shall be received into any Asylum upon or under any Certificate signed by any Medical Officer of such Asylum.

Power to Two Visitors of any Asylum, being Justices, to order Removal of Pauper Lunatics to or from such Asylum.

LXXVII. It shall be lawful for any Two of the Visitors of any Asylum, being Justices, by an Order in Writing under their Hands and Seals, to order any Pauper Lunatic chargeable to any Parish or Union within the County or Borough or any County or Borough to which such Asylum wholly or in part belongs, or to such County, and who may be confined in any other Asylum, or in any registered Hospital or licensed House, to be removed to such first-mentioned Asylum ; and it shall be lawful for any Two of the Visitors of any Asylum, being Justices, in manner aforesaid to order any Pauper Lunatic to be removed from such Asylum to some other Asylum, or to some registered Hospital or licensed House ; but no such Lunatic shall be removed as last aforesaid without the Consent in Writing of Two of the Commissioners in Lunacy, except to an Asylum within or belonging wholly or in part to the County within which the Asylum from which the Lunatic is removed is situate, or the County in some Parish of which the Lunatic may have been adjudged to be settled, or a registered Hospital or licensed House within any such County as aforesaid, or an Asylum, registered Hospital, or licensed House into which the Lunatic can be received under a subsisting Contract for the Reception of Lunatics therein ; and it shall be lawful for the Justices making any such Order in and by the same to direct or require any Overseer or Relieving or other Officer of the Parish, Union, or County to which such Lunatic is chargeable, or to authorize any other Person to execute the same; and every such Order and Consent shall be made and given respectively in Duplicate, and One Duplicate shall be delivered to and left with the Superintendent or Proprietor of the Asylum, Hospital, or licensed House from which the Patient is removed, and the other shall be delivered to and left with the Superintendent or Proprietor of the Asylum, Hospital, or licensed House to which the Patient is removed, and such Order, with such Consent in Writing (where such Consent is required) shall be a sufficient Authority for the Removal of such Patient, and also for his Reception into the Asylum, Hospital, or licensed House to which he is ordered to be removed : Provided always, that no Person shall be removed under any such Order without a Medical Certificate, signed by the Medical Officer of the Asylum, or the Medical Practitioner, or One of the Medical Practitioners, keeping, residing in, or visiting the Hospital or licensed House from which such Person is ordered to be removed, certifying that he is in a fit Condition of bodily Health to be removed in pursuance of such Order; and the Superintendent or Proprietor of such Asylum, Hospital, or licensed House shall, at the Time of delivering the Person ordered to be removed to the Overseer, Officer, or Person having the Execution of the Order for Removal, deliver to such Overseer or Officer, free of any Charge for the same, the Certificate of such Medical Officer, and also a Copy (certified under the Hand of such Superintendent or Proprietor to be a true Copy) of the Order and Certificate under which such Person was received into and detained in such Asylum, Hospital, or licensed House, and the said Certificate and certified Copies, with One Duplicate of the Order for Removal, shall be delivered by such Overseer, Officer, or Person to the Superintendent or Proprietor of the Asylum, Hospital, or licensed House to which such Person is ordered to be removed, or any other Officer of such Asylum, Hospital, or licensed House into whose Care such Person is delivered.

Pauper Lunatics not to be received into any other than the County or Borough Asylum without Endorsement of Order by a Visitor, and Orders not compulsory on Hospitals or licensed Houses.

LXXVIII. Provided always, that no Lunatic being a Pauper shall be received under any Order made by virtue of this Act into any Asylum, other than an Asylum belonging wholly or in part to the County or Borough in which the Parish or Place from which such Lunatic is sent, or the Parish in which he is adjudged to be settled, is situate, except there be a subsisting Contract for the Reception of Lunatics of such County or Borough therein, or such Borough otherwise contributes to such Asylum, unless such Order be endorsed by a Visitor of such Asylum ; and it shall not be compulsory on the Superintendent of any registered Hospital or the Proprietor of any licensed House to receive any Lunatic under any such Order, except in pursuance of any subsisting Contract.

Discharge of Lunatics from Asylums.

LXXIX. It shall be lawful for any Three of the Visitors of any Asylum, by Writing under their Hands and Seals, to order the Discharge of any Person detained in such Asylum, whether such Person be recovered or not, and also for any Two of such Visitors, with the Advice in Writing of the Medical Officer of such Asylum, to discharge any Person detained therein, or to permit any such Person to be absent from the Asylum upon Trial for such Period as such Visitors think fit; and it shall be lawful for such Visitors to make such Allowance to such last-mentioned Person, not exceeding what would be the Charge for such Person in the Asylum, which Allowance, and no greater Sum, shall be charged for him and be payable as if he were actually in the Asylum ; and in case any Person so allowed to be absent on Trial for any Period do not return at the Expiration of such Period, and a Medical Certificate as to his State of Mind, certifying that his Detention in an Asylum is no longer necessary, be not sent to the Visitors, he may, at any Time within Fourteen Days after the Expiration of such Period, be retaken, as herein provided in the Case of an Escape.

APPENDIX.

Overseers and Relieving Officers to remove Lunatics upon Notice of Discharge, and to be liable to a Penalty for Refusal or wilful Neglect.

LXXX. When the Visitors of any Asylum shall order a Pauper Lunatic confined therein to be discharged therefrom, it shall be lawful for them, when they shall see Occasion, to send Notice in Writing, signed by their Clerk, through the Post or otherwise, of their Intention to discharge such Lunatic, to the Overseers of the Parish wherein it shall have been adjudged that such Lunatic is settled, or, if no such Adjudication shall have been made, to the Overseers of the Parish from which such Lunatic shall have been sent to such Asylum, unless such Lunatic shall be chargeable to the Common Fund of any Union, and in any such last-mentioned Case to some One Relieving Officer of such Union ; and upon Receipt of such Notice the Overseers or Relieving Officers respectively shall cause such Lunatic, upon his Discharge, to be forthwith removed to their Parish, or to the Workhouse of the Union at the Cost and Charge of their Parish or of the Common Fund of the Union, as the Case shall require ; and any Overseer or Relieving Officer who shall refuse or wilfully neglect to remove such Lunatic from the said Asylum within the Space of Seven Days after such Notice shall have been sent to him shall be guilty of an Offence against this Act, and shall forfeit for such Offence any Sum not exceeding Ten Pounds, to be recovered as other Penalties imposed by this Act are recoverable.

Visitors may discharge a Lunatic on the Undertaking of a Relative or Friend that he shall no longer be chargeable, and shall be taken care of.

LXXXI. Where Application is made to the Committee of Visitors of any Asylum by any Relative or Friend of a Pauper Lunatic confined therein, requiring that he may be delivered over to the Custody and Care of such Relative or Friend, it shall be lawful for any Two of the Visitors aforesaid, if they think fit, and upon the Undertaking in Writing of such Relative or Friend to the Satisfaction of such Visitors that such Lunatic shall be no longer chargeable to any Union, Parish, or County, and shall be properly taken care of, and shall be prevented from doing Injury to himself or others, to discharge such Lunatic.

Commissioners in Lunacy may order Removal of Lunatics.

LXXXII. It shall be lawful for the Commissioners in Lunacy, or any Two of them, by Writing under their Hands and Seals, to order and direct the Removal of any Lunatic from any Asylum, registered Hospital, or licensed House to any other Asylum, registered Hospital, or licensed House ; and every such Order shall be made in Duplicate, and One Duplicate shall be delivered to and left with the Superintendent or Proprietor of the Asylum, Hospital, or licensed House from which the Patient is removed, and the other shall be delivered to and left with the Superintendent or Proprietor of the Asylum, Hospital, or licensed House to which the Patient is removed, and such Order shall be a sufficient Authority for the Removal of such Patient, and also for his Reception into the Asylum, Hospital, or licensed House to which he is ordered to be removed.

The Person who signed the Order for the Reception of a private Patient may order his Discharge or Removal.

LXXXIII. If and when any Person who signed the Order on which any Patient (not being a Pauper) was received into any Asylum (whether or not such Patient have since been removed under any Order made under this Act or otherwise to any other Asylum) shall by Writing under his Hand direct that such Patient be discharged or removed, then and in such Case such Patient shall forthwith be discharged or removed as the Person who signed the Order for his Reception may direct.

Provision where the Person who signed the Order for Reception is dead or incapable of acting.

LXXXIV. If the Person who signed the Order on which any Patient (not being a Pauper) was received into any Asylum be dead, or be incapable, by reason of Insanity, Absence from *England*, or otherwise, of giving an Order for the Discharge or Removal of such Patient, then the Person who made the last Payment on account of such Patient, or the Husband or Wife, or (if there be no Husband, or the Husband or Wife be incapable as aforesaid,) the Father, or (if there be no Father, or he be incapable as aforesaid,) the Mother of such Patient, or if there be no Mother, or she be incapable as aforesaid, then any One of the nearest of Kin for the Time being of such Patient, may, by Writing under his or her Hand, give such Direction as aforesaid for the Discharge or Removal of such Patient, and thereupon such Patient shall be forthwith discharged or removed accordingly.

Patient not to be discharged where certified to be dangerous, without Visitors' Consent. Not to prevent Transfer under Control of an Attendant.

LXXXV. Provided always, That no Patient shall be discharged under either of the Two last foregoing Provisions of the Medical Officer of the Asylum in which such Patient is certified in Writing under his Hand that in the Opinion of such Medical Officer such Patient is dangerous and unfit to be at large, together with the Grounds on which such Opinion is founded, unless Two of the Visitors of such Asylum, being Justices, shall, after such Certificate shall have been produced to them, give their Consent in Writing to such Patient being so discharged; provided that nothing in this Enactment shall prevent the Transfer of any Patient so certified to be dangerous and unfit to be at large from any Asylum to any other Asylum, or to any registered Hospital or licensed House, but in such Case the Patient shall be placed under the Control of an Attendant belonging to the Asylum, Hospital, or House from or to which he is about to be removed for the Purpose of such Removal, and shall remain under such Control until such Time as the Removal has been duly effected.

Provision authorizing Transfer of private and single Patients.

LXXXVI. Any Person, having Authority to order the Discharge of any Patient (not being a Pauper) from any Asylum, registered Hospital, or licensed House, or of any single Patient, may, with the previous Consent in Writing of Two of the Commissioners, direct, by an Order in Writing under his Hand, the Removal of such Patient to any Asylum, registered Hospital, or licensed House, or to the Care or Charge of any Person mentioned or named in such Order; and every such Order and Consent shall be made and given respectively in Duplicate, and One of the Duplicates shall be delivered to and left with the Superintendent or Proprietor of the Asylum, Hospital, or House from which or the Person from whose Care or Charge the Patient is ordered to be removed, and the other Duplicate shall be delivered to and left with the Superintendent or Proprietor of the Asylum, Hospital, or House into which or the Person into whose Care or Charge the Patient is ordered to be removed; and such Order for Removal, together with such Consent in Writing, shall be a sufficient Authority for the Removal of such Patient, and also for his Reception into the Asylum, registered Hospital, or licensed House into which or by the Person into whose Care or Charge he is ordered to be removed: Provided always, that a Copy of the Order and Certificates upon which such Patient was received into the Asylum, Hospital, or House from which he is removed, or as a single Patient, by the Person from whose Care he is removed, certified under the Hand of the Superintendent or Proprietor of such Asylum, Hospital, or House, or of such Person as last aforesaid, to be a true Copy, shall be furnished by him free of Expense, and shall be delivered, with One Duplicate of the said Order of Removal and Consent, to the Superintendent or Proprietor of the Asylum, Hospital, or House to which, or to the Person to whose Care or Charge, such Patient is removed.

Orders and Medical Certificates may be amended.

LXXXVII. If after the Reception of any Lunatic into any Asylum it appear that the Order or the Medical Certificate, or (if more than One) both or either of the Medical Certificate, upon which he was received, is or are in any respect incorrect or defective, such Order and Medical Certificate or Certificates may be amended by the Person or Persons signing the same at any Time within Fourteen Days next after the Reception of such Lunatic; provided nevertheless, that no such Amendment shall have any Force or Effect unless the same shall receive the Sanction of One or more of the Commissioners in Lunacy.

Persons received into Asylums, &c. may be detained till Removal or Discharge, and in case of Escape may be retaken within Fourteen Days.

LXXXVIII. Every Person received into any Asylum, registered Hospital, or licensed House under such Order as is required by this Act, accompanied by the requisite Medical Certificate, may be detained therein until he be removed or discharged as authorized by this Act, and in case of Escape may, by virtue of such Order and Certificate or Certificates, be retaken at any Time within Fourteen Days after his Escape by the Superintendent or Proprietor of such Asylum, Hospital, or House, or any Officer or Servant, belonging thereto, or any other Person authorized in Writing in this Behalf by such Superintendent or Proprietor, and conveyed to and received and detained in such Asylum, Hospital, or House.

Every Clerk receiving a Lunatic into an Asylum to make an Entry thereof, and to transmit a Copy of the Order and Certificate of Medical Officer of the Asylum to Commissioners in Lunacy.

LXXXIX. The Clerk of every Asylum shall, immediately on the Admission of any Person as a Lunatic into such Asylum, make an Entry with respect to such Lunatic in a Book to be kept for that Purpose, to be called "The Register of Patients," according to the Form, and containing the Particulars specified in the Schedule (G.) No. 1. to this Act, except as to the Form of Disorders, the Entry as to which is to be supplied by the Medical Officer of the Asylum within One Month after the Admission of the Patient, and after the Second and before the End of the Seventh clear Day from the Day of the Admission of any Person as a Lunatic into any Asylum shall transmit to the Commissioners in Lunacy a Copy of the Order and Statement and Certificate or Certificates on which such Lunatic has been so received, together with a Statement, to be made and signed by the Medical Officer of the Asylum, not sooner than Two clear Days after such Admission, according to the Form in the said Schedule (F.) No. 4. to this Act annexed; and any Clerk omitting so to make such Entry, or to transmit such Copy and Statement within the Time aforesaid, and every Medical Officer omitting to make or sign such Statement, shall for every such Offence forfeit any Sum not exceeding Twenty Pounds.

Weekly Journal and Case Book to be kept in every Asylum.

XC. In every Asylum the Medical Officer thereof shall once in every Week enter in a Book to be kept for that Purpose, to be called "The Medical Journal," a Statement according to the Form in the said Schedule (G.) No. 3., showing the Number of Patients of each Sex then in such Asylum, the Christian Name and Surname of every Patient who is or has been under Restraint or in Seclusion since the last Entry, and when and for what Period and Reasons, and in case of Restraint by what Means, and the Christian Name and Surname of every Patient under Medical Treatment, and for what, if any, bodily Disorder, and every Death, Injury, and Violence which shall have happened to or affected any Patient since the then last preceding Entry, and shall also enter into a Book to be called "The Case Book," as soon as may be after the Admission of any Patient, the mental State and bodily Condition of every Patient at the Time of his Admission, and also the History from Time to Time of his Case whilst he shall continue in the Asylum; and such Books shall from Time to Time be regularly laid before the Visitors for their Inspection and Signature, and every Medical Officer omitting to make such Entries, or any of them, shall for every such Offence forfeit any Sum not exceeding Twenty Pounds.

S

APPENDIX.

Copies of Entries made by Commissioners visiting Asylums to be sent to the Office of Commissioners.

XCI. The Clerk of every Asylum shall, within Three Days after every Visit to such Asylum of Two or more of the Commissioners in Lunacy, transmit to the Office, of such Commissioners a true and perfect Copy of any Entries of any Remarks or Observations made by such Visiting Commissioners in any of the Books of such Asylum, and every such Clerk omitting to transmit as aforesaid any such Copy shall for every such Offence forfeit any Sum not exceeding Ten Pounds.

In case of the Death of a Lunatic the Cause of Death to be stated, and sent to the Registrar of Deaths, the Commissioners in Lunacy, and Relieving Officer or Overseers.

XCII. In case of the Death of any Patient in any Asylum a Notice and Statement according to the Form in Schedule (F.) No. 5. of the Death and Cause of the Death of such Patient, and the Name of any Person or Persons who was or were present at the Death, shall be drawn up and signed by the Clerk and Medical Officer of such Asylum, and a Copy thereof shall be by the Clerk transmitted to the Registrar of Deaths for the District and to the Commissioners in Lunacy within Forty-eight Hours of the Death of such Patient, and also to the Relieving Officer or the Overseers of the Union or Parish to which such Lunatic (if a Pauper) was chargeable, and if not a Pauper to the Person who shall have signed the Order for the Admission of the Lunatic, or who made the last Payment on account of such Lunatic; and every Clerk or Medical Officer who neglects or omits to draw up, sign, or transmit such Notice or Statement as aforesaid, within the Time aforesaid, shall respectively forfeit and pay any Sum not exceeding Twenty Pounds.

Entries to be made of Deaths, Discharges, and Removals, and Notice given to the Commissioners in case of the Discharge, Removal, Escape, and Recapture of every Lunatic.

XCIII. The Clerk of every Asylum shall, within Three clear Days after the Death, Discharge, or Removal of any Patient, make an Entry thereof in the said Register of Patients, and also in a Book to be kept for that Purpose according to the Form and containing the Particulars in the Schedule (G.) No. 2. to this Act, and shall also, within Three clear Days after the Discharge, Removal, Escape, or Recapture of any Patient, transmit a written Notice of such Discharge or Removal, according to the Form in the said Schedule (F.) No. 5., or of such Escape or Recapture, to the Commissioners in Lunacy; and every such Clerk who neglects or omits to make such Entry as aforesaid, or transmit such Notice as aforesaid within the Time aforesaid, shall forfeit and pay any Sum not exceeding Ten Pounds; and every such Clerk who shall knowingly and wilfully in such Entry untruly set forth any of the Particulars required shall be guilty of a Misdemeanor.

AS TO EXPENSE OF MAINTENANCE AND REMOVAL, &c. OF PAUPER AND OTHER LUNATICS.

How Justices are to proceed where it appears to them that the Lunatic has Property applicable to his Maintenance.

XCIV. Where any Lunatic shall be sent to an Asylum, registered Hospital, or licensed House, under any Order made by virtue of the Authority herein-before given to Two Justices, if it appear to such Justices that such Lunatic hath an Estate applicable to his Maintenance, and more than sufficient to maintain his Family (if any), it shall be lawful for such Justices to make an Application in Writing under their Hands and Seals to the nearest known Relative or Friend of such Lunatic, for the Payment of the Charges of the Examination, Removal, Lodging, Maintenance, Clothing, Medicine, and Care of such Lunatic; and in case such Charges be not paid within One Month after such Application, it shall be lawful for the same or any other Justices, by an Order under their Hands and Seals, to direct a Relieving Officer or Overseer of the Parish from which such Lunatic shall be sent, or where any Property of such Lunatic shall be, to seize so much of the Money, and to seize and sell so much of the Goods and Chattels, and take and receive so much of the Rents and Profits of the Lands and Tenements of such Lunatic, and of any other Income of such Lunatic, as may be necessary to pay the Charges of the Examination, Removal, Lodging, Maintenance, Clothing, Medicine, and Care of such Lunatic, accounting for the same to the same or any other Justices, such Charges having been first proved to the Satisfaction of the said Justices, and the Amount set forth in such Order; and if any Trustee or other Person having the Possession, Custody or Charge of any Property of such Lunatic, or if the Governor and Company of the Bank of *England*, or any other Body or Person having in their or his Hands any Stock, Interest, Dividend, or Annuity belonging or due to such Lunatic, pay the whole or any Part thereof to any Overseer or Relieving Officer, to defray the Charges set forth in such Order the Receipt of such Overseer or Relieving Officer shall be a good Discharge to such Trustee, Governor, and Company, or other Body or Person as aforesaid: Provided always, that, notwithstanding it may appear to the said Justices that such Lunatic hath such Estate as aforesaid it shall be lawful for such Justices, in the meantime and until such Charges as aforesaid shall be paid, in pursuance of such Application or Order as aforesaid to make an Order on the Guardians of the Union or Parish, or the Overseers of the Parish, from which such Lunatic shall be sent for Confinement, for Payment of the Charges of the Removal, Lodging, Maintenance, Clothing, Medicine, and Care of such Lunatic; and such Guardians or Overseers shall be reimbursed such Charges under any Order to be made as aforesaid for Payment of such Charges, out of the Property of the Lunatic, unless the same be sooner repaid by some Relative or Friend of such Lunatic in pursuance of such Application as aforesaid.

Every Pauper Lunatic to be chargeable to the Parish from which he is sent till otherwise adjudged.

XCV. When any Pauper Lunatic is confined under the Provisions of this Act he shall, for the Purposes of this Act, be chargeable to the Parish from which, or at the Instance of some Officer or Officiating Clergyman of which, he has been sent, unless and until such Parish shall have established, under the Provisions herein contained, that such Lunatic is settled in some other Parish, or that it cannot be ascertained in what Parish such Lunatic is settled; and every Pauper Lunatic who is chargeable to any Parish shall, whilst he resides in an Asylum, registered Hospital, or licensed House, be deemed for the Purposes of his Settlement to be residing in the Parish to which he is chargeable.

Justices to make an Order upon the Officers of Unions and Parishes for Maintenance of Lunatics.

XCVI. It shall be lawful for the Justice by whom any Pauper Lunatic is sent to an Asylum, registered Hospital, or licensed House under the Powers of this Act, or for any Two Justices of the County or Borough in which the Asylum, registered Hospital, or licensed House in which any Pauper Lunatic is confined is situate, or from any Part of which any Pauper Lunatic has been sent, or for any Two Justices being Visitors of such Asylum or licensed House, to make an Order upon the Guardians of the Union or Parish or the Overseers of the Parish (if not in a Union or under a Board of Guardians) from which, or at the Instance of any Officer or Officiating Clergyman of which, such Lunatic is or has been sent for Confinement, for Payment to the Treasurer, Officer, or Proprietor of the Asylum, registered Hospital, or licensed House of the reasonable Charges of the Lodging, Maintenance, Medicine, Clothing, and Care of such Lunatic in such Asylum, Hospital, or House, and any such Order may be retrospective or prospective, or partly retrospective and partly prospective; and the Guardians or Overseers on whom such Order shall be made shall from Time to Time pay to the said Treasurer, Officer, or Proprietor the Charges aforesaid.

s 2

Two Justices may inquire into and adjudge the Settlement of a Lunatic, and order Payment of Maintenance, &c. accordingly.

XCVII. It shall be lawful for any Two Justices for the County or Borough in which any Asylum, registered Hospital, or licensed House in which any Pauper Lunatic is or has been confined is situate, or to which such Asylum, wholly or in part belongs, or from any Part of which any Pauper Lunatic is or has been sent for Confinement, at any Time to inquire into the last legal Settlement of such Pauper Lunatic, and if satisfactory Evidence can be obtained as to such Settlement in any Parish, such Justices shall, by Order under their Hands and Seals, adjudge such Settlement accordingly, and order the Guardians of the Union to which the Parish in which such Lunatic is adjudged to be settled belongs, or of such Parish in case such Parish be in a Union or be under a Board of Guardians, and if not, then the Overseers of such Parish, to pay to the Guardians of any Union or Parish, or the Overseers of any Parish, all Expenses incurred by or on behalf of such Union or Parish in or about the Examination of such Lunatic, and the bringing him before a Justice or Justices, and his Conveyance to the Asylum, Hospital, or House, and of all Monies paid by such last-mentioned Guardians or Overseers to the Treasurer, Officer, or Proprietor of the Asylum, Hospital, or House, for the Lodging, Maintenance, Medicine, Clothing, and Care of such Lunatic, and incurred within Twelve Calendar Months previous to the Date of such Order, and, if such Lunatic is still in Confinement, also to pay to the Treasurer, Officer, or Proprietor of the Asylum, Hospital, or House the reasonable Charges of the future Lodging, Maintenance, Medicine, Clothing, and Care of such Lunatic ; and the Guardians or Overseers on whom any such Order is made shall immediately pay to the Guardians or Overseers to whom the same are ordered to be paid the Amount of the Expenses and Monies by such Order directed to be paid to them, and from Time to Time pay to the said Treasurer, Officer, or Proprietor of the Asylum, Hospital, or House the future Charges aforesaid.

If Settlement cannot be ascertained, a Pauper Lunatic may be made chargeable to the County.

XCVIII. If any Pauper Lunatic be not settled in the Parish by which, or at the Instance of some Officer or Officiating Clergyman of which, he is sent to any Asylum, registered Hospital, or licensed House, and it cannot be ascertained in what Parish such Pauper Lunatic is settled, and if a Relieving Officer of such first-mentioned Parish, or of the Union in which the same is situate, or the Overseers of such first-mentioned Parish, shall give Ten Days' Notice to the Clerk of the Peace of the County in which such Lunatic was found to appear for such County before Two Justices thereof, at a Time and Place to be appointed in such Notice, it shall be lawful for such Two Justices, or any Two or more Justices of such County, upon the Appearance of such Clerk of the Peace, or any one on his Behalf, or, in case of his Non-appearance, upon Proof of his having been served with such Notice, to inquire into the Circumstances of the Case, and to adjudge such Pauper Lunatic to be chargeable to such County, and to order the Treasurer of such County to pay to the Guardians of any Union or Parish or the Overseers of any Parish all Expenses incurred by or on behalf of such Union or Parish in or about the Examination of such Lunatic, and the bringing him before a Justice or Justices, and his Conveyance to the Asylum, Hospital, or House, and all Monies paid by such Guardians or Overseers to the Treasurer, Officer, or Proprietor of the Asylum, Hospital, or House, for the Lodging, Maintenance, Medicine, Clothing, and Care of such Lunatic, and incurred within Twelve Calendar Months previous to the Date of such Order, and (if such Lunatic is still in Confinement) also to pay to the Treasurer, Officer, or Proprietor of the Asylum, Hospital, or House the reasonable Charges of the future Lodging, Maintenance, Medicine, Clothing, and Care of such Lunatic ; and every such Treasurer of a County on whom any such Order is made shall, out of any Monies which may come into his Hands by virtue of his Office, immediately pay to such Guardians or Overseers the Amount of the Expenses and Monies by such Order directed to be paid to them, and from Time to Time pay to the said Treasurer, Officer, or Proprietor of the Asylum,

Hospital, or House, the future Charges aforesaid: Provided always, that such Justices may direct such Inquiry to be made to ascertain the Parish in which any Pauper Lunatic is settled as they think fit, and delay adjudging such Pauper Lunatic to be chargeable to any County until such further Inquiry has been made: Provided also, that every County to which any Pauper Lunatic is adjudged to be chargeable as aforesaid may at any Time thereafter inquire as to the Parish in which such Lunatic is settled, and may procure such Lunatic to be adjudged to be settled in any Parish.

Provision for the Reimbursement to a County of Monies paid on account of a Lunatic afterwards adjudged to belong to any Parish.

XCIX. If, after any Pauper Lunatic has been sent to an Asylum, registered Hospital, or licensed House as aforesaid, and has been adjudged to be chargeable to a County, such County procure such Lunatic to be adjudged to be settled in any Parish, it shall be lawful for any Two Justices of the County or Borough in which the Asylum, registered Hospital, or licensed House in which such Lunatic is confined is situate, or from any Part of which such Lunatic was sent for Confinement, or for any Two Justices being Visitors of such Asylum or licensed House, to make an Order upon the Guardians of the Union to which such Parish belongs, or of any such Parish, if such Parish be in a Union or be under a Board of Guardians, or if not, then upon the Overseers of such Parish, for Payment to the Treasurer of the said County of all Expenses and Monies paid by such Treasurer as herein-before is provided, and of all Monies paid by such Treasurer to the Treasurer, Officer, or Proprietor of the Asylum, Hospital, or House, for the Lodging, Maintenance, Medicine, Clothing, and Care of such Lunatic, and incurred within Twelve Calendar Months previous to such Order, and (if such Lunatic is still in Confinement) and also for Payment to the Treasurer or Officer or Proprietor of the Asylum, Hospital, or House of the reasonable Charges of the future Lodging, Maintenance, Medicine, Clothing, and Care of such Lunatic; and such Guardians or Overseers shall immediately pay to the Treasurer of such County the Amount of the Expenses and Monies by such Order directed to be paid to him, and from Time to Time pay to the said Treasurer, Officer, or Proprietor of the Asylum, Hospital, or House the future Charges aforesaid.

Justices to make Orders out of their respective Jurisdictions.

C. It shall be lawful for any Justices herein-before authorized to make any such Order as aforesaid upon the Guardians of any Union or Parish, or upon the Overseers of any Parish, to make such Order upon such Guardians or Overseers, although such Union or Parish be not within the Jurisdiction of such Justices.

Order for Payment of Charges of Maintenance in Asylums, &c. to extend to any Asylum, &c. to which the Lunatic may be removed.

CI. Where any Order has been made for the Payment of the future Charges of the Lodging, Maintenance, Medicine, Clothing, and Care of any Lunatic in any Asylum, registered Hospital, or licensed House, such Order shall extend to and be applicable in respect of the Charges of the Lodging, Maintenance, Medicine, Clothing, and Care of such Lunatic in any Asylum, registered Hospital, or licensed House to which he may be removed under the Powers of this or any other Act, in like Manner as if such Charges had by such Order been directed to be paid to the Treasurer or an Officer or the Proprietor of the Asylum, registered Hospital, or licensed House in which such Lunatic may for the Time being be confined.

The Costs of Pauper Lunatics who are irremovable to be borne by the Parish wherein they were exempt from Removal or by the Common Fund in Unions. Section 5. of 12 & 13 Vict. c. 103. repealed.

CII. Provided always, That all the Expenses incurred since the Twenty-ninth Day of *September* One thousand eight hundred and fifty-three, or hereafter to be incurred, in and about the Examination, bringing before a Justice or Justices, Removal, Lodging, Maintenance, Medicine, Clothing, and Care of a Pauper Lunatic heretofore or hereafter removed to an Asylum, registered Hospital, or licensed House under the Authority of this or any other Act, who would, at the Time of his being conveyed to such Asylum, Hospital, or House, have been exempt from Removal to the Parish of his Settlement or the Country of his Birth by reason of some Provision in the Act of the Session holden in the Ninth and Tenth Years of Her Majesty, Chapter Sixty-six, shall be paid by the Guardians of the Parish wherein such Lunatic shall have acquired such Exemption if such Parish be subject to a separate Board of Guardians, or by the Overseers of such Parish where the same is not subject to such separate Board, and where such Parish shall be comprised in any Union the same shall be paid by the Guardians, and be charged to the Common Fund of such Union so long as the Cost of the Relief of Paupers rendered irremovable by the last-mentioned Act shall continue to be chargeable upon the Common Funds of Unions; and no Order shall be made under any Provision contained in this or any other Act upon the Parish of the Settlement in respect of any such Lunatic Pauper during the Time that the above-mentioned Charges are to be paid and charged as herein provided; and Section Five of the Act of the Session holden in the Twelfth and Thirteenth Years of Her Majesty, Chapter One hundred and three, shall be repealed.

Guardians and Overseers may pay Charges without Orders of Justices.

CIII. Provided also, That any Guardians or Overseers who would be liable under any Provision contained in this Act to have an Order made upon them for the Payment of any Money may pay the same without any such Order being made, and may charge the same to such Account as they could have done if such Order had been made.

Lunatic's Property to be available for his Maintenance.

CIV. If it appear to any Justice or Justices by this Act authorized to make any Order for the Payment of Money for the Maintenance of any Lunatic that such Lunatic has an Estate, Real or Personal, applicable to his Maintenance, and more than sufficient to maintain his Family, if any, he or they shall, by an Order under his or their Hand and Seal or Hands and Seals, direct the Overseers of the Parish, or a Relieving Officer of the Parish or Union, or the Treasurer or some other Officer of the County to which such Lunatic is chargeable, or in which any Property of the Lunatic may be, or an Officer of the Asylum in which the Lunatic may be, to seize so much of any Money, and to seize and sell so much of the Goods and Chattels, and to take and receive so much of the Rents and Profits of the Lands and Tenements of such Lunatic and other Income of such Lunatic as may be necessary to pay the Charges of the Examination, bringing before a Justice or Justices, Removal, Lodging, Maintenance, Clothing, Medicine, and Care of such Lunatic, accounting for the same to such Justice or Justices, such Charges having been first proved to the Satisfaction of such Justice or Justices, and the Amount set forth in such Order; and if any Trustee or other Person having the Possession, Custody, or Charge of any Property of such Lunatic, or if the Governor and Company of the Bank of *England*, or any other Body or Person having in their or his Hands any Stock, Interest, Dividend, or Annuity belonging or due to such Lunatic, pay any Money according to any such Order, or pay any Money without any such Order, to the Guardians of any Union or Parish, or to any Overseer of any Parish not in a Union or under a Board of Guardians, or to the Treasurer of any County, or any other Officer of any County authorized to receive the same, to defray the Charges paid or incurred by or on behalf of such Parish, Union, or County for the Examination, bringing before a Justice or Justices, Removal, Lodging, Maintenance, Clothing, Medicine, and Care of such Lunatic, the Receipt of the Person authorized to receive such Money under such Order, or of such Guardians, Overseer, or Treasurer, or other Officer, shall be a good Discharge to such Trustee, Governor, and Company, or other Body or Person as aforesaid.

APPENDIX. 263

Liability of Relations of Pauper not to be affected.

CV. The Liability of any Relation or Person to maintain any Lunatic shall not be taken away or affected where such Lunatic is sent to or confined in any Asylum, registered Hospital, or licensed House by any Provision herein contained concerning the Maintenance of such Lunatic.

Persons aggrieved by Refusal of an Order may appeal to the Sessions.

CVI. If any Person feel aggrieved by any Refusal of an Order of any Justice or Justices as aforesaid, such Person may appeal to the next General or Quarter Sessions of the Peace for the County or Borough where the Matter of Appeal has arisen, the Person so appealing having given to the Justice or Justices against whom such Appeal is made Fourteen clear Days' Notice of such Appeal, and such Sessions are hereby authorized and required to hear and determine the Matter of such Appeal in a summary Way, and their Determination shall be final and conclusive.

Party obtaining Order of Adjudication to send Copy thereof and Statement of Grounds to the Parish or County affected.

CVII. The Overseers of any Parish, and the Guardians of any Union or Parish, and the Clerk of the Peace of any County obtaining any Order under this Act adjudging the Settlement of any Lunatic to be in any Parish, shall, within a reasonable Time after such Order has been made, send or deliver, by Post or otherwise, to the Overseers or Guardians of the Parish in which such Lunatic is adjudged to be settled, a Copy or Duplicate of such Order, and also a Statement in Writing under their or his Hands or Hand, or where they are the Guardians of a Union or Parish under the Hands of any Three or more of such Guardians, stating the Description and Address of the Overseers, Guardians, or Clerk of the Peace obtaining such Order, and the Place of Confinement of the Lunatic, and setting forth the Grounds of such Adjudication, including the Particulars of any Settlement or Settlements relied upon in support thereof; and on the Hearing of any Appeal against any such Order it shall not be lawful for the Respondents to go into or give Evidence of any other Grounds in support of such Order than those set forth in such Statement.

Appeal against Order of Adjudication.

CVIII. If the Guardians of any Union or Parish, or the Overseers of any Parish, feel aggrieved by any such Order as aforesaid adjudging the Settlement of any Lunatic, they or he may appeal against the same to the next General Quarter Sessions of the Peace for the County in behalf of which such Order has been obtained, or in which the Union or Parish obtaining such Order is situate, or in case such Parish or Union extend into several Jurisdictions, then to the next General Quarter Sessions of the Peace for the County or Borough in which the Asylum, registered Hospital, or licensed House in which such Lunatic is or has been confined is situate, and such Sessions upon hearing the said Appeal shall have full Power finally to determine the Matter.

Copy of Depositions to be furnished on Application.

CIX. The Clerk to the Justices making any Order adjudging the Settlement of any Lunatic, or the Clerk of the Peace in the Case herein-after provided for, shall keep the Depositions upon which such Order was made, and shall within Seven Days furnish a Copy of such Depositions to any Party authorized to appeal against such Order, if such Party apply for such Copy, and pay for the same at the Rate of Twopence for every Folio of Seventy-two Words; provided that no Omission or Delay in furnishing such Copy of the Depositions shall be deemed or construed to be any Ground of Appeal against the Order: Provided also, that on the Trial of any Appeal against any such Order, no such Order shall be quashed or set aside either wholly or in part on the Ground that such Depositions do not furnish sufficient Evidence to support, or that any Matter therein contained or omitted raises an Objection to the Order, or Grounds on which the same was made: Provided also, that if the Justices who make any such Order have not any Clerk, they shall send or deliver the Depositions to the Clerk of the Peace of the County or Borough to the General Quarter Sessions whereof the Appeal against such Order is given by this Act, and the Party obtaining such Order shall, in such Statement of Grounds of Adjudication as aforesaid, state that such Justices have not any Clerk.

No Appeal if Notice not given within a certain Time after Notice of Order.

CX. No Appeal shall be allowed against any such Order if Notice in Writing of such Appeal be not sent or delivered by Post or otherwise to the Party on whose Application the Order was obtained within the Space of Twenty-one Days after the sending or Delivery, as herein-before directed, of a Copy or Duplicate of such Order and such Statement as herein-before mentioned, unless within such Period of Twenty-one Days a Copy of the Depositions shall have been applied for as aforesaid by the Party intending to appeal, in which Case a further Period of Fourteen Days after the sending of such Copy shall be allowed for the giving of such Notice of Appeal.

Grounds of Appeal to be stated.

CXI. In every Case where Notice of Appeal against such Order is given the Appellant shall, with such Notice, or Fourteen Days at least before the First Day of the Sessions at which such Appeal is intended to be tried, send or deliver by Post or otherwise to the Respondent a Statement in Writing under their or his Hands or Hand, or where the Appellants are the Guardians of any Union or Parish, under the Hands of any Three or more of such Guardians, of the Grounds of such Appeal; and it shall not be lawful for the Appellant on the Hearing of any Appeal to go into or give Evidence of any other Grounds of Appeal than those set forth in such Statement.

As to the Sufficiency of Statement of Grounds of Adjudication or Appeal. Power to amend Statement.

CXII. Upon the Hearing of any Appeal against any such Order no Objection whatever on account of any Defect in the Form of setting forth any Ground of Adjudication or Appeal in any such Statement shall be allowed, and no Objection to the Reception of legal Evidence offered in support of any such Ground alleged to be set forth in any such Statement shall prevail unless the Court be of opinion that such alleged Ground is so imperfectly or incorrectly set forth as to be insufficient to enable the Party receiving the same to inquire into the Subject of such Statement, and to prepare for Trial: Provided always, that in all Cases where the Court is of opinion that any such Objection to such Statement or to the Reception of Evidence ought to prevail, it shall be lawful for such Court, if it so think fit, to cause any such Statement to be forthwith amended by some Officer of the Court, or otherwise, on such Terms as to Payment of Costs to the other Party, or postponing the Trial to another day in the same Sessions, or to the next subsequent Sessions, or both Payment of Costs and Postponement, as to such Court appears just and reasonable.

Power for Court to amend Order on account of Omission or Mistake. Proviso.

CXIII. If, upon the Trial of any Appeal against any such Order or upon the Return to a Writ of Certiorari, any Objection be made on account of any Omission or Mistake in the drawing up of such Order, and it be shown to the Satisfaction of the Court that sufficient Grounds were in proof before the Justices making such Order to have authorized the drawing up thereof free from the said Omission or Mistake, it shall be lawful for the Court, upon such Terms as to Payment of Costs as it think fit, to amend such Order and to give Judgment as if no such Omission or Mistake had existed: Provided always, that no Objection on

APPENDIX. 265

account of any Omission or Mistake in any such Order brought up upon a Return to a Writ of Certiorari shall be allowed, unless such Omission or Mistake have been specified in the Rule for issuing such Writ of Certiorari.

Party making frivolous or vexatious Statement of Grounds liable to pay Costs.

CXIV. If either of the Parties to the said Appeal shall have included in the Statement of Grounds of Adjudication or of Appeal sent to the opposite Party any Ground or Grounds in support of the Order or of Appeal which in the Opinion of the Court determining the Appeal, is or are frivolous and vexatious, such Party shall be liable, at the Discretion of the said Court, to pay the whole or any Part of the Costs incurred by the other Party in disputing any such Ground or Grounds.

Party losing Appeal to pay such Costs as Court may direct.

CXV. Upon every such Appeal the Court before whom the same is brought shall and may, if they think fit, order and direct the Party against which the same is decided to pay to the other such Costs and Charges as may to such Court appear just and reasonable, and shall certify the Amount thereof.

Decisions of Courts upon hearing Appeals to be final.

CXVI. The Decision of the Court upon the Hearing of any Appeal against any such Order, as well upon the Sufficiency and Effect of the Statement of the Grounds in support of the Order and Appeal, and of the Copy or Duplicate of the Order sent to the Appellant Parish or County, as upon the amending or refusing to amend the Order as aforesaid, or the Statement of Grounds, shall be final, and shall not be liable to be reviewed in any Court by means of a Writ of Certiorari or Mandamus or otherwise.

Abandonment of Orders.

CXVII. In any Case in which an Order has been made as aforesaid, and a Copy or Duplicate thereof sent as herein required, it shall and may be lawful for the Party who has obtained such Order, whether any Notice of Appeal against such Order has or has not been given, and whether any Appeal has or has not been entered, to abandon such Order, by Notice in Writing under the Hand or Hands of such Party, or, where such Order has been obtained by the Guardians of any Union, under the Hands of any Three or more of such Guardians, to be sent by Post or delivered to the Appellant or the Party entitled to appeal, and thereupon the said Order and all Proceedings consequent thereon shall become and be null and void to all Intents and Purposes as if the same had not been made, and shall not be in any way given in Evidence, in case any other Order for the same Purposes shall be obtained: Provided always, that in all Cases of such Abandonment the Party so abandoning shall pay to the Appellant or the Party entitled to Appeal the Costs which he has incurred by reason of such Order and of all subsequent Proceedings thereon; which Costs the proper Officer of the Court before whom any such Appeal (if it had not been abandoned) might have been brought shall, upon Application, tax and ascertain at any Time, whether the Court be sitting or not, upon Production to him of such Notice of Abandonment, and upon Proof to him that such Reasonable Notice of Taxation, together with a Copy of the Bill of Costs, has been given to the Overseers, Guardians, or Clerk of the Peace abandoning such Order, as the Distance between the Parties shall in his Judgment require; and thereupon the Sum allowed for Costs, including the usual Costs of Taxation, which such Officer is hereby empowered to charge and receive, shall be endorsed upon the said Notice of Abandonment, and the said Notice so endorsed shall be filed among the Records of the said Court.

Provisions of this Act as to Expenses to extend to Pauper Lunatics sent to Asylums under any other Act, &c.

CXVIII. The Provisions of this Act for and concerning the Payment of Expenses incurred or to be incurred in relation to Pauper Lunatics shall be applicable with respect to Persons confined as Pauper Lunatics sent to any Asylum, registered Hospital, or licensed House under any other Act authorizing their Reception therein as Pauper Lunatics, and (save as herein otherwise provided concerning any Lunatic who shall appear to have an Estate, Real or Personal, applicable to his Maintenance) with respect to all other Lunatics sent to any Asylum, registered Hospital, or licensed House under any Order of a Justice or Justices made under this Act, or the Acts hereby repealed, or any of them, as if such last-mentioned Lunatics were at the Time of being so sent actually chargeable to the Parish from which they have been or shall be sent.

In Cases of Inquiries and Appeals Guardians and Officers interested to have Access to the Lunatic.

CXIX. In every Case of an Inquiry, Investigation, Dispute, or Appeal as to the Parish in which a Pauper Lunatic is settled, the Guardians, Clerks of the Guardians, Relieving Officers, and Overseers of every Union including any Parish, or of any Parish, which Parish respectively is interested in such Inquiry, Investigation, Dispute, or Appeal, and every Person duly authorized by them respectively, and the Clerk of the Peace of any County interested in such Inquiry, Investigation, Dispute, or Appeal, and every Person duly authorized by such Clerk of the Peace, shall at all reasonable Times be allowed free Access, in the Presence of the Medical Attendant, to the Lunatic, to examine him as to the Premises.

Expenses of the Burial, Removal, or Discharge of a Pauper.

CXX. On the Death, Discharge, or Removal of any Pauper from any Asylum, registered Hospital, or licensed House, the necessary Expenses attending the Burial, Discharge, or Removal of such Pauper shall be borne by the Union or Parish (if any) to which such Pauper is chargeable, as herein-before provided, or if such Pauper be chargeable to a County as herein-before provided, then by such County, and shall be paid by the Guardians of such Union or Parish, or by the Overseers of such Parish if not in a Union or under a Board of Guardians, or by the Treasurer of such County.

Money ordered to be paid by any Clerk, Overseer, Relieving Officer, or Treasurer to be levied (in case of Neglect to pay) by Distress or Action.

CXXI. If any Overseer, or any Treasurer of any County, upon whom any Order of Justices for the Payment of Money under the Provisions of this Act or of any Act hereby repealed is made, shall refuse or neglect for the Space of Twenty Days next after due Notice of such Order to pay the Money so ordered to be paid, the said Money, together with the Expenses of recovering the same, shall be recovered by Distress and Sale of the Goods of the Overseer or Treasurer so refusing or neglecting, by Warrant under the Hands and Seals of any Two Justices hereby authorized to make the Order for Payment of the Money aforesaid, or by an Action at Law, or by any other Proceeding in any Court of competent Jurisdiction, against such Overseer or Treasurer; and if the Guardians upon whom any such Order is made refuse or neglect for such Time as aforesaid to pay the Money so ordered to be paid, the same, together with the Expenses of recovering the same, may be recovered by an Action at Law or by any other Proceeding in any such Court; and in case of any such Action or Proceeding no Objection shall be taken to any Default or Want of Form in any Order of Admission or Maintenance, or in any Certificate or Adjudication under this Act, if such Order or Adjudication shall not have been appealed against, or if appealed against shall have been affirmed.

MISCELLANEOUS.

Medical Men signing false Certificates, and Persons not being Medical Men giving Certificates as such, guilty of Misdemeanor.

CXXII. Any Physician, Surgeon, or Apothecary who shall sign any Certificate contrary to any of the Provisions herein contained shall for every such Offence forfeit any Sum not exceeding Twenty Pounds; and any Physician, Surgeon, or Apothecary who shall falsely state or certify anything in any Certificate under this Act, and any Person who shall sign any Certificate under this Act, in which he shall be described as a Physician, Surgeon, or Apothecary, not being a Physician, Surgeon, or Apothecary respectively within the Meaning of this Act, shall be guilty of a Misdemeanor.

Penalty on Officers or Servants ill-treating Lunatics.

CXXIII. If any Superintendent, Officer, Nurse, Attendant, Servant, or other Person employed in any Asylum strike, wound, ill-treat, or wilfully neglect any Lunatic confined therein, he shall be guilty of a Misdemeanor, and shall be subject to Indictment for every such Offence, or to forfeit for every such Offence, on a summary Conviction thereof before Two Justices, any Sum not exceeding Twenty Pounds nor less than Two Pounds.

Penalty on Officers, &c. allowing Lunatics to escape or be at large without Permission.

CXXIV. If any Superintendent, Officer, or Servant in any Asylum shall, through wilful Neglect or Connivance, permit any Patient in any Case to quit or escape from such Asylum, or be at large without such Order as in this Act mentioned (save in the Case of temporary Absence authorized under the Regulations of the Committee of Visitors), or shall secrete, or abet or connive at the Escape of any such Person, he shall for every such Offence forfeit and pay any Sum not more than Twenty Pounds nor less than Two Pounds.

Visitors may sue and be sued in the Name of their Clerk, whose Removal shall not abate Action.

CXXV. Every Committee of Visitors may sue and be sued in the Name of their Clerk ; and no Action brought or commenced by or against any such Committee of Visitors in the Name of their Clerk shall abate or be discontinued by the Death or Removal of such Clerk, but the Clerk for the Time being to the Visitors shall always be deemed Plaintiff or Defendant in such Action, as the Case may be.

Secretary of Commissioners in Lunacy and Clerks to Visitors may prosecute for Offences.

CXXVI. It shall be lawful for the Secretary of the Commissioners in Lunacy, by their Order, to prosecute or proceed against any Person for any Offence against this Act, and for the Clerk to any Committee of Visitors of any Asylum, by their Order, to prosecute or proceed against any Person for any Offence against this Act committed by any Officer or Servant belonging thereto or employed therein ; and such Secretary or Clerk acting as the Prosecutor or Complainant in any such Prosecution or Proceeding shall be competent to be a Witness therein, in the same Manner as if he were not such Prosecutor or Complainant ; and no such Prosecution or Proceeding shall abate or be discontinued by reason of the Death or Removal of such Secretary or Clerk, but his Successor shall come and be in his Place.

Penalties to be recovered in manner provided by 11 & 12 Vict. c. 43. Application of Penalties.

CXXVII. All Penalties and Forfeitures imposed by this Act shall and may be recovered summarily before Two Justices in manner provided by the Act of the Twelfth Year of Her Majesty, "to facilitate the Performance of the Duties of Justices of the Peace out of Sessions, within *England* and *Wales*, with respect to summary Convictions and Orders ;" and such Penalties and Forfeitures, when recovered upon Proceedings taken by the Secretary of the Commissioners, shall be paid to such Secretary, and be applied and accounted for by him in like Manner as

Money received for Licences for the Reception of Lunatics granted by the said Commissioners, and when recovered upon Proceedings taken by the Clerk to any Committee of Visitors of any Asylum shall be paid to the Treasurer of such Asylum to be by him applied for the Purposes of such Asylum in such Manner as such Committee may think fit and direct, and in all other Cases shall be paid to the Treasurer of the County or Borough for which the Justices by whom the Person convicted of such Offence have acted in such Conviction.

Power of Appeal to the Quarter Sessions.

CXXVIII. Any Person who thinks himself aggrieved by any Order or Determination of any Justices under this Act, other than Orders adjudicating as to the Settlement of any Lunatic Pauper, and providing for his Maintenance, may, within Four Calendar Months after such Order or Determination made or given, appeal to the General or Quarter Sessions, the Persons appealing having first given at least Fourteen clear Days' Notice in Writing of such Appeal and the Nature and Matter thereof to the Person appealed against, and forthwith after such Notice entering into a Recognizance before some Justice of the Peace, with Two sufficient Sureties, conditioned to try such Appeal, and to abide the Order and Award of the said Court thereupon ; and the said General or Quarter Sessions, upon Proof of such Notice and Recognizance having been given and entered into, shall in a summary Way hear and determine such Appeal, or, if they think proper, adjourn the Hearing thereof until the next General or Quarter Sessions, and if they see Cause may reduce any Penalty or Forfeiture to not less than One Fourth of the Amout imposed by this Act, and may order any Money to be returned which shall have been levied in pursuance of such Order or Determination, and may also award such further Satisfaction to be made to the Party injured, or such Costs to either of the Parties, as they shall judge reasonable and proper; and all such Determinations of the said General or Quarter Sessions shall be final, binding, and conclusive upon all Parties to all Intents and Purposes whatsoever.

Council of every Borough to exercise the same Duties, &c. of erecting Asylums as are conferred upon Justices, &c.

CXXIX. The Council of every Borough which shall within Six Months after the passing of this Act, by Writing under their Common Seal, give Notice to One of Her Majesty's Principal Secretaries of State of the Intention of such Council to take upon itself the Duties, Powers, and Authorities herein-before imposed or conferred upon or given to the Justices of the Borough, shall from and after the giving of such Notice be subject to and have and exercise all the Duties, Powers, and Authorities of and for erecting and providing Asylums and carrying into execution the Purposes of this Act which by this Act are imposed or conferred upon or given to the Justices of such Borough, or upon any Committee of Visitors to be appointed as directed by this Act, and all Liabilities and Contracts incurred or entered into by such Justices or Committee on behalf of such Borough under this Act, or any Act hereby repealed, shall thereupon become transferred to and obligatory upon such Council to the same Extent as they would have been binding or obligatory on such Justices or Committee, and all Matters and Things which in this Act are required to be done at any General or Quarter Sessions, or at any Meeting of the Justices of such Borough, may and shall thenceforth be done at any Meeting of the Council of such Borough, and all Notices which by this Act are required to be given to or by the Clerk of the Peace shall and may thenceforth be given to or by the Town Clerk of such Borough.

Committee appointed by Council to have same Powers as Committee of Visitors.

CXXX. It shall and may be lawful for the Council of any such Borough to confer upon any Committee to be appointed by such Council such of the Powers and Authorities which by this Act are conferred upon any Committee of Visitors to be appointed thereunder, as to such Council shall seem fit.

APPENDIX. 269

Every City, Town, Liberty, &c. not being a Borough within the Meaning of this Act, to be annexed to and rated as Part of the County within which the same is situate.

CXXXI. Every City, Town, Liberty, Parish, Place, or District, not being a Borough or Part of a Borough within the Meaning of this Act, shall for all the Purposes of this Act be annexed to and be treated and rated as Part of the County within which the same is situate, or if such City, Town, Liberty, Parish, Place, or District be situate partly in One County and partly in another, then to and as Part of such one of the same Counties as such City, Town, Liberty, Parish, Place, or District may have been annexed to under the said Act of the Eighth and Ninth Years of Her Majesty, hereby repealed, or if not already so annexed, then to and as Part of such one of the same Counties as One of Her Majesty's Principal Secretaries of State shall by Writing under his Hand and Seal direct, and shall contribute rateably to the Expenses of the Asylum of the County to which it is or shall be so annexed, whether such Asylum have been provided before or after the passing of this Act, and shall for the Purposes of this Act be within the Jurisdiction of the Justices of such County ; and in every Case in which any such City, Town, Liberty, Parish, Place, or District as aforesaid is or shall be annexed to a County in which an Asylum has been or shall have been already erected or provided, and such City, Town, Liberty, Parish, Place, or District shall not have contributed as provided by Law towards the Expenses incurred in erecting or providing such Asylum, the present or any future Committee of Visitors of such Asylum shall, as soon as conveniently may be after the passing of this Act, or after such Annexation, fix a Sum to be paid by the City, Town, Liberty, Parish, Place, or District so annexed towards the Expenses then already incurred in erecting or providing such Asylum, in due Proportion to the Population of such City, Town, Liberty, Parish, Place, or District, and of the County to which it shall be annexed, according to the last Returns under the Authority of Parliament, and the same shall be paid by every such City, Town, Liberty, Parish, Place, or District to the Treasurer of such Asylum, and shall be levied and raised by such City, Town, Liberty, Parish, Place, or District by a Rate to be made therein in the same Manner as any Rate to be made therein for the Purpose of levying or raising any other Monies hereby directed to be levied and raised for the Purposes of this Act ; and the Justices for the County to which such City, Town, Liberty, Parish, Place, or District is or shall be annexed as aforesaid in General or Quarter Sessions, are hereby authorized and required to make such Rate as aforesaid ; and the Sum so paid by such City, Town, Liberty, Parish, Place, or District shall be applied by the Treasurer of the Asylum to whom the same shall have been paid in such Manner as the Committee of Visitors shall direct, according to the Provisions and for carrying into execution the Purposes of this Act.

Interpretation of Terms. 48 G. 3. c. 96. 9 G. 4. c. 40.

CXXXII. In this Act the Words and Expressions following shall have the several Meanings hereby assigned to them, unless there be something in the Subject or Context repugnant to such Construction ; (that is to say,)

"County" shall mean every County, Riding, and Division of a County, County of a City, County of a Town, and shall include every City, Town, Parish, Place, or District by this Act annexed to a County for the Purposes hereof :

"Borough" shall mean every Borough Town and City Corporate having a Quarter Sessions, Recorder, and Clerk of the Peace :

"Parish" shall mean any Parish, Township, Vill, Tithing, Extra-parochial Place, or Place maintaining its own Poor :

"Union" shall mean a Union of Parishes formed under the Act of the Fifth Year of King *William* the Fourth, intituled *An Act for the Amendment and better Administration of the Laws relating to the Relief of the Poor in England and Wales*, or under the Act of the Twenty-second Year of King *George* the Third, intituled *An Act for the better Relief and Employment of the Poor*, or incorporated or united for the Relief or Maintenance of the Poor under any Local Act :

"Lunatic" shall mean and include every Person of unsound Mind, and every Person being an Idiot:

"Pauper" shall mean every Person maintained wholly or in part by or chargeable to any Parish, Union, or County:

"Justice" shall mean Justice of the Peace:

"Officiating Clergyman of the Parish" shall include the Chaplain of the Workhouse of the same Parish, or of the Workhouse of a Union to which such Parish belongs:

"Guardians" shall mean Guardians, Governors, Directors, Managers, or acting Guardians, entitled to act in the ordering of Relief to the Poor from Poor Rates:

"Overseer" shall mean Overseer of the Poor of any Parish, or any Person acting as such:

"Relieving Officer" and "Clerk of the Guardians" shall respectively mean such Relieving Officer and Clerk of the Guardians, and any Persons acting as such respectively:

"Clerk of the Peace" shall mean every Clerk of the Peace, and every Person acting as such, or any Deputy duly appointed:

"Physician," "Surgeon," and "Apothecary" shall respectively mean a Physician, Surgeon, and Apothecary duly authorized or licensed to practise as such by or as a Member of some College, University, Company, or Institution legally established, and qualified to grant such Authority or Licence, in some Part of the United Kingdom, or having been in Practice as an Apothecary in *England* or *Wales* on or before the Fifteenth Day of *August* One thousand eight hundred and fifteen, and being in actual Practice as a Physician, Surgeon, or Apothecary:

"Treasurer of the Borough" shall mean every Officer who has the Custody of any Monies raised by a Borough Rate:

"Treasurer of the County" shall mean every Officer who has the Custody of any County Rate, or of any Rate of any City, Town, Parish, Place, or District by this Act annexed to a County for the Purposes hereof:

"County Rate" shall mean a County Rate and any Funds assessed upon or raised in or belonging to any County in the Nature of County Rates, and applicable to the Purposes to which County Rates are applicable:

"Borough Rate" shall mean a Borough Fund or Rate, and any Fund assessed upon or raised in or belonging to any Borough in the Nature of Borough Rates, and applicable to the Purposes to which Borough Rates are applicable:

"Asylum" shall mean any Asylum, House, Building, or Place already erected or provided under the Provisions of an Act passed in the Forty-eighth Year of King *George* the Third, Chapter Ninety-six, or an Act of the Ninth Year of King *George* the Fourth, Chapter Forty, or the said Acts hereby repealed, or any of them, or subject to the Provisions of the said Acts or any of them, or to be erected or provided under the Provisions of this Act.

Nothing to affect Provisions of 39 & 40 G. 3. c. 94. 1 & 2 Vict. c. 14., or 3 & 4 Vict. c. 54. as to Criminal Lunatics.

CXXXIII. Nothing in this Act shall affect the Provisions of any of the following Acts; (that is to say,) an Act of the Session holden in the Thirty-ninth and Fortieth Years of King *George* the Third, Chapter Ninety-four; an Act of the Session holden in the First and Second Years of Her Majesty, Chapter Fourteen; and an Act of the Session holden in the Third and Fourth Years of Her Majesty, Chapter Fifty-four; or any other Provisions relating to Criminal Lunatics.

Commencement of Act.

CXXXIV. This Act shall commence and come into operation on the First Day of *November* One thousand eight hundred and Fifty-three.

APPENDIX. 271

Extent of Act.
CXXXV. This Act shall extend only to *England* and *Wales*.

Short Title.
CXXXVI. This Act may be cited as "The Lunatic Asylums Act, 1853."

SCHEDULES referred to by the foregoing Act.

SCHEDULE (A.)

FORM of AGREEMENT for uniting under the foregoing Act for the Purpose of erecting or providing an Asylum for the Reception of Lunatics.

IT is agreed this   Day of   by and between the Committees of Justices of the Peace for the County [*or* Counties] and the Borough [*or* Boroughs] of   and the Committee of the Subscribers of the Lunatic Hospital of   [*as the Case may be*], severally appointed to treat for the uniting of the said County and Borough [*or* Counties and Boroughs] [and   Lunatic Hospital, *as the Case may be*], for the Purposes of an Act passed in the   Year of Her Majesty Queen Victoria, intituled "An Act" [*here insert the title of this Act*], that the said County [*or* Counties] and Borough [*or* Boroughs, and the said Lunatic Hospital, *as the Case may be*], shall henceforth be united for the Purposes of the said Act; and that an Asylum for the Reception of Lunatics with all necessary Buildings, Courts, Yards, and Outlets, shall be immediately provided and properly fitted up and accommodated for the Purposes mentioned in the said Act: and that the necessary Expenses attending the providing, building, fitting up, Repairs, and Maintenance of the said Asylum shall be defrayed by the said County [*or* Counties] and Borough [*or* Boroughs and Lunatic Hospital], so united, in the following Proportions; (that is to say,)

The County of   Five Ninths of the said Expenses.
The Borough of   Two Ninths of the same.
The Lunatic Hospital of   Two Ninths of the same [*as the Case may be*].

And it is further agreed, that the Committee of Visitors to superintend the Building, Erection, and Management of the said Asylum shall be appointed in the following Proportions; the Justices of the Peace for the said County of   shall appoint*   , the Justices of the Peace for the Borough of   shall appoint*   , and the Subscribers to the said Lunatic Hospital of   shall appoint*   , and the Proportions in which the said Committee of Visitors are to be appointed as aforesaid may be from Time to Time varied, with the Consent in Writing under the Hands of the greater Number of Visitors of the said County and Borough [*or* each of the said Counties and Boroughs], and of the greater Number of Visitors appointed by the said Body of Subscribers, and with the Consent of the Commissioners in Lunacy: And hereunto we, the Undersigned, being the major Part of each of the Committees of Justices of the Peace for the said County and Borough [*or* Counties and Boroughs] respectively, and the major Part of the Committee of Subscribers to the said Lunatic Hospital, do, on behalf of the said County and Borough [*or* Counties and Boroughs] and Lunatic Hospital, set our Hands and Seals, this   Day of   in the Year   .

* *Insert in these Blanks either the Number or the Proportion of Visitors; and where the Number of the Committee of Visitors is not fixed in the Agreement, but only the Proportions, a Provision shall be made by the Agreement for fixing from Time to Time the Number of such Committee.*

272   APPENDIX.

SCHEDULE (B.)

FORM of MORTGAGE and CHARGE upon the County or Borough Rates for securing the Money borrowed.

WE,   the Chairman of the Court of Quarter Sessions of the Peace of the County of   holden at   the Day of   and Two other of Her Majesty's Justices of the Peace for the said County, assembled in the said Court [or We, the Mayor and Council of the Borough of   as the Case shall be], in pursuance of the Powers to us given by an Act passed in the   Year of Her Majesty Queen Victoria, intituled "An Act" [here insert the Title of this Act], do hereby mortgage and charge all the Rates and Funds to be raised and paid within the said County [or Borough, as the Case may be], under the Description of County Rates [or Borough Fund or Rates], with the Payment of the Sum of which   of   , hath advanced and paid towards defraying the Expenses of purchasing Lands, and for building and repairing, &c. [as the Case shall be] a Lunatic Asylum for the said County [or Borough, or the united Counties and Boroughs of, &c., as the Case may be], and we do hereby grant and confirm the same Rates and Funds unto the said   his Executors, Administrators, and Assigns, for securing the Repayment of the said Sum of   and Interest for the same after the Rate of   per Centum per Annum, and do order the Treasurer for such County [or Borough, &c., as the Case shall be] to pay the Interest of the said Sum of   half-yearly, as the same shall become due, until the Principal shall be discharged, at the Times and in the Manner agreed upon between the said   and the said Justices [or the said Mayor and Council, as the Case may be], pursuant to the Directions of the said Act.

SCHEDULE (C.) No. 1.

NAMES of all Pauper Lunatics in the Asylum at   for the County [or Borough, &c., as the Case may be,] of on the   Day of   18   .

Names of those chargeable to a Parish.	Date of Admission.	Names of those chargeable to County.	Date of Admission.	Names of Criminals.

This is a correct Return

(Signed)

Dated   Clerk of the Asylum.

SCHEDULE (C.) No. 2.

NAMES of all private Lunatics in the Asylum at   for the County [*or* Borough, &c., *as the Case may be,*] of
on the   Day of   18 .

Names.	Date of Admission.

This is a correct List.

(Signed)

Dated   Clerk of the Asylum.

SCHEDULE (D.)

FORM OF ANNUAL RETURN.

A TRUE LIST of all LUNATICS, IDIOTS, and other Persons of unsound Mind, chargeable to the Common Fund, or to the Parishes comprised within [such Part of] the   Union [as is situate] [*or* to the Parish of
] in the County of   specifying the Names, Sex, and Age of each, and whether dangerous or otherwise, and for what Length of Time they have been supposed to be of unsound Mind, and where detained, or how otherwise disposed of.*

Name.	Age.	Sex.	Parish to which chargeable.	Where maintained.					Weekly Cost of Maintenance and Clothing.	Whether Lunatic or Idiot.	Dangerous to himself or others.	Of dirty Habits.	For what Length of Time supposed to be of unsound Mind.	Observations.
				In a County or Borough Asylum, and what Asylum, and when sent thither.	In a registered Hospital or licensed House, and where, and when sent thither.	In the Workhouse.	In Lodgings, or boarded out, and where, and with whom, by Name	Residing with Relatives, and where, and with whom, by Name.						

Signed by me this   Day of   18 .
                                    A.B.
        Clerk to the Board of Guardians of the said Union,
                [*or* Overseer of the said Parish].

---

* *Lunatics chargeable to the Common Fund, who are in the Workhouse, should be entered as in the County where the Workhouse is situate; and those who are not in the Workhouse, as in the County in which they reside.*

SCHEDULE (E.)

County of
 Union [or Parish] of
 District of

QUARTERLY LIST of LUNATIC PAUPERS within the   District of the
Union of   [or the Parish of   ]
in the County or Borough of   not in any Asylum, registered
Hospital, or licensed House.

Name.	Sex.	Age.	Form of mental Disorder.	Duration of present Attack of Insanity, and, if idiotic, whether or not from Birth.	Where and with whom resident.	Date of Visit.	In what Condition, and if ever restrained, why, by what Means, and how often.

I declare that I have personally examined the several Persons whose Names are specified in this List, on the Days set opposite to their Names, and that they are all [or all except A.B., C.D., and E.F.] properly taken care of, and may properly remain out of an Asylum, and that these are the only Pauper Lunatics, to the best of my Knowledge, in the   District of the Union [or in the Parish] of   who are not in an Asylum, registered Hospital, or duly licensed House.

(Signed)   A.B.
Medical Officer of the   District
of the Union [or Parish] of

 Dated the   Day of   One thousand eight hundred
and   .

SCHEDULE (F.) No. 1.

ORDER for the RECEPTION of a PAUPER PATIENT.

I, C.D. [in the Case of a single Justice of the Peace, or in the Case of Two Justices, or of a Clergyman and Relieving Officer, &c., We C.D. and E.F.], the Undersigned, having called to my [or our] Assistance a Physician, [or Surgeon, or Apothecary, as the Case may be,] and having personally examined A.B., a Pauper, [omit the Words "a Pauper" when the Lunatic is not a Pauper,] and being satisfied that the said A.B. is a Lunatic [or an Idiot, or a Person of unsound Mind,] [add, where the Lunatic is sent as being wandering at large, the Words "wandering at large," and in the Case of a Lunatic sent by virtue of the Authority given to Two Justices, add "not under proper Care and Control," or "and is cruelly treated [or neglected] by the Person having the Care or Charge of him," as may appear to the Justices to be the Case], and a proper Person to be taken charge of and detained under Care and Treatment, hereby direct you to receive

APPENDIX. 275

the said *A.B.* as a Patient into your Asylum [or Hospital, or House]. Subjoined is a Statement respecting the said *A.B.*

      (Signed) *C.D.*
      * A Justice of the Peace for the City or Borough
      of   [or an or the Officiating
      Clergyman of the Parish of   ]
      (Signed) *E.F.*
      The Relieving Officer of the Union or Parish
      of   [or an Overseer of the
      Parish of   ].

   * To be signed by Two Justices, where required by the foregoing Act.

Dated the   Day of   One thousand eight hundred

To   Superintendent of the Asylum for the
 County of or the Lunatic Hospital of
     or Proprietor of the licensed
House of [*describing the Asylum, Hospital,*
or *House.*]

NOTE.—*Where the Order directs the Lunatic to be received into any Asylum other than an Asylum of the County or Borough in which the Parish or Place from which the Lunatic is sent is situate, or into a registered Hospital or licensed House, it should state that the Justice or Justices or other Persons making the Order is or are satisfied that there is no Asylum of such County or Borough, or that the Asylum or Asylums thereof is or are full; or (as the Case may require) the special Circumstances by reason whereof the Lunatic cannot conveniently be taken to an Asylum for such first-mentioned County or Borough.*

STATEMENT.

[*If any Particulars in this Statement be not known, the Fact to be so stated.*]

Name of Patient, and Christian Name, at Length.
Sex and Age.
Married, single, or widowed.
Condition of Life, and previous Occupation (if any).
The Religious Persuasion, as far as known.
Previous Place of Abode.
Whether First Attack.
Age (if known) on First Attack.
When and where previously under Care and Treatment.
Duration of existing Attack.
Supposed Cause.
Whether subject to Epilepsy.
Whether suicidal.
Whether dangerous to others.
Parish or Union to which the Lunatic is chargeable (if a Pauper or destitute Lunatic).
Name and Christian Name and Place of Abode of the nearest known Relative of the Patient, and Degree of Relationship (if known).

  I certify that to the best of my Knowledge the above Particulars are correctly stated.
      (Signed)
      [*In the Case of a Pauper, to be signed by the
       Relieving Officer or Overseer.*]

T 2

APPENDIX.

SCHEDULE (F.) No. 2.

ORDER for the RECEPTION of a PRIVATE PATIENT.

I, the Undersigned, hereby request you to receive A.B., a Lunatic, [or an Idiot, or a Person of unsound Mind,] as a Patient into your Asylum. Subjoined is a Statement respecting the said A.B.

(Signed) Name.
Occupation (if any).
Place of Abode.
Degree of Relationship (if any), or other Circumstance of Connexion with the Patient.

Dated this　　　　　Day of　　　　　One thousand eight hundred and
To　　　　　Superintendent of the Asylum for the
County [or Borough] of　　　　　[describing the Asylum].

STATEMENT.

[*If any of the Particulars in this Statement be not known, the Fact to be so stated.*]
Name of Patient, with Christian Name at Length.
Sex and Age.
Married, single, or widowed.
Condition of Life, and previous Occupation (if any).
The Religious Persuasion, as far as known.
Previous Place of Abode.
Whether First Attack.
Age (if known) on First Attack.
When and where previously under Care and Treatment.
Duration of existing Attack.
Supposed Cause.
Whether subject to Epilepsy.
Whether suicidal.
Whether dangerous to others.
Whether found lunatic by Inquisition, and Date of Commission or Order for Inquisition.
Special Circumstances (if any) preventing the Patient being examined, before Admission, separately by Two Medical Practitioners.

(Signed) Name.

Where the Person signing the Statement is not the Person who signs the Order, the following Particulars concerning the Person signing the Statement are to be added ; viz.

Occupation (*if any*).
Place of Abode.
Degree of Relationship (*if any*) or other Circumstances of Connexion with the Patient.

SCHEDULE (F.) No. 3.

FORM of MEDICAL CERTIFICATE.

I, the Undersigned [*here set forth the Qualification entitling the Person certifying to practise as a Physician, Surgeon, or Apothecary, ex. gra.*, " being a Fellow of the Royal College of Physicians in London,"] and being in actual Practice as a [Physician, Surgeon, or Apothecary, *as the Case may be*], hereby certify, That I, on the　　　　　Day of　　　　　at
[*here insert the Street and Number of the House (if any) or other like Particulars*], in the County of　　　　　, [*in any Case where more than One Medical Certificate is required by this Act, here insert* separately from any other Medical Practitioner,] personally examined A.B. of

[insert *Residence and Profession* or *Occupation, if any*], and that the said *A. B.* is a [Lunatic, *or* an Idiot, *or* a Person of unsound Mind], and a proper Person to be taken charge of and detained under Care and Treatment, and that I have formed this Opinion upon the following Grounds ; viz.
1. Facts indicating Insanity observed by myself [*here state the Facts*].
2. Other Facts (*if any*) indicating Insanity communicated to me by others [*here state the Information, and from whom*].

(Signed)

Place of Abode

Dated this   Day of   One thousand eight hundred and   .

SCHEDULE (F.) No. 4.

NOTICE OF ADMISSION.

I hereby give you Notice, That *A. B.* was admitted into this Asylum as a Private [*or* Pauper] Patient on the   Day of
and I hereby transmit a Copy of the Order and Statement and Medical Certificates [*or* Certificate] on which he was received.

[*If a private Patient be received upon One Certificate only, the special Circumstances which have prevented the Patient from being examined by Two Medical Practitioners to be here stated, as in the Statement accompanying the Order for Admission.*]

Subjoined is a Statement with respect to the mental and bodily Condition of the above-named Patient.

(Signed)
Clerk of   Asylum.

Dated the   Day of   One thousand eight hundred and   .

STATEMENT.

I have this Day [*some Day not less than Two clear Days after the Admission of the Patient*] seen and examined   the Patient mentioned in the above Notice, and hereby certify that with respect to mental State he [*or* she]   and that with respect to bodily Health and Condition he [*or* she]

(Signed)
Medical Officer of   Asylum.

Dated the   Day of   One thousand eight hundred and

SCHEDULE (F.) No. 5.

FORM OF NOTICE OF DISCHARGE, REMOVAL, OR DEATH.

I hereby give you Notice, That   Pauper [*or* a Private] Patient, admitted into this Asylum on the   Day of   was discharged therefrom recovered [*or* relieved, *or* not improved,] *or* was removed to [*mentioning the Asylum, &c.*] relieved [*or* not improved], by the Authority of   *or* died therein in the presence of   ], on the
Day of   .

(Signed)
Clerk of the   Asylum.

Dated the   Day of   One thousand eight hundred and   .

In case of Death, add, "I certify that the apparent Cause of Death of the said   [as ascertained by post mortem examination (*if so*)] was   ."

(Signed)
Medical Officer of the   Asylum.

SCHEDULE (G.) No. 1.
REGISTRY OF ADMISSIONS.
REGISTER OF PATIENTS.*

Date of last previous Admission, if any.	No. in Order of Admission.	Date of Admission.	Christian and Surname at Length.	Sex. M.	Sex. F.	Age.	Married.	Single.	Widowed.	Condition of Life, and previous Occupation.	Previous Place of Abode.	County, Union, or Parish to which chargeable.	By whose Authority sent.	Date of Medical Certificates, and by whom signed.	Form of mental Disorder.	Supposed Cause of Insanity.	Bodily Condition, and Name of Disease, if any.	Epileptics.	Congenital Idiots.	Years.	Months.	Weeks.	Number of previous Attacks.	Age on First Attack.	Date of Discharge, Removal, or Death.	Recovered.	Relieved.	Not improved.	Died.	Observations.
	1	1846: Jan. 3	William Johnson		1	23		1		Carpenter					Melancholia						4		2	17	1846: Sept. 1	1				
	2																													
	3																													
	4	1848: June 9	William Johnson		1	25		1												7			3		1848: Dec. 2	1				
	5																													
	6																													
	7	1852: May 6	William Johnson		1	29	1													3			4		1853: June 8					
	8																													

* In the Case of an Asylum receiving both private and Pauper Patients, a separate Register in the above Form to be kept for each Class.

APPENDIX. 279

SCHEDULE (G.) No. 2.

REGISTER of DISCHARGES, REMOVALS, and DEATHS.*

Date of Death, Discharge, or Removal.	Date of last Admission.	No. in Register of Patients.	Christian and Surname at Length.	Sex.		Discharged.						Removal, and to what Asylum, registered Hospital, or licensed House.				Died.		Assigned Cause of Death.	Age at Death.		Observations.
						Recovered.		Relieved.		Not improved.		Relieved.		Not improved.							
				M.	F.	M.	F.	M.	F.	M.	F.	M.	F.	M.	F.	M.	F.		M.	F.	
1846: Sept. 1	1846: Jan. 3	1	William Johnson	1	-	1															
1849: Dec. 2	1849: June 9	4	William Johnson	1	-	1															
1853: June 8	1852: May 6	7	William Johnson	1	-											1	-	Phthisis	27		

* In the Case of an Asylum receiving both private and Pauper Patients, a separate Register in the above Form to be kept for each class.

SCHEDULE (G.) No. 3.

Form of Medical Journal.*

Date.	Number of Patients.		Patients who are, or since the last Entry have been, under Restraint or in Seclusion, when and for what Period, and Reasons, and, in case of Restraint, by what Means.	Patients under Medical Treatment, and for what, if any, bodily Disorder.		Deaths, Injuries, and Violence to Patients since the last Entry.
	M.	F.	Males. Females.	Males.	Females.	

* In the Case of an Asylum receiving both Pauper and private Patients, a separate Journal to be kept in the above Form for each Class.

ANNO DECIMO NONO & VICESIMO VICTORIÆ REGINÆ.

CAP. LXXXVII.

An Act to amend the Lunatic Asylums Act, 1853.

[29th *July* 1856.]

BE it enacted by the Queen's most Excellent Majesty, by and with the Advice and Consent of the Lords Spiritual and Temporal, and Commons, in this present Parliament assembled, and by the Authority of the same, as follows:

Recorder to appoint Two Justices to be Members of the Committee of Justices of the County for providing an Asylum.

I. Where a Committee is or shall hereafter be appointed to provide an Asylum for any County under the Lunatics Asylum Act, 1853, the Recorder of every Borough now or hereafter annexed to such County for the Purposes of the said Act shall, at the General or Quarter Sessions next after such Appointment as aforesaid, or where such Committee has been already appointed, shall, at the General or Quarter Sessions next after the passing of this Act, appoint Two Justices of such Borough to be Members of such Committee.

APPENDIX.

ANNO DECIMO OCTAVO & DECIMO NONO VICTORIÆ REGINÆ.

CAP. CV.

An Act to amend the Lunatic Asylums Act, 1853, and the Acts passed in the Ninth and Seventeenth Years of Her Majesty, for the Regulation of the Care and Treatment of Lunatics.

[14th *August* 1855.]

BE it enacted by the Queen's most Excellent Majesty, by and with the Advice and Consent of the Lords Spiritual and Temporal, and Commons, in this present Parliament assembled, and by the Authority of the same, as follows:

Any single County or Borough may unite with the Subscribers to a Hospital, and any Committee of Visitors of an existing Asylum may so unite.

I. Section Three of the Lunatic Asylums Act, 1853, shall extend to empower the Justices of any One County or Borough to authorize any Committee of Justices elected for such County or Borough thereunder to treat and enter into an Agreement for uniting with the Subscribers to any such Hospital as therein mentioned, and it shall not be necessary that any other County or Borough be a Party to such Agreement; and Section Five of the said Act shall extend to empower any such Committee of Visitors as therein mentioned to enter into an Agreement for uniting with the Subscribers to any such Hospital alone.

The Proportion of Expenses between any County and Borough may be fixed with reference to Accommodation likely to be required.

II. When Two or more Committees agree to unite under the Lunatic Asylums Act, 1853, or under that Act as amended by this Act, the Proportion in which the Expenses of carrying into execution the Purposes of the said Act shall be charged upon and raised by each County and Borough so uniting may be calculated and fixed according to the Extent of the Accommodation which in the Judgment of the Committees entering into such Agreement will be required for the Pauper Lunatics of such County and Borough respectively; and the Power in Section Sixteen of the Lunatic Asylums Act, 1853, of repealing or altering the Stipulations of any Agreement for uniting, shall extend to authorize the Alteration thereof by readjusting the Proportions in which the Expenses aforesaid shall be charged on each County and Borough and the Subscribers (if any) uniting, or any of the said Parties, and, where the Committee of Visitors think fit, by fixing as aforesaid, according to the probable Extent of Accommodation required, the Proportion in which each County and Borough is to contribute to such Expenses; and where the Proportions of any Contributions are fixed according to the probable Extent of Accommodation required as aforesaid the Agreement shall specify that such Proportions are fixed according to that Basis.

Agreements for uniting to be hereafter entered into to stipulate for Contribution by Counties and Boroughs according to their relative Populations for the Time being, where not fixed according to foregoing Provision.

III. Where an Agreement for uniting is hereafter entered into under the Lunatic Asylums Act, 1853, or under that Act as amended by this Act, and the Proportion in which the Expenses of carrying the Purposes of the said Act into execution are to be charged upon each County and Borough is not fixed, under the foregoing Provision, with reference to the probable Extent of Accommodation required, the Agreement shall stipulate that such Expenses, or, where any Committee of Subscribers of a Lunatic Hospital are a Party to the Agreement, then that the aggregate Amount to be contributed by the Counties and Boroughs towards such Expenses, shall be from Time to Time charged upon and raised by the Counties and Boroughs in proportion to their respective Populations as stated in the last Return for the Time being made of the same under the Authority of Parliament, and such Agreement shall be varied from the Form in Schedule (A.) to the Lunatic Asylums Act, 1853, accordingly.

Where Expenses are to be contributed in proportion to Population, the same to be ascertained by last Census for the Time being.

IV. Where an Agreement for uniting has been already entered into under the Lunatic Asylums Act, 1853, or any former Act, the Expenses of carrying into execution any such Act, or, where any Committee of Subscribers is a Party to the Agreement, the aggregate Amount to be contributed by such Counties and Boroughs, shall be from Time to Time charged upon and raised by the Counties and Boroughs united in proportion to their respective Populations as stated in the last Return for the Time being made of the same under the Authority of Parliament, save where such Expenses are adjusted and fixed under the foregoing Provision according to the probable Extent of Accommodation required.

Where there is a Dissolution of a Union a new Asylum to be provided.

V. To the Intent that due Provision may be made for the Reception and Care of the Pauper Lunatics of Counties and Boroughs Parties to Unions upon the Dissolution of such Unions, the Justices of every County and Borough united (either alone or with any Subscribers) shall, before any Dissolution of their Union takes effect, at a General or Quarter Sessions for such County, or at a Special Meeting of the Justices of such Borough, (as the Case may require,) elect a Committee to provide an Asylum for their County or Borough, and authorize such Committee to proceed for that Purpose in manner by the Lunatic Asylums Act, 1853, provided in the Case of a County or Borough not having an Asylum ; and all the Provisions of the said Act and this Act applicable to a Committee elected to provide an Asylum in the Case of a County or Borough not having an Asylum shall be applicable to the Committee elected under this Provision.

Provisions to apply to Councils of Boroughs where they have taken upon themselves the Execution of the Lunatic Asylums Act, 1853.

VI. Where the Council of a Borough has taken upon itself, under the Lunatic Asylums Act, 1853, or the Act of the Session holden in the Eighth and Ninth Years of Her Majesty, Chapter One hundred and twenty-six, the Duties, Powers, and Authorities imposed or conferred upon or given to the Justices of the Borough, such Council shall be subject to and have and exercise the Duties, Powers, and Authorities by this Act imposed or conferred upon the Justices of a Borough, or any Committee elected by them ; and such Council may confer upon any Committee appointed by them such of the said Duties, Powers, and Authorities as under this Act are or may be conferred upon a Committee elected by the Justices of a Borough; and where the Council of a Borough had before the Commencement of the Lunatic Asylums Act, 1853, taken upon itself under the said Act of the Eighth and Ninth Years of Her Majesty, Chapter One hundred and twenty-six, the Duties, Powers, and Authorities imposed or conferred upon or given to the Justices of the Borough, such Council shall, from the Commencement of the Lunatic Asylums Act, 1853, be deemed to have been subject to and to have had the Duties, Powers, and Authorities by that Act imposed or conferred upon the Justices of a Borough, or any Committee elected by them, and to have been authorized to confer upon any Committee appointed by such Council such of the said Duties, Powers, and Authorities as under such Act may be conferred upon a Committee elected by the Justices of a Borough.

Places becoming Boroughs after the Commencement of the Lunatic Asylums Act, 1853, to be deemed Boroughs annexed to the Counties in which they are situate.

VII. Any Place which has become a Borough within the Definition contained in Section One hundred and thirty-two of the Lunatic Asylums Act, 1853, since, the Commencement of that Act, shall, from and after the passing of this Act, be deemed to be a Borough annexed to the County in which the same is situate, and any Place which after the passing of this Act becomes a Borough within such Definition shall, from and after the Time of becoming such Borough, be deemed a Borough so annexed, and the Provisions contained in Section Nine in the Lunatic Asylums Act, 1853, for the Appointment of Two Justices of a Borough annexed thereunder to a County to be Members of the Committee of Visitors of the Asylum of such County, and in relation to the Contribution by such Borough to the Expenses of the Asylum of such County, shall extend to any Borough annexed under this Enactment.

Powers given by Sect. 77. of Lunatic Asylums Act, 1853, to Visitors of an Asylum to order Removal of Pauper Lunatics extended.

VIII. The Power given by Section Seventy-seven of the Lunatic Asylums Act, 1853, to any Two of the Visitors of any Asylum, being Justices, to order any Pauper Lunatic chargeable to any Parish or Union within the County or Borough, or any County or Borough to which such Asylum wholly or in part belongs, or to any such County, and who may be confined in any other Asylum, or in any registered Hospital or licensed House, to be removed to such first-mentioned Asylum, shall be extended so as to authorize such Visitors to order any Pauper Lunatic chargeable to any Parish or Union within any County or Borough, or to any County for the Reception of the Pauper Lunatics whereof into such first mentioned Asylum there is a subsisting Contract, and who may be confined as aforesaid, to be removed to such first-mentioned Asylum, and also to order any such Pauper Lunatic as herein-before mentioned to be removed from such first-mentioned Asylum to any Asylum, registered Hospital, or licensed House, subject nevertheless to the Restriction contained in Section Seventy-eight of the Lunatic Asylums Act, 1853.

Powers of Commissioners and Visitors to continue applicable to a House which has been licensed after Expiration of Licence, while any Patients are therein.

IX. The Powers of the Commissioners and Visitors under the Lunatic Asylums Act, 1853, and the Acts of the Eighth and Ninth Years of Her Majesty, Chapter One hundred, and the Sixteenth and Seventeenth Years of Her Majesty, Chapter Ninety-six, with reference to any licensed House and the Inmates thereof, and all Powers and Provisions of the said Acts having reference to the Discharge, Removal, and Transfer of such Inmates, shall, after the Expiration or Revocation of any Licence granted in respect of such House, continue in force for all Purposes, so long as any Lunatics are detained therein, in the same Manner as if the Licence subsisted.

Contracts under Forty-second Section of Lunatic Asylums Act, 1853, may be renewed.

X. Whereas Doubts have been entertained whether under the Forty-second Section of the Lunatic Asylums Act, 1853, a Contract for the Reception of Pauper Lunatics thereby authorized can be renewed : Be it declared and enacted, That upon or after the Expiration or other Determination of any Contract for any of the Purposes of the said Section it shall be lawful for every Committee of Visitors, under and subject to the several Provisions of the said Act applicable thereto, from Time to Time to enter into a new Contract for any of the Purposes mentioned in the said Section with the Committee of Visitors of any Asylum, or with the Subscribers to any Hospital registered or the Proprietor of any House licensed for the Reception of Lunatics, and for the Committee of Visitors of any Asylum, or the Subscribers to any registered Hospital or the Proprietor of any licensed House, to contract with any Committee of Visitors accordingly.

Provision for Burial of Pauper Lunatics.

XI. Where the Visitors of Lunatic Asylums for Counties and Boroughs in *England*, or any of their Officers duly authorized in that Behalf, shall undertake the Burial of any Pauper Lunatic, and the Burial cannot take place in the Parish where the Death shall have taken place by reason of the public Burial Ground of such Parish having been closed, and no other having been provided, or where, in consequence of the crowded State of such Burial Ground, the Visitors as aforesaid are of opinion that the Burial of such dead Body therein would be improper, it shall be lawful to bury such Body in a public Burial Ground of or in some other Parish as near as conveniently may be to the Parish wherein the Death shall have taken place, with the Consent of the Minister and Churchwardens of such Parish: Provided, that in all Cases of Burial under the Direction of the Visitors or their Officers as aforesaid the Fee or Fees payable by the Custom of the Place where the Burial may be, or under the Provisions of any Act of Parliament shall be paid by the said Visitors for the Burial of each such Body to the Person or Persons who by such Custom or under such Act of Parliament shall be entitled to receive such Fee or Fees.

Power to enter into Agreements with Cemetery Company or Burial Board.

XII. The Visitors of Lunatic Asylums in *England* may from Time to Time enter into Agreements with the Proprietors of any Cemetery established under the Authority of Parliament, or with any Burial Board duly constituted under the Statutes in that Behalf for the Burial of the dead Bodies of any Pauper Lunatics which such Visitors may undertake to bury; and thereupon the Burial of any such Body, under the Directions of the said Visitors or their Officer, in such Cemetery, or in the Burial Ground of such Burial Board, shall be lawful: Provided, however, that no such Agreement shall be valid unless made in such Form and with such Stipulations as the Commissioners in Lunacy shall approve.

Committee of Visitors may convey Land for Burial Ground for Lunatics, &c. dying in the Asylum.

XIII. And whereas it is expedient that Burial Grounds should be provided for Persons dying in any County or Borough Lunatic Asylum built or to be built under the Authority of any Act of Parliament for the Reception of Pauper Lunatics: Be it therefore enacted, That it shall be lawful for every Committee of Visitors of any County or Borough Lunatic Asylum, or for any Trustees or Trustee in whom any Land shall be vested for the Purposes of an Asylum, with the previous Consent of One of Her Majesty's Principal Secretaries of State under his Hand, to give, grant, and convey to Her Majesty's Commissioners for building new Churches, and it shall be lawful for them to accept, any Portion not exceeding Two Statute Acres of any Land which belongs to or has been or may be purchased for any such Asylum, for the Purpose of Consecration as a Burial Ground for Pauper or other Lunatics or Officers or Servants dying in such Asylum, and that in all such Cases the Freehold of every Burial Ground, of which Her Majesty's said Commissioners shall accept a Conveyance under the Provisions of this Act for the Purpose of Consecration, shall. after the same Burial Ground shall have been consecrated, vest in the Visitors or Trustees or Trustee, as the Case may be, for the Time being of the County or Borough Lunatic Asylum to which such Burial Ground shall belong, and be for ever thereafter exclusively appropriated for the Burial of Pauper and other Lunatics dying in such Asylum, and of the Officers and Servants belonging to such Asylum and dying therein; and that from and after the Consecration of such Land the Incumbent of the Parish in which such Burial Ground is situate shall not be entitled to any Fee for the Interment therein of any Pauper or other Lunatic dying in such Asylum, or of any of the Officers and Servants belonging to such Asylum and dying therein.

Pauper Lunatics, whose Settlements cannot be ascertained, where found in a Borough which does not contribute to the County Expenditure, to be chargeable to such Borough.

XIV. And whereas Doubts are entertained as to the Chargeability of Pauper Lunatics found in Boroughs whose Settlements cannot be ascertained, and it is expedient to remove such Doubts:

Section Three of the Act of the Session holden in the Twelfth and Thirteenth Years of Her Majesty, Chapter Eighty-two, shall be repealed; and where any Pauper Lunatic is not settled in the Parish by which or at the Instance of some Officer or Officiating Clergyman of which he is sent to an Asylum, registered Hospital, or licensed House, and it cannot be ascertained in what Parish such Pauper Lunatic is settled, and such Lunatic was found in a Borough having a separate Court of Quarter Sessions of the Peace, and which is not liable, under the Act of the Session holden in the Fifth and Sixth Years of King *William* the Fourth, Chapter Seventy-six, Section One hundred and seventeen, to the Payment of a Proportion of the Sums expended out of the County Rate, such Lunatic may be adjudged to be chargeable to such Borough by any Two Justices of such Borough; and it shall not be lawful for any Justices to adjudge such Lunatic to be chargeable to any County, nor to make any Order upon the Treasurer of any County for the Payment of any Expenses whatsoever incurred or to be incurred in respect of the said Lunatic; and all the Provisions in the Lunatic Asylums Act, 1853, as to the Mode of determining that a Pauper Lunatic is chargeable to a County, and as to the Order to be made for the Maintenance of such Pauper Lunatic, shall extend and be applied to such Borough, as fully and effectually, to all Intents and Purposes, as if all the said Provisions were repeated and re-enacted in this Act, and made applicable to such Borough, in the same Manner in all respects as though for the Purposes of this Provision such Borough were a separate and distinct County.

Seals of Commissioners, Visitors, and Justices, to Orders, &c. dispensed with.

XV. In all Cases in which, under the Lunatic Asylums Act, 1853, or the Act of the Session holden in the Eighth and Ninth Years of Her Majesty, Chapter One hundred, or the Act of the Session holden in the Sixteenth and Seventeenth Years of Her Majesty, Chapter Ninety-six, any Order or other Instrument is required to be under the Hand and Seal or Hands and Seals of any Visitor or Visitors, Justice or Justices, it shall be sufficient for such Order or Instrument to be signed only; and all such Orders and Instruments as aforesaid which have been signed before the passing of this Act, and have not had a Seal or Seals affixed to them, as by Law required, shall be and be deemed to have been valid and sufficient to justify any Proceeding thereon or thereunder.

So much of Sect. 6. of 16 & 17 Vict. c. 96. as requires personal Examination of Patients, repealed.

XVI. So much of Section Six of the said Act of the Sixteenth and Seventeenth Years of Her Majesty, Chapter Ninety-six, as requires such Assent as therein mentioned of Two of the Commissioners not to be given until after such Commissioners have by personal Examination of the Patient satisfied themselves of his Desire to remain, shall be repealed.

Consent of Committee of Management of any Hospital sufficient to authorize a Patient being sent to any Place for Health.

XVII. The Superintendent of any registered Hospital may, with the Consent in Writing of Two Members of the Committee having the Management or Government of such Hospital, send or take, under proper Control, any Patient to any specified Place for any definite Time for the Benefit of his Health; and any such Consent, and any Consent under Section Eighty-six of the said Act of the Eighth and Ninth Years of Her Majesty, Chapter One hundred, may be from Time to Time renewed and the Place varied.

Detention of Lunatics after Expiration of or Revocation of Licence a Misdemeanor.

XVIII. If after the Lapse of Two Months from the Expiration of any Licence for the Use of any House for the Reception of Lunatics which has not been renewed, or if after the Revocation of any such Licence there be in any such House Two or more Lunatics, every Person keeping such House, or having the Care and Charge of such Lunatics, shall be guilty of a Misdemeanor.

Act to be read with the Acts amended as One Act.

XIX. This Act, so far as the same amends or affects the said Acts of the Eighth and Ninth Years of Her Majesty, Chapter One hundred, and of the Sixteenth and Seventeenth Years of Her Majesty, Chapter Ninety-six, or either of them, shall be read and construed together with the said Acts as One Act, and the Provision contained in Section One hundred and six of the said Act of the Eighth and Ninth Years of Her Majesty shall extend to Offences against this Act; and this Act, so far as the same amends or affects the Lunatic Asylums Act, 1853, shall be read and construed therewith as One Act.

ANNO VICESIMO QUINTO & VICESIMO SEXTO VICTORIÆ REGINÆ.

CAP. LXXXVI.

An Act to amend the Law relating to Commissions of Lunacy and the Proceedings under the same, and to provide more effectually for the visiting of Lunatics, and for other Purposes.

[7th *August* 1862.]

WHEREAS it is expedient to amend the Law relating to Commissions of Lunacy and the Proceedings under the same, and to provide more effectually for the visiting of Persons found lunatic by Inquisition, and to make the other Provisions herein-after contained with respect to certain Officers in Lunacy, and otherwise: Be it therefore enacted by the Queen's most Excellent Majesty, by and with the Advice and Consent of the Lords Spiritual and Temporal, and Commons, in this present Parliament assembled, and by the Authority of the same, as follows:

Short Title.

1. This Act may be cited as "The Lunacy Regulation Act, 1862."

INTERPRETATION.

Act to be construed as Part of Lunacy Regulation Act, 1853. 16 & 17 *Vict.* c. 70.

2. In this Act, unless there be something in the Subject Matter or Context repugnant to the Construction, the following Terms shall have the Meanings herein-after assigned to them:

The Expression "the Lord Chancellor intrusted as aforesaid" and the several other Expressions and Words mentioned and referred to in the Second Section of the Act passed in the Session of Parliament holden in the Sixteenth and Seventeenth Years of the Reign of Her present Majesty, Chapter Seventy, intituled *An Act for the Regulation of Proceedings under Commissions of Lunacy, and the Consolidation and Amendment of the Acts respecting Lunatics so found by Inquisition and their Estates,* shall be read and construed according to the Interpretations thereof contained in the said Section:

And generally the Provisions of the said Act (except so far as the same are altered by or inconsistent with this Act) shall extend and apply to the several Cases and Matters provided for by this Act, in the same Way as if this Act had been incorporated with and had formed Part of the said Act.

Nature and Limit of Inquiry under Commissions of Lunacy.

3. The Inquiry to be made under every Order for Inquiry or Commission of Lunacy or Issue shall be confined to the Question whether or not the Person who is the Subject of the Inquiry is at the Time of such Inquiry of unsound Mind, and incapable of managing himself or his Affairs, and no Evidence as to anything

APPENDIX. 287

done or said by such Person, or as to his Demeanour or State of Mind at any Time being more than Two Years before the Time of the Inquiry, shall be receivable in Proof of Insanity on any such Inquiry, or on the Trial of any Traverse of an Inquisition, unless the Judge or Master shall otherwise direct.

Inquiries before a Jury to be made by means of an Issue to one of the Superior Courts of Common Law.

4. Wherever, under the said Act, the Lord Chancellor intrusted as aforesaid shall order an Inquiry before a Jury, he may by his Order direct an Issue to be tried in One of Her Majesty's Superior Courts of Common Law at *Westminster*, and the Question in such Issue shall be, whether the alleged insane Person is of unsound Mind and incapable of managing himself or his Affairs; and the Provisions of the said Act with respect to Commissions of Lunacy, and Orders for Inquiry to be tried by a Jury, and the Trial thereof, and the Constitution of the Jury, shall apply to any Issue to be directed as aforesaid, and the Trial thereof, and subject thereto such Issue and the Trial thereof shall be regulated by the Act of the Eighth and Ninth Years of the Reign of Her present Majesty, Chapter One hundred and nine, intituled *An Act to amend the Law concerning Games and Wagers*, and the Verdict upon any such Issue finding the alleged insane Person to be of unsound Mind and incapable of managing himself or his Affairs, shall have the same Force to all Intents and Purposes as an Inquisition under a Commission of Lunacy, finding a Person to be of unsound Mind and incapable of managing himself or his Affairs, returned into the Court of Chancery.

Reference in other Acts to Inquisition to apply to Verdict on Issue.

5. Where in any Act of Parliament, Order, Rule of Court, or Instrument Reference is made to a Commission of Lunacy, or the Inquisition thereon, the Issue hereby authorized to be directed, and the Verdict thereon, operating as an Inquisition, shall be deemed to be intended by or comprehended in the Reference.

Examination of alleged Lunatic on the holding of the Inquisition.

6. On the Trial of every such Issue as last aforesaid the alleged insane Person shall, if he is within the Jurisdiction, be examined before the taking of the Evidence is commenced, and at the Close of the Proceedings, before the Jury consult as to their Verdict, unless the presiding Judge shall otherwise direct; and such Examinations of the alleged insane Person shall take place either in open Court or in private as such Judge shall direct.

No Traverse of an Inquisition made by One of the Judges of the Superior Courts and by a Jury to be granted, but new Trial may be ordered by the Lord Chancellor.

7. No Person shall be entitled to a Traverse of any Inquisition made under any such Order as aforesaid upon the Oath of a Jury; but it shall be lawful for the Lord Chancellor intrusted as aforesaid, if he shall think fit, upon a Petition being presented to him within Three Months next after the Trial of any such Issue, to order that a new Trial shall be had of such Issue or a new Inquiry made as to the Insanity of such Person, subject to such Directions and upon such Conditions as to the Lord Chancellor intrusted as aforesaid may seem proper.

Sections One hundred and forty-eight, One hundred and forty-nine, and One hundred and fifty of the said Act (which Sections relate to Petitions and Orders for the Traverse of Inquisitions) shall not apply to any Case coming within the last preceding Section of this Act.

Section One hundred and fifty-one of the said Act shall apply to all Proceedings taken, Orders made, and Things done, pending a new Trial or new Inquiry or the Petition for the same, in the same Manner as is provided by the said Section with respect to such Matters pending a Traverse or the Petition for the same.

Demand of Inquiry by Jury.

8. And with Reference to Inquiries before the Master without a Jury, and the Right of the alleged Lunatic to demand an Inquiry by a Jury, be it enacted, upon the Hearing of any Petition for Inquiry it shall be lawful for the alleged Lunatic, by himself, his Council or Solicitor, orally, or by Petition addressed to the Lord Chancellor intrusted as aforesaid, to demand an Inquiry by a Jury, and such Demand shall have the same Effect as if made by Notice filed with the Registrar in accordance with the Provisions of the said Act.

Demand of Inquiry by Jury may be withdrawn.

9. Upon such Hearing the alleged Lunatic may, by himself, his Counsel or Solicitor, orally, or by Petition as aforesaid, withdraw any Notice of demanding an Inquiry by a Jury previously filed by him.

Commission may be superseded on Conditions.

10. And with respect to the superseding of Commissions, be it enacted, That if it shall appear to the Lord Chancellor that it is not expedient or for the Benefit of the Lunatic that the Commission should be unconditionally superseded, but that same should be superseded on Terms and Conditions, he may, upon the Consent of the Lunatic and such other Persons, if any, whose Consent he may deem necessary, order the Commission to be superseded upon such Terms and Conditions as he shall think proper ; and all the Provisions contained in " The Lunacy Regulation Act, 1853," in relation to the superseding of the Commission in Cases where a Traverse has been applied for, and to the Proceedings for the fulfilling of such Terms and Conditions, shall apply to all Cases in which the Commission shall be superseded upon Terms and Conditions under the Provisions herein contained.

Lord Chancellor may order Costs.

11. It shall be lawful for the Lord Chancellor intrusted as aforesaid to order the Costs, Charges, and Expenses of and incidental to the Presentation of any Petition for a Commission in the Nature of a Writ de lunatico inquirendo, or for any Order of Inquiry under "The Lunacy Regulation Act, 1853," and of and incidental to the Prosecution of any Inquiry, Inquisition, Issue, Traverse or other Proceeding consequent upon such Commission or Order, to be paid either by the Party or Parties who shall have presented such Petition, or by the Party or Parties opposing such Petition, or out of the Estate of the alleged Lunatic, or partly in one way and partly in another, as the Lord Chancellor intrusted as aforesaid shall in each Case think proper ; and such Order shall have the same Force and Effect as Orders for the Payment of Money made by the High Court of Chancery in Cases within its Jurisdiction.

AS TO PROPERTY OF INSANE PERSONS WHEN OF SMALL AMOUNT.

Power to Lord Chancellor, where Property of Lunatic does not exceed 1000l. *in Value, or* 50l. *per Annum, to apply it for his Benefit in a summary Manner, without Inquisition.*

In order that the Property of insane Persons when the same is of small Amount may be applied for their Benefit in a summary and inexpensive Manner, be it enacted as follows :

12. Where, by the Report of One of the Masters in Lunacy or of the Commissioners in Lunacy, or by Affidavit or otherwise, it is established to the Satisfaction of the Lord Chancellor intrusted as aforesaid that any Person is of unsound Mind and incapable of managing his Affairs, and that his Property does not exceed One thousand Pounds in Value, or that the Income thereof does not exceed Fifty Pounds *per Annum*, the Lord Chancellor intrusted as aforesaid may, without directing any Inquiry under a Commission of Lunacy, make such Order as he may consider expedient for the Purpose of rendering the Property of such Person, or

the Income thereof, available for his Maintenance or Benefit or for carrying on his Trade or Business: Provided nevertheless, that the alleged insane Person shall have such personal Notice of the Application for such Order as aforesaid as the Lord Chancellor shall by General Order to be made as after mentioned direct.

Power to sell Land or other Property of Lunatic for his Benefit.

13. For the Purpose of giving effect to any such Order as is mentioned in the last preceding Section the Lord Chancellor intrusted as aforesaid may order any Land, Stock, or other Property of such Person as aforesaid to be sold, charged by way of Mortgage, or otherwise disposed of, and a Conveyance, Transfer, Charge, or other Disposition thereof to be executed or made by any Person on his Behalf, and may order the Proceeds of any such Sale, Charge, or other Disposition, or the Dividends or Income of such Land, Stock, or Property, to be paid to any Relative of such insane Person, or to such other Person as it may be considered proper to trust with the Application thereof, to be by him applied in the Maintenance or for the Benefit of the insane Person, or of him and his Family, either at the Discretion of such Relative or Person, or in such Manner, and subject to such Control, and with or without such Security for the Application thereof, as the Lord Chancellor intrusted as aforesaid may direct; and for the Purpose above mentioned the Lord Chancellor intrusted as aforesaid shall have all the same Powers with respect to the Transfer, Sale, and Disposition of, and otherwise respecting, the Real and Personal Property of such Person as aforesaid as if he had been found lunatic by Inquisition.

Power to make General Orders, to carry into effect the Objects of the last preceding Section.

14. The Lord Chancellor may from Time to Time make such General Orders as he may think fit for regulating the Procedure to be adopted and the Duties to be performed by the Masters and Officers in Lunacy for obtaining such Reports as aforesaid, and for carrying the Objects of the Two last preceding Sections into effect, and for vesting in such Masters and Officers such Powers as the Lord Chancellor may consider expedient for the Purposes aforesaid.

Power to apply Property of Persons acquitted on the Ground of Insanity for their Benefit.

15. Where any Person has, on the Trial of any Indictment, been acquitted on the Ground of Insanity, it shall be lawful for the Lord Chancellor intrusted as aforesaid, on being satisfied by Affidavit or otherwise of the continued Insanity of such Person, and of his being still in Confinement, to make any such Order with respect to the Property of such Person, and the Application thereof for his Maintenance or Benefit, or that of his Family, or for carrying on his Trade or Business, as is mentioned in the Three last preceding Sections of this Act.

CHARGING ORDERS.

Extending Powers of charging Lunatic's Property for his Maintenance, Debts, and Costs.

And for the Purpose of extending the Powers over the Property of Lunatics given by Section One hundred and sixteen of the said Act, be it enacted as follows:

16. Where it appears to the Lord Chancellor intrusted as aforesaid to be for the Lunatic's Benefit, he may by Order direct any Estate or Interest of the Lunatic in Land or Stock, either in possession, reversion, remainder, contingency, or expectancy, and either existing or which may exist at any future Time, to stand and be charged with any Monies advanced or to be advanced, or due or to become due, to any Person for or in respect of any of the Purposes or Matters mentioned in the said Section, and either with or without Interest on such Monies; and he may also by Order direct any such Estate and Interest to be dealt with and disposed of in such Manner as he shall consider expedient for any of the Purposes

U

aforesaid, or for securing any Monies advanced or to be advanced for such Purposes or any of them, and with or without Interest for the same ; and every Charge and Disposition directed or made by or in pursuance of any such Order shall be valid and effectual to all Intents and Purposes and shall take effect accordingly, subject only to any prior Charge to which the Estate or Interest affected thereby may at the Date of such Order be subject.

GENERAL.

All Deeds, Transfers, Payments, &c. made in pursuance of this Act to be valid and binding. Indemnity to the Bank of England, &c.

17. Every Conveyance, Transfer, Charge, or other Disposition made or executed by virtue of this Act, and every Payment made in pursuance of this Act, shall be valid to all Intents, and binding upon all Persons whomsoever ; and this Act shall be a full Indemnity and Discharge to the Governor and Company of the Bank of *England*, their Officers and Servants, and all other Persons respectively, for all Acts and Things done or permitted to be done in pursuance thereof, or of any Order of the Lord Chancellor intrusted as aforesaid made or purporting to be made under this Act ; and such Acts and Things respectively shall not be questioned or impeached in any Court of Law or Equity to their Detriment.

Power to Masters to summon Witnesses.

18. To give further and better Effect to the Fifty-fifth, Fifty-sixth, and Sixtieth Sections of the said Act, respecting the Attendance of Witnesses before the said Masters, the Masters may in the Matter of any Lunatic or alleged Lunatic compel by Summons the Attendance of any Person to give Evidence before them, whether such Person has or has not previously given Evidence by Affidavit ; and every Person so summoned shall be bound to attend as required by the Summons, and give Evidence before the said Masters in like Manner as is provided by the Sixtieth Section of the said Act in the Case of Persons who have given Evidence by Affidavit.

VISITING.

Duties of Visitors.

And with respect to the visiting of Lunatics, be it enacted as follows :
19. It shall be the Duty of the Visitors to visit Persons of unsound Mind within the Meaning of this Act at such Times and in such Rotation and Manner, and to make such Inquiries and Investigations as to their Care and Treatment and mental and bodily Health, and the Arrangements for their Maintenance and Comfort, and otherwise respecting them, as the Lord Chancellor shall by General Orders, or as the Lord Chancellor intrusted as aforesaid shall by Special Order in any particular Case from Time to Time direct.

All Lunatics to be visited Four Times a Year.

20. Provided always, That from and after the First Day of *October* next every Lunatic shall be personally visited and seen by One of the said Visitors Four Times at least in every Year, and such Visits shall be so regulated as that the Interval between successive Visits to any such Lunatic shall in no Case exceed Four Months : Provided always, that Lunatics who are resident in licensed Houses, Asylums, or registered Hospitals shall not necessarily be visited by any of the said Visitors more than once in the Year, unless the Lord Chancellor intrusted as aforesaid shall otherwise direct.

Visitors also to visit alleged Lunatics, and make a Report, &c. to the Lord Chancellor.

21. The Visitors shall also visit such Persons alleged to be insane, and shall make such Inquiries and Reports in reference to them as the Lord Chancellor intrusted as aforesaid may direct, and at the Expiration of every Six Calendar Months they shall report to the Lord Chancellor the Number of Visits which they shall have

made, the Number of Patients they shall have seen, and the Number of Miles they shall have travelled during such Months, and shall on the First Day of *January* in each Year make a Return to the Lord Chancellor of all Sums received by them for travelling Expenses, or upon any other Account ; and a Copy of such Reports, showing the Number of Visits made, the Number of Patients seen, and the Number of Miles travelled, and also a Copy of such Return of Sums received for travelling Expenses, or upon any other Account, shall be laid before Parliament on or before the First Day of *February* in each Year, if Parliament be then sitting, and if not, within Twenty-one Days next after the Commencement of the next Session of Parliament.

Sections 104 and 105 of 16 & 17 Vict. c. 70. repealed.

22. Sections One hundred and four and One hundred and five of the said Act (which Sections relate to the visiting of Lunatics) are hereby repealed.

OFFICERS IN LUNACY.

Power to the Lord Chancellor to allow Pensions to present Visitors, if desirous of retiring.

23. The Lord Chancellor may, if he shall so think fit, on a Petition presented to him for that Purpose, order Annuities, not exceeding One Half of their respective Salaries, to be paid to the present Medical Visitors or either of them, in case they or either of them shall be desirous of retiring from the Offices held by them, they having already attained the respective Ages of Seventy-eight and Eighty-one Years, and having served as such Medical Visitors for Twenty-eight and Twenty Years respectively.

Visitors to hold Office during good Behaviour. Visitors to receive Salaries, but not to practise in their Professions.

24. The Medical Visitors to be hereafter appointed and the Legal Visitor shall hold their Offices during their good Behaviour, but may be removed therefrom by the Lord Chancellor in Case of Misconduct or Neglect in the Discharge of their Duties, or of their being disabled from performing the same, and they shall receive Salaries of Fifteen hundred Pounds each, and shall not be in any way engaged in the Practice of their respective Professions.

Clerks to the Visitors.

25. Such Clerks to the Visitors may from Time to Time be appointed by the Lord Chancellor and at such Salaries as the Lord Chancellor, with the Approbation of the Commissioners of Her Majesty's Treasury, shall from Time to Time direct : So much of Section Twenty-three of the said Act as refers to the Clerk of the Secretary to the said Visitors is hereby repealed.

Superannuation Allowances to Officers in Lunacy.

26. The Lord Chancellor may, if he shall so think fit, order to be paid to any Officer who has served for Twenty Years in any Office or Offices in Lunacy, and who shall be above Sixty Years of Age and shall be desirous of retiring, or who is disabled by permanent Infirmity from the Performance of his Duties, such Superannuation Allowance not exceeding Two Thirds of the Salary payable to such Officer or Person at the Time of his Resignation, as the Lord Chancellor, with the Approbation of the Commissioners of Her Majesty's Treasury may approve.

Payment of Pensions and Salaries.

27. All Annuities and Salaries ordered to be paid in pursuance of this Act shall be payable out of " The Suitors Fee Fund Account" mentioned in the said Act, and at the Times and in the Manner directed by the Twenty-fifth Section of the said Act.

APPENDIX.

Registrar to hold Office during good Behaviour.

28. The Registrar in Lunacy shall hold his Office during good Behaviour, and may be removed therefrom by the Lord Chancellor in case of Misconduct or Neglect in the Discharge of his Duties or his being disabled from performing the same.

ORDERS.

Office Copies of Orders to be acted upon by Accountant General and others.

And with respect to Orders in Lunacy, be it enacted as follows :

29. The Accountant General and all other Persons, and the Governor and Company of the Bank of *England,* shall act upon all Office Copies of Orders in Lunacy purporting to be signed by the Registrar in Lunacy, and sealed with the Seal of his Office, in the same Manner as such Persons are by Section One hundred and one of the said Act required to act upon Office Copies of Reports confirmed by Fiat.

ANNO VICESIMO QUINTO & VICESIMO SEXTO
VICTORIÆ REGINÆ.

CAP. CXI.

An Act to amend the Law relating to Lunatics.

[7th *August* 1862.]

WHEREAS it is expedient to amend the Law relating to Lunatics, other than those found Lunatics by Inquisition, or Lunatics convicted of Crime, or acquitted on the Ground of Insanity : Be it enacted by the Queen's most Excellent Majesty, by and with the Advice and Consent of the Lords Spiritual and Temporal, and Commons, in this present Parliament assembled, and by the Authority of the same, as follows ; (that is to say,)

PRELIMINARY.

Interpretation of Terms.

1. In the Construction and for the Purposes of this Act (if not inconsistent with the Context or Subject Matter) the following Terms shall have the respective Meanings hereinafter assigned to them ; that is to say,

" Lunacy Act, Chapter One hundred," shall mean an Act passed in the Session holden in the Eighth and Ninth Years of the Reign of Her present Majesty, Chapter One hundred, and intituled *An Act for the Regulation of the Care and Treatment of Lunatics :*

"Lunacy Act, Chapter Ninety-six," shall mean an Act passed in the Session holden in the Sixteenth and Seventeenth Years of the Reign of Her present Majesty, Chapter Ninety-six, intituled *An Act to amend an Act passed in the Ninth Year of Her Majesty, for the Regulation of the Care and Treatment of Lunatics :*

" Lunacy Act, Chapter Ninety-seven," shall mean an Act passed in the Session holden in the Sixteenth and Seventeenth Years of the reign of Her present Majesty, Chapter Ninety-seven, intituled *An Act to consolidate and amend the Laws for the Provision and Regulation of Lunatic Asylums for Counties and Boroughs, and for the Maintenance and Care of Pauper Lunatics, in England :*

" The Lunacy Acts" shall include the Three Acts above mentioned and this Act :

" Asylum" shall have the same meaning as it has in the Lunacy Act, Chapter Ninety-seven :

" Registered Hospital" shall mean any Hospital registered for the Reception of Lunatics.

APPENDIX.

Construction of Act.

2. This Act shall be construed as One Act with the Lunacy Acts, Chapters One hundred, Ninety-six, and Ninety-seven, and Words defined by the said Acts or any of them shall have the same Meaning in this Act.

Short Title.

3. This Act may be cited for all Purposes as the "Lunacy Acts Amendment Act, 1862."

ESTABLISHMENT OF COUNTY ASYLUMS.

Plans, &c. of Visitors, when not approved by the Quarter Sessions, to be submitted to Secretary of State.

4. Whereas by Section Thirty-one of the Lunacy Act, Chapter Ninety-seven, it is provided, "that the said Visitors shall from Time to Time make their Report to the General or Quarter Sessions of the County or Borough, Counties or Boroughs, for which they (or such of them as have not been elected by Subscribers, as therein mentioned,) have been elected, of the several Plans, Estimates, and Contracts which have been agreed upon, and of the Sum or Sums of Money necessary to be raised and levied for defraying the Purchase Monies and Expenses thereof on the County or Borough, or, in the Case of such Union as therein mentioned, on each or every of the Counties or Boroughs; which Plans, Estimates, and Contracts shall be subject to the Approbation of the Court or Courts of General or Quarter Sessions of such County or Counties, and of the Justices of such Borough or Boroughs, before the same are completed or carried into execution" (save in the Case therein mentioned):

Where a Plan, Estimate, or Contract agreed upon by any Committee of Visitors on behalf of a Union of Counties, or of a Union of Counties and Boroughs, is disapproved of by One or more but not all of the Courts of General or Quarter Sessions, or other Bodies of Justices whose Approbation is required, in pursuance of the said Enactment, each Court of General or Quarter Sessions or Body of Justices disapproving of the same shall, within Four Months after such Plan, Estimate, or Contract is reported to them, or where the same has been reported to them before the passing of this Act, then within One Month after the holden of the first Court of General or Quarter Sessions of the County or the First Meeting of the Justices of the Borough after the passing of this Act, as the Case may be, set forth their Objections, with any Observations they may think fit in relation thereto, in a Report in Writing, and forthwith transmit the same to One of Her Majesty's Principal Secretaries of State, and the Secretary of State shall cause such Inquiries to be made in relation to the Matter as he may deem proper, and shall by Writing under his Hand direct the Plan, Estimate, or Contract in question, with or without any Alteration therein, or such other Plan, Estimate, or Contract, for the like Purpose as he may think fit, to be proceeded with and carried into execution. The Decision of the Secretary of State, given in pursuance of this Section, shall be final, and shall be acted upon without further Report or Approval.

Estimates to accompany Plans.

5. Together with every Plan for building, or providing or enlarging or improving, any Asylum for Pauper Lunatics, which is to be submitted to the Commissioners in Lunacy, under Section Forty-five of the said Lunacy Act, Chapter Ninety-seven, an Estimate of the Cost and Expense of carrying such Plan into execution shall be also submitted to the said Commissioners.

Excess of Payment may be paid to a Building and Repair Fund.

6. Where the Committee of Visitors enter into any Agreement for the Reception into the County Asylum of Pauper Lunatics belonging to a County or Borough which has not contributed to the erecting or providing such Asylum, and think fit under the Lunacy Act, Chapter Ninety-seven, Section Fifty-four, to fix a greater weekly Sum than is charged by them in respect of Lunatics sent from or

settled in some Place, Parish, or Borough which has contributed to the building or providing such Asylum, they may, if they think fit, pay over the Excess created by the Payment of such greater weekly Sum to a Building and Repair Fund, to be applied by them to the altering, repairing, or improving such Asylum, and shall annually submit to the General or Quarter Sessions a detailed Statement of the Manner in which such Fund has been expended.

Provision as to Contract for Reception of Lunatics.

7. Where any Contract has been made by a Committee of Visitors of any County or Borough under the Lunacy Act, Chapter Ninety-seven, Section Forty-two, for the Reception into any Asylum, Hospital, or licensed House of the whole or a Portion of the Pauper Lunatics of such County or Borough, it shall be lawful for the Justices of such County or Borough, so long as such Contract is subsisting, to defray out of the County or Borough Rate so much of the weekly Charge agreed upon for each Pauper Lunatic received therein as may, in the Opinion of such Committee of Visitors, represent the Sum due for the Use of such Asylum, Hospital, or licensed House, not exceeding, however, One Fourth of the whole of such weekly Charge, in exoneration to that Extent of the Union to which the Maintenance of any such Pauper Lunatic may be chargeable.

Provision for Care of chronic Lunatics.

8. It shall be lawful for the Visitors of any Asylum and the Guardians of any Parish or Union within the District for which the Asylum has been provided, if they shall see fit, to make Arrangements, subject to the Approval of the Commissioners and the President of the Poor Law Board, for the Reception and Care of a limited Number of chronic Lunatics in the Workhouse of the Parish or Union, to be selected by the Superintendent of the Asylum, and certified by him to be fit and proper so to be removed.

Lunatics in Asylum.

9. The Committee of Visitors of any Asylum may provide Accommodation for the Burial of Pauper Lunatics dying in the Asylum by acquiring a new Burial Ground, or by enlarging any existing Burial Ground ; they may purchase for the Purposes aforesaid any Land, and may grant any Land when purchased, or any Land already belonging to them, to any Person or Body of Persons, to be held on trust for a new Burial Ground or as Part of an existing Burial Ground, or they may themselves hold such Land on trust as a new Burial Ground or as Part of an existing Burial Ground ; they may also contribute any Sums of Money to any Person or Body of Persons on condition of such Person or Body of Persons agreeing to provide Accommodation for the Burial of such Paupers as aforesaid in any Burial Ground ; they may also take Steps for the Consecration of any new Burial Ground or enlarged Burial Ground, or any Part thereof, and in the Case of a new Burial Ground they may provide for the Appointment of a Chaplain therein ; they may enter into any Agreements necessary for carrying into effect the Powers conferred by this Section, but the Exercise of such Powers shall be subject to the Restrictions following :

Firstly, That not more than Two Statute Acres shall in the Case of any One Asylum be purchased or granted as a new Burial Ground, or for an Enlargement of an existing Burial Ground :

Secondly, That the Sanction of the Court of General or Quarter Sessions and of One of Her Majesty's Principal Secretaries of State shall be given to any Plan that may be proposed by any Visitors for carrying into effect this Section.

All Expenses incurred by any Visitors in providing Accommodation for the Burial of Pauper Lunatics, in pursuance of this Act, shall be deemed to be Monies, Costs, and Expenses payable for the Purposes of the Lunacy Act, Chapter Ninety-seven, and may be defrayed accordingly.

8 & 9 Vict. c. 18, incorporated.

10. All the Provisions of "The Lands Clauses Act, 1845," except the Pro-

visions of that Act "with respect to the Purchase and taking of any Lands otherwise than by Agreement," "with respect to the Recovery of Forfeitures, Penalties, and Costs," "with respect to Lands acquired by the Promoters of the Undertaking, under the Provisions of the Lands Clauses Consolidation Act, 1845, or the Special Act, or any Act incorporated therewith, but which shall not be required for the Purposes thereof," "and with respect to the Provision to be made for affording Access to the Special Act by all Parties interested," shall be incorporated with this Act; and for the Purposes of this Act the Expression "the Promoters of the Undertaking," wherever used in the said Lands Clauses Consolidation Act, shall mean any such Committee of Visitors as aforesaid.

Taking on Lease additional Lands for Use of Asylum.

11. It shall be lawful for any Committee of Visitors, with the Sanction of the Court of General or Quarter Sessions, to hire or take on Lease, from Year to Year or for any Term of Years, at such Rent, and upon such Terms, and under such Covenants as they think fit, any Land or Buildings, either for the Employment or Occupation of the Patients in the Asylum, or for the temporary Accommodation of any Pauper Lunatics for whom the Accommodation in the Asylum may be inadequate.

The Restrictions in Section Thirty-three of the Lunacy Act, Chapter Ninety-seven, as to the Term for which the Committee of Visitors are thereby authorized to take a Lease, or to rent Land, shall not apply to Land or Buildings to be hired or taken under this Provision.

The Land and Buildings so to be hired or taken shall, while used for the Purposes of this Section, be deemed Part of the Asylum, and all existing Provisions as to the Asylum or Part of the Asylum shall be applicable thereto accordingly.

Superannuation of Officers in Asylum.

12. The Power vested in the Visitors of an Asylum of granting an Annuity by way of Superannuation to any Person that has been an Officer or Servant in such Asylum for not less than Twenty Years, under Section Fifty-seven of the Lunacy Act, Chapter Ninety-seven, may be exercised by them when any such Person has been an Officer or Servant for not less than Fifteen Years, in the same Manner as if the Time of such Service had been Twenty Years ; and in calculating the Amount of Superannuation regard may be had, if the Visitors think fit, to the Value of the Lodgings, Rations, or other Allowances enjoyed by the Person superannuated : Provided, that no Annuity by way of Superannuation granted by the Visitors of any Asylum under the Provisions of this Act, or of the Lunacy Act, Chapter Ninety-seven, shall be chargeable on or payable out of the Rates of any County until such Annuity shall have been confirmed by a Resolution of the Justices of such County in General or Quarter Sessions assembled.

Provision for Superannuation of Matrons.

13. Where the Offices of Superintendent and Matron of any Asylum are held by Man and Wife, and an Order has been made under the Lunacy Act, Chapter Ninety-seven, granting an Annuity by way of Superannuation to the Superintendent, it shall be lawful for the Committee of Visitors of such Asylum, if they think fit to do so, and if the Matron has been an Officer in the Asylum for not less than Twenty Years, to grant to her such Annuity by way of Superannuation as they in their Discretion think proportionate to her Merits and Time of Service, although she may not have become incapable of executing her Office from Sickness, Age, or Infirmity ; and every Annuity granted in pursuance of this Section shall be payable out of the Rates lawfully applicable to the building or repairing of such Asylum: Provided, firstly, that the annual Amount by way of Superannuation paid to any Matron under this Section shall not exceed Two Thirds of the Salary payable at the Time of her Retirement; secondly, that no such Superannuation shall be granted unless Notice of the Meeting at which the same is to be granted, and of the Intention to determine thereat the Question of such Superannuation, have been given in such Manner and so long before the Time appointed for such

Meeting as is provided in the said Act with respect to Notices of Meetings of Committees of Visitors, nor unless Three Visitors concur in and sign the Order granting the same; thirdly, if any such Matron as aforesaid at any Time thereafter is appointed to any Public Office, or to any Office under the Lunacy Act, in respect of which she receives a Salary, the Payment of the Compensation awarded to her under this Act shall be suspended so long as she receives such Salary, if the Amount thereof is greater than the Amount of Compensation, or, if not, shall be diminished by the Amount of such Salary.

LICENSED HOUSES.

Inspection by Commissioners before Licence granted by Justices.

14. Before the Grant by the Justices of a Licence for the Reception of Lunatics to a House which has not been previously licensed for that Purpose, the Notice given by the Applicant, and the Plan and Statements accompanying the same, or Copies of such Notice, Plan, and Statements respectively, shall be transmitted by the Applicant to the Commissioners, and the Commissioners shall inspect or cause to be inspected the House and Land or Appurtenances proposed to be included in the Licence, and shall ascertain, with reference as well to the Situation as to the Structure, Arrangements, and Condition of the Premises, whether the same are suitable for the Reception of the Patients proposed to be received therein, and the Commissioners shall transmit to the Clerk of the Peace for the County or Borough a Report in reference to such Application; and no Licence shall be granted by the Justices of the County or Borough, in pursuance of such Application, until the Report of the Commissioners with reference thereto has been received by the said Clerk of the Peace, and taken into consideration by the Justices in General or Quarter or Special Sessions assembled.

Where a Licence is granted by the Justices of a County or Borough in respect of a House not previously licensed, such Licence shall, as nearly as conveniently may be, be according to the Form in the Schedule marked A. to this Act, instead of in the Form prescribed by the Lunacy Act, Chapter One hundred.

Notice of Alterations to be given to the Commissioners.

15. Before the Consent of any Visitors is given to any Addition or Alteration being made in or about any licensed House, or the Appurtenances, the Notice of the proposed Addition or Alteration, and Plan thereof, and accompanying Description given to the Clerk of the Peace, or Copies thereof respectively, shall be transmitted by him to the Commissioners, who shall, after making or causing to be made such Inquiries or Inspection (if any) as they may deem proper, transmit to the said Clerk of the Peace a Report stating their Approval or Disapproval thereof; and the Visitors shall not consent to such Addition or Alteration until they have received and considered such Report.

Provision as to non-resident Proprietors.

16. Whereas by the Second Section of the Lunacy Act, Chapter Ninety-six, it is enacted, "that no Person having, after the passing of the Lunacy Act, Chapter One hundred, received for the First Time a Licence for the Reception of Lunatics, or thereafter receiving for the First Time such Licence, shall receive a Licence unless he resides on the Premises licensed, and no Two or more Persons having, after the passing of the last-mentioned Act, received for the First Time a joint Licence for the Reception of Lunatics, or thereafter receiving for the First Time such joint Licence, shall receive such Licence unless they or One of them should reside on the Premises licensed:" And whereas it is expedient that in the licensed Houses to which the said Section does not apply, by reason of the Proprietor or Proprietors thereof having first received a Licence prior to the Date mentioned in the said Section, the following Provision shall be made : Be it enacted,

That in all Cases of licensed Houses, where the Proprietor or Proprietors thereof have first received their Licence or Licences before the Date of the

APPENDIX. 297

passing of the Lunacy Act, Chapter One hundred, the Physician, Surgeon, or Apothecary required by Act of Parliament to reside in or visit such House shall be approved, in the Case of a House licensed by the Commissioners, by the Commissioners, and in the Case of a House licensed by Justices, by the Justices ; and any Proprietor of a licensed House to which this Section applies who permits any Physician, Surgeon, or Apothecary who has not been approved by the Commissioners, or by the Justices, as the Case may be, to reside in or visit at such House in such Capacity as aforesaid for a Period exceeding One Calendar Month, shall incur a Penalty not exceeding Five Pounds for every Day beyond such Month during which such Physician, Surgeon, or Apothecary so resides or visits ; the above-mentioned Period of One Month shall be reckoned in the Case of a Physician, Surgeon, or Apothecary so resident or visiting at the Time of the passing of this Act from the Date of the passing thereof, and in the Case of any fresh Appointment of any such Physician, Surgeon, or Apothecary as aforesaid from the Date of such Appointment.

Penalty on infringing Terms of Licence.

17. If any Person empowered by Licence issued under the Lunacy Act, Chapter One hundred, to employ his House and Premises for the Reception of Lunatics receives into his House any Patients beyond the Number specified in his Licence, or fails to comply with the Regulations of his Licence in respect of the Sex of the Patients to be received, or the Class of Patients, whether Private or not, to be received, he shall, in respect of each Patient received in contravention of his Licence, incur a Penalty not exceeding Fifty Pounds.

Extension of Powers to take Boarders in Houses.

18. It shall be lawful for the Proprietor or Superintendent of any licensed House, with the previous Assent in Writing of Two or more of the Commissioners, or in the Case of a House licensed by Justices of Two or more of the Visitors, to entertain and keep in such House as a Boarder for such Time as may be specified in the Assent any Person who may have been within Five Years immediately preceding the giving of such Assent a Patient in any Asylum, Hospital, or licensed House, or under Care as a single Patient.

ADMISSION AND VISITATION OF PATIENTS.

Provision for sending Pauper Lunatics to Asylums.

19. Whereas by the Sixty-seventh Section of the Lunacy Act, Chapter Ninety-seven, it is amongst other things enacted as follows : "That every Relieving Officer of any Parish within a Union or under a Board of Guardians, and every Overseer of a Parish of which there is no Relieving Officer, who shall have Knowledge either by such Notice or otherwise that any Pauper resident in such Parish is or is deemed to be a Lunatic and a proper Person to be sent to an Asylum, shall within Three Days after obtaining such Knowledge give Notice thereof to some Justice of the County or Borough within which such Parish is situate :" Now be it enacted, That the said Section shall be construed as if the Words "and a proper Person to be sent to an Asylum" had been omitted in the said recited Enactment.

Lunatics proper to be sent to Asylums.

20. No Person shall be detained in any Workhouse, being a Lunatic or alleged Lunatic, beyond the Period of Fourteen Days, unless in the Opinion, given in Writing, of the Medical Officer of the Union or Parish to which the Workhouse belongs such Person is a proper Person to be kept in a Workhouse, nor unless the Accommodation in the Workhouse is sufficient for his Reception, and any Person detained in a Workhouse in contravention of this Section shall be deemed to be a proper Person to be sent to an Asylum within the Meaning of Section Sixty-seven of the Lunacy Act, Chapter Ninety-seven ; and in the event of any Person being detained in a Workhouse in contravention of this Section, the Medical Officer shall

for all the Purposes of the Lunacy Act, Chapter Ninety-seven, be deemed to have Knowledge that a Pauper resident within his District is a Lunatic, and a proper Person to be sent to an Asylum, and it shall be his Duty to act accordingly, and further to sign such Certificate as is contained in Schedule F. to the said Act, No. 3, with a view to more certainly securing the Reception into an Asylum of such Pauper Lunatic as aforesaid.

Amendment of Form of List as respects Pauper Lunatics in Workhouses.

21. The List of Lunatic Paupers required by Section Sixty-six of the Lunacy Act, Chapter Ninety-seven, to be made out by the Medical Officer, shall be in the Form in the Schedule marked B. hereto, and not in the Form required by the said Section, and shall, as respects such of the Lunatics therein mentioned as may be in any Workhouse, state whether, in the Opinion of the Medical Officer, the Workhouse is or not sufficient for the Accommodation of the Lunatics detained therein, and whether or not the Lunatics detained therein are proper Persons to be kept in a Workhouse.

Order for Reception and Medical Visitation of Persons found lunatic by Inquisition.

22. When a Person has been found lunatic by Inquisition an Order, signed by the Committee appointed by the Lord Chancellor, and having annexed thereto an Office Copy of the Order appointing such Committee, shall be a sufficient Authority for the Reception of such Person into any Asylum, Hospital, licensed House, or other House, without any further Order or any such Medical Certificates as are required by Section Ninety of the Lunacy Act, Chapter One hundred, and Section Four and Eight of the Lunacy Act, Chapter Ninety-six, and the Provisions of the Section Ninety of the Lunacy Act, Chapter One hundred, as to the Visitation of every single Patient once in every Two Weeks by a Physician, Surgeon, or Apothecary, shall not apply to any Person found lunatic by Inquisition as aforesaid.

Persons signing Orders for Admission to have seen Patient within One Month.

23. No Order for the Reception of a Private Patient into any Asylum or registered Hospital, licensed or other House, made in pursuance of the Lunacy Acts, Chapters Ninety-six and Ninety-seven, or either of them, shall authorize the Reception of such Patient after the Expiration of One Calendar Month from its Date, nor unless the Person subscribing such Order has himself seen the Patient within One Month prior to its Date, nor unless a Statement of the Time and Place when such Person last saw the Patient is added to such Order.

Certain Persons prohibited from signing Orders for Admission.

24. The following Persons shall be prohibited from signing any Certificate or Order for the Reception of any Private Patient into any licensed or other House:
First, Any Person receiving any Per-centage on or otherwise interested in the Payments to be made by or on account of any Patient received into a licensed or other House:
Second, Any Medical Attendant as defined by the Lunacy Act, Chapter One hundred.

Relative of Pauper to be named in Order of Admission.

25. Where an Order is made, in pursuance of the Lunacy Acts or any of them, for the Reception of any Private or Pauper Lunatic into any Asylum, registered Hospital, or licensed House, there shall be inserted in every such Order, wherever it be possible, the Name and Address of One or more of the Relations of the Lunatic ; and in the event of his Death it shall be the Duty of the Clerk of such Asylum, the Superintendent of such Hospital, and the Proprietor or Superintendent of such licensed House, to send by Post Notice of his Death in a prepaid Letter addressed to such Relation or One of such Relations.

APPENDIX.

Same Order and Certificates to justify Detention as Pauper or Private Patient.

26. The Order and Certificate required by Law for the Detention of a Patient as a Pauper shall extend to authorize his Detention, although it may afterwards appear that he is entitled to be classified as a Private Patient; and the Order and Certificates required by Law for the Detention of a Patient as a Private Patient shall authorize his Detention, although it may afterwards appear that he ought to be classified as a Pauper Patient.

Provision as to defective Certificates.

27. Where any Medical Certificate upon which a Patient has been received into any Asylum, registered Hospital, licensed or other House, or either of such Certificates, is deemed by the Commissioners incorrect or defective, and the same are or is not duly amended to their Satisfaction within Fourteen Days after the Reception by the Superintendent or Proprietor of such Asylum, registered Hospital, or licensed or other House of a Direction or Writing from the Commissioners requiring Amendment of the same, the Commissioners or any Two of them may, if they see fit, make an Order for the Patient's Discharge.

Transmission of Documents to Commissioners on Admission of Patient.

28. The Documents required by the Lunacy Act, Chapter One hundred, Sections Fifty-two and Ninety, and the Lunacy Act, Chapter Ninety-seven, Section Eighty-nine, to be sent to the Commissioners in Lunacy, after Two clear Days, and before the Expiration of Seven clear Days from the Day on which any Private Patient has been received into any licensed House, registered Hospital, or Asylum, shall, with the Exception of the Statement now required to be subjoined to the Notice of Admission into any Asylum, Hospital, or licensed House, be transmitted to the said Commissioners within One clear Day from the Day on which any Patient has been received into any such House, Hospital, or Asylum as aforesaid, and the said Sections shall, so far as relates to the said Documents, other than the said Statement, be construed as if the Words "One clear Day" were substituted therein for the Words "after Two clear Days, and before the Expiration of Seven clear Days;" nevertheless the said excepted Statement shall be transmitted as heretofore, save that it shall be separate from the said Notice, and shall refer to the Order of Admission by the Date thereof, instead of referring to it as the above Notice, and the Words referring to the said Statement as being subjoined shall be omitted in the said Notice.

Visits by Commissioners.

29. Every licensed House may be visited at any Time, and, if situate within their immediate Jurisdiction, shall be visited Twice at least in every Year by any One or more of the Commissioners, in addition to the Visits now required to be made by Two at least of the Commissioners; and if not within the immediate Jurisdiction of the Commissioners, may be visited at any Time, and shall be visited twice at least in every Year by One or more of the Visitors, in addition to the Visits now required to be made by Two at least of the Visitors.

Every Commissioner visiting alone shall have the same Powers as Two Commissioners would have under Section Sixty-one of the Lunacy Act, Chapter One hundred; and all the Provisions of the said Act contained in Sections Sixty-three, Sixty-four, Sixty-five, Sixty-six, and Sixty-seven shall apply to a Commissioner or Visitor visiting alone, as the Case may be, in the same Manner as they would apply under the said Act to Two or more Commissioners or Two or more Visitors visiting together.

Single Commissioner to Visit Asylums and Gaols.

30. Any One or more of the Commissioners may at any Time visit every Asylum and Hospital for Lunatics, and every Gaol in which there may be, or alleged to be, any Lunatic, in addition to the Visits now required or empowered to

be made by Two at least of the Commissioners, and every Commissioner so visiting alone shall have the same Powers as Two or more Commissioners would perform and have, in the Case of an Asylum or Gaol, in pursuance of the One hundred and tenth Section of the Lunacy Act, Chapter One hundred, and in the Case of a Hospital in pursuance of Section Sixty-one of the Lunacy Act, Chapter One hundred.

Power to remove Lunatic from Workhouse to Asylum.

31. Where upon the Visitation of any Workhouse by any Two or more of the Commissioners in Lunacy it appears to them that any Lunatic or alleged Lunatic therein is not a proper Person to be kept in a Workhouse, they may by an Order under their Hands direct such Lunatic to be received into an Asylum, and any Order so made shall have the same Effect, and be obeyed by the same Persons, and subject them to the same Penalties, in case of Disobedience, as an Order made by a Justice for the Reception of a Lunatic into an Asylum under the Sixty-seventh Section of the Lunacy Act, Chapter Ninety-seven : Provided always, that it shall be lawful for the Guardians of the Union or Parish to which any Workhouse belongs to appeal against such Order at any Time within One Calendar Month from the making thereof to Her Majesty's Principal Secretary of State for the Home Department, who shall thereupon exercise the Power given to him by Section One hundred and thirteen of the Lunacy Act, Chapter One hundred, save that he shall not appoint thereunder the Commissioners who made the Order appealed against, or either of them ; and the Order in the Matter of the Secretary of State, made upon the Report of the special Visitation, shall be binding on all Parties concerned.

Removal of single Pauper Patients to Asylums.

32. Any Two or more of the Commissioners in Lunacy may visit any Pauper Lunatic or alleged Lunatic not in an Asylum, Hospital, licensed House, or Workhouse, and may, if they think fit so to do, call to their Assistance a Physician, Surgeon, or Apothecary, and examine such Pauper : and if such Physician, Surgeon, or Apothecary sign a Certificate with respect to such Pauper, according to the Form in Schedule F. No. 3, annexed to the Lunacy Act, Chapter Ninety-seven, and the Commissioners are satisfied that such Pauper is a Lunatic, and a proper Person to be taken charge of and detained under Care and Treatment, they may, by an Order under their Hands, direct such Lunatic or alleged Lunatic to be received into an Asylum, and any Order so made shall have the same Effect, and be obeyed by the same Persons, and subject them to the same Penalties in case of Disobedience, as an Order made by a Justice for the Reception of a Lunatic into an Asylum under the Sixty-seventh Section of the Lunacy Act, Chapter Ninety-seven.

Effect of Order for Removal.

33. The Order made by any Two or more of the Commissioners in Lunacy in pursuance of this Act may authorize the Admission of a Lunatic not only into any Asylum of the County or Borough in which the Parish or Place from which the Lunatic is sent is situate, but also into any other Asylum for the Reception of Pauper Lunatics of such County or Borough, and also into any Asylum for any other County or Borough, or any Hospital registered or House licensed for the Reception of Lunatics, under the same Circumstances and subject to the same Conditions under which an Order of the Justice or Justices may authorize such Admission in pursuance of Section Seventy-two of the Lunacy Act, Chapter Ninety-seven.

Statement of Condition of Pauper Lunatics to be transmitted to Guardians.

34. The Superintendent of every Asylum shall, once at the least in each Half Year, transmit to the Guardians of every Union, and of every Parish under a Board of Guardians, and the Overseers of every Parish not in a Union nor under a Board of Guardians, a Statement of the Condition of every Pauper Lunatic chargeable to such Union or Parish.

APPENDIX.

Amendment of Sect. 61. of 8 & 9 Vict. c. 100.

35. The Inquiries authorized to be made under Section Sixty-four of the Lunacy Act, Chapter One hundred, or under Section Ninety-two of the same Act, and the Provisions amending the same, may include Inquiries as to the Monies paid to the Superintendent or Proprietor on account of any Lunatic under the Care of such Superintendent or Proprietor.

Copies of Entries of Commissioners and Visitors.

36. The Proprietor of every licensed House within the Jurisdiction of Visitors appointed by Justices shall, within Three Days after a Visit by the Visiting Commissioners or Commissioner, transmit a true and perfect Copy of the Entries made by them or him in the Visitors' Book, the Patients' Book, and the Medical Visitation Book, respectively, distinguishing the Entries in the several Books, to the Clerk of the Visitors as well as to the Commissioners, and the Copies so transmitted to the Clerk of the Visitors of all such Entries in the Visitors' Book relating to any such licensed House, and made since the Grant or last Renewal of the Licence thereof, shall be laid before the Justices on taking into consideration the Renewal of the Licence to the House to which such Entries relate ; and every such Proprietor as aforesaid who shall omit to transmit as herein-before mentioned a true and perfect Copy of every or any such Entry as aforesaid shall for every such Omission forfeit a Sum not exceeding Ten Pounds.

Visiting Committee to enter Observations in a Book respecting Dietary, Accommodation, &c. of Lunatics in Workhouses.

37. The Visiting Committee of every Union, and of every Parish under a Board of Guardians, and the Overseers of every Parish not in a Union nor under a Board of Guardians, shall once at the least in each Quarter of a Year enter in a Book to be provided and kept by the Master of the Workhouse such Observations as they may think fit to make respecting the Dietary, Accommodation, and Treatment of the Lunatics or alleged Lunatics for the Time being in the Workhouse of their Union or Parish, and the Book containing the Observations made in pursuance of this Section by the Visiting Guardians or Overseers shall be laid by the Master before the Commissioner or Commissioners on his or their next Visit.

MISCELLANEOUS CLAUSES.

Patients may be permitted to be absent on Trial from Hospitals and Private Houses.

38. Section Eighty-six of the Lunacy Act, Chapter One hundred, and Section Seventeen of the Act Eighteenth and Nineteenth *Victoria*, Chapter One hundred and five, shall extend to authorize the Proprietor or Superintendent of any licensed House or Hospital, with such Consent, and to be given on such Approval as thereby required, to permit any Patient to be absent from such Hospital or House upon Trial for such Period as may be thought fit :

Two of the Commissioners, as regards any Hospital or any licensed House, and Two of the Committee of Governors of any Hospital, and Two of the Visitors of any licensed House, as regards any licensed House within the Jurisdiction of Visitors, may of their own Authority permit any Pauper Patient therein to be absent from such Hospital or House upon Trial for such Period as they may think fit, and may make or order to be made an Allowance to such Pauper not exceeding what would be the Charge for him in such Hospital or House, which Allowance shall be charged for him and be payable as if he were actually in such Hospital or House, but shall be paid over to him, or for his Benefit, as the said Commissioners or Visitors may direct :

In case any Person so allowed to be absent on Trial for any Period do not return at the Expiration thereof, and a Medical Certificate as to his State of Mind certifying that his Detention as a Lunatic is no longer necessary be not sent to the Proprietor or Superintendent of such licensed House or Hospital, he may at any Time within Fourteen Days after the Expiration of the same Period be retaken as in the Case of an Escape.

APPENDIX.

Penalty on Officer conniving at the Escape of Lunatics.

39. If any Officer or Servant in any Hospital or licensed House through wilful Neglect or Connivance permits any Patient to escape from such Hospital or licensed House, or secretes or abets or connives at the Escape of any Patient from such Hospital or licensed House, he shall for every such Offence incur a Penalty not exceeding Twenty Pounds.

Correspondence of Private Patients.

40. Every Letter written by a Private Patient in any Asylum, Hospital, or licensed House, or by any single Patient, and addressed to the Commissioners in Lunacy or Committee, or in the Case of Houses within the Jurisdiction of Visitors to the Visitors or any of them, shall, unless special Regulations to the contrary have been given by such Commissioners or Visitors, be forwarded unopened.

Every Letter written by a Private Patient in any Asylum, Hospital, or licensed House, or by any single Patient, and addressed to any Person other than the Commissioners or Committee or Visitors or One of them, shall be forwarded to the Person to whom it is addressed, unless the Superintendent in the Case of an Asylum or Hospital, the Proprietor in the Case of a licensed House, and the Person having the Charge of a single Patient in the Case of a single Patient, prohibit the forwarding of such Letter, by Endorsement to that Effect under his Hand on the Letter, in which Case he shall lay all Letters so endorsed before the Visiting Commissioners, Committee, or Visitors, as the Case may be, on their next Visit.

Any Superintendent, Proprietor, or Person in charge of a single Patient failing to comply with the Provisions of this Section as to laying any Letter before the Commissioners or Committee or Visitors that is not forwarded to the Address of the Person to whom it is directed, or being privy to the Detention by any other Person of any Letter detained in contravention of this Section, shall incur a Penalty not exceeding Twenty Pounds in respect of each Offence ; and any Person detaining any Letter in contravention of this Section shall incur, in respect of each Letter so detained, a Penalty not exceeding Twenty Pounds.

Statement as to Condition of single Patients.

41. Every Person having the Care or Charge of a single Patient shall in addition to the Notice required to be given by the Ninetieth Section of the Lunacy Act, Chapter One hundred, before the Expiration of Seven clear Days from the Day on which he has taken the Patient under his Care or Charge, transmit to the Commissioners a Statement of the Condition of the Patient, according to the Form in Schedule F. annexed to the said last-mentioned Act, such Statement to be signed by the Physician, Surgeon, or Apothecary visiting the Patient in pursuance of the Ninetieth Section of the Lunacy Act, Chapter One hundred.

If any Person having the Care or Charge of a single Patient fails to transmit such Statement as aforesaid within such Time as is required by this Section he shall be guilty of a Misdemeanor.

Commissioners empowered to prescribe Forms, &c. of Medical Visitation Book.

42. In the Case of single Patients the Commissioners may from Time to Time make Regulations as to the Form of and the Particulars to be entered in the "Medical Visitation Book," required to be kept by the Ninetieth Section of the Lunacy Act, Chapter One hundred, and if the Person having the Care or Charge of a single Patient fails to comply with the Regulations so made he shall in respect of each Offence incur a Penalty not exceeding Five Pounds.

Discharge of a Private Patient.

43. If there be no Person capable or qualified, under Section Seventy-two or Section Seventy-three of the said Lunacy Act, Chapter One hundred, to direct the Discharge or Removal of any such Patient as therein mentioned from any registered Hospital or licensed House, the Commissioners may order the Discharge or Removal of such Patient, as they may think fit.

APPENDIX. 303

Report to Coroner of Death of single Patient.

44. The Superintendent of every Asylum, and every Person having the Care or Charge of a single Patient, shall, in the event of the Death of any Patient, transmit to the Coroner of the County or Borough the same Statement as is required by Law to be transmitted in the Case of the Death of any Patient in any Hospital or licensed House, and if such Coroner, after receiving such Statement, thinks that any reasonable Suspicion attends the Cause and Circumstances of the Death of such Patient, he shall summon a Jury to inquire into the Circumstances of such Death.

Any Superintendent or Person in charge who makes default in complying with the Requisitions of this Section shall be guilty of a Misdemeanor.

Chargeability of Pauper Lunatics whose Settlements cannot be ascertained where found in certain Boroughs.

45. Section Fourteen of the Act of the Session holden in the Eighteenth and Nineteenth Years of Her Majesty, Chapter One hundred and five, shall be repealed, and in lieu thereof be it enacted, Where any Pauper Lunatic is not settled in the Parish by which or at the Instance of some Officer or Officiating Clergyman of which he is sent to an Asylum, registered Hospital, or licensed House, and it cannot be ascertained in what Parish such Pauper Lunatic is settled, and such Lunatic is found in a Borough which has a separate Court of Sessions of the Peace, and is not liable, under the Act of the Session holden in the Fifth and Sixth Years of King *William* the Fourth, Chapter Seventy-six, Section One hundred and seventeen, to the Payment of a Proportion of the Sums expended out of the County Rate, or is found in any Borough which under the Act of the Session holden in the Twelfth and Thirteenth Years of Her Majesty, Chapter Eighty-two, is exempted from Liability to contribute to the Payment of the Expenses incurred for maintaining Pauper Lunatics chargeable to the County in which such Borough is situate, such Lunatic shall be adjudged to be chargeable to the Borough in which he is found; and it shall not be lawful for any Justices to adjudge such Lunatic to be chargeable to any County, nor to make any Order upon the Treasurer of any County for the Payment of any Expenses whatsoever incurred or to be incurred in respect of such Lunatic.

All the Provisions in the Lunacy Act, Chapter Ninety-seven, as to the Mode of determining that a Pauper Lunatic is chargeable to a County, and as to the Orders to be made for Payment of Expenses and other Monies in respect of such Lunatic, and for the Repayment thereof to the Treasurer of a County, shall extend to the Case of a Borough to which a Lunatic is made chargeable under this Section as if the said Provisions were re-enacted in this Act, and such Borough were therein mentioned or referred to instead of a County.

Amendment of 8 & 9 Vict. c. 100. s. 100. as to Power of administering Oaths.

46. Any Two or more Commissioners or Visitors, in exercise of the Powers given to them by the One hundredth Section of the Lunacy Act, Chapter One hundred, may, if they think fit, examine on Oath any Person appearing before them as a Witness, notwithstanding a Summons may not have been served on him in pursuance of the said Section.

Definition of Physician, Surgeon, or Apothecary.

47. The Term Physician, Surgeon, or Apothecary, wherever used in the Lunacy Acts, shall mean a Person registered under "The Medical Act," passed in the Session holden in the Twenty-first and Twenty-second Years of the Reign of Her present Majesty, Chapter Ninety.

Part of Sect. 132. of 16 & 17 Vict. c. 97. repealed.

48. So much of Section One hundred and thirty-two of the Lunacy Act, Chapter Ninety-seven, as enacts that in that Act, unless there be something in the Subject

or Context repugnant to such Construction, the Word "County" shall mean a County of a City or County of a Town, shall, except with respect to the City of *London*, be repealed, and all the Provisions of the said Act and of the Acts amending the same shall be read and construed accordingly.

SCHEDULE A.

Form of Licence.

Know all Men, That we, the Undersigned Justices of the Peace, acting in and for   in General [*or* Quarter *or* Special] Sessions assembled, do hereby certify that *A.B.* of   in the Parish of in the County of   hath delivered to the Clerk of the Peace a Plan and Description of a House and Premises proposed to be licensed for the Reception of Lunatics, situate at   in the County of and which has not been previously licensed for that Purpose, and hath applied to us for a Licence thereof: And whereas the Particulars of the said Application have been transmitted to the Commissioners in Lunacy, and their Report in reference to the said Application has been received, and has been taken into consideration by us; and we, having considered and approved the Application, do hereby authorize and empower the said *A.B.* (be intending *or* not intending to reside therein) to use and employ the said House and Premises for the Reception of   Male [*or* Female, *or*   Male and   Female] Lunatics, of whom not more than   shall be Private Patients, for the Space of Calendar Months from this Date.

Given under our Hand and Seals, this   Day of   in the Year of our Lord One thousand eight hundred and

Witness, *Y.Z.*, *Clerk of the Peace.*

SCHEDULE B.

County of
Union [*or* Parish of]
District of

Quarterly List of Lunatic Paupers within the District of the Union of [*or* the Parish of   ], in the County or Borough of , not in any Asylum, Registered Hospital, or Licensed House.

Name.	Sex.	Age.	Form of Mental Disorder.	Duration of present Attack of Insanity, and if idiotic, whether or not from Birth.	Resident in	Non-Resident in Workhouse, where and with whom resident.	Date of Visit.	In what Condition, and, if ever restrained, why, and by what Means, and how often.

I declare that I have personally examined the several Persons whose Names are specified in the above List on the Days set opposite their Names; and I certify, firstly, with respect to those appearing by the above List to be in the Workhouse, that the Accommodation in the Workhouse is sufficient for their Reception, and that they are all [or all except A.B. and C.D.] proper Patients to be kept in the Workhouse; and, secondly, with respect to those appearing by the above List to be resident elsewhere than in the Workhouse, that they are all [or all except A.B. and C.D.] properly taken care of, and may properly remain out of an Asylum.

I declare that the Persons in the above List are to the best of my Knowledge the only Pauper Lunatics in the District of the Union of   [or in the Parish of   ] who are not in an Asylum, registered Hospital, or duly licensed House.

                    (Signed)   A.B.,
                    Medical Officer of the   District
                    of the Union [or Parish] of
    Dated the   Day of   One thousand eight hundred and   .

ANNO DECIMO QUINTO & DECIMO SEXTO
VICTORIÆ REGINÆ.

CAP. LVI.

An Act for regulating the Qualifications of Pharmaceutical Chemists.
[30th *June* 1852.]

WHEREAS it is expedient for the safety of the public that persons exercising the business or calling of pharmaceutical chemists in *Great Britain* should possess a competent practical knowledge of pharmaceutical and general chemistry and other branches of useful knowledge : And whereas certain persons desirous of advancing chemistry and pharmacy, and of promoting a uniform system of educating those who should practise the same, formed themselves into a society called " The Pharmaceutical Society of *Great Britain*," which said society was on the eighteenth day of *February* one thousand eight hundred and forty-three incorporated by Royal Charter, whereby it was provided that the said society should consist of members who should be chemists and druggists who were or had been established on their own account at the date of the said charter, or who should have been examined in such manner as the council of the said society should deem proper, or who should have been certified to be duly qualified for admission, or who should be persons elected as superintendents by the council of the said society : And whereas it is expedient to prevent ignorant and incompetent persons from assuming the title of or pretending to be pharmaceutical chemists or pharmaceutists in *Great Britain*, or members of the said Pharmaceutical Society, and to that end it is desirable that all persons before assuming such title should be duly examined as to their skill and knowledge by competent persons, and that a register should be kept by some legally authorized officer of all such persons : And whereas for the purpose aforesaid, and for extending the benefits which have already resulted from the said Charter of Incorporation, it is desirable that additional powers should be granted for regulating the qualifications of persons who may carry on the business of pharmaceutical chemists : Be it enacted by the Queen's most excellent Majesty, by and with the advice and consent of the Lords Spiritual and Temporal, and Commons, in this present Parliament assembled, and by the authority of the same,

    Charter, dated 18th Feb. 1843, confirmed, save as altered.

I. That the said Charter of Incorporation granted to the said society on the eighteenth day of *February* one thousand eight hundred and forty-three, save and except such part or parts thereof as are hereby altered, varied, or repealed, shall

X

be and the same is hereby confirmed and declared to be in full force and virtue, and shall be as good and effectual to all intents and purposes as if this Act had not been passed.

Power to Council to alter Byelaws, provided they are approved by a General Meeting of Society and the Secretary of State.

II. The council of the said Pharmaceutical Society shall be and the same are hereby authorized and empowered to alter and amend the byelaws of the said society made and established under or in pursuance of the said Charter of Incorporation, and to make and establish such new or additional byelaws as they shall deem proper and necessary for the purposes contemplated by the said charter or by this Act: Provided always, that all such original byelaws, and all altered, amended, or additional byelaws, shall be confirmed and approved by a special general meeting of the members of the said Pharmaceutical Society, and by one of Her Majesty's principal Secretaries of State: Provided also, that the existing byelaws of the said society shall continue in force until the next annual meeting of the said society to be held in the month of *May* one thousand eight hundred and fifty-three.

Manner of Voting.

III. At all meetings of the said society at which votes shall be given for the election of officers, all members entitled to vote may give their votes either personally, or, in cases of residence exceeding five miles from the General Post Office, *Saint Martin's le Grand, London*, by voting papers authorized by writing, in a form to be defined in the byelaws of the said society, or in a form to the like effect, such voting papers being transmitted under cover to the secretary not less than five clear days prior to the day on which the election is to take place.

Council to appoint Registrar, &c.

IV. The council of the said Pharmaceutical Society shall, within three calendar months after the passing of this Act, appoint a fit and proper person as a registrar under this Act, and the council of the said society shall have the power to remove the said registrar, or any future registrar to be appointed under this Act, from the said office, and from time to time to appoint a new registrar in the room of any registrar who may die, or retire, or be removed from office as aforesaid, and also to appoint and remove from time to time a deputy registrar, and such clerks and other subordinate officers as may be requisite for carrying out the purposes of this Act, and also to pay suitable salaries to the said registrar, deputy registrar, clerks, and officers.

Registrar to make Registers of Members of Society, &c., and to keep an Index and Books, as may be required.

V. The registrar to be appointed under or by virtue of this Act shall from time to time make out and maintain a complete register of all persons being members of the said society, and also of all persons being associates and apprentices or students respectively, according to the terms of the Charter of Incorporation, and shall keep a proper index of the register, and all such other registers and books as may be required by the council of the said society, and may be necessary for giving effect to the byelaws of the said society and to the provisions of this Act.

All Members, Associates, &c. of the Society at the passing of this Act entitled to be registered.

VI. All such persons as shall at the time of the passing of this Act be members, associates, apprentices, or students of the said Pharmaceutical Society of *Great Britian*, according to the terms of the said Charter of Incorporation, shall be registered as pharmaceutical chemists, assistants, and apprentices or students respectively.

APPENDIX.

Registrar bound to certify, &c., and his Certificate good Evidence in Absence of the contrary.

VII. The registrar to be appointed under or by virtue of this Act shall be bound, on the application of any person paying one shilling, to certify under his hand whether or no any person whose name and address shall be furnished to him appears in the said register or is a member of the Pharmaceutical Society of *Great Britian* or not; and the certificate of such registrar, signed by the said registrar, and countersigned by the president or two members of the council of the said society, shall, in the absence of evidence to the contrary, be sufficient evidence of the facts therein stated up to the date of the said certificate.

Persons appointed under Charter or Byelaws, or this Act, to conduct Examinations. Power to Examiners to grant Certificates.

VIII. All such persons as shall from time to time be appointed under or in pursuance of the said Charter of Incorporation or the byelaws thereof, or under this Act, shall be and the same are hereby declared to be fit and proper persons to conduct all such examinations as are provided for or contemplated by this Act, and shall respectively have full power and authority and are hereby authorized and empowered to examine all persons who shall present themselves for examination under the provisions of this Act in their knowledge of the Latin language, in botany, in materia medica, and in pharmaceutical and general chemistry, and such other subjects as may from time to time be determined by any byelaw; provided always, that such examinations shall not include the theory and practice of medicine, surgery, or midwifery; and the said examiners are hereby empowered to grant or refuse to such persons, as in their discretion may seem fit, certificates of competent skill and knowledge and qualification to exercise the business or calling of pharmaceutical chemists, or, as the case may require, to be engaged or employed as students, apprentices, or assistants respectively.

Examiners to be appointed for Scotland.

IX. And to enable the said society to provide for the examination in *Scotland* of such students, apprentices, or assistants in *Scotland* as may desire to be examined there, it shall be lawful for the council of the society, and they are hereby required, to appoint such fit and proper persons in *Scotland*, to meet in *Edinburgh* or *Glasgow*, or such other place or places as the council may think desirable, and to conduct there all such examinations as are provided for and contemplated by this Act, with such and the like powers and authorities in respect thereof as are herein conferred, and to grant to the persons to be so examined such and the like certificates as are herein-before specified and referred to, or to refuse the same; and all the provisions of this Act shall be equally applicable to the examiners, examinations, and parties examined in *Scotland* as to the examiners, examinations, and parties examined in *England*.

Persons who have obtained Certificates entitled to be registered; Persons registered as Pharmaceutical Chemists eligible to be Members; Eligibility of Persons registered as Assistants, &c.

X. Every such person who shall have been examined by the persons appointed as aforesaid, and shall have obtained a certificate of qualification from them, shall be entitled to be registered by the registrar according to the provisions of this Act, upon payment of such fee or fees as shall be fixed by the byelaws; and every such person duly registered as a pharmaceutical chemist shall be eligible to be elected as a member of the said society; and every such person duly registered as an assistant shall be eligible for admission as an associate of the said society; and every such person duly registered as a student or apprentice to a pharmaceutical chemist shall be eligible for admission into the said society; according to the byelaws thereof.

Exceptions.

XI. That no person who is a member of the medical profession, or who is practising under right of a degree of any university, or under a diploma or licence of a medical or surgical corporate body, shall be entitled to be registered under this Act; and if any registered pharmaceutical chemist shall obtain such diploma or licence, his name shall not be retained on the said register during the time that he is engaged in practice as aforesaid.

Persons not duly registered shall not assume or use Name or Title of Pharmaceutical Chemist, or any Name, Title, or Sign implying the same. Penalty on Persons offending. Penalties, how recoverable in England and Wales; in Scotland.

XII. From and after the passing of this Act, it shall not be lawful for any person, not being duly registered as a pharmaceutical chemist according to the provisions of this Act, to assume or use the title of pharmaceutical chemist or pharmaceutist in any part of *Great Britain*, or to assume, use, or exhibit any name, title, or sign implying that he is registered under this Act, or that he is a member of the said society; and if any person not being duly registered under this Act, shall assume or use the title of pharmaceutical chemist or pharmaceutist, or shall use, assume, or exhibit any name, title, or sign implying that he is a person registered under this Act, or that he is a member of the said society, every such person shall be liable to a penalty of five pounds; and such penalty may be recovered by the registrar to be appointed under this Act, in the name and by the authority of the council of the said society in manner following ; (that is to say,)

In *England* or *Wales*, by plaint under the provisions of any Act in force for the more easy recovery of small debts and demands :

In *Scotland*, by action before the Court of Session in ordinary form, or by summary action before the sheriff of the county, or in the royal burghs before the magistrates of the burghs where the offence may be committed or the offender resides, who, upon proof of the offence or offences either by confession of the party offending or by the oath or affirmation of one or more credible witnesses, shall convict the offender, and find him liable in the penalty or penalties aforesaid, as also in expenses; and it shall be lawful for the sheriff or magistrate, in pronouncing such judgment for the penalty or penalties and costs, to insert in such judgment a warrant, in the event of such penalty or penalties and costs not being paid, to levy and recover the amount of the same by poinding :

Provided always, that it shall be lawful to the sheriff or magistrate, in the event of his dismissing the action and assoilzieing the defender, to find the complainer liable in expenses ; and any judgment so to be pronounced by the sheriff or magistrate in such summary application shall be final and conclusive, and not subject to review, by advocation, suspension, reduction, or otherwise.

Limitation for Recovery of Penalties, &c.

XIII. Provided always, that no action or other proceeding for any offence under this Act shall be brought after the expiration of six months from the commission of such offence; and in every such action or proceeding the party who shall prevail shall recover his full costs of suit or of such other proceedings.

Application of Penalties.

XIV. All and every sums and sum of money which shall arise from any conviction and recovery of penalties for offences incurred under this Act shall be paid as the commissioners of Her Majesty's Treasury shall direct.

Registrar falsifying Register, &c. guilty of a Misdemeanor.

XV. If any registrar under this Act shall wilfully make or cause to be made any falsification in any matters relating to any register or certificate aforesaid, every such offender shall be deemed guilty of a misdemeanor.

Any Person wilfully procuring false Certificate guilty of a Misdemeanor.

XVI. If any person shall wilfully procure by any false or fraudulent means a certificate purporting to be a certificate of registration under this Act, or shall fraudulently exhibit a certificate purporting to be a certificate of membership of the Pharmaceutical Society, every such person so offending shall be adjudged guilty of a misdemeanor.

ANNO TRICESIMO PRIMO & TRICESIMO SECUNDO VICTORIÆ REGINÆ.

CAP. CXXI.

An Act to regulate the Sale of Poisons, and alter and amend the Pharmacy Act, 1852.   [31st *July* 1868.]

15 & 16 *Vict. c.* 56.

WHEREAS it is expedient for the safety of the public that persons keeping open shop for the retailing, dispensing, or compounding of poisons, and persons known as chemists and druggists, should possess a competent practical knowledge of their business, and to that end that from and after the day herein named all persons not already engaged in such business should, before commencing such business, be duly examined as to their practical knowledge, and that a register should be kept as herein provided, and also that the Act passed in the fifteenth and sixteenth years of the reign of Her present Majesty, intituled *An Act for regulating the Qualification of Pharmaceutical Chemists*, herein-after described as the Pharmacy Act, should be amended :

Be it enacted by the Queen's most excellent Majesty, by and with the advice and consent of the Lords Spiritual and Temporal, and Commons, in this present Parliament assembled, and by authority of the same, as follows :

Persons selling or compounding Poisons, or assuming the Title of Chemist and Druggist, to be qualified.

1. From and after the thirty-first day of *December* one thousand eight hundred and sixty-eight it shall be unlawful for any person to sell or keep open shop for retailing, dispensing, or compounding poisons, or to assume or use the title "chemist and druggist," or chemist or druggist, or pharmacist, or dispensing chemist or druggist, in any part of *Great Britain*, unless such person shall be a pharmaceutical chemist, or a chemist and druggist within the meaning of this Act, and be registered under this Act, and conform to such regulations as to the keeping, dispensing, and selling of such poisons as may from time to time be prescribed by the Pharmaceutical Society with the consent of the Privy Council.

Articles named in Schedule (A.) to be deemed Poisons within the Meaning of this Act.

2. The several articles named or described in the Schedule (A.) shall be deemed to be poisons within the meaning of this Act, and the council of the Pharmaceutical Society of *Great Britain* (herein-after referred to as the Pharmaceutical Society) may from time to time, by resolution, declare that any article in such resolution named ought to be deemed a poison within the meaning of this Act ; and thereupon the said society shall submit the same for the approval of the Privy Council, and if such approval shall be given, then such resolution shall be advertised in the *London Gazette*, and on the expiration of one month from such advertisement the article named in such resolution shall be deemed to be a poison within the meaning of this Act.

APPENDIX.

Chemists and Druggists within the Meaning of this Act.

3. Chemists and druggists within the meaning of this Act shall consist of all persons who at any time before the passing of this Act have carried on in *Great Britain* the business of a chemist and druggist, in the keeping of open shop for the compounding of the prescriptions of duly qualified medical practitioners, also of all assistants and associates who before the passing of this Act shall have been duly registered under or according to the provisions of the Pharmacy Act, and also of all such persons as may be duly registered under this Act.

Apprentices and Assistants to be registered.

4. Any person who at the time of the passing of this Act shall be of full age, and shall produce to the registrar, on or before the thirty-first day of *December* one thousand eight hundred and sixty-eight, certificates according to Schedule (E.) to this Act that he had been for a period of not less than three years actually engaged and employed in the dispensing and compounding of prescriptions as an assistant to a pharmaceutical chemist, or to a chemist and druggist as defined by clause three of this Act, shall, on passing such a modified examination as the council of the Pharmaceutical Society with the consent of the Privy Council may declare to be sufficient evidence of his skill and competency to conduct the business of a chemist and druggist, be registered as a chemist and druggist under this Act.

Registration of Chemists and Druggists.

5. The persons who at the time of the passing of this Act shall have been duly admitted pharmaceutical chemists, or shall be chemists and druggists within the meaning of the Act, shall be entitled to be registered under the Act without paying any fee for such registration: Provided, however, as regards any such chemist and druggist, that his claim to be registered must be by notice in writing, signed by him, and given to the registrar, with certificates according to the Schedules (C.) and (D.) to this Act; and provided also, that for any such registration of a chemist and druggist, unless it be duly claimed by him on or before the thirty-first day of *December* one thousand eight hundred and sixty-eight, the person registered shall pay the same fee as persons admitted to the register after examination under this Act.

Examiners under Pharmacy Act to be the Examiners under this Act.
Certificate of competent Skill, &c.

6. All such persons as shall from time to time have been appointed to conduct examinations under the Pharmacy Act shall be and are hereby declared to be examiners for the purposes of this Act, and are hereby empowered and required to examine all such persons as shall tender themselves for examination under the provisions of this Act; and every person who shall have been examined by such examiners, and shall have obtained from them a certificate of competent skill and knowledge and qualification, shall be entitled to be registered as a chemist and druggist under this Act; and the examination aforesaid shall be such as is provided under the Pharmacy Act for the purposes of a qualification to be registered as assistant under that Act, or as the same may be varied from time to time by any byelaw to be made in accordance with the Pharmacy Act as amended by this Act; provided that no person shall conduct any examination for the purposes of this Act until his appointment has been approved by the Privy Council; and such appointment and approval shall not in any case be in force for more than five years; moreover it shall be the duty of the said Pharmaceutical Society to allow any officer appointed by the said Privy Council to be present during the progress of any examination held for the purposes of this Act.

Application of Fees to Purpose of Pharmaceutical Society.

7. Upon every such examination and registration as aforesaid such fees shall be payable as shall from time to time be fixed and determined by any byelaw to be made in accordance with the Pharmacy Act as amended by this Act, and shall be paid to the treasurer of the said society for the purposes of the said society.

APPENDIX. 311

Registrar under Pharmacy Act to be so under this Act.

8. The registrar appointed or to be appointed under or by virtue of the Pharmacy Act shall be registrar for the purposes of this Act.

Council of Pharmaceutical Society to make Orders for regulating Register to be kept.

9. The council of the Pharmaceutical Society shall, with all convenient speed after the passing of this Act, and from time to time as occasion may require, make orders or regulations for regulating the register to be kept under this Act as nearly as conveniently may be in accordance with the form set forth in the Schedule (B.) to this Act or to the like effect, and such register shall be called the Register of Chemists and Druggists.

Duty of Registrar to make and keep Register.

10. It shall be the duty of the registrar to make and keep a correct register, in accordance with the provisions of this Act, of all persons who shall be entitled to be registered under this Act, and to erase the names of all registered persons who shall have died, and from time to time to make the necessary alterations in the addresses of the persons registered under this Act: to enable the registrar duly to fulfil the duties imposed upon him, it shall be lawful for the registrar to write a letter to any registered person, addressed to him according to his address on the register, to inquire whether he has ceased to carry on business or has changed his residence, such letter to be forwarded by post as a registered letter, according to the Post Office regulations for the time being, and if no answer shall be returned to such letter within the period of six months from the sending of the letter, a second, of similar purport, shall be sent in like manner, and if no answer be given thereto within three months from the date thereof it shall be lawful to erase the name of such person from the register: Provided always, that the same may be restored by direction of the council of the Pharmaceutical Society should they think fit to make an order to that effect.

Notice of Death of Pharmaceutical Chemist or Chemist and Druggist to be given by Registrars.

11. Every registrar of deaths in *Great Britain*, on receiving notice of the death of any Pharmaceutical chemist, or chemist and druggist, shall forthwith transmit by post to the registrar under the Pharmacy Act a certificate under his own hand of such death, with the particulars of the time and place of death, and on the receipt of such certificate the said registrar under the Pharmacy Act shall erase the name of such deceased pharmaceutical chemist, or chemist and druggist, from the register, and shall transmit to the said registrar of deaths the cost of such certificate and transmission, and may charge the cost thereof as an expense of his office.

Evidence of Qualification to be given before Registration.

12. No name shall be entered in the register, except of persons authorized by this Act to be registered, nor unless the registrar be satisfied by the proper evidence that the person claiming is entitled to be registered; and any appeal from the decision of the registrar may be decided by the council of the Pharmaceutical Society; and any entry which shall be proved to the satisfaction of such council to have been fraudulently or incorrectly made may be erased from or amended in the register by order in writing of such council.

Annual Register to be published and be Evidence.

13. The registrar shall, in the month of *January* in every year, cause to be printed, published, and sold a correct register of the names of all pharmaceutical chemists, and a correct register of all persons registered as chemists and druggists, and in such registers respectively the names shall be in alphabetical order accord

ing to the surnames, with the respective residences, in the form set forth in Schedule (D) to this Act, or to the like effect, of all persons appearing on the register of pharmaceutical chemists, and on the register of chemists and druggists, on the thirty-first day of *December* last preceding, and such printed registers shall be called "The Registers of Pharmaceutical Chemists and Chemists and Druggists," and a printed copy of such registers for the time being, purporting to be so printed and published as aforesaid, or any certificate under the hand of the said registrar, and countersigned by the president or two members of the council of the Pharmaceutical Society, shall be evidence in all courts, and before all justices of the peace and others, that the persons therein specified are registered according to the provisions of the Pharmacy Act or of this Act, as the case may be, and the absence of the name of any person from such printed register shall be evidence, until the contrary shall be made to appear, that such person is not registered according to the provisions of the Pharmacy Act or of this Act.

Penalty on wilful Falsification of Register, or for obtaining Registration by false Representation.

14. Any registrar who shall wilfully make or cause to be made any falsification in any matter relating to the said registers, and any person who shall wilfully procure or attempt to procure himself to be registered under the Pharmacy Act or under this Act, by making or producing or causing to be made or produced any false or fraudulent representation or declaration, either verbally or in writing, and any person aiding or assisting him therein, shall be deemed guilty of a misdemeanor in *England*, and in *Scotland* of a crime or offence punishable by fine or imprisonment, and shall on conviction thereof be sentenced to be imprisoned for any term not exceeding twelve months.

Protection of Titles, and Restrictions on Sale of Poisons.

15. From and after the thirty-first day of *December* one thousand eight hundred and sixty-eight any person who shall sell or keep an open shop for the retailing, dispensing, or compounding poisons, or who shall take, use, or exhibit the name or title of chemist and druggist, or chemist or druggist, not being a duly registered pharmaceutical chemist, or chemist and druggist, or who shall take, use, or exhibit the name or title pharmaceutical chemist, pharmaceutist, or pharmacist, not being a pharmaceutical chemist, or shall fail to conform with any regulation as to the keeping or selling of poisons made in pursuance of this Act, or who shall compound any medicines of the British Pharmacopœia except according to the formularies of the said Pharmacopœia, shall for every such offence be liable to pay a penalty or sum of five pounds, and the same may be sued for, recovered, and dealt with in the manner provided by the Pharmacy Act for the recovery of penalties under that Act ; but nothing in this Act contained shall prevent any person from being liable to any other penalty, damages, or punishment to which he would have been subject if this Act had not passed.

Reserving Rights of certain Persons.

16. Nothing herein-before contained shall extend to or interfere with the business of any legally qualified apothecary or of any member of the Royal College of Veterinary Surgeons of *Great Britain*, nor with the making or dealing in patent medicines, nor with the business of wholesale dealers in supplying poisons in the ordinary course of wholesale dealing ; and upon the decease of any pharmaceutical chemist or chemist and druggist actually in business at the time of his death it shall be lawful for any executor, administrator, or trustee of the estate of such pharmaceutical chemist or chemist and druggist to continue such business if and so long only as such business shall be *bonâ fide* conducted by a duly qualified assistant, and a duly qualified assistant within the meaning of this clause shall be a pharmaceutical chemist or a chemist and druggist registered by the registrar under the Pharmacy Act or this Act; Provided always, that registration under this Act shall not entitle any person so registered to practice medicine or surgery, or any branch of medicine or surgery.

APPENDIX.

Regulations to be observed in the Sale of Poisons.

17. It shall be unlawful to sell any poison, either by wholesale or by retail, unless the box, bottle, vessel, wrapper, or cover in which such poison is contained be distinctly labelled with the name of the article and the word poison, and with the name and address of the seller of the poison ; and it shall be unlawful to sell any poison of those which are in the first part of Schedule (A.) to this Act, or may hereafter be added thereto under Section two of this Act, to any person unknown to the seller, unless introduced by some person known to the seller ; and on every sale of any such article the seller shall, before delivery, make or cause to be made an entry in a book to be kept for that purpose, stating, in the form set forth in Schedule (F.) to this Act, the date of the sale, the name and address of the purchaser, the name and quantity of the article sold, and the purpose for which it is stated by the purchaser to be required, to which entry the signature of the purchaser and of the person, if any, who introduced him shall be affixed ; and any person selling poison otherwise than is herein provided shall, upon a summary conviction before two justices of the peace in *England* or the Sheriff in *Scotland*, be liable to a penalty not exceeding five pounds for the first offence, and to a penalty not exceeding ten pounds for the second or any subsequent offence, and for the purposes of this section the person on whose behalf any sale is made by any apprentice or servant shall be deemed to be the seller ; but the provisions of this section, which are solely applicable to poisons in the first part of the Schedule (A.) to this Act, or which require that the label shall contain the name and address of the seller, shall not apply to articles to be exported from *Great Britain* by wholesale dealers, nor to sales by wholesale to retail dealers in the ordinary course of wholesale dealing, nor shall any of the provisions of this section apply to any medicine supplied by a legally qualified apothecary to his patient, nor apply to any article when forming part of the ingredients of any medicine dispensed by a person registered under this Act ; provided such medicine be labelled in the manner aforesaid, with the name and address of the seller, and the ingredients thereof be entered, with the name of the person to whom it is sold or delivered, in a book to be kept by the seller for that purpose ; and nothing in this Act contained shall repeal or affect any of the provisions of an Act of the Session holden in the fourteenth and fifteenth years in the reign of Her present Majesty, intituled *An Act to regulate the Sale of Arsenic.*

Chemists and Druggists in Business prior to passing of Act eligible for Election as Members of Pharmaceutical Society.

18. Every person who at the time of the passing of this Act is or has been in business on his own account as a chemist and druggist as aforesaid, and who shall be registered as a chemist and druggist, shall be eligible to be elected and continue a member of the Pharmaceutical Society according to the byelaws thereof ; but no person shall, in right of membership acquired pursuant to this clause, be placed on the register of pharmaceutical chemists, nor, save as is herein-after expressly provided, be eligible for election to the council of the Pharmaceutical Society.

Council of Pharmaceutical Society.

19. Every person who is or has been in business on his own account as a chemist and druggist as aforesaid at the time of the passing of this Act, and who shall become a member of the Pharmaceutical Society, shall be eligible for election to the council of the Pharmaceutical Society ; but the said council shall not at any time contain more than seven members who are not on the register of pharmaceutical chemists.

Chemists and Druggists registered eligible to be elected Associates, and, being in Business, have the Privilege of voting in the Society, on paying the same Subscriptions as Members.

20. Every person who shall have been registered as a chemist and druggist under this Act by reason of having obtained a certificate of qualification from the board of examiners shall be eligible to be elected an associate of the Pharmaceu-

tical Society, and every such person so elected and continuing as such associate, being in business on his own account, shall have the privilege of attending all meetings of the said society and of voting thereat, and otherwise taking part in the proceedings of such meetings, in the same manner as members of the said society: Provided always, that such associates contribute to the funds of the said society the same fees or subscriptions as members contribute for the time being under the byelaws thereof.

Voting Papers for Election of Council.

21. At all meetings of the Pharmaceutical Society at which votes shall be given for the election of officers all or any of the votes may be given either personally or by voting papers in a form to be defined in the byelaws of the said society, or in a form to the like effect, such voting papers being transmitted under cover to the secretary not less than one clear day prior to the day on which the election is to take place.

Benevolent Fund way be applied to past Members and Associates, also to Pharmaceutical Chemists and registered Chemists and Druggists.

22. And whereas by the Charter of Incorporation of the said Pharmaceutical Society it is provided that the council of the said society shall have the sole control and management of the real and personal property of the said society, subject to the byelaws thereof, and shall make provision thereout, or out of such part thereof as they shall think proper, for the relief of the distressed members or associates of the said society, and their widows and orphans, subject to the regulations and byelaws of the said society: And whereas, for extending the benefits which have resulted from the said provision in the said Charter of Incorporation, it is desirable that additional power should be granted to the said council: Be it enacted, that from and after the passing of this Act the said council may make provision out of the real and personal property aforesaid, and out of any special fund known as the benevolent fund, not only for the relief of the distressed members or associates of the said society and their widows and orphans, subject to the said regulations and byelaws, but also for all persons who may have been and have ceased to be members or associates of the said society, or who may be or have been duly registered as " pharmaceutical chemists" or " chemists and druggists," and the widows and orphans of such persons, subject to the regulations and byelaws of the said society.

Registration under " Medical Act."

23. Persons registered under " The Medical Act," shall not be or continue to be registered under this Act.

Adulteration of Food or Drink Act to extend to Medicines.

24. The provisions of the Act of the twenty-third and twenty-fourth of *Victoria*, chapter eighty-four, intituled *An Act for preventing the Adulteration of Articles of Food or Drink*, shall extend to all articles usually taken or sold as medicines, and every adulteration of any such article shall be deemed an admixture injurious to health; and any person registered under this Act who sells any such article adulterated shall, unless the contrary be proved, be deemed to have knowledge of such adulteration.

Acts of Privy Council.

25. On and after the passing of this Act all powers vested by the Pharmacy Act in one of Her Majesty's principal Secretaries of State shall be vested in the Privy Council, and the seventh section of the Public Health Act, 1858, shall apply to all proceedings and acts of the Privy Council herein authorized.

Power to Privy Council to erase Names of Persons from Register.

26. The Privy Council may direct the name of any person who is convicted of any offence against this Act which in their opinion renders him unfit to be on the register under this Act to be erased from such register, and it shall be the duty of the registrar to erase the same accordingly.

APPENDIX. 315

Extent of Act.
27. This Act shall not extend to *Ireland*.

Short Title.
28. This Act may be cited as the Pharmacy Act, 1868.

SCHEDULES.

SCHEDULE (A.)

Part 1.

Arsenic and its preparations.
Prussic acid.
Cyanides of potassium and all metallic cyanides.
Strychnine and all poisonous vegetable alkaloids and their salts.
Aconite and its preparations.
Emetic tartar.
Corrosive sublimate.
Cantharides.
Savin and its oil.
Ergot of rye and its preparations.

Part 2.

Oxalic acid.
Chloroform.
Belladonna and its preparations.
Essential oil of almonds unless deprived of its prussic acid.
Opium and all preparations of opium or of poppies.

SCHEDULE (B.)

Name.	Residence.	Qualification.
A.B.	*Oxford Street, London.*	In business prior to Pharmacy Act, 1868.
C.D.	*George Street, Edinburgh.*	Examined and certified.
E.F.	*Cheapside, London.*	Assistant prior to Pharmacy Act, 1868.

SCHEDULE (C.)

DECLARATION BY A PERSON WHO WAS IN BUSINESS AS A CHEMIST AND DRUGGIST IN GREAT BRITAIN BEFORE THE PHARMACY ACT, 1868.

To the Registrar of the Pharmaceutical Society of Great Britain.

I   , residing at   in the county of   hereby declare that I was in business as a chemist and druggist, in the keeping of open shop for the compounding of the prescriptions of duly qualified medical practitioners at   in the county of   , on or before the   day of   186 .

                                  Signed   (*Name.*)
   Dated this   day of   18 .

SCHEDULE (D.)

DECLARATION TO BE SIGNED BY A DULY QUALIFIED MEDICAL PRACTITIONER, OR MAGISTRATE, RESPECTING A PERSON WHO WAS IN BUSINESS AS A CHEMIST AND DRUGGIST IN GREAT BRITAIN BEFORE THE PHARMACY ACT, 1868.

To the Registrar of the Pharmaceutical Society of Great Britain.
I   , residing at   in the county of   ,
hereby declare that I am a duly qualified medical practitioner [or magistrate], and that to my knowledge   , residing at   in the county of   , was in business as a chemist and druggist, in the keeping of open shop for the compounding of the prescriptions of duly qualified medical practitioners before the   day of   186 .

(*Signed*)

SCHEDULE (E.)

DECLARATIONS TO BE SIGNED BY AND ON BEHALF OF ANY ASSISTANT CLAIMING TO BE REGISTERED UNDER THE PHARMACY ACT, 1868.

To the Registrar of the Pharmaceutical Society of Great Britain.
I hereby declare that the undersigned   , residing at   in the county of   , had for three years immediately before the passing of the Pharmacy Act, 1868, been employed in dispensing and compounding prescriptions as an assistant to a pharmaceutical chemist or chemist and druggist, and attained the age of twenty-one years.
As witness my hand, this   day of   186 .

      A. B., duly qualified Medical Practitioner.
      C. D., Pharmaceutical Chemist.
      E. F., Chemist and Druggist.
      G. H., Magistrate.

(*To be signed by one of the four parties named.*)

I hereby declare that I was an assistant to   of   in the county of   in the year   , and was for three years immediately before the passing of this Act actually engaged in dispensing and compounding prescriptions, and that I had attained the full age of twenty-one years at the time of the passing of the Pharmacy Act, 1868.

      N. O., Assistant.

SCHEDULE (F.)

Date.	Name of Purchaser.	Name and Quantity of Poison sold.	Purpose for which it is required.	Signature of Purchaser.	Signature of Person introducing Purchaser.

APPENDIX. 317

ANNO DECIMO SEXTO & DECIMO SEPTIMO
VICTORIÆ REGINÆ.

CAP. C.

An Act further to extend and make compulsory the practice of Vaccination. [20th *August* 1853.]

3 & 4 *Vict.* c. 29. 4 & 5 *Vict.* c. 32.

WHEREAS an Act was passed in the fourth year of the reign of Her present Majesty, intituled *An Act to extend the Practice of Vaccination:* and whereas an Act was passed in the fifth year of the same reign, intituled *An Act to amend an Act to extend the Practice of Vaccination:* and whereas it is expedient that the practice of vaccination should be still further extended: Be it therefore enacted by the Queen's most excellent Majesty, by and with the advice and consent of the Lords Spiritual and Temporal, and Commons, in this present Parliament assembled, and by the authority of the same, as follows :

Parishes or Unions to be divided into Districts for the Purpose of Vaccination, and Places appointed for the Performance of Vaccination.

I. Within six weeks after the passing of this Act the guardians of every parish or union, and the overseers of every parish in which relief to the poor shall not be administered by guardians, in *England* and *Wales*, shall, subject to the approval of the Poor Law Board, divide such parish or union, if need be, into convenient districts for the purpose of affording increased facilities for the vaccination of the poor, and shall appoint a convenient place in each such district for the performance of such vaccination, and shall take the most effectual means for giving from time to time to all persons resident within such district due notice of the days and hours at which the medical officer or practitioner contracted with for such purpose will attend at such place to vaccinate all persons not already successfully vaccinated who may then appear there, and also of the days and hours at which such medical officer or practitioner will attend at such place to inspect the progress of such vaccination in the persons so vaccinated.

Parents and Guardians of Children born after 1st *August,* 1853, *to have such Children vaccinated within Three or Four Months after Birth.*

II. The father or mother of every child born in *England* or *Wales* after the first day of *August* in the year of our Lord one thousand eight hundred and fifty-three shall within three calendar months after the birth of the said child, or in the event of the death, illness, absence, or inability of the father and mother, then the person who shall have the care, nurture, or custody of the said child shall within four calendar months after the birth of such child take or cause to be taken the said child to the medical officer or practitioner appointed in the union or parish in which the said child is resident according to the provisions of the first-recited Act for the purpose of being vaccinated, unless he shall have been previously vaccinated by some duly qualified medical practitioner, and the vaccination duly certified, and the said medical officer or practitioner so appointed shall and he is hereby required thereupon, or as soon after as it may conveniently and properly be done, to vaccinate the said child.

Children to be taken for Inspection by Medical Officer on Eighth Day after the Operation.

III. Upon the eighth day following the day on which any child has been vaccinated as aforesaid the father or mother, or other person having the care, nurture, or custody of the said child, shall again take or cause to be taken the said child to the medical officer or practitioner by whom the operation was performed, in order that such medical officer or practitioner may ascertain by inspection the result of such operation.

APPENDIX.

Certificate of successful Vaccination to be delivered.

IV. Upon and immediately after the successful vaccination of any child the medical officer or practitioner who shall have performed the operation shall deliver to the father or mother of the said child, or to the person who shall have the care, nurture, or custody of the said child, a certificate under his hand, according to the form of Schedule herein-after inserted, marked (A.), that the said child has been successfully vaccinated, and shall also transmit a duplicate of the said certificate to the registrar of births and deaths of the sub-district in which the operation was performed; and such certificate shall, without further proof, be admissible as evidence of the successful vaccination of such child in any information or complaint which shall be brought against the father or mother of the said child, or against the person who shall have the care, nurture, or custody of such child as aforesaid, for noncompliance with the provisions of this Act.

If the Child be not in a fit State for Vaccination, the Medical Officer to deliver a Certificate to that Effect, to be in force for Two Months.

V. If any medical officer or practitioner shall be of opinion that any child is not in a fit and proper state to be successfully vaccinated, he shall thereupon and immediately deliver, without fee or reward, to the father or mother of such child, or the person having the care, nurture, or custody of the said child, a certificate under his hand according to the form of Schedule herein-after inserted, marked (B.), that the child is in an unfit state for successful vaccination, and such certificate shall remain in force for two calendar months from its delivery as aforesaid; and the father or mother of the said child, or the person having the care, nurture, or custody of the said child, shall, unless they shall within each succeeding period of two months have obtained from a medical officer or practitioner a renewal of such certificate, within two months next after the delivery of the said certificate as aforesaid, and if the said child be not vaccinated at or by the termination of such period of two months then during each succeeding period of two calendar months until such child has been successfully vaccinated, take or cause to be taken to the said medical officer or practitioner such child to be vaccinated by him; and if the said medical officer or practitioner deem the said child to be then in a fit and proper state for successful vaccination, he shall forthwith vaccinate it accordingly, and shall deliver to the father or mother of such child, or person having the care, nurture, or custody of such child a certificate under his hand according to the form of Schedule herein-after inserted, marked (A.), that such child has been successfully vaccinated; but if the said medical officer or practitioner be of opinion that the child is still in an unfit state for successful vaccination, then he shall again deliver to the father or mother of such child, or person having the care, nurture, or custody of the said child a certificate under his hand, according to the said form of Schedule (B.), that the child is still in an unfit state for successful vaccination; and the said medical officer or practitioner, so long as such child remains in an unfit state for vaccination and unvaccinated, shall at the expiration of every succeeding period of two calendar months deliver, if required, to the said father or mother of such child, or person having the care, nurture, or custody of such child, a fresh certificate under his hand according to the said form of schedule; and the production of such certificate shall be a sufficient defence against any complaint which shall be brought against the said father or mother, or person having the care, nurture, or custody of such child for noncompliance with the provisions of this Act.

Rules of Payment for Vaccination in Contracts by Guardians or Overseers.

VI. In all contracts to be hereafter made under the provisions of the first-recited Act by any guardians or overseers of the poor with any medical officers or practitioners for the vaccination of the persons resident in their respective unions or parishes the sums contracted to be paid shall not be less than the following rates; that is to say, for every person successfully vaccinated at the residence of such medical officer or practitioner, or within two miles therefrom by the nearest public road, a sum not less than one shilling and sixpence, and for every person successfully vaccinated at any place more than two miles distant from such residence any sum not less than two shillings and sixpence.

Child's Incapacity to receive the Vaccine Disease to be certified.

VII. In the event of any medical practitioner acting under the provisions of this Act being of opinion that any child that has been vaccinated by him is insusceptible of the vaccine disease, he shall deliver to the father or mother, or person having the care, nurture, or custody of such child, a certificate under his hand according to the form of Schedule herein-after inserted, marked (D.); and the production of such certificate shall be a sufficient defence against any complaint which may be brought against the said father, mother, or person having the care, nurture, or custody of such child for noncompliance with the provisions of this Act.

Registrars of Births and Deaths to keep a Register of Cases of successful Vaccination, of which Searches and Extracts may be made.

VIII. The registrar of births and deaths in every sub-district in which the operation has been performed shall keep a register of the persons of whose successful vaccination a certificate shall have been transmitted to him as above provided by the said medical officer or practitioner, and shall at all reasonable times allow searches to be made of any such register book in his keeping, and shall give a copy, certified under his hand, of any entry or entries in the same, on payment of the fee of one shilling for each search, and sixpence for each certificate.

Notice to be given of the Requirement of Vaccination, and on Failure of Parent or Guardian to comply therewith, Penalty.

IX. The registrar of births and deaths in every sub-district shall, on or within seven days after the registration of the birth of any child not already vaccinated within the said sub-district, give notice in writing in manner herein-after directed, and according to the form of Schedule herein-after inserted, marked (C.), to the father or mother of such child, or in the event of the death, illness, absence, or inability from sickness or otherwise of the father and mother, then to the person upon whom the care, nurture, or custody of such child shall have devolved, that it is the duty of such father or mother, or person having the care, nurture, or custody of such child as aforesaid, to take care that the said child shall be vaccinated in the manner directed by this Act, and shall together therewith deliver to such person a notice of the days, hours, and places within the district of such registrar at which the medical officer or practitioner as herein-before provided will attend for the purpose of vaccination; and if after such notice the father or mother of the said child, or the person so having as aforesaid the care, nurture, or custody of the said child, shall not cause such child to be vaccinated, or shall not on the eighth day after the vaccination has been performed take or cause to be taken such child for inspection according to the provisions in this Act respectively contained, then such father or mother, or person having the care, nurture, or custody of such child as aforesaid, so offending, shall forfeit a sum not exceeding twenty shillings.

Fee to Registrar.

X. A fee of threepence shall be paid to such registrar for each child vaccinated in respect of which he shall have performed the duties required in this Act; and he shall keep a book, to be provided as herein-after directed, containing a minute of his having duly given such notice as herein-before directed; and the said fee shall be payable in the same manner as the fee now payable to such registrar for registering the birth of such child as aforesaid is paid.

Registrar General to provide Books and Forms for carrying out the Provisions of this Act.

XI. The Registrar General for *England* and *Wales* shall and he is empowered and directed, within two months after the passing of this Act, to frame and provide such books, forms, and regulations as he may deem requisite for carrying into full effect the provisions of this Act, and shall transmit the same to the superintendent registrars of each district in *England* and *Wales*, who shall deliver to the medical officers so appointed as aforesaid, and other duly qualified medical practitioners in the said district, such of the said books, forms, and regulations as they may require for the performance of the duties imposed upon them by this Act ; and the expenses to be incurred by the Registrar General under the provisions of this Act shall be defrayed in the same manner as the expenses under the Act of the sixth and seventh years of King *William* the Fourth, chapter eighty-five.

Recovery of Penalties.

XII. All penalties by this Act imposed shall be recoverable before any two justices of the peace for the county, city, borough, or place where the offence may have been committed ; and the provisions of the Act of the twelfth year of Her present Majesty, chapter forty-three, shall be applicable to the recovery of such penalties.

Application of Penalties.

XIII. All penalties recovered under this Act shall be applied in aid of the funds applicable to the relief of the poor in the parish or place maintaining its own poor wherein the offence may have been committed.

SCHEDULES referred to by this Act.

SCHEDULE (A.)

I, the undersigned, hereby certify, that the child of
aged of the parish of in the county of
has been successfully vaccinated by me.
Dated this day of 185 .
 (Signed) *A.B.*
 Surgeon of the union *or* parish
 (*or other medical practitioner,
 as the case may be*).

SCHEDULE (B.)

I, the undersigned, hereby certify, that I am of opinion that
the child of of the parish of in the county
of aged is not now in a fit and proper state
to be successfully vaccinated, and I do hereby postpone the vaccination until
the day of
Dated this day of 185 .
 (Signed) *A.B.*
 Surgeon of the union *or* parish
 (*or other medical practitioner,
 as the case may be*).

SCHEDULE (C.)

I, the undersigned, hereby give you notice, and require you to have *C.D.* vaccinated within three (or four, *as the case may be, according to the second section of this Act,*) months after the birth, pursuant to the provisions and directions of the Act of the 16 Victoria, cap. . As witness my hand this day of 185 .
 J.B.
 Registrar of births and deaths for
 the sub-district (*as
 the case may be*).

APPENDIX. 321

SCHEDULE (D.)

I, the undersigned, hereby certify, that I am of opinion, that the child of   of the parish of   in the county of   is insusceptible of the vaccine disease.
Dated this   day of   18 .
Signed, *A. B.*, surgeon of the union or parish of   (*or other medical practitioner, as the case may be*).

ANNO TRICESIMO & TRICESIMO PRIMO VICTORIÆ REGINÆ.

CAP. LXXXIV.

An Act to consolidate and amend the Laws relating to Vaccination.
[12th *August* 1867.]

WHEREAS it is expedient to consolidate and amend the statutes relating to vaccination in *England:* Be it therefore enacted by the Queen's most excellent Majesty, by and with the advice and consent of the Lords Spiritual and Temporal, and Commons, in this present Parliament assembled, and by the authority of the same, as follows:

Acts and Parts of Acts herein named repealed on and after January 1, 1868.

1. From and after the day when this Act shall come into operation as hereinafter provided, the statute of the third and fourth years of the reign of Her Majesty, chapter twenty-nine, that of the fourth and fifth years of the same reign, chapter thirty-two, that of the sixteenth and seventeenth years of the same reign, chapter one hundred, the seventh section of the statute of the twenty-first and twenty-second years of the same reign, chapter twenty-five, the second section of the statute of the twenty-first and twenty-second years of the same reign, chapter ninety-seven, and the statute of the twenty-fourth and twenty-fifth years of the same reign, chapter fifty-nine, shall be repealed,—

Exceptions.

Except in regard to the divisions and districts of unions and parishes previously made, and to all contracts under the said statutes then in force, and to all Acts and proceedings duly commenced under the same, and not then completed, and except in regard to all liabilities and responsibilities incurred under the same, all which shall remain in as full force as if the same statutes had not been repealed, unless they be in any respect inconsistent with the provisions herein contained.

Guardians to divide Unions and Parishes into Vaccination Districts, or to consolidate or alter them, subject to Approval of the Poor Law Board.

2. The guardians of every union or parish where the same shall not have been divided into districts for the purpose of vaccination shall, unless such union or parish respectively shall be of so limited an area as not to require subdivision, in which case the same shall be treated as a vaccination district within the meaning hereof, forthwith divide the union or parish for which they act into districts for the performance of vaccination; and when the Poor Law Board shall by their order require any districts for the time being to be consolidated or otherwise altered, the guardians shall proceed to consolidate or alter the same, and they shall in every such case of division, consolidation, or alteration report their proposal to the Poor Law Board for their approval, which board shall approve or disapprove of the same as they see fit; and the guardians of every union or parish may, with like

Y

approval, from time to time as they shall find it requisite, alter the districts heretofore formed or hereafter to be formed for the purpose of vaccination.

If the Board do not approve, another Scheme to be prepared; when approved, Guardians to contract for Performance of Vaccination.

3. If the said board disapprove of the proposal the guardians shall forthwith proceed to prepare another, and submit the same to the said board for approval, and so on from time to time as shall be requisite until their proposal shall be approved, and when the said board shall have approved of the same the guardians shall enter into a contract with some duly registered medical practitioner for the performance of vaccination of all persons resident within each district; and every such medical practitioner shall be termed the public vaccinator of the district; and as and when the contracts now existing shall determine the guardians shall enter into others, with such modifications as the circumstances shall render necessary, subject to the like approval of the Poor Law Board as aforesaid.

Qualification of Vaccinator to be prescribed by Lords of the Privy Council, and other Regulations to be prescribed by them.

4. No person shall be appointed a public vaccinator, or act as deputy for a public vaccinator who shall not possess the qualification heretofore prescribed by the Lords of Her Majesty's Council, or such as shall be from time to time hereafter prescribed by them, except when such Lords shall upon sufficient cause sanction any departure from their directions; and all such regulations as the said Lords have heretofore made or shall hereafter make, which they are hereby authorized to make, to secure the efficient performance of vaccination or the provision and supply of vaccine lymph by the public vaccinator, and all such directions or regulations as the said Lords acting under any Act for the prevention of diseases may issue in relation to smallpox, shall be duly observed by the several persons to whom they apply; and the said Lords may from time to time cause such inquiries to be made relating to the observance of such regulations and to the execution of this Act as to them shall seem fit, and shall direct how any money hereafter to be provided by Parliament for or towards defraying the expenses of the national vaccine establishment, or otherwise providing for the supply of vaccine lymph, shall be applied.

As to Allowances to Public Vaccinators.

5. On reports made to the Lords of Her Majesty's Council with regard to the number and quality of the vaccinations performed in the several vaccination districts of *England*, or any of them, the said Lords may from time to time, out of monies provided by Parliament, and under regulations to be approved by the Lords Commissioners of Her Majesty's Treasury, authorize to be paid to any public vaccinators, in addition to the payments received by them from guardians or overseers, further payments not exceeding in any case the rate of one shilling for each child whom the vaccinator has successfully vaccinated during the time to which the award of the said Lords of the Council relates.

As to Fees payable for Vaccination.

6. Every such contract for vaccination shall provide for payment in respect only of the successful vaccination of persons, and so that the rate of payment for primary vaccinations shall be not less than the following; that is to say, for every such vaccination done at an appointed station situated at or within one mile from the residence of the vaccinator, or in the workhouse of the union or parish, not less than one shilling and sixpence; and for every such vaccination done at any station over one mile and under two miles distant from his residence, not less than two shillings; and for every such vaccination done at any station over two miles distant from his residence, not less than three shillings; such distance being measured according to the nearest public carriage road; but in respect of successful vaccinations performed elsewhere than at a station or in the workhouse as aforesaid, the payment shall be according to the terms specified in the contract as approved of by the Poor Law Board.

APPENDIX. 323

Conditions may be imposed in the Contracts to secure due Vaccination of Persons.

7. The guardians shall, with the consent of the Poor Law Board, make stipulations and conditions in their contracts to secure the due vaccination of persons, the observance of the provisions of this Act with regard to the transmission of the certificate of successful vaccination, and the fulfilment of all other provisions of this Act on the part of the public vaccinator, and shall provide all stations at which the vaccination shall be appointed to be performed other than the surgery or residence of the public vaccinator.

Provision for Re-vaccination.

8. The provisions of the contracts entered into before this Act comes into operation shall not after the thirty-first day of *December* next apply to the cases of persons who having been previously successfully vaccinated shall be re-vaccinated, but if the Lords of Her Majesty's Council shall have issued or shall hereafter issue regulations in respect of the re-vaccination of persons who may apply to be re-vaccinated, which such Lords are hereby authorized to do, the guardians shall pay in respect of every case of successful re-vaccination performed in conformity with such regulations under such contracts or under new contracts entered into after the date hereof a sum amounting to two-thirds of the fee payable upon each case of successful primary vaccination.

Contract not valid unless approved of by the Poor Law Board, who may determine the same at any Time.

9. No contract for vaccination entered into under the provisions of this Act shall be valid until the same shall have been approved of by the Poor Law Board, and such board may, at their discretion, upon the application of the Lords of Her Majesty's Council or otherwise, at any time after the same shall have been approved of by them, determine it either forthwith or at a future day.

No Payment to be made out of the Poor Rate or any other public Fund unless the Poor Law Board have approved of the Contract.

10. No payment in respect of vaccination shall be made out of the common fund of any union, or out of the poor rate of any parish, or out of any other public or parochial fund, where the Poor Law Board shall not have approved of a contract for the performance thereof, or after they shall have determined any such contract; and every payment made contrary hereto shall be disallowed by the auditor in the accounts of every board of guardians, or of the overseers, or of any officer who shall have made the same.

No Public Vaccinator to be paid for Vaccination out of his District.

11. Where a district shall have been or shall be assigned to a vaccinator, he shall not be entitled to be paid a fee in respect of the vaccination or re-vaccination of any child or other person resident out of his district, except in case of a vacancy in the office of vaccinator in any adjoining district, or of the default of the vaccinator therein, of which default notice shall have been given to him in writing by the guardians, or when a relieving officer of his union or parish shall in writing refer any child to him for vaccination.

Provision for Districts in particular Places of scanty Population.

12. The guardians may with the consent of the Poor Law Board provide in districts where the population is scanty or much scattered, or where some peculiar circumstances may render it expedient for them to do so, for the attendance of the public vaccinator at the appointed places after intervals exceeding three months; and if by reason of such intervals the vaccination of any child cannot be performed within the respective periods herein prescribed, no parent or other person who would otherwise be liable shall be liable to any penalty in respect of a neglect to procure the vaccination during any such period; but every such parent or other

Y 2

person shall be bound to procure such vaccination to be performed at the time and place so appointed before the commencement of the next interval, unless it be otherwise performed by a medical practitioner as herein provided, or unless the child shall be certified to be then in an unfit state for or insusceptible of vaccination.

Guardians to give Notice of Alteration in Districts.

13. When the guardians make any alteration in a vaccination district, or otherwise in the local arrangements for vaccination, they shall give public notice of such alteration by printed papers to be affixed in the districts affected by such alteration for one month prior to the alteration taking effect.

The Registrar-General to provide Forms.

14. The Registrar-General for *England* and *Wales* shall, when he shall deem it necessary, from time to time as occasion shall require, after the passing of this Act, frame and provide appropriate books, forms, and regulations for the use and guidance of the registrars in the exercise of their duties therein prescribed, and also such forms as shall be required for the use of the public vaccinators and the signature of the medical practitioners under the provisions of this Act, and shall transmit the same to all registrars of births and deaths, who shall retain such as relate to themselves, and distribute among the vaccinators within their respective districts such as relate to them without any fee or reward.

Registrar of Births to deliver Notice of Vaccination to Parent or other Person registering Birth.

15. The registrar of births shall, on or within seven days after the registration with him of the birth of any child not already vaccinated, give a notice, according to the form in the schedule hereto annexed marked A., or to the like effect, to the parent, or, in the event of the death, illness, absence, or inability of the parent, to the person having the custody of such child, if known to him, requiring such child to be duly vaccinated according to the provisions of this Act, and specifying the days, hours, and places where the public vaccinator of the vaccination district wherein such child resides, or the vaccinator of any station duly authorized by the Lords of Her Majesty's Council, will attend for the purpose of performing the operation, to which notice forms according to those in the said schedule marked B., C., and D., and also the address of the registrar giving the notice, shall be attached in such form as the Registrar-General shall deem most convenient.

Parent or other Person to procure the Vaccination of Child within Three Months.

16. The parent of every child born in *England* shall within three months after the birth of such child, or where, by reason of the death, illness, absence, or inability of the parent or other cause, any other person shall have the custody of such child, such person shall, within three months after receiving the custody of such child, take it or cause it to be taken to the public vaccinator of the vaccination district in which it shall be then resident, according to the provisions of this or any other Act, to be vaccinated, or shall within such period as aforesaid cause it to be vaccinated by some medical practitioner; and the public vaccinator to whom such child shall be so brought is hereby required, with all reasonable despatch, subject to the conditions herein-after mentioned, to vaccinate such child.

Provision for Inspection of Vaccination.

17. Upon the same day in the following week when the operation shall have been performed by the public vaccinator such parent or other person, as the case may be, shall again take the child or cause it to be taken to him or to his deputy that he may inspect it, and ascertain the result of the operation, and if he see fit, take from such child lymph for the performance of other vaccinations; and in the event of the vaccination being unsuccessful such parent or other person shall, if the vaccinator so direct, cause the child to be forthwith again vaccinated and inspected as on the previous occasion.

Provision for the Unfitness of the Child for Vaccination.

18. If any public vaccinator or medical practitioner shall be of opinion that the child is not in a fit and proper state to be successfully vaccinated he shall forthwith deliver to the parent or other person having the custody of such child a certificate under his hand according to the form in the schedule hereto annexed marked B., or to the like effect, that the child is then in a state unfit for successful vaccination, which certificate shall remain in force for two months, and shall be renewable for successive periods of two months until a public vaccinator or medical practitioner shall deem the child to be in a fit state for successful vaccination, when the child shall, with all reasonable despatch, be vaccinated, and the certificate of successful vaccination duly given if warranted by the result.

Provision for successive Certificates.

19. At or before the end of each successive period the parent or such person as aforesaid shall take or cause the child to be taken to some public vaccinator or medical practitioner, who shall then examine the child, and give the certificate according to the said form B., so long as he deems requisite under the circumstances of the case.

Provision for Insusceptibility of successful Vaccination.

20. If any such public vaccinator or medical practitioner shall find that a child whom he has three times unsuccessfully vaccinated is insusceptible of successful vaccination, or that a child brought to him for vaccination has already had the smallpox, he shall deliver to the parent or other person as aforesaid a certificate under his hand according to the form in the schedule hereunto annexed marked C., or to the like effect, and the parent or such person as aforesaid shall thenceforth not be required to cause the child to be vaccinated.

Certificate of successful Vaccination to be transmitted to the Registrar, and a Duplicate given to the Parent.

21. Every public vaccinator who shall have performed the operation of vaccination upon any child, and have ascertained that the same has been successful, shall, within twenty-one days after the performance of the operation, transmit by post or otherwise a certificate according to form D. in the same schedule, or to the like effect, certifying that the said child has been successfully vaccinated, to the registrar of births and deaths in the district within which the birth was registered, but if such district be not known to him, or if the birth of the child shall not have been registered, to the registrar within whose district the operation shall have been performed, and upon request shall deliver a duplicate thereof to the parent or other person as aforesaid.

No Fee to be charged for Certificate.

22. No fee or remuneration shall be charged by the public vaccinator to the parent or other person for any such certificate or duplicate certificate as aforesaid, nor for any vaccination done under his contract, nor shall he be entitled to payment under his contract for any vaccination in respect of which he shall have been paid by the parent or other person for whom or on whom it is performed; and if he should have received payment under his contract he shall not be entitled to recover payment for the vaccination from any other person.

Parent, &c. to transmit Certificate of successful Vaccination by Medical Practitioner to Registrar of District.

23. Where the vaccination shall be successfully performed by a medical practitioner not being a public vaccinator, the parent or other person as aforesaid causing the child to be vaccinated shall submit a certificate according to the said form marked D. to such medical practitioner, to be filled up and signed by him, and shall within twenty-one days after the performance of the operation transmit the

same so signed, by post or otherwise, to the registrar of the district where the birth of such child was registered, or if such child shall not have been registered, or the district of the registration shall not be known to such parent or other person, to the registrar of the district in which the operation shall have been performed.

Registrar to keep Books and Register of Vaccination to be open to Searches. Fees for Searches and Copies. Proviso.

24. Every registrar shall keep a book in which he shall enter minutes of the notices of vaccination given by him as herein required, and also register the certificates transmitted to him as herein provided, and shall at all reasonable times allow searches to be made therein, and upon demand give a copy under his hand or under that of his deputy of any entry in the same, on payment of a fee of sixpence for each search and threepence for each copy ; and every registrar shall receive a fee of one penny in respect of every child whose birth he shall have registered, and in respect of whom he shall give the notice as aforesaid, and another fee of threepence in respect of every such child whose certificate he shall have registered as herein provided, and he shall receive a fee of one penny in respect of each child whose certificate he shall have registered without having registered the birth : Provided that no fee shall be charged for any search made by a public vaccinator, or any officer of the guardians, authorized by them to make such search, or any inspector appointed by the Poor Law Board or the Lords of Her Majesty's Council.

Registrar to be paid Fees by the Boards of Guardians.

25. The registrar shall make out an account of the fees to which he shall be entitled under this Act at the usual quarter days of the year, and submit the same to the guardians of the union or parish for which he acts, and they shall, after examining the same and comparing with the register of successful vaccinations kept by him and finding the account to be correct, forthwith pay the amount of the same out of the funds in their possession.

Vaccination declared to be not Parochial Relief so as to disqualify.

26. It is hereby declared, that the vaccination, or the surgical or medical assistance incident to the vaccination, of any person in a union or parish, heretofore or hereafter performed or rendered by a public vaccinator, shall not be considered to be parochial relief, alms, or charitable allowance to such person or his parent, and no such person or his parent shall by reason thereof be deprived of any right or privilege, or be subject to any disability or disqualification.

Half-yearly Proceedings by Registrars and Guardians.

27. The registrar of each district shall, within one week after the first day of *January* and the first day of *July* in each year, make a list of all cases in which certificates of vaccination have not been duly received by him during the last preceding half year, and shall submit the same to the next meeting of the guardians of the union or parish wherein he acts, and the said guardians shall forthwith make inquiry into the circumstances of the cases contained in the list, and if they find that the provisions of the Act have been neglected shall cause proceedings to be taken against the persons in default.

Power to Guardians to pay certain Expenses out of their Funds.

28. The guardians of any union or parish may pay out of their funds all reasonable expenses incurred by them in causing notices to be printed and circulated as to the provisions of this Act, and in and about inquiries and reports as to the state of smallpox or vaccination in their union or parish, and in taking measures to prevent the spread of smallpox and to promote vaccination upon any actual or expected outbreak of that disease therein, and may pay any officer appointed by them to prosecute persons charged with offences against this Act, or otherwise to enforce its provisions.

APPENDIX.

Penalty on Parent, &c. neglecting to procure Vaccination of the Child.

29. Every parent or person having the custody of a child who shall neglect to take such child or to cause it to be taken to be vaccinated, or after vaccination to be inspected, according to the provisions of this Act, and shall not render a reasonable excuse for his neglect, shall he guilty of an offence, and be liable to be proceeded against summarily, and upon conviction to pay a penalty not exceeding twenty shillings.

Penalty on Vaccinator and Parent neglecting to transmit Certificate, and Persons signing false Certificates.

30. Every public vaccinator, parent or person, as the case shall require, who shall neglect to transmit any certificate required of him by the provisions of this Act completely filled up and legibly written to the registrar within the time herein specified, and every public vaccinator who shall refuse to deliver the duplicate to the parent or other person, on request, and every medical practitioner who shall refuse to fill up and sign the certificate of successful vaccination when submitted to him as aforesaid, shall be liable to pay upon a summary conviction a penalty not exceeding twenty shillings; and every person who shall wilfully sign a false certificate or duplicate under this Act shall be guilty of a misdemeanor, and punishable accordingly.

Justices may make an Order for the Vaccination of any Child under 14 Years. Penalty for Disobedience. Proviso for Costs to Person improperly summoned.

31. If any registrar, or any officer appointed by the guardians to enforce the provisions of this Act, shall give information in writing to a justice of the peace that he has reason to believe that any child under the age of fourteen years, being within the union or parish for which the informant acts, has not been successfully vaccinated, and that he has given notice to the parent or person having the custody of such child to procure its being vaccinated, and that this notice has been disregarded, the justice may summon such parent or person to appear with the child before him at a certain time and place, and upon the appearance, if the justice shall find, after such examination as he shall deem necessary, that the child has not been vaccinated, nor has already had the smallpox, he may, if he see fit, make an order under his hand and seal directing such child to be vaccinated within a certain time; and if at the expiration of such time the child shall not have been so vaccinated, or shall not be shown to be then unfit to be vaccinated, or to be insusceptible of vaccination, the person upon whom such order shall have been made shall be proceeded against summarily, and, unless he can show some reasonable ground for his omission to carry the order into effect, shall be liable to a penalty not exceeding twenty shillings:

Provided that if the justice shall be of opinion that the person is improperly brought before him, and shall refuse to make any order for the vaccination of the child, he may order the informant to pay to such person such sum of money as he shall consider to be a fair compensation for his expenses and loss of time in attending before the justice.

Penalty upon Persons inoculating with Smallpox.

32. Any person who shall after the passing of this Act produce or attempt to produce in any person by inoculation with variolous matter, or by wilful exposure to variolous matter, or to any matter, article, or thing impregnated with variolous matter, or wilfully by any other means whatsoever produce, the disease of smallpox in any person, shall be guilty of an offence, and shall be liable to be proceeded against summarily, and upon conviction to be imprisoned for any term not exceeding one month.

11 & 12 Vict. c. 43., except Sect. 11., Sect. 59 of 7 & 8 Vict. c. 101., and Sect. 9 of 28 & 29 Vict. c. 79. to apply to these Proceedings.

33. The statute of the eleventh and twelfth *Victoria*, chapter forty-three, except section eleven, shall apply to all proceedings to be taken under this Act; and the

justices for the county, city, borough, or other place where the offence shall have been committed shall have jurisdiction to hear and determine the complaint, and where a union or parish shall be comprised in several jurisdictions the complaint as to any matter arising in such union or parish may be heard and determined in any one of such jurisdictions ; and all prosecutions undertaken by the guardians or their officers or any registrar under this Act shall be deemed to be within the operation of the seventh and eighth *Victoria*, chapter one hundred and one, section fifty-nine, and the Union Chargeability Act of 1865, section nine.

Notice not to be proved by Prosecutors. Certificates to be Defence.

34. In any prosecution for neglect to procure the vaccination of a child, it shall not be necessary in support thereof to prove that the defendant had received notice from the registrar or any other officer of the requirements of the law in this respect ; but if the defendant produce any such certificate as herein-before described, or the register of vaccinations kept by the registrar as herein-before provided, in which the certificate of successful vaccination of such child shall be duly entered, the same shall be a sufficient defence for him, except in regard to the certificate marked B., when the time specified therein for the postponement of the vaccination shall have expired before the time when the information shall have been laid.

Interpretation of Terms.

35. The word "parent" shall include the father and mother of a legitimate child and the mother of an illegitimate child; "medical practitioner" shall mean a registered medical practitioner ; and the several words herein contained shall be construed, except where any inconsistency would ensue from such construction, in the same manner as in the several Acts for the amendment of the law for the relief of the poor.

Sect. 7 of 21 & 22 Vict. c. 97. to apply to Acts of Privy Council.

36. The seventh section of the Public Health Act, 1858, shall apply to all the proceedings and acts of the Lords of Her Majesty's Council herein authorized.

Commencement of Act. Short Title.

37. This Act shall come into operation on the first day of *January* next, and may be cited as " The Vaccination Act of 1867."

SCHEDULE OF FORMS.

A.

I, the undersigned hereby give you notice to have the child (*insert name*), whose birth is now registered, vaccinated within three months from the date of its birth, pursuant to the provisions and directions of the Vaccination Act ; and that in default of your doing so you will be liable to the penalties thereby imposed for neglect of those provisions.

If you intend to apply to the public vaccinator of your district I have to inform you that he will attend at   on
at the hour of

You are required to produce to the public vaccinator or medical practitioner who may be applied to the forms herewith supplied for him to fill up and sign ; and if the operation be performed by a medical practitioner who is not the public vaccinator, you must transmit to me by post or otherwise the certificate signed by him within twenty-one days after the performance of the operation, or you will be liable to a penalty of twenty shillings, to be recovered on a summary conviction.

Dated this   day of   18 .
     (Signed)   *C.D.*,
           registrar of births and deaths for the sub-district of
              in the   union *or* parish.

B.

I, the undersigned, hereby certify, that I am of opinion that the child of   of   in the parish or township of   in the county or borough of aged   is not now in a fit and proper state to be successfully vaccinated. I do hereby postpone the vaccination until the   day of (a).

Dated this   day of   18 .

  (Signed) *A.B.*,
    public vaccinator of the   union
    or parish.
    or *A.B.*,   of
    medical practitioner (*i.e.* M.D., L.A.C., or F.R.C.S.,
    or otherwise, as the case may be).

Mem.—*This is to be kept by the parent or other person to whom it is given.*

(a) *This must not exceed two calendar months from the date of the certificate.*

C.

I, the undersigned, hereby certify, that I have   times unsuccessfully vaccinated   the child of of   in the parish or township of   in the county or borough of   aged   , [or that the child has already had smallpox, *as the case may be*,] and I am of opinion that such child is insusceptible of successful vaccination.

Dated this   day of   18 .

  (Signed) *A.B.*,
    public vaccinator of the   union
    or parish.
    or *A.B.*,   of
    medical practitioner (*i.e.* M.D., L.A.C., or F.R.C.S., or
    otherwise, as the case may be).

Mem.—*This is to be kept by the parent or other person to whom it is given.*

D.

I, the undersigned, hereby certify, that   the child of   aged   of   in the parish or township of   in the county or borough of   has been successfully vaccinated by me.

Dated this   day of   18 .

  (Signed) *A.B.*,
    public vaccinator of the   union
    or parish.
    or *A.B.*,   of
    medical practitioner (*i.e.* M.D., L.A.C., or F.R.C.S., or
    otherwise, as the case may be).

NOTICE.—*This certificate is to be transmitted within twenty-one days from the performance of the operation by the public vaccinator to the registrar of the district in which the birth was registered, or, if that be not known to him, to the registrar of* the district in which the operation was performed. A duplicate is to be given to the parent, or other person, procuring the vaccination, if required.

When the vaccination is performed by a medical practitioner, not the public vaccinator of the district, he is to fill up and sign this certificate, and the parent or such other person is within the same time to transmit it to the registrar with whom the birth was registered, or, if his district be not known to such parent or other person, to the registrar of the district in which the operation was performed.

The transmission may be by post or otherwise.

In each case the Vaccination Act of 1867 imposes a penalty of twenty shillings for default.

ANNO VICESIMO NONO VICTORIÆ REGINÆ.

CAP. XXXV.

An Act for the better Prevention of Contagious Diseases at certain Naval and Military Stations. [11th *June* 1866.]

BE it enacted by the Queen's most excellent Majesty, by and with the advice and consent of the Lords Spiritual and Temporal, and Commons, in this present Parliament assembled, and by the authority of the same, as follows :

PRELIMINARY.

Short Title.

1. This Act may be cited as The Contagious Diseases Act, 1866.

Intrepretation of Terms.

2. In this Act—

The term "contagious disease" means venereal disease, including gonorrhœa :

The term "police" means Metropolitan Police or other police or constabulary authorized to act in any part of any place to which this Act applies :

The term "superintendent" includes inspector :

The term "chief medical officer" means the principal physician or surgeon for the time being attached to or doing duty at a hospital, or the house surgeon or resident surgeon of the hospital :

The term "justice" means a justice of the peace having jurisdiction in the county, borough, or place where the matter requiring the cognizance of a justice arises, or in any part of any place to which this Act applies :

The term "two justices" means two or more justices assembled and acting together, and includes any police or stipendiary magistrate or other justice having by law for any purpose the powers of two justices.

Act to commence from Sept. 30, 1866, and then 27 & 28 Vict. c. 85. to cease to operate, except, &c.

3. This Act shall commence from and immediately after the thirtieth day of September one thousand eight hundred and sixty-six, and on the commencement of this Act the Contagious Diseases Prevention Act, 1864, shall cease to operate ; but the discontinuance of that Act by this Act shall not affect the validity or invalidity of anything done or suffered before the commencement of this Act ; and that discontinuance or anything in this Act shall not apply to or in respect of any offence, act, or thing committed or done or omitted before the commencement of this Act ; and every such offence, act, or thing shall after and notwithstanding the commencement of this Act have the same consequences and effect in all respects as if the Contagious Diseases Prevention Act, 1864, had not been discontinued.

Every order of a justice under the said Act shall remain in force as if this Act had not been passed.

APPENDIX. 331

Every hospital certified under the said Act shall continue to be a certified hospital, for the purposes of this Act, for three months after the commencement of this Act, unless before the expiration of that time the certificate is withdrawn or the hospital is certified under this Act; and every hospital certified under this Act shall be deemed a certified hospital for the purposes of the said Act, as long as the operation thereof continues for any purpose under this Act.

EXTENT OF ACT.

Act to extend only to places in Schedule.

4. The places to which this Act applies shall be the places mentioned in the first schedule to this Act, the limits of which places shall for the purposes of this Act be such as are defined in that schedule.

EXPENSES OF EXECUTION OF ACT.

Expenses of Act to be defrayed by Admiralty.

5. Expenses incurred in the execution of this Act shall be paid under the direction of the Lord High Admiral of the United Kingdom or the commissioners for executing the office of Lord High Admiral (hereafter in this Act styled the Admiralty) and of such one of Her Majesty's principal Secretaries of State as Her Majesty thinks fit for the time being to intrust with the seals of the War Department (hereafter in this Act styled the Secretary of State for War) out of money to be provided by Parliament for that purpose.

VISITING SURGEONS.

Appointment of Visiting Surgeons and Assistants.

6. The Admiralty or the Secretary of State for War may, on the commencement of this Act, appoint a medical officer for each of the places to which this Act applies, to be, during pleasure, visiting surgeon there for the purposes of this Act, and may from time to time, on the death, resignation, or removal from office of any visiting surgeon, appoint another such officer in his stead.

The Admiralty or the Secretary of State for War may, from time to time as occasion requires, appoint a medical officer to be the assistant of any such visiting surgeon; and every such assistant shall have the like powers and duties as the visiting surgeon to whom he is appointed assistant.

A notice of the appointment of every such visiting surgeon and of every such assistant shall be published in the *London* or *Dublin Gazette* according as the place for which he is appointed is in *England* or in *Ireland*.

A copy of the Gazette containing such a notice shall be conclusive evidence of the appointment.

INSPECTOR OF HOSPITALS.

Appointment of Inspector and Assistant Inspector of Certified Hospitals.

7. The Admiralty and the Secretary of State for War shall, on the commencement of this Act, appoint a medical officer to be, during pleasure, inspector of certified hospitals under this Act, and shall from time to time, on the death, resignation, or removal from office of any such inspector, appoint another such officer in his stead.

The Admiralty and the Secretary of State for War may, from time to time as occasion requires, appoint a medical officer to be an assistant inspector of certified hospitals under this Act, which assistant shall have the like powers and duties as the inspector.

A notice of the appointment of every such inspector and of every such assistant shall be published in the *London Gazette*.

A copy of the Gazette containing such a notice shall be conclusive evidence of the appointment.

APPENDIX.

CERTIFIED HOSPITALS.
Power to Admiralty, &c. to provide Hospitals, and certify them.

8. The Admiralty or the Secretary of State for War may from time to time provide any buildings or parts of buildings as hospitals for the purposes of this Act, and any building or part of a building so provided and certified in writing by the Admiralty or Secretary of State for War (as the case may be) to be so provided shall be deemed a certified hospital under this Act; and every certified hospital so provided shall be placed under the control or management of such persons as to the Admiralty or the Secretary of State for War from time to time seem fit.

Power to certify other Hospitals.

9. The Admiralty or the Secretary of State for War may from time to time, on such application or with such consent as to them or him seems requisite, and on the report of the inspector of certified hospitals, certify in writing any building or part of a building (not provided as a hospital by the Admiralty or Secretary of State for War) to be useful and efficient as a hospital for the purposes of this Act, and thereupon that building or part of a building shall be deemed a certified hospital under this Act.

Inspection of Certified Hospitals.

10. The inspector of certified hospitals shall from time to time visit and inspect every certified hospital.

Power to withdraw Certificate.

11. The Admiralty or the Secretary of State for War may at any time, by declaration in writing, declare the certificate relative to any certified hospital withdrawn as from a time specified in the declaration, and thereupon the same shall cease to be a certified hospital as from the time so specified.

Provision for Moral and Religious Instruction.

12. A hospital shall not be certified under this Act unless at the time of the granting of a certificate adequate provision is made for the moral and religious instruction of the women detained therein under this Act; and if at any subsequent time it appears to the Admiralty or the Secretary of State for War that in any such hospital adequate provision for that purpose is not made, the certificate of that hospital shall be withdrawn.

Certificate and Declaration of Withdrawal to be gazetted.

13. Every certificate and every declaration of withdrawal of a certificate relative to any hospital under this Act shall be published in the *London* or *Dublin Gazette*, according as the hospital to which the certificate or declaration relates is in *England* or in *Ireland*.

A copy of the Gazette containing any such certificate or declaration shall be conclusive evidence of such certificate or declaration.

Every certificate proved to have been made shall be presumed to be in force until the withdrawal thereof is proved.

Power to make Regulations for Certified Hospitals. A printed Copy of Regulations to be Evidence.

14. The managers or persons having the control or management of each certified hospital shall make regulations for the management and government of the hospital, as far as regards women authorized by this Act to be detained therein for medical treatment, or being therein under medical treatment for a contagious disease, such regulations not being inconsistent with the provisions of this Act, and may from time to time alter any such regulations; but all such regulations, and all alterations thereof, shall be subject to the approval in writing of the Admiralty or the Secretary of State for War.

A printed copy of regulations purporting to be regulations of a certified hospital so approved, such copy being signed by the inspector of certified hopitals, or the chief medical officer of the hospital, shall be evidence of the regulations of the hospital, and of the due making and approval thereof, for the purposes of this Act.

PERIODICAL MEDICAL EXAMINATIONS.

On Information, Justice may issue Notice to Woman who is a common Prostitute.

15. Where an information on oath is laid before a justice by a superintendent of police, charging to the effect that the informant has good cause to believe that a woman therein named is a common prostitute, and either is resident within the limits of any place to which this Act applies, or, being resident within five miles of those limits, has, within fourteen days before the laying of the information, been within those limits for the purpose of prostitution, the justice may, if he thinks fit, issue a notice thereof addressed to such woman, which notice the superintendent of police shall cause to be served on her:

Provided that nothing in this Act contained shall apply or extend, in the case of *Woolwich*, to any woman who is not resident within one of the parishes of *Woolwich, Plumstead,* or *Charlton.*

Power to Justice to order periodical Medical Examination.

16. In either of the following cases, namely,—

If the woman on whom such a notice is served appears herself, or by some person on her behalf, at the time and place appointed in the notice, or at some other time and place appointed by adjournment;—

If she does not so appear, and it is shown (on oath) to the justice present that the notice was served on her a reasonable time before the time appointed for her appearance, or that reasonable notice of such adjournment was given to her (as the case may be),—

The justice present, on oath being made before him substantiating the matter of the information to his satisfaction, may, if he thinks fit, order that the woman be subject to a periodical medical examination by the visiting surgeon for any period not exceeding one year, for the purpose of ascertaining at the time of each such examination whether she is affected with a contagious disease; and thereupon she shall be subject to such a periodical medical examination, and the order shall be a sufficient warrant for the visiting surgeon to conduct such examination accordingly.

The order shall specify the time and place at which the woman shall attend for the first examination.

The superintendent of police shall cause a copy of the order to be served on the woman.

Voluntary Submission by Woman.

17. Any woman, in any place to which this Act applies, may voluntarily, by a submission in writing signed by her in the presence of and attested by the superintendent of police, subject herself to a periodical medical examination under this Act for any period not exceeding one year.

Power to make Regulations as to Examinations.

18. For each of the places to which this Act applies, either the Admiralty or the Secretary of State for War (but not both for any one place) may from time to time make regulations respecting the times and places of medical examinations under this Act at that place, and generally respecting the arrangements for the conduct there of those examinations; and a copy of all such regulations from time to time in force for each place shall be sent by the Admiralty or the Secretary of State for War (as the case may be) to the clerk of the peace, town clerk (if any), clerk of the justices, visiting surgeon, and superintendent of police.

APPENDIX.

Visiting Surgeon to prescribe Times, &c.

19. The visiting surgeon, having regard to the regulations aforesaid and to the circumstances of each case, shall at the first examination of each woman examined by him, and afterwards from time to time as occasion requires, prescribe the times and places at which she is required to attend again for examination; and he shall from time to time give or cause to be given to each such woman notice in writing of the times and places so prescribed.

DETENTION IN HOSPITAL.

Certificate of Visiting Surgeon.

20. If on any such examination the woman examined is found to be affected with a contagious disease, she shall thereupon be liable to be detained in a certified hospital subject and according to the provisions of this Act, and the visiting surgeon shall sign a certificate to the effect that she is affected with a contagious disease, naming the certified hospital in which she is to be placed; and he shall sign that certificate in triplicate, and shall cause one of the originals to be delivered to the woman and the others to the superintendent of police.

Placing in Certified Hospital for Treatment.

21. Any woman to whom any such certificate of the visiting surgeon relates may, if she thinks fit, proceed to the certified hospital named in that certificate, and place herself there for medical treatment, but if after the certificate is delivered to her she neglects or refuses to do so, the superintendent of police, or a constable acting under his orders, shall apprehend her, and convey her with all practicable speed to that hospital, and place her there for medical treatment, and the certificate of the visiting surgeon shall be a sufficient authority to him for so doing.

The reception of a woman in a certified hospital by the managers or persons having the control or management thereof shall be deemed to be an undertaking by them to provide for her care and treatment, lodging, clothing, and food, during her detention in the hospital.

Detention in Hospital.

22. Where a woman certified by the visiting surgeon to be affected with a contagious disease places herself, or is placed as aforesaid, in a certified hospital for medical treatment, she shall be detained there for that purpose by the chief medical officer of the hospital until discharged by him by writing under his hand.

The certificate of the visiting surgeon, one of the three originals whereof shall be delivered by the superintendent of police to the chief medical officer, shall, when so delivered, be sufficient authority for such detention.

Power to transfer to another Certified Hospital.

23. The inspector of certified hospitals may, if in any case it seems to him expedient, by order in writing signed by him, direct the transfer of any woman detained in a certified hospital for medical treatment from that certified hospital to another named in the order.

Every such order shall be made in triplicate, and one of the originals shall be delivered to the woman and the others to the superintendent of police.

Every such order shall be sufficient authority for the superintendent of police or any person acting under his orders to transfer the woman to whom it relates from the one hospital to the other, and to place her there for medical treatment; and she shall be detained there for that purpose by the chief medical officer of the hospital until discharged by him by writing under his hand.

The order of the inspector of certified hospitals, one of the originals whereof shall be delivered by the superintendent of police to the chief medical officer of the hospital to which the transfer is made, shall when so delivered be sufficient authority for such detention.

APPENDIX.

Limitation of Detention.

24. Provided always, that any woman shall not be detained under any one certificate for a longer time than three months, unless the chief medical officer of the hospital in which she is detained, and the inspector of certified hospitals, or visiting surgeon for the place whence she came or was brought, conjointly certify that her further detention for medical treatment is requisite (which certificate shall be in duplicate, and one of the originals thereof shall be delivered to the woman); and in that case she may be further detained in the hospital in which she is at the expiration of the said period of three months by the chief medical officer until discharged by him by writing under his hand ; but so that any woman be not detained under any one certificate for a longer time in the whole than six months.

Power for Woman detained to apply to Justice for Discharge.

25. If any woman detained in any hospital considers herself entitled to be discharged therefrom, and the chief medical officer of the hospital refuses to discharge her, such woman shall on her request be conveyed before a justice, who, if he is satisfied upon reasonable evidence that she is free from a contagious disease, shall discharge her from such hospital, and such order of discharge shall have the same effect as the discharge of the chief medical officer.

During Conveyance to Certified Hospital, &c., Woman deemed to be in legal Custody.

26. Every woman conveyed or transferred under this Act to a certified hospital shall, while being so conveyed or transferred thither, and also while detained there, be deemed to be legally in the custody of the person conveying, transferring, or detaining her, notwithstanding that she is for that purpose removed out of one into or through another jurisdiction, or is detained in a jurisdiction other than that in which the certificate of the visiting surgeon was made.

Expenses of Woman's Return Home.

27. Every woman shall, on her discharge from the hospital, be sent to the place of her residence, if she so desires, without expense to herself.

REFUSAL TO BE EXAMINED, &c.

Punishment of Women for refusing to be examined, &c.

28. In the following cases, namely,—

If any woman subjected by order of a justice under this Act to periodical medical examination at any time temporarily absents herself in order to avoid submitting herself to such examination on any occasion on which she ought so to submit herself, or refuses or wilfully neglects to submit herself to such examination on any such occasion ;

If any woman authorized by this Act to be detained in a certified hospital for medical treatment quits the hospital without being discharged therefrom by the chief medical officer thereof by writing under his hand (the proof whereof shall lie on the accused) ;

If any woman authorized by this Act to be detained in a certified hospital for medical treatment, or any woman being in a certified hospital under medical treatment of a contagious disease, refuses or wilfully neglects while in the hospital to conform to the regulations thereof approved under this Act ;

Then and in every such case such woman shall be guilty of an offence against this Act, and on summary conviction shall be liable to imprisonment, with or without hard labour, in the case of a first offence for any term not exceeding one month, and in the case of a second or any subsequent offence for any term not exceeding three months ; and in the case of the offence of quitting the hospital without being discharged as aforesaid the woman may be taken into custody without warrant by any constable.

APPENDIX.

Effect of Order of Imprisonment for Absence, &c. from Examination.

29. If any woman is convicted of and imprisoned for the offence of absenting herself or of refusing or neglecting to submit herself to examination as aforesaid, the order subjecting her to periodical medical examination shall be in force after and notwithstanding her imprisonment, unless the surgeon or other medical officer of the prison, or a visiting surgeon appointed under this Act, at the time of her discharge from imprisonment, certifies in writing to the effect that she is then free from a contagious disease (the proof of which certificate shall lie on her), and in that case the order subjecting her to periodical examination shall, on her discharge from imprisonment, cease to operate.

Effect on Order of Imprisonment for quitting Hospital, &c.

30. If any woman is convicted of and imprisoned for the offence of quitting a hospital without being discharged, or of refusing or neglecting while in a hospital to conform to the regulations thereof as aforesaid, the certificate of the visiting surgeon under which she was detained in the hospital shall continue in force, and on the expiration of her term of imprisonment she shall be sent back from the prison to that certified hospital, and shall (notwithstanding anything in this Act) be detained there under that certificate as if it were given on the day of the expiration of her term of imprisonment, unless the surgeon or other medical officer of the prison, or a visiting surgeon appointed under this Act, at the time of her discharge from imprisonment, certifies in writing to the effect that she is then free from a contagious disease (the proof of which certificate shall lie on her), and in that case the certificate under which she was detained, and the order subjecting her to periodical medical examination, shall, on her discharge from imprisonment, cease to operate.

Penalty on Woman discharged uncured conducting herself as Prostitute.

31. If on any woman leaving a certified hospital a notice in writing is given to her by the chief medical officer of the hospital to the effect that she is still affected with a contagious disease, and she is afterwards in any place for the purpose of prostitution without having previously received from a visiting surgeon appointed under this Act a certificate in writing endorsed on the notice or on a copy thereof certified by the chief medical officer of the hospital (proof of which certificate shall lie on her) to the effect that she is then free from a contagious disease, she shall be guilty of an offence against this Act, and on summary conviction before two justices shall be liable to be imprisoned with or without hard labour, in the case of a first offence for any term not exceeding one month, and in the case of a second or any subsequent offence for any term not exceeding three months.

DURATION OF ORDER.

Order to operate whenever Woman is resident in any Place where Order made, &c.

32. Every order under this Act subjecting a woman to periodical medical examination shall be in operation and enforceable, in manner in this Act provided, as long as and whenever from time to time the woman to whom it relates is resident within the limits of the place to which this Act applies wherein the order was made, or within five miles of those limits, but not in any case for a longer period than one year ; and where the chief medical officer of a certified hospital, on the discharge by him of any woman from the hospital, certifies that she is free from a contagious disease (proof of which certificate shall lie on her), the order subjecting her to periodical examination shall thereupon cease to operate.

RELIEF FROM EXAMINATION.

Application for Relief from Examination.

33. If any woman subjected to a periodical medical examination under this Act (either on her own submission or under the order of a justice), desiring to be

relieved therefrom, and not being under detention in a certified hospital, makes application in writing in that behalf to a justice, the justice shall appoint by notice in writing a time and place for the hearing of the application, and shall cause the notice to be delivered to the applicant, and a copy of the application and of the notice to be delivered to the superintendent of police.

Order for Relief from Examination on Discontinuance of Prostitution, &c.

34. If on the hearing of the application it is shown, to the satisfaction of a justice, that the applicant has ceased to be a common prostitute, or if the applicant, with the approval of the justice, enters into a recognizance, with or without sureties, as to the justice seems meet, for her good behaviour during three months thereafter, the justice shall order that she be relieved from periodical medical examination.

Forfeiture of Recognizance by Return to Prostitution.

35. Every such recognizance shall be deemed to be forfeited if at any time during the term for which it is entered into the woman to whom it relates is (within the limits of any place to which this Act applies) in any public thoroughfare, street, or place for the purpose of prostitution, or otherwise (within those limits) conducts herself as a common prostitute.

PENALTIES FOR HARBOURING, &c.

Penalties for permitting Prostitute having contagious Disease to resort to any House, &c., for Prostitution.

36. If any person, being the owner or occupier of any house, room, or place within the limits of any place to which this Act applies, or being a manager or assistant in the management thereof, having reasonable cause to believe any woman to be a common prostitute and to be affected with a contagious disease, induces or suffers her to resort to or be in that house, room, or place for the purpose of prostitution, he shall be guilty of an offence against this Act, and on summary conviction thereof before two justices shall be liable to a penalty not exceeding twenty pounds, or, at the discretion of the justices, to be imprisoned for any term not exceeding six months, with or without hard labour.

Provided that a conviction under this enactment shall not exempt the offender from any penal or other consequences to which he may be liable for keeping or being concerned in keeping a bawdy house or disorderly house, or for the nuisance thereby occasioned.

PROCEDURE, &c.

Application of 11 & 12 Vict. c. 43. and 14 & 15 Vict. c. 93. to this Act.

37. All proceedings under this Act before and by justices shall be had in *England* according to the provisions of the Act of the Session of the eleventh and twelfth years of Her Majesty (chapter forty-three), " to facilitate the performance of the duties of justices of the peace out of Sessions within *England* and *Wales* with respect to summary convictions and orders," and in *Ireland* according to the provisions of the Petty Sessions (*Ireland*) Act, 1851, as far as those provisions respectively are not inconsistent with any provision of this Act, and save that the room or place in which a justice sits to inquire into the truth of the statements contained in any information or application under this Act against or by a woman shall not unless the woman so desires, be deemed an open court for that purpose ; and, unless the woman otherwise desires, the justice may, in his discretion, order that no person have access to or be or remain in that room without his consent or permission.

Forms in Second Schedule to be used.

38. The forms of certificates, orders, and other instruments given in the Second Schedule to this Act, or forms to the like effect, with such variations and additions as circumstances require, may be used for the purposes therein indicated and according to the directions therein contained, and instruments in those forms shall (as regards the form thereof) be valid and sufficient.

Z

APPENDIX.

Instruments may be in Print, &c.

39. Any certificate, order, notice, or other instrument made or issued for the purposes of this Act may be partly in print and partly in writing.

Presumption as to Signatures of Justices, &c.

40. In any proceeding under this Act any notice, order, certificate, copy of regulations, or other instrument purporting to be signed by a justice, superintendent of police, visiting surgeon, assistant visiting surgeon, surgeon or other medical officer of a prison, chief medical officer of a certified hospital, or the inspector or an assistant inspector of certified hospitals, or by any person in Her Majesty's service or in that of the Admiralty, shall on production be received in evidence, and shall be presumed to have been duly signed by the person, and in the character by whom and in which it purports to be signed, until the contrary is shown.

Mode of Service.

41. Every notice, order, or other instrument by this Act required to be served on a woman shall be served by delivery thereof to some person for her at her usual place of abode, or by delivery thereof to her personally.

Limitation of Actions, &c.

42. Any action or prosecution against any person for anything done in pursuance or execution or intended execution of this Act shall be laid and tried in the county where the thing was done, and shall be commenced within three months after the thing done, and not otherwise.

Notice in writing of every such action and of the cause thereof shall be given to the intended defendant one month at least before the commencement of the action.

In any such action the defendant may plead generally that the act complained of was done in pursuance or execution or intended execution of this Act, and give this Act and the special matter in evidence at any trial to be had thereupon.

The plaintiff shall not recover if tender of sufficient amends is made before action brought, or if a sufficient sum of money is paid into court after action brought, by or on behalf of the defendant.

If a verdict passes for the defendant, or the plaintiff becomes nonsuit, or discontinues the action after issue joined, or if, on demurrer or otherwise, judgment is given against the plaintiff, the defendant shall recover his full costs as between attorney and client, and shall have the like remedy for the same as any defendant has by law for costs in other cases.

Though a verdict is given for the plaintiff, he shall not have costs against the defendant unless the judge before whom the trial is had certifies his approbation of the action.

SCHEDULES.

THE FIRST SCHEDULE.

Names of Places.	Limits of Places.
Portsmouth - - - -	The limits of the municipal borough of Portsmouth, and of the residue of the island of Portsea, and of the parish of Alverstoke, and of the township of Landport.
Plymouth and Devonport	The limits of the following places; namely,— The municipal borough of Plymouth. The parliamentary borough of Devonport. The parish of Laira. The tithing of Pennycross or Western Peveril. The tithing of Compton Gifford. Torpoint in the county of Cornwall, within the distance of half a mile from the Ferry Gate.

APPENDIX.

THE FIRST SCHEDULE—*continued.*

Names of Places.	Limits of Places.
Woolwich	The limits of the parishes of Woolwich, Plumpstead, and Charlton.
Chatham	The limits of the following parishes; namely,— Chatham, Gillingham, St. Nicholas, Rochester, St. Margaret, Rochester, The Precincts, Rochester, Brompton, New Brompton, Strood, and Frindsbury, and of the hamlet of Grange, otherwise Grench.
Sheerness	The limits of the the parish of Minster, and of the township of Queenborough.
Aldershot	The limits of the following parishes; namely,— Purbright, Ash, Compton, Pepper Harrow, Frimley, Puttenham, Seal, and Tongham, } in the county Surrey. Elstead, Farnham, Bisley, Aldershot, Yateley, Crondall, Dogmersfield, Winchfield, Hartley Wintney, } in the county of Hants. Cove, Eversley, Farnborough, Binstead, Bentley Sandhurst, in the county of Berks.
Windsor	The limits of the following parishes; namely,— New Windsor, Old Windsor, } in the county of Berks. Clewer, Eton, in the county of Bucks.
Colchester	The limits of the following parishes or ecclesiastical districts; namely,— All Saints. St. Botolph. St. Giles. St. James. St. John. St. Leonard.

THE FIRST SCHEDULE—continued.

Name of Places.	Limits of Places.
Colchester (continued)	St. Martin. St. Mary at the Walls. St. Mary Magdalene. St. Nicholas. St. Peter. St. Runwald. The Holy Trinity.
Shorncliffe	The limits of the following parishes; namely,— Cheriton. Hythe. Folkstone.
The Curragh	The limits of the following parishes; namely,— Kilcullen. Kildare. Ballysax. Great Conwell. Morristown-beller.
Cork	The limits of the borough of Cork for municipal purposes.
Queenstown	The limits of the town of Queenstown for the purposes of town improvement.

THE SECOND SCHEDULE.

FORMS.

(A.)

Gazette Notice of Appointments.

London   18 .

THE Lords Commissioners of the Admiralty have [or the Secretary of State for War has] appointed *R.S.* to be visiting surgeon [or assistant visiting surgeon] for [*Portsmouth*, or the Lords Commissioners of the Admiralty and the Secretary of State for War have appointed *P.T.* to be inspector (or assistant inspector) of certified hospitals] under the Contagious Diseases Act, 1866.

(B.)

Certificate for Hospital provided by Admiralty, &c.

THE CONTAGIOUS DISEASES ACT, 1866.

IN pursuance of the above-mentioned Act, it is hereby certified by the Commissioners for executing the office of Lord High Admiral of the United Kingdom [or by Her Majesty's Principal Secretary of State intrusted with the Seals of the War Department], that the following building [or part of a building], namely [*here describe generally the building or part of building*], has been provided by the said

Lords Commissioners [*or* Secretary of State] as a hospital for the purposes of the said Act.

 Dated this   day of   18 .
  By order of the Lords Commissioners of the Admiralty.
        (Signed) *C.P.*,
           Secretary of the Admiralty.
  [*Or*
  By order of the Secretary of State for War.
        (Signed) *E.L.*,
           Under-Secretary of State.]

(C.)

Certificate for Hospital not provided by Admiralty, &c.

THE CONTAGIOUS DISEASES ACT, 1866.

 IN pursuance of the above-mentioned Act, it is hereby certified by the Commissioners for executing the office of Lord High Admiral of the United Kingdom [*or* by Her Majesty's Principal Secretary of State intrusted with the seals of the War Department], that the following building [*or* part of a building], namely [the Lock Wards of the Portsmouth, Portsea, and Gosport Hospital, *or as the case may be*], is useful and efficient as a hospital for the purposes of the said Act.

 Dated this   day of   18 .
  By order of the Lords Commissioners of the Admiralty.
        (Signed) *C.P.*,
           Secretary of the Admiralty.
  [*Or*
  By order of the Secretary of State for War.
        (Signed) *E.L.*,
           Under-Secretary of State.]

(D.)

Declaration of Withdrawal of Certificate.

THE CONTAGIOUS DISEASES ACT, 1866.

 IN pursuance of the above-mentioned Act, it is hereby declared by the Commissioners for executing the office of Lord High Admiral of the United Kingdom [*or* by Her Majesty's Principal Secretary of State intrusted with the seals of the War Department], that the certificate under the said Act, dated the   day of   , constituting the   hospital [*or as the case may be*] a certified hospital under the said Act, has been and the same is hereby withdrawn as from the   day of   18 .

 Dated this   day of   18 .
  By order of the Lords Commissioners of the Admiralty.
        (Signed) *C.P.*,
           Secretary of the Admiralty.
  [*Or*
  By order of the Secretary of State for War.
        (Signed) *E.L.*,
           Under-Secretary of State.]

(E.)

Information.

 } THE information of *C.D.* of   , superintendent o
to wit. } police for   [*or as the case may be*], under the Contagious Diseases Act, 1866, taken this   day of
186 , before the undersigned, one of Her Majesty's justices of the peace in and for

the said [*county*] of   who says he has good cause to believe that *A.B.* is a common prostitute, and is resident within the limits of a place to which the said Act applies, that is to say, at   in the [*county*] of [*or* is a common prostitute, and being resident within five miles of a place to which the said Act applies, that is to say, at   in the county of   , was within fourteen days before the laying of this information, that is to say, on the   day of   , within those limits, that is to say, at   in the county of   , for the purpose of prostitution].

Taken and sworn before me the day and year first above mentioned.

(Signed)   *L.M.*

(F.)

Notice for Attendance of Woman.

To *A.B.* of

TAKE notice, that an information, a copy whereof is subjoined hereto, has been laid before me, and that, in accordance with the provisions of the Act therein mentioned, the truth of the statements therein contained will be inquired into before me, or some other justice, at   , on the   day of   , at   o'clock in the   noon.

You are therefore to appear before me or such other justice at that place and time, and to answer to what is stated in the said information.

You may appear yourself, or by any person on your behalf.

If you do not appear, you may be ordered, without further notice, to be subject to a periodical medical examination by the visiting surgeon under the said Act.

If you prefer it, you may, by a submission in writing signed by you in the presence of the superintendent of police [*or as the case may be*], and attested by him, subject yourself to such a periodical examination.

If you do so before the time above appointed for your appearance, it will not be necessary for you to appear then before a justice.

Dated this   day of

(Signed)   *L.M.*,

justice of the peace for

[*Subjoin copy of information.*]

(G.)

Order subjecting Woman to Examination.

              } BE it remembered, that on the   day of   ,
to wit.   } in pursuance of the Contagious Diseases Act, 1866, I, one of Her Majesty's justices of the peace in and for the said [*county*] of   , do order that *A.B.*,   of   , be subject to a periodical medical examination by the visiting surgeon for [*Portsmouth, or as the case may be*] for   calendar months from this day, for the purpose of ascertaining at the time of each such examination whether she is affected with a contagious disease within the meaning of the said Act, and that she do attend for the first examination at   on the   day of   , at   o'clock in the   noon.

(Signed)   *L.M.*

(H.)

Voluntary Submission to Examination.

THE CONTAGIOUS DISEASES ACT, 1866.

I *A.B.*   of   , in pursuance of the above-mentioned Act, by this submission, voluntarily subject myself to a periodical

medical examination by the visiting surgeon for [*Portsmouth, or as the case may be*] for   calendar months from the date hereof.
Dated this   day of   18 .
                                                         (Signed)   *A.B.*
Witness,
X.Y.,
superintendent of police for   [*or as the case may be*].

(J.)

Notice by Visiting Surgeon to Woman of Times, &c. of Examination.

To *A.B.* of
Take notice, that in pursuance of the Contagious Diseases Act, 1866, you are required to attend for medical examination as follows :
[*Here state times and places of examination.*]
Dated this   day of   18 .
                                    (Signed)   *E.F.*,
                                    visiting surgeon for [*Portsmouth*].

(K.)

Certificate of Visiting Surgeon.

IN pursuance of the Contagious Diseases Act, 1866, I hereby certify that I have this day examined *A.B.*   of   , and that she is affected with a contagious disease within the meaning of that Act; and the certified hospital in which she is to be placed under the said Act is the   hospital.
Dated this   day of   18 .
                                    (Signed)   *E.F.*,
                                    visiting surgeon for [*Portsmouth*].

(L.)

Order by Inspector of Certified Hospitals for Transfer.

BY virtue of the power in this behalf vested in me by the Contagious Diseases Act, 1866, I hereby order that *A.B.*   of   , now detained under that Act in the certified hospital of   for medical treatment, be transferred thence to the certified hospital of   .
Dated this   day of   18 .
                                    (Signed)   *M.N.*,
                                    inspector of certified hospitals.

(M.)

Certificate for Detention beyond Three Months.

THE CONTAGIOUS DISEASES ACT, 1866.

WE, the undersigned, hereby certify that the further detention for medical treatment of *A.B.* of   , now an inmate of this hospital, is requisite.
Dated this   day of   18   , at the   hospital.
                                   (Signed)   *M.N.*,
                                inspector of certified hospitals,
                                [*or as the case may be*],
                                      G.H.,
                                   chief medical officer.

(N.)

Discharge from Hospital.

IN pursuance of the Contagious Diseases Act, 1866, I hereby discharge *A.B.* of   from this hospital [*add according to the fact,* and certify that she is now free from a contagious disease].
Dated this   day of   18   , at the   hospital.
(Signed)   *G.H.*,
chief medical officer.

(O.)

Certificate on Discharge from Imprisonment.

THE CONTAGIOUS DISEASES ACT, 1866.

WHEREAS under the above-mentioned Act *A.B.*   of   was on the   day of   convicted of the offence of   and has since been imprisoned for that offence in the gaol of   and is now discharged from imprisonment therein : Now in pursuance of the said Act I hereby certify that she is now free from a contagious disease.
Dated this   day of   .
R.O.,
surgeon of the gaol of
[or *E.F.*,
visiting surgeon for *Portsmouth*].

(P.)

Notice to Woman leaving Hospital.

THE CONTAGIOUS DISEASES ACT, 1866.

To *A.B.*

As you are now leaving this hospital, I hereby, in pursuance of the above-mentioned Act, give you notice that you are still affected with a contagious disease.
Dated this   day of   .
(Signed)   *G.H.*,
chief medical officer.

Note.—The above-mentioned Act provides as follows :—
If on any woman leaving a certified hospital a notice [*set out section of Act*].

(Q.)

Certificate on last foregoing Notice or Copy.

IN pursuance of the within-mentioned Act, I hereby certify that the within-named woman is now free from a contagious disease.
Dated this   day of   .
(Signed)   *E.F.*,
visiting surgeon for [*Portsmouth*].

(R.)

Application to be relieved from Examination.

To *L.M.*, Esq., and others, Her Majesty's justices of the peace for the [*county*] of   .
I *A.B.*   of   , being in pursuance of the Contagious Diseases Act, 1866, subject to a periodical medical examination on my own submission [*or* under the order of *L.M.*, Esq., *as the case may be*], dated the   day of   , do hereby apply to be relieved therefrom.
Dated this   day of   18   .
(Signed)   *A.B.*
Witness, *G.W.*

APPENDIX. 345

CHAP. XCVI.

An Act to amend the Contagious Diseases Act, 1866.
[11th *August*, 1869.]

BE it enacted by the Queen's most excellent Majesty, by and with the advice and consent of the Lords Spiritual and Temporal, and Commons, in this present Parliament assembled, and by the authority of the same, as follows :

Short Title.

1. This Act may be cited as the Contagious Diseases Act, 1869.

Construction of Act.

2. This Act shall be construed as one with the Contagious Diseases Act, 1866 (in this Act referred to as the principal Act), and with the Act of the Session of the thirty-first and thirty-second years of the reign of Her present Majesty, chapter eighty, and those Acts and this Act may be cited together as the Contagious Diseases Acts, 1866 to 1869.

Temporary Detention of Women.

3. Any woman who, on attending for examination or being examined by the visiting surgeon, is found by him to be in such a condition that he cannot properly examine her, shall, if such surgeon has reasonable grounds for believing that she is affected with a contagious disease, be liable to be detained in a certified hospital, subject and acording to the provisions of the Contagious Diseases Acts, 1866 to 1869, until the visiting surgeon can properly examine her, so that she be not so detained for a period exceeding five days. The visiting surgeon shall sign a certificate to the effect that she was in such a condition that he could not properly examine her, and that he has reasonable grounds to believe that she is affected with a contagious disease, and shall name therein the certified hospital in which she is to be placed ; and such certificate shall be signed and otherwise dealt with in the same manner, and have the same effect, except as regards duration, as a certificate under the principal Act.

If the reason that the visiting surgeon cannot examine the woman is that she is drunk, she may be detained upon an order of the visiting surgeon for a period not exceeding twenty-four hours in any place named in the order where persons accused of being drunk and disorderly or of offences punishable summarily are usually detained, and the gaoler or the keeper of such place shall upon the receipt of such order receive and detain the woman accordingly.

Notice by Justice to Woman being a Common Prostitute.

4. Where an information on oath is laid before a justice by a superintendent of police, charging to the effect that the informant has good cause to believe that a woman therein named is a common prostitute, and either is resident within the limits of any place to which this Act applies, or, being resident within ten miles of those limits, or having no settled place of abode, has, within fourteen days before the laying of the information, either been within those limits for the purpose of prostitution, or been outside of those limits for the purposes of prostitution in the company of men resident within those limits, the justice may, if he thinks fit, issue a notice thereof addressed to such woman, which notice the superintendent of police shall cause to be served on her:

Provided that nothing in the Contagious Diseases Acts, 1866 to 1869, shall extend, in the case of Woolwich, to any woman who is not resident within the limits specified in the first schedule to this Act.

Section fifteen of the principal Act is hereby repealed, and the foregoing enactment in this section is substituted for it ; provided that all proceedings taken and acts done under the section hereby repealed shall, notwithstanding, remain of full effect, and shall, if necessary, be continued as if they had been taken and done under this section.

Duration of Order.

5. Any order for subjecting a woman to periodical medical examination shall be in operation and enforceable as long as and whenever such woman is resident within ten miles of the limits of the place where the order was made, instead of within five miles, as prescribed by section thirty-two of the principal Act.

Effect of Voluntary Submission by Women.

6. Where any woman, in pursuance of the principal Act, voluntarily subjects herself by submisson in writing to a periodical medical examination under that Act, such submission shall, for all the purposes of the Contagious Diseases Acts, 1866 to 1869, have the same effect as an order of a justice subjecting the woman to examination ; and all the provisions of the principal Act respecting the attendance of the woman for examination, and her absenting herself to avoid examination, and her refusing or wilfully neglecting to submit herself for examination, and the force of the order subjecting her to examination after imprisonment for such absence, refusal, or neglect, shall apply and be construed accordingly.

Duration of Detention.

7. A woman may be detained for a further period not exceeding three months, in addition to the six months allowed under section twenty-four of the principal Act, if such certificate as is required by that section (to the effect that her further detention for medical treatment is requisite) is given at the expiration of such six months ; so, nevertheless, that any woman be not detained under one certificate for a longer time in the whole than nine months.

Custody of Orders of Discharge.

8. Where an order is made discharging a woman from any hospital, or where a certificate is given, under section thirty of the principal Act, that a woman is free from a contagious disease, such order and certificate shall be delivered to the superintendent of police, and retained by him.

Application to Surgeon for Relief from Examination.

9. Any woman subjected, either on her own submission or under the order of a justice, to a periodical medical examination under the principal Act, who desires to be relieved therefrom, and is not under detention in a certified hospital, may make application in writing in that behalf to the visiting surgeon.

The visiting surgeon shall cause a copy of such application to be delivered to the superintendent of police, and if, after a report from such superintendent, he is satisfied by such report or other evidence that the applicant has ceased to be a common prostitute, may, by order under his hand, direct that she be relieved, and she shall thereupon be relieved, from periodical medical examination.

Such order shall be in triplicate ; one copy shall be delivered to the woman, and two copies shall be delivered to the superintendent of police, who shall communicate one copy to the justice (if any) who made the order subjecting the woman to a periodical medical examination, or to his successor in office.

The provisions of this section shall be in addition to and not in substitution for the provisions of the principal Act, for relieving a woman from examination.

Places to which Act extends.

10. The places to which the Contagious Diseases Acts, 1866 to 1869, apply, shall be the places mentioned in the first schedule to this Act, the limits of which places shall, for the purposes of the said Acts, be such as are defined in that schedule.

APPENDIX. 347

Forms in Second Schedule to be used.

11. The forms of certificates, orders, and other instruments given in the second schedule to this Act, or forms to the like effect, with such variations and additions as circumstances require, may be used for the purposes therein indicated, and according to the directions therein contained, and instruments in those forms shall (as regards the form thereof) be valid and sufficient.

Repeal of Parts of 29 & 30 Vict. c. 35.

12. Sections four and thirty-eight of the principal Act, and the two schedules to that Act, are hereby repealed.

As to Settlement of Child born in Certified Hospital.

13. The settlement of a child born of the body of a mother while detained in a certified hospital shall be the same as if such hospital were a house licensed for the public reception of pregnant women under the Act of the thirteenth year of King George III., chapter eighty-two.

FIRST SCHEDULE.

Names of Places.	Limits of Places.
Aldershot	The limits of the following parishes; namely, Pirbright, Ash, Compton, Peper Harow, Frimley, Puttenham, Seal, Tongham, Elstead, Farnham, Bisley, in the county of Surrey. Aldershot, Yateley, Crondall, Dogmersfield, Winchfield, Hartley Wintney, Cove, Eversley, Farnborough, Binsted, Bentley, in the county of Hants. Sandhurst, in the county of Berks.
Canterbury	The limits of the following parishes or ecclesiastical districts; namely, St. Andrew. All Saints. St. Alphage. St. Mary Bredin. St. Mary Bredman. St. George-the-Martyr. St. Mary Magdalen.

APPENDIX.

First Schedule—*continued.*

Names of Places.	Limits of Places.
Canterbury (*continued*) -	St. Margaret. St. Mildred. St. Mary, Northgate. St. Martin. St. Paul. St. Peter. The Archbishop's Palace. St. Dunstan. Christ Church. St. Gregory. Staplegate. Westgate Within. Westgate Without. St. Augustine. Old Castle.
Chatham - - - - -	The limits of the following parishes and places ; namely, Chatham. Gillingham. St. Nicholas, Rochester. St. Margaret, Rochester. The Precincts, Rochester. Brompton. New Brompton. Strood. Frindsbury, and The hamlet of Grange, otherwise Grench.
Colchester - - - - -	The limits of the following parishes or ecclesiastical districts ; namely, All Saints. St. Botolph. St. Giles. St. James. St. John. St. Leonard. St. Martin. St. Mary at the Walls. St. Mary Magdalene. St. Nicholas. St. Peter. St. Runwald. The Holy Trinity. St. Andrew's, Greenstead. Lexden. St. Michaels, Mile End.
Dover - - - - - -	The limits of the parishes of— Buckland. Charlton. Hougham. St. Mary's. St. James's. Eastcliff (extra-parochial). Guston.

FIRST SCHEDULE—*continued.*

Names of Places.	Limits of Places.
Gravesend	The limits of the parishes of— Gravesend. Milton. Northfleet. Den'on. Chalk.
Maidstone	The limits of the parishes of— Maidstone. Barming. East Farleigh. Loose. Boughton Monchelsea. Allington, and The hamlet of Tovil.
Plymouth and Devonport	The limits of the following places; namely, The municipal borough of Plymouth. The parliamentary borough of Devonport. The district of Laira. The tithing of Pennycross or Western Peveril. The tithing of Compton Gifford. Torpoint in the county of Cornwall, within the distance of half a mile from the Ferry Gate. Ivy Bridge. The parishes of Plympton St. Maurice and Plympton St. Mary. Dartmouth.
Portsmouth	The limits of the following places and parishes; namely, The municipal borough of Portsmouth. The residue of the island of Portsea. The parish of Alverstoke. The township of Landport.
Sheerness	The limits of the parish of Minster, of the township of Queenborough, and of the Isle of Grain.
Shorncliffe	The limits of the following parishes; namely, Cheriton. Hythe. Folkstone. Walmer. Deal. Sholden. Mongeham. Ringwold. Ripple.
Southampton	The limits of the municipal borough of Southampton.
Winchester	The limits of the parliamentary borough of Winchester.
Windsor	The limits of the following parishes; namely, New Windsor, Old Windsor, Clewer, } in the county of Berks. Eton, Datchet, Upton, } in the county of Bucks.

APPENDIX.

First Schedule—*continued.*

Names of Places.	Limits of Places.
Woolwich - - - - -	The limits of the following parishes and places; namely, Woolwich. Plumstead. Charlton. St. Paul, } Deptford. St. Nicholas, Hamlet of Hatcham. St. Alphage, Greenwich.
	IRELAND.
The Curragh - - - -	The limits of the following parishes; namely, Kilcullen. Kildare. Ballysax. Great Conwell. Morristown-beller.
Cork - - - - - -	The limits of the borough of Cork for municipal purposes.
Queenstown - - - -	The limits of the town of Queenstown for the purposes of town improvement.

SECOND SCHEDULE.

Forms.

(A.)

Gazette Notice of Appointments.

London,   18 .

The Lords Commissioners of the Admiralty have [*or* the Secretary of State for War has] appointed *R.S.* to be visiting surgeon [*or* assistant visiting surgeon] for [*Portsmouth*, *or* the Lords Commissioners of the Admiralty and the Secretary of State for War have appointed *P.S.* to be inspector (*or* assistant inspector) of certified hospitals] under the Contagious Diseases Acts, 1866 to 1869.

(B.)

Certificate for Hospital provided by Admiralty, &c.

The Contagious Diseases Acts, 1866 to 1869.

In pursuance of the above-mentioned Acts, it is hereby certified by the Commissioners for executing the office of Lord High Admiral of the United Kingdom [*or* by Her Majesty's Principal Secretary of State intrusted with the seals of the War Department], that the following building [*or* part of a building], namely [*here describe generally the building or part of building*], has been provided by the said Lords Commissioners [*or* Secretary of State] as a hospital for the purposes of the said Acts.

Dated this   day of   18 .

By order of the Lords Commissioners of the Admiralty.

(Signed)   *C.P.*,

Secretary of the Admiralty.

[*Or*,

By order of the Secretary of State for War.

(Signed)   *E.L.*,

Under-Secretary of State.]

APPENDIX. 351

(C.)

Certificate for Hospital not provided by Admiralty, &c.

THE CONTAGIOUS DISEASES ACTS, 1866 to 1869.

IN pursuance of the above-mentioned Acts, it is hereby certified by the Commissioners for executing the office of Lord High Admiral of the United Kingdom [or by Her Majesty's Principal Secretary of State intrusted with the seals of the War Department], that the following building [or part of a building], namely [the Lock Wards of the Portsmouth, Portsea, and Gosport Hospital, or *as the case may be*], is useful and efficient as a hospital for the purposes of the said Acts.

Dated this   day of   18   .
By order of the Lords Commissioners of the Admiralty.
                            (Signed)   *C.P.*,
                                          Secretary of the Admiralty.
                [*Or*
By order of the Secretary of State for War.
                            (Signed)   *E.L.*,
                                          Under-Secretary of State.]

(D.)

Declaration of Withdrawal of Certificate.

THE CONTAGIOUS DISEASES ACTS, 1866 to 1869.

IN pursuance of the above-mentioned Acts, it is hereby declared by the Commissioners for executing the office of Lord High Admiral of the United Kingdom [or by Her Majesty's Principal Secretary of State intrusted with the seals of the War Department], that the certificate under the said Acts, dated the day of   , constituting the   hospital [or *as the case may be*] a certified hospital under the said Acts, has been and the same is hereby withdrawn as from the   day of   18   .

Dated this   day of   18   .
By order of the Lords Commissioners of the Admiralty.
                            (Signed)   *C.P.*,
                                          Secretary of the Admiralty.
                [*Or*
By order of the Secretary of State for War.
                            (Signed)   *E.L.*,
                                          Under-Secretary of State.]

(E.)

Information.

         } THE information of *C.D.* of   , superintendent
to wit, } of police for   [or *as the case may be*]
under the Contagious Diseases Acts, 1866 to 1869, taken this
day of   186 , before the undersigned, one of Her Majesty's justices of the peace in and for the said [*county*] of
who says he has good cause to believe that *A.B.* is a common prostitute, and is resident within the limits of a place to which the said Acts apply, that is to say, at   in the [*county*] of   [or is a common prostitute, and being resident within fifteen miles of a place to which the said Act applies, that is to say, at   in the county of   ], was within fourteen days before the laying of this information, that is to say, on the   day of   , within those limits [or outside of those limits], that is to say, at
in the county of   for the purpose of prostitution [in the company of men resident within those limits].

Taken and sworn before me the day and year first above mentioned.
                            (Signed)   *L.M.*

(F.)
Notice for Attendance of Woman.
To *A.B.* of

TAKE notice, that an information, a copy whereof is subjoined hereto, has been laid before me, and that, in accordance with the provisions of the Acts therein mentioned, the truth of the statements therein contained will be inquired into before me, or some other justice, at , on the day of , at o'clock in the noon.

You are therefore to appear before me or such other justice at that place and time, and to answer to what is stated in the said information.

You may appear yourself, or by any person on your behalf.

If you do not appear, you may be ordered, without further notice, to be subject to a periodical medical examination by the visiting surgeon under the said Acts.

If you prefer it, you may, by a submission in writing signed by you in the presence of the superintendent of police [*or as the case may be*], and attested by him, subject yourself to such a periodical examination.

If you do so before the time above appointed for your appearance, it will not be necessary for you to appear then before a justice.

Dated this day of

(Signed) *L.M.*,
justice of the peace for
[*Subjoin copy of information.*]

(G.)
Order subjecting Woman to Examination.

to wit. } BE it remembered, that on the day of in pursuance of the Contagious Diseases Acts, 1866 to 1869, I, one of Her Majesty's justices of the peace in and for the said [*county*] of do order that *A.B.* of be subject to a periodical medical examination by the visiting surgeon for [*Portsmouth, or as the case may be*] for calendar months from this day, for the purpose of ascertaining at the time of each such examination whether she is affected with a contagious disease within the meaning of the said Acts, and that she do attend for the first examination at on the day of at o'clock in the noon.

(Signed) *L.M.*

(H.)
Voluntary Submission to Examination.

THE CONTAGIOUS DISEASES ACTS, 1866 to 1869.

I *A.B.* of , in pursuance of the above-mentioned Acts, by this submission, voluntarily subject myself to a periodical medical examination by the visiting surgeon for [*Portsmouth, or as the case may be*] for calendar months from the date hereof.

Dated this day of 18 .

(Signed) *A.B*

Witness,
X.Y.,
superintendent of police for [*or as the case may be*].

(J.)
Notice by Visiting Surgeon to Woman of Times, &c. of Examination.
To *A.B.* of

TAKE notice, that in pursuance of the Contagious Diseases Acts, 1866 to 1869, you are required to attend for medical examination as follows :

[*Here state times and places of examination.*]
Dated this day of 18 .

(Signed) *E.F.*,
visiting surgeon for [*Portsmouth*].

(K.)
Certificate of Visiting Surgeon.

IN pursuance of the Contagious Diseases Acts, 1866 to 1869, I hereby certify that I have this day examined *A.B.*, of , and that she is affected with a contagious disease within the meaning of those Acts; and the certified hospital in which she is to be placed under the said Acts is the hospital.

Dated this day of 18 .
(Signed) *E.F.*,
visiting surgeon for [*Portsmouth*].

(L.)
Certificate of Visiting Surgeon where Woman cannot properly be examined.

I HEREBY certify that *A.B.*, on being examined by me this day, in pursuance of the Contagious Diseases Acts, 1866 to 1869, was in such a condition that I could not properly examine her, and I have reasonable ground to believe that she is affected with contagious disease within the meaning of those Acts, and the certified hospital in which she is to be placed under the said Acts is the hospital.

Dated this day of 18 .
(Signed) *E.F.*,
visiting surgeon for [*Portsmouth*].

(M.)
Order of Visiting Surgeon for temporary Detention of Woman.

I HEREBY certify that *A.B.*, on attending this day for examination, in pursuance of the Contagious Diseases Acts, 1866 to 1869, was drunk, so that I could not properly examine her, and I have reasonable ground to believe that she is affected with contagious disease within the meaning of those Acts, and I hereby order that she be detained in the lock-up [*or as the case may be*], at in accordance with the said Acts.

Dated this day of 18 .
(Signed) *E.F.*,
visiting surgeon for [*Portsmouth*].

(N.)
Order by Inspector of Certified Hospitals for Transfer.

BY virtue of the power in this behalf vested in me by the Contagious Diseases Acts, 1866 to 1869, I hereby order that *A.B.* of , now detained under those Acts in the certified hospital of for medical treatment, be transferred thence to the certified hospital of

Dated this day of 18 .
(Signed) *M.N.*,
inspector of certified hospitals.

(O.)
Certificate for Detention beyond Three Months.
THE CONTAGIOUS DISEASES ACTS, 1866 to 1869.

WE, the undersigned, hereby certify that the further detention for medical treatment of *A.B.*, of , now an inmate of this hospital, is requisite.

Dated this day of 18 , at the hospital.
(Signed) *M.N.*,
inspector of certified hospitals,
[*or as the case may be.*]
G.H.,
chief medical officer.

(P.)

Discharge from Hospital.

IN pursuance of the Contagious Diseases Acts, 1866 to 1869, I hereby discharge *A.B.* of   from this hospital [*add according to the fact*, and certify that she is now free from a contagious disease].
Dated this   day of   18 , at the   hospital.
(Signed)   *G.H.*,
chief medical officer.

(Q.)

Certificate on Discharge from Imprisonment.

THE CONTAGIOUS DISEASES ACTS, 1866 to 1869.

WHEREAS under the above-mentioned Acts, *A.B.*   of   was on the   day of   convicted of the offence of   and has since been imprisoned for that offence in the gaol of   and is now discharged from imprisonment therein : Now in pursuance of the said Acts, I hereby certify that she is now free from a contagious disease.
Dated this   day of   .
R.O.,
surgeon of the gaol of
[*or E.F.*,
visiting surgeon for *Portsmouth*].

(R.)

Notice to Woman leaving Hospital.

THE CONTAGIOUS DISEASES ACTS, 1866 to 1869.

To *A.B.*

As you are now leaving this hospital, I hereby, in pursuance of the above-mentioned Acts, give you notice that you are still affected with a contagious disease.
Dated this   day of   .
(Signed)   *G.H.*,
chief medical officer.

Note.—The above-mentioned Acts provide as follows :—
If on any woman leaving a certified hospital a notice [*set out section of Act*].

(S.)

Certificate on last foregoing Notice or Copy.

IN pursuance of the within-mentioned Acts, I hereby certify that the within-named woman is now free from a contagious disease.
Dated this   day of   .
(Signed)   *E.F.*,
visiting surgeon for [*Portsmouth*].

(T.)

Application to be relieved from Examination.

To *L.M.*, Esq., and others, Her Majesty's justices of the peace for the [*county*] of   [*or* to *N.O.*, Esq., visiting surgeon for *Portsmouth, or as the case may be*].
I *A.B.*   of   , being in pursuance of the Contagious Diseases Acts, 1866 to 1869, subject to a periodical medical examination on my own submission [*or* under the order of *L.M.*, Esq., *as the case may be*], dated the   day of   , do hereby apply to be relieved therefrom.
Dated this   day of   18 .
(Signed)   *A.B.*
Witness, *G.W.*

ANNO DECIMO OCTAVO & DECIMO NONO VICTORIÆ REGINÆ.

CAP. CXVI.

An Act for the better Prevention of Diseases.

[14th *August* 1855.]

WHEREAS the provisions of "The Nuisances Removal and Diseases Prevention Act, 1848," amended by "The Nuisances Removal and Diseases Prevention Amendment Act, 1849," in so far as the same relate to the prevention or mitigation of epidemic, endemic, or contagious diseases, are defective, and it is expedient to substitute other provisions more effectual in that behalf: Be it therefore enacted by the Queen's most excellent Majesty, by and with the advice and consent of the Lords Spiritual and Temporal, and Commons, in this present Parliament assembled, and by the authority of the same, as follows:

Short Title.

I. This Act may be cited for all purposes as the "Diseases Prevention Act, 1855."

Local Authority for Execution of Act.

II. The local authority for executing this Act shall be the local authority acting in execution of any general Act in force for the time being for the removal of nuisances.

Expenses of Act.

III. The expenses incurred in execution of this Act shall be borne out of the rates or funds administered by such local authority, under the provisions and for the purposes of any such general Act as is referred to in the preceding section.

Power of Entry.

IV. The local authority and their officers shall have power of entry for the purposes of this Act, and for executing or superintending the execution of the regulations and directions of the General Board issued under this Act.

Power to Privy Council to issue Orders that Provisions herein contained for Prevention of Diseases may be put in force.

V. Whenever any part of *England* appears to be threatened with or is affected by any formidable epidemic, endemic, or contagious disease, the Lords and others of Her Majesty's most honourable Privy Council, or any three or more of them (the Lord President of the Council or one of Her Majesty's Principal Secretaries of State being one), may, by order or orders to be by them from time to time made, direct that the provisions herein contained for the prevention of diseases be put in force in *England*, or in such parts thereof as in such order or orders respectively may be expressed, and may from time to time, as to all or any of the parts to which any such order or orders extend, and in like manner, revoke or renew any such order; and, subject to revocation and renewal as aforesaid, every such order shall be in force for six calendar months, or for such shorter period as in such order shall be expressed; and every such order of Her Majesty's Privy Council, or of any members thereof, as aforesaid, shall be certified under the hand of the Clerk in Ordinary of Her Majesty's Privy Council, and shall be published in the *London Gazette;* and such publication shall be conclusive evidence of such order, to all intents and purposes.

Power to General Board of Health to issue Regulations to carry out such Provisions. Local Extent and Duration of Regulations of General Board.

VI. From time to time after the issuing of any such order as aforesaid, and whilst the same continues in force, the General Board of Health may issue directions and regulations, as the said Board think fit—

A A 2

For the speedy interment of the dead :
For house-to-house visitation :
For the dispensing of medicines, guarding against the spread of disease, and affording to persons afflicted by or threatened with such epidemic, endemic, or contagious diseases such medical aid and such accommodation as may be required :
And from time to time, in like manner, may revoke, renew, and alter any such directions and regulations as to the said Board appears expedient, to extend to all parts in which the provisions of this Act for the prevention of disease shall for the time being be put in force under such orders as aforesaid, unless such directions and regulations be expressly confined to some of such parts, and then to such parts as therein are specified ; and (subject to the power of revocation and alteration herein contained) such directions and regulations shall continue in force so long as the said provisions of this Act shall under such order be applicable to the same parts.

Publication of such Regulations.

VII. Every such direction and regulation as aforesaid, when issued, shall be published in the *London Gazette*, and the Gazette in which such direction or regulation was published shall be conclusive evidence of the direction or regulation so published, to all intents and purposes.

The Local Authority to see to the Execution of such Regulations, &c.

VIII. The local authority shall superintend and see to the execution of such directions and regulations, and shall appoint and pay such medical or other officers or persons, and do and provide all such acts, matters, and things, as may be necessary for mitigating such disease, or for superintending or aiding in the execution of such directions and regulations, or for executing the same, as the case may require.

The Local Authority may direct Prosecutions for violating the Same.

IX. The local authority may from time to time direct any prosecutions or legal proceedings for or in respect of the wilful violation or neglect of any such direction and regulation.

Orders of Council, Directions, and Regulations to be laid before Parliament.

X. Every order of Her Majesty's Privy Council, and every direction and regulation of the General Board of Health, under this Act, shall be laid before both Houses of Parliament, forthwith upon the issuing thereof if Parliament be then sitting, and if not then within fourteen days next after the commencement of the then next Session of Parliament.

Order in Council may extend to Parts and Arms of the Sea.

XI. Orders in Council issued in pursuance of this Act for putting in force the provisions for the prevention of disease in the said Nuisances Removal and Diseases Prevention Acts contained, in *Great Britain*, may extend to parts and arms of the sea lying within the jurisdiction of the Admiralty ; and the Board of Health for *England* may issue under this Act directions and regulations for cleansing, purifying, ventilating, and disinfecting, and providing medical aid and accommodation, and preventing disease in ships and vessels, as well upon arms and parts of the sea aforesaid as upon inland waters.

Medical Officer of Unions and others entitled to Costs of attending Sick on Board Vessels, when required by Orders of General Board of Health.

XII. Whenever, in compliance with any regulation of the General Board of Health, which they may be empowered to make under this Act, any medical officer appointed under and by virtue of the laws for the time being for the relief of the poor shall perform any medical service on board of any vessel, such medical officer shall be entitled to charge extra for any such service, at the general rate of his allowance for his services for the union or place for which he is appointed, and such charges shall be payable by the captain of the vessel, on behalf of the owners, together with any reasonable expenses for the treatment of the sick; and if such services shall be rendered by any medical practitioner who is not a union or parish officer, he shall be entitled to charges for any service rendered on board, with extra remuneration on account of distance, at the same rate as those which he is in the habit of receiving from private patients of the class of those attended and treated on shipboard, to be paid as aforesaid; and in case of dispute in respect of such charges, such dispute may, where the charges do not exceed twenty pounds, be determined summarily, at the place where the dispute arises, as in case of seamen's wages not exceeding fifty pounds, according to the provisions of the law in that behalf for the time being in force; and any justice before whom complaint is made shall determine summarily as to the amount which is reasonable, according to the accustomed rate of charge within the place for attendance on patients of the like class or condition as those in respect of whom the charge is made.

Authentication of Directions and Regulations of General Board of Health.

XIII. The directions and regulations of the General Board of Health under this enactment shall be under the seal of the said Board, and the hand of the president or two or more members thereof; and any copy of such regulations purporting to bear such seal and signature, whether the said signature and seal be respectively impressed and written, or printed only, shall be evidence in all proceedings in which such regulations may come in question.

Penalty for obstructing Execution of Act.

XIV. Whoever wilfully obstructs any person acting under the authority or employed in the execution of this Act, and whosoever wilfully violates any direction or regulation issued by the General Board of Health as aforesaid, shall be liable for every such offence to a penalty not exceeding five pounds, to be appropriated in or towards the defraying the expenses of executing this Act.

Certain Provisions of Nuisances Removal Act to apply to this Act.

XV. The provisions of any general Act in force for the removal of nuisances, with regard to the service of notices, the proof of orders or resolutions of the local authority, and the recovery of penalties, shall extend and apply to this Act.

ANNO DECIMO OCTAVO & DECIMO NONO VICTORIÆ REGINÆ.

CAP. CXXI.

An Act to consolidate and amend the Nuisances Removal and Diseases Prevention Acts, 1848 and 1849. [14th *August* 1855.]

11 & 12 *Vict.* c. 123.   12 & 13 *Vict.* c. 111.

WHEREAS the provisions of "The Nuisances Removal and Diseases Prevention Act, 1848," amended by "The Nuisances Removal and Diseases Prevention Amendment Act, 1849," are defective, and it is expedient to repeal the said Acts as far as relates to *England*, and to substitute other provisions more effectual in that behalf: Be it therefore enacted by the Queen's most excellent Majesty, by and with the advice and consent of the Lords Spiritual and Temporal, and Commons, in this present Parliament assembled, and by the authority of the same, as follows:

APPENDIX.

Recited Acts repealed as far as relates to England, except as to Proceedings commenced.

I. From and after the passing of this Act the said Acts are by this section repealed as far as relates to *England:* Provided always, that all proceedings commenced or taken under the said Acts, and not yet completed, may be proceeded with under the said Acts; and all contracts or works undertaken by virtue of the said Acts shall continue and be as effectual as if the said Acts had not been repealed.

Interpretation of certain Terms used in this Act.

II. In this Act the following words and expressions have the meanings by this section herein-after assigned to them, unless such meanings be repugnant to or inconsistent with the context; (that is to say,) the word "place" includes any city, borough, district under the Public Health Act, parish, township, or hamlet, or part of any such city, borough, district, town, parish, township, or hamlet; the word "guardians" includes the directors, wardens, overseers, governors, or other like officers having the management of the poor for any parish or place where the matter or any part of the matter requiring the cognizance of any such officer arises; the word "borough," and the expressions "mayor, aldermen, and burgesses," "council," and "borough fund," have respectively the same meanings as in the Acts for the Regulation of Municipal Corporations, and shall also respectively mean, include, and apply to any royal borough, royal town, or other town having a warden, high bailiff, borough reeve, or other chief officer, and burgesses or inhabitants, however designated, associated with him in the government or management thereof, or any town or place having a governing body therein in the nature of a corporation or otherwise, and to the chief officers and governing bodies of such boroughs, towns, and places, and to the funds and property under the management of or at the disposal of such chief officers and governing bodies; the expression "Improvement Act" means an Act for regulating and managing the police of, and for draining, cleansing, paving, lighting, watching, and improving a place, and an Act for any of those purposes; the word "owner" includes any person receiving the rents of the property in respect of which that word is used from the occupier of such property on his own account, or as trustee or agent for any other person, or as receiver or sequestrator appointed by the Court of Chancery or under any order thereof, or who would receive the same if such property were let to a tenant; the word "premises" extends to all messuages, lands, or tenements, whether open or enclosed, whether built on or not, and whether public or private; the word "parish" includes every township or place separately maintaining its poor or separately maintaining its own highways; the expression "quarter sessions" means the court of general or quarter sessions of the peace for a county, riding, or division of a county, city, or borough; the word "person," and words applying to any person or individual, apply to and include corporations, whether aggregate or sole; and the expression "two justices" shall, in addition to its ordinary signification, mean one stipendiary or police magistrate acting in any police court for the district.

PART I.

CONSTITUTION OF LOCAL AUTHORITY, EXPENSES, DESCRIPTION OF NUISANCES, AND POWERS OF ENTRY.

And with respect to the constitution of the local authority for the execution of this Act, the expenses of its execution, the description of nuisances that may be dealt with under it, and the powers of entry for the purposes of the Act, be it enacted thus:

The Local Authority to execute this Act in Places as herein stated.

III. The following bodies shall respectively be the local authority to execute this Act in the districts hereunder stated in *England:*

In any place within which the Public Health Act is or shall be in force, the local board of health:

In any other place wherein a council exists or shall exist, the mayor, aldermen, and burgesses by the council, except in the city of *London* and the liberties thereof, where the local authority shall be the commissioners of sewers for the time being; and except in the city of *Oxford* and borough of *Cambridge*, where the local authority shall be the commissioners acting in execution of the local Improvements Acts in force respectively in the said city and borough :

In any place in which there is no local board of health or council, and where there are or shall be trustees or commissioners under an Improvement Act, such trustees or commissioners :

In any place within which there is no such local board of health nor council, body of trustees, or commissioners, and where there is or shall be a board for the repair of the highways of such place, that board :

In any place where there is no such local board of health, council, body of trustees, or commissioners, nor highway board, a committee for carrying this Act into execution, by the name of "The Nuisances Removal Committee," of which the surveyor or surveyors of highways for the time being of such place shall be *ex officio* a member or members, may be annually chosen by the vestry on the same day as the overseers or surveyors of highways, and the first of such committees may be chosen at a vestry to be specially held for that purpose ; and such committee may consist of such number of members as the vestry shall determine, not being more than twelve, exclusive of such surveyor or surveyors, and of such committee three shall be a quorum :

In any place wherein there is no such local board of health, council, body of trustees, or commissioners, highway board, or committee appointed as aforesaid, and wherein there is or shall be a board of inspectors for lighting and watching under the Act 3 & 4 W. 4. c. 90., that board with the surveyor of highways :

In any place in which there is no such local board of health, council, body of trustees, or commissioners, nor highway board, nor committee appointed as aforesaid, nor board of inspectors for lighting and watching, the guardians and overseers of the poor and the surveyors of the highways in and for such place.

As to filling up Vacancies.

IV. On any vacancy in such nuisances removal committee arising from death, change of residence, or otherwise, notice shall be given by the committee to the churchwardens, who shall forthwith summon a meeting of the vestry, and fill up such vacancy by election ; and until such vacancy is filled up the remaining members of the committee may act in all respects as if their number was complete.

Power to Local Authority to appoint Committees.

V. The local authority may appoint any committee of their own body to receive notices, take proceedings, and in all or certain specified respects execute this Act, whereof two shall be a quorum ; and such local authority, or their committee, may, in each particular case, by order in writing under the hand of the chairman of such body or committee, empower any officer or person to make complaints and take proceedings on their behalf.

As to the Execution of this Act in Extra-parochial Places.

VI. In extra-parochial places not comprised within the jurisdiction of any of the local authorities aforesaid, and having a population of not less than two hundred persons, the local authority for the execution of this Act shall be a nuisance removal committee, elected annually by the householders within the extra-parochial place :

The first election of such committee shall take place at a meeting of such householders summoned for that purpose by the churchwardens of the adjacent place having the largest common boundary with such extra-parochial place ; and

APPENDIX.

Subsequent elections shall be held annually on some day in *Easter* week at meetings summoned by the chairman of the local authority for the year preceding:

Extra-parochial places not so comprised as aforesaid, and having a population of less than two hundred persons, shall for the purpose of this Act be attached to and form part of the adjacent place having the largest common boundary with the extra-parochial place, and notice of vestry meetings for the election of a local authority under and for the purposes of this Act shall be given in such extra-parochial places, and the householders within such places may attend such vestry meetings, and vote on such elections.

As to defraying Expenses of Executing this Act.

VII. All charges and expenses incurred by the local authority in executing this Act, and not recovered, as by this Act provided, may be defrayed as follows: to wit,

Out of general district rates, where the local authority is a local board of health ;

Out of the borough fund or borough rate, where the local authority is the mayor, aldermen, and burgesses by the council, or if there be an Improvement Act for the borough administered by the council, then out of rates levied thereunder applicable to the purposes of such Improvement Act ; or in the city of *London* and the liberties thereof, any rates or funds administered by the commissioners of sewers for the said city and liberties ;

Provided always, that in the city of *Oxford* and borough of *Cambridge* such expenses shall be deemed annual charges and expenses of cleansing the streets of the said city and borough respectively, and shall be so payable ;

Out of the rates levied for purposes of improvement under any Improvement Act, where the local authority is a body of trustees or commissioners acting in execution of the powers of such an Act ;

Out of highway rates, or any fund applicable in aid or in lieu thereof, where the local authority is a highway board, or a nuisance removal committee ;

Out of the rates for lighting and watching, where the local authority is a board of inspectors appointed for lighting and watching ;

And if there be no such rates or funds, or if the local authority be the guardians and surveyors of highways, then out of the rates or funds applicable to the relief of the poor of the parish or place wherein such rates or funds are collected or arise, if such parish or place be co-extensive with the district within which the charges and expenses are incurred, but if such parish or place be now or hereafter shall be partly comprised within and partly without the limits of a place where a local authority other than a highway board, nuisance removal committee, inspectors of watching and lighting, and surveyors or guardians and surveyors, exists or shall exist, all the charges and expenses incurred in the district comprising that part of the parish or place which is excluded from such limits shall be defrayed out of any highway rate or rates, or any funds applicable in lieu thereof, collected or raised within the part so excluded ; and if there be more than one highway rate collected within such district, the local authority shall settle the proportion in which the respective parties or places liable thereto shall bear such charges and expenses ; and if any portion of such excluded part be exempt from such highway rate or rates, then all the charges and expenses incurred in the whole of such excluded part shall be defrayed out of any district police rate or other rate which may by the Act 12 & 13 Vict. cap. 65. be raised and assessed upon such excluded part :

And when the local authority has not control of such rates or funds, the officer or person having the custody or control thereof shall pay over the amount to the local authority on the order of two justices, directed to such officer or person ; and on neglect or refusal to pay the sum specified in such order for six days after the service thereof, the same may, by warrant under the hands of the same or any two justices, be levied by distress and sale of the goods and chattels of the officer or person in default, and such levy shall include the cost of such distress and sale :

In extra-parochial places having a population of not less than two hundred persons,

out of a rate assessed by the local authority on all such property in the place as would be assessable to highway rate if such rate were levied therein :
In extra-parochial places having a population of less than two hundred persons, out of a similar rate assessed by the surveyor of highways of the adjacent p'ace having the largest common boundary with such extra-parochial place :
And the local authority in the first case, and the surveyor of highways in the second, may levy and collect the sums so assessed, in the same manner, and with the same remedies in case of any default in payment thereof, and with the same right of appeal against the amount of such assessment reserved to the person assessed, as are provided by the law in force for the time being with regard to rates for the repair of highways.

What are deemed Nuisances under this Act.

VIII. The word "nuisances" under this Act shall include—
Any premises in such a state as to be a nuisance or injurious to health :
Any pool, ditch, gutter, watercourse, privy, urinal, cesspool, drain, or ashpit so foul as to be a nuisance or injurious to health :
Any animal so kept as to be a nuisance or injurious to health :
Any accumulation or deposit which is a nuisance or injurious to health :
Provided always, that no such accumulation or deposit as shall be necessary for the effectual carrying on of any business or manufacture shall be punishable as a nuisance under this section, when it is proved to the satisfaction of the justices that the accumulation or deposit has not been kept longer than is necessary for the purposes of such business or manufacture, and that the best available means have been taken for protecting the public from injury to health thereby.

Power to Local Authority to appoint a Sanitary Inspector, and allow him a proper Salary.

IX. The local authority shall for the purposes of this Act appoint or employ, or join with other local authorities in appointing or employing, a sanitary inspector or inspectors, and may appoint a convenient place for his or their office, and may allow to every such person on account of his employment a proper salary or allowance ; and where local authorities join in such appointment or employment they may apportion among themselves the payment of such salary or allowance : Provided always, that where the local authority has already appointed an officer who executes the duties of such inspector under any Improvement Act, it shall not be necessary to appoint any other inspector under this Act, but the inspector acting in execution of the Improvement Act shall have all the powers, authorities, and privileges granted to any inspector appointed under this Act.

Notice of Nuisances to be given to Local Authority, &c. to ground Proceedings.

X. Notice of nuisance may be given to the local authority by any person aggrieved thereby, or by any of the following persons : the sanitary inspector or any paid officer under the said local authority ; two or more inhabitant householders of the parish or place to which the notice relates ; the relieving officer of the union or parish ; any constable or any officer of the constabulary or police force of the district or place ; and in case the premises be a common lodging-house any person appointed for the inspection of common lodging-houses ; and the local authority may take cognizance of any such nuisance after entry made as hereinafter provided, or in conformity with any Improvement Act under which the inspector has been appointed.

Power of Entry to Local Authority or their Officer.

XI. The local authority shall have power of entry for the following purposes of this Act, and under the following conditions :—
1. To ground proceedings.
For this purpose, when they or any of their officers have reasonable grounds for believing that a nuisance exists on any private premises, demand may be made by them or their officer, on any person having custody of the premises, of admission to inspect the same at any hour between nine in the morning and six in the even-

ing; and if admission be not granted, any justice having jurisdiction in the place may, on oath made before him of belief in the existence of the nuisance, and after reasonable notice of the intended application to such justice being given in writing to the party on whose premises the nuisance is believed to exist, by order under his hand, require the person having the custody of the premises to admit the local authority or their officer; and if no person having custody of the premises can be discovered, any such justice may and shall, on oath made before him of belief in the existence of such nuisance, and of the fact that no person having custody of the premises can be discovered, by order under his hand authorize the local authority or their officers to enter the premises between the hours aforesaid.

2. To examine premises where nuisances exist to ascertain the course of drains, and to execute or inspect works ordered by justices to be done under this Act.

For these purposes, whenever, under the provisions of this Act, a nuisance has been ascertained to exist, or when an order of abatement or prohibition under this Act has been made, or when it becomes necessary to ascertain the course of a drain, the local authority may enter on the premises, by themselves or their officers, between the hours aforesaid, until the nuisance shall have been abated or the course of the drain shall have been ascertained, or the works ordered to be done shall have been completed, as the case may be.

3. To remove or abate a nuisance in case of noncompliance with or infringement of the order of justices, or to inspect or examine any carcase, meat, poultry, game, flesh, fish, fruit, vegetables, corn, bread, or flour, under the powers and for the purposes of this Act.

For this purpose the local authority or their officer may from time to time enter the premises where the nuisance exists, or the carcase, meat, poultry, game, flesh, fish, fruit, vegetables, corn, bread, or flour is found, at all reasonable hours, or at all hours during which business is carried on on such premises, without notice.

PART II.

WITH REGARD TO REMOVAL OF NUISANCES.

Proceedings by Local Authority before Justices in the Case of Nuisances likely to recur, &c. If proved to Justices that Nuisance exists, &c., they shall issue Order for Abatement, &c.

With regard to the removal of nuisances, be it enacted thus:

XII. In any case where a nuisance is so ascertained by the local authority to exist, or where the nuisance in their opinion did exist at the time when the notice was given, and, although the same may have been since removed or discontinued, is in their opinion likely to recur or to be repeated on the same premises or any part thereof, they shall cause complaint thereof to be made before a justice of the peace; and such justice shall thereupon issue a summons requiring the person by whose act, default, permission, or sufferance the nuisance arises or continues, or, if such person cannot be found or ascertained, the owner or occupier of the premises on which the nuisance arises, to appear before any two justices in petty sessions assembled, at the usual place of meeting, who shall proceed to inquire into the said complaint; and if it be proved to their satisfaction that the nuisance exists, or did exist at the time when the notice was given, or, if removed or discontinued since the notice was given, that it is likely to recur or to be repeated, the justices shall make an order in writing under their hands and seals on such person, owner, or occupier for the abatement or discontinuance and prohibition of the nuisance as herein-after mentioned, and shall also make an order for the payment of all calls incurred up to the time of hearing or making the order for abatement or discontinuance or prohibition of the nuisance.

Justices Order for Abatement. Prohibitive Order against future Nuisance.

XIII. By their order the justices may require the person on whom it is made to provide sufficient privy accommodation, means of drainage or ventilation, or to make safe and habitable, or to pave, cleanse, whitewash, disinfect, or purify the premises which are a nuisance or injurious to health, or such part thereof as the justices may direct in their order, or to drain, empty, cleanse, fill up, amend, or remove the injurious pool, ditch, gutter, watercourse, privy, urinal, cesspool, drain, or ashpit which is a nuisance or injurious to health, or to provide a substitute for that complained of, or to carry away the accumulation or deposit which is a nuisance or injurious to health, or to provide for the cleanly and wholesome keeping of the animal kept so as to be a nuisance or injurious to health, or if it be proved to the justices to be impossible so to provide, then to remove the animal, or any or all of these things (according to the nature of the nuisance), or to do such other works or acts as are necessary to abate the nuisance complained of, in such manner and within such time as in such order shall be specified; and if the justices are of opinion that such or the like nuisance is likely to recur, the justices may further prohibit the recurrence of it, and direct the works necessary to prevent such recurrence, as the case may in the judgment of such justices require; and if the nuisance proved to exist be such as to render a house or building, in the judgment of the justices, unfit for human habitation, they may prohibit the using thereof for that purpose until it is rendered fit for that purpose in the judgment of the justices, and on their being satisfied that it has been rendered fit for such purpose they may determine their previous order by another declaring such house habitable, from the date of which other order such house may be let or inhabited.

Penalty for Contravention of Order of Abatement; and of Prohibition. Local Authority may enter and remove or abate Nuisance.

XIV. Any person not obeying the said order for abatement shall, if he fail to satisfy the justices that he has used all due diligence to carry out such order, be liable for every such offence to a penalty of not more than ten shillings *per* day during his default; and any person knowingly and wilfully acting contrary to the said order of prohibition shall be liable for every such offence to a penalty not exceeding twenty shillings *per* day during such contrary action; and the local authority may, under the powers of entry given by this Act, enter the premises to which the order relates, and remove or abate the nuisance condemned or prohibited, and do whatever may be necessary in execution of such order, and charge the cost to the person on whom the order is made, as herein-after provided.

Appeal against Order of Prohibition.

XV. Any such order of prohibition may be appealed against as provided in this Act.

Appeal against Order of Abatement when structural Works are required.

XVI. When it shall appear to the justices that the execution of structural works is required for the abatement of a nuisance, they may direct such works to be carried out under the direction or with the consent or approval of any public board, trustees, or commissioners having jurisdiction in the place in respect of such works; and if within seven days from the date of the order the person on whom it is made shall have given notice to the local authority of his intention to appeal against it as provided in this Act, and shall have entered into recognizances to try such appeal as provided by this Act, and shall appeal accordingly, no liability to penalty shall arise, nor shall any work be done nor proceedings taken under such order until after the determination of such appeal, unless such appeal cease to be prosecuted.

If Person causing Nuisance cannot be found, Local Authority to execute Order at once.

XVII. Whenever it appears to the satisfaction of the justices that the person by whose act or default the nuisance arises, or the owner or occupier of the premises is not known or cannot be found, then such order may be addressed to and executed by such local authority, and the cost defrayed out of the rates or funds applicable to the execution of this Act.

Manure, &c. to be sold.

XVIII. Any matter or thing removed by the local authority in pursuance of this enactment may be sold by public auction, after not less than five days' notice by posting bills distributed in the locality, unless in cases where the delay would be prejudicial to health, when the justices may direct the immediate removal, destruction, or sale of the matter or thing ; and the money arising from the sale retained by the local authority, and applied in payment of all expenses incurred under this Act with reference to such nuisance, and the surplus, if any, shall be paid, on demand, by the local authority, to the owner of such matter or thing.

Costs and Expenses of Works to be paid by Person on whom Order is made, or Owner or Occupier.

XIX. All reasonable costs and expenses from time to time incurred in making a complaint, or giving notice, or in obtaining an order of justices under this Act, or in carrying the same into effect under this Act, shall be deemed to be money paid for the use and at the request of the person on whom the order is made, or if the order be made on the local authority, or if no order be made, but the nuisance be proved to have existed when the complaint was made or the notice given, then of the person by whose act or default the nuisance was caused ; and in case of nuisances caused by the act or default of the owner of premises, the said premises shall be and continue chargeable with such costs and expenses, and also with the amount of any penalties incurred under this Act, until the same be fully discharged, provided that such costs and expenses shall not exceed in the whole one year's rack-rent of the premises ; and such costs and expenses, and penalties, together with the charges of suing for the same, may be recovered in any county or superior court, or, if the local authority think fit, before any two justices of the peace ; and the said justices shall have power to divide such costs, expenses, and penalties, between the persons by whose act or default the nuisance arises, in such manner as they shall consider reasonable ; and if it appear to them that a complaint made under this Act is frivolous or unfounded, they may order the payment by the local authority or person making the complaint of the costs incurred by the person against whom the complaint is made, or any part thereof.

Proceedings before Justices to recover Expenses.

XX. Where any costs, expenses, or penalties are due under or in consequence of any order of justices made in pursuance of this Act as aforesaid, any justice of the peace, upon the application of the local authority, shall issue a summons requiring the person from whom they are due to appear before two justices at a time and place to be named therein ; and upon proof to the satisfaction of the justices present that any such costs, expenses, or penalties are so due, such justices, unless they think fit to excuse the party summoned upon the ground of poverty or other special circumstances, shall, by order in writing under their hands and seals, order him to pay the amount to the local authority at once, or by such instalments as the justices think fit, together with the charges attending such application and the proceedings thereon ; and if the amount of such order, or any instalment thereof, be not paid within fourteen days after the sum is due, the same may, by warrant of the said or other justices, be levied by distress and sale.

Surveyors of Highways to cleanse Ditches, &c., paying Owners, &c. for Damages.

XXI. All surveyors and district surveyors may make, scour, cleanse, and keep open all ditches, gutters, drains, or watercourses in and through any lands or grounds adjoining or lying near to any highway, upon paying the owner or occupier of such lands or grounds, provided they are not waste or common, for the damages which he shall thereby sustain, to be settled and paid in such manner as the damages for getting materials in enclosed lands or grounds are directed to be settled and paid by the law in force for the time being with regard to highways.

APPENDIX.   365

Power to Local Authority to cover and improve open Ditches, &c.

XXII. Whenever any ditch, gutter, drain, or watercourse used or partly used for the conveyance of any water, filth, sewage, or other matter from any house, buildings, or premises is a nuisance within the meaning of this Act, and cannot, in the opinion of the local authority, be rendered innocuous without the laying down of a sewer or of some other structure along the same or part thereof or instead thereof, such local authority shall and they are hereby required to lay down such sewer or other structure, and to keep the same in good and serviceable repair; and they are hereby declared to have the same powers as to entering lands for the purposes thereof, and to be entitled to recover the same penalties in case of interference as are contained in the sixty-seventh and sixty-eighth sections of the Act passed in the fifth and sixth years of the reign of King *William* IV., intituled "An Act for consolidating and amending the Laws relating to Highways in *England*;" and such local authority are hereby authorized and empowered to assess every house, building, or premises then or at any time thereafter using for the purposes aforesaid the said ditch, gutter, drain, watercourse, sewer, or other structure, to such payment, either immediate or annual, or distributed over a term of years, as they shall think just and reasonable, and, after fourteen days' notice at the least left on the premises so assessed, to levy and collect the sum and sums so assessed in the same manner, and with the same remedies in case of default in payment thereof, as highway rates are by the law in force for the time being leviable and collectable, and with the same right and power of appeal against the amount of such assessments reserved to the person or persons so assessed as by the law for the time being in force shall be given against any rate made for the repair of the highways; and the provisions contained in this section shall be deemed to be part of the law relating to highways in *England*: Provided always, that where such ditch, gutter, drain, or watercourse shall, as to parts thereof, be within the jurisdiction of different local authorities, this enactment shall apply to each local authority only as to so much of the works hereby required, and the expenses thereof, as is included within the respective jurisdiction of that authority: Provided also, that such assessment shall in no case exceed a shilling in the pound on the assessment to the highway rate, if any.

Penalty for causing Water to be corrupted by Gas Washings.

XXIII. Any person or company engaged in the manufacture of gas who shall at any time cause or suffer to be brought or to flow into any stream, reservoir, or aqueduct, pond, or place for water, or into any drain communicating therewith, any washing or other substance produced in making or supplying gas, or shall wilfully do any act connected with the making or supplying of gas whereby the water in any such stream, reservoir, aqueduct, pond, or place for water shall be fouled, shall forfeit for every such offence the sum of two hundred pounds.

Penalty to be sued for in Superior Courts within Six Months.

XXIV. Such penalty may be recovered, with full costs of suit, in any of the superior courts, by the person into whose water such washing or other substance shall be conveyed or shall flow, or whose water shall be fouled by any such act as aforesaid, or if there be no such person, or in default of proceedings by such person, after notice to him from the local authority of their intention to proceed for such penalty, by the local authority; but such penalty shall not be recoverable unless it be sued for during the continuance of the offence, or within six months after it shall have ceased.

Daily Penalty during the Continuance of the Offence.

XXV. In addition to the said penalty of two hundred pounds (and whether such penalty shall have been recovered or not), the person or company so offending shall forfeit the sum of twenty pounds (to be recovered in the like manner) for each day during which such washing or other substance shall be brought or shall flow as aforesaid, or during which the act by which such water shall be fouled shall continue, after the expiration of twenty-four hours from the time when notice of the offence shall have been served on such person or company by the local authority, or the person into whose water such washing or other substance shall be brought or flow, or whose water shall be fouled thereby, and such penalty shall be paid to the parties from whom such notice shall proceed; and all monies recovered by a local authority under this or the preceding section shall, after payment of any damage caused by the act for which the penalty is imposed, be applied towards defraying the expenses of executing this Act.

Penalty on Sale of unwholesome Meat, &c.

XXVI. The sanitary inspector may at all reasonable times inspect and examine any carcase, meat, poultry, game, flesh, fish, fruit, vegetables, corn, bread, or flour exposed for sale, or in the course of or on their way to slaughtering, dressing, or preparation for sale or use, or landed from any ship or vessel in any port in *England*; and in case any such carcase, meat, poultry, game, flesh, fish, fruit, vegetables, corn, bread, or flour appear to him to be unfit for such food, the same may be seized; and if it appear to a justice that any such carcase, meat, poultry, game, flesh, fish, fruit, vegetables, corn, bread, or flour, is unfit for the food of man, he shall order the same to be destroyed or to be so disposed of as to prevent its being exposed for sale or used for such food; and the person to whom such carcase, meat, poultry, game, flesh, fish, fruit, vegetables, corn, bread, or flour belongs, or in whose custody the same is found, shall be liable to a penalty not exceeding ten pounds for every carcase, fish, or piece of meat, flesh, or fish, or any poultry or game, or for the parcel of fruit, vegetables, corn, bread, or flour so found.

As to Nuisances arising in Cases of noxious Trades, Businesses, Processes, or Manufactures.

XXVII. If any candle-house, melting-house, melting-place, or soaphouse, or any slaughterhouse, or any building or place for boiling offal or blood, or for boiling, burning, or crushing bones, or any manufactory, building, or place used for any trade, business, process, or manufacture causing effluvia, be at any time certified to the local authority by any medical officer, or any two legally qualified medical practitioners, to be a nuisance or injurious to the health of the inhabitants of the neighbourhood, the local authority shall direct complaint to be made before any justice, who may summon before any two justices in petty sessions assembled at their usual place of meeting the person by or in whose behalf the work so complained of is carried on, and such justices shall inquire into such complaint, and if it shall appear to such justices that the trade or business carried on by the person complained against is a nuisance, or causes any effluvia injurious to the health of the inhabitants of the neighbourhood, and that such person shall not have used the best practicable means for abating such nuisance or preventing or counteracting such effluvia, the person so offending (being the owner or occupier of the premises, or being a foreman or other person employed by such owner or occupier), shall, upon a summary conviction for such offence, forfeit and pay a sum of not more than five pounds nor less than forty shillings, and upon a second conviction for such offence, the sum of ten pounds, and for each subsequent conviction a sum double the amount of the penalty imposed for the last preceding conviction, but the highest amount of such penalty shall not in any case exceed the sum of two hundred pounds: Provided always, that the justices may suspend their final determination in any such case, upon condition that the person so complained against shall undertake to adopt, within a reasonable time, such means as the said justices shall judge to be practicable and order to be carried into effect for abating such nuisance, or mitigating or preventing the injurious effects of such effluvia, or shall give notice of appeal in the manner provided by this Act, and shall enter into recognizances to try such appeal, and shall appeal accordingly: Provided always, that the provisions herein-before contained shall not extend or be applicable to any place without the limits of any city, town, or populous district.

APPENDIX. 367

Reference to Superior Court at the Option of the Party complained against.

XXVIII. Provided also, that if, upon his appearance before such justices, the party complained against object to have the matter determined by such justices, and enter into recognizances, with sufficient sureties to be approved by the justices, to abide the event of any proceedings at law or in equity that may be had against him on account of the subject matter of complaint, the local authority shall thereupon abandon all proceedings before the justices, and shall forthwith take proceedings at law or in equity in Her Majesty's superior courts for preventing or abating the nuisance complained of.

On Certificate of Medical Officer to Local Authority that House is overcrowded, Proceedings may be taken to abate the same.

XXIX. Whenever the medical officer of health, if there be one, or, if none, whenever two qualified medical practitioners, shall certify to the local authority that any house is so overcrowded as to be dangerous or prejudicial to the health of the inhabitants, and the inhabitants shall consist of more than one family, the local authority shall cause proceedings to be taken before the justices to abate such overcrowding, and the justices shall thereupon make such order as they may think fit, and the person permitting such overcrowding shall forfeit a sum not exceeding forty shillings.

Local Authority to order Costs of Prosecutions to be paid out of the Rates.

XXX. The local authority may, within the area of their jurisdiction, direct any proceedings to be taken at law or in equity in cases coming within the purview of this Act, and may order proceedings to be taken for the recovery of any penalties, and for the punishment of any persons offending against the provisions of this Act, or in relation to appeals under this Act, and may order the expenses of all such proceedings to be paid out of the rates or funds administered by them under this Act.

PART III.

AS TO PROCEDURE UNDER THIS ACT.

Service of Notices, Summonses, and Orders.

And with regard to procedure under this Act, be it enacted, that—

XXXI. Notices, summonses, and orders under this Act may be served by delivering the same to or at the residence of the persons to whom they are respectively addressed, and where addressed to the owner or occupier of premises they may also be served by delivering the same or a true copy thereof to some person upon the premises, or if there be no person upon the premises who can be so served, by fixing the same upon some conspicuous part of the premises, or if the person shall reside at a distance of more than five miles from the office of the inspector then by a registered letter through the post.

Proof of Resolutions of Local Authority.

XXXII. Copies of any orders or resolutions of the local authority or their committee, purporting to be signed by the chairman of such body or committee, shall, unless the contrary be shown, be received as evidence thereof, without proof of their meeting, or of the official character or signature of the person signing the same.

As to Proceedings taken against several Persons for the same Offence.

XXXIII. Where proceedings under this Act are to be taken against several persons in respect of one nuisance caused by the joint act or default of such persons, it shall be lawful for the local authority to include such persons in one complaint, and for the justices to include such persons in one summons, and any order made in such a case may be made upon all or any number of the persons included in the summons, and the costs may be distributed as to the justices may appear fair and reasonable.

One or more joint Owners or Occupiers may be proceeded against alone.

XXXIV. In case of any demand or complaint under this Act to which two or more persons, being owners or occupiers of premises or partly the one or partly the other, may be answerable jointly or in common or severally, it shall be sufficient to proceed against any one or more of them without proceeding against the others or other of them; but nothing herein contained shall prevent the parties so proceeded against from recovering contribution in any case in which they would now be entitled to contribution by law.

Designation of " Owner" or " Occupier."

XXXV. Whenever in any proceeding under this Act, whether written or otherwise, it shall become necessary to mention or refer to the owner or occupier of any premises, it shall be sufficient to designate him as the "owner" or "occupier" of such premies, without name or further description.

Penalty for obstructing Execution of this Act.

XXXVI. Whoever refuses to obey an order of justices under this Act for admission on premises of the local authority or their officers, or wilfully obstructs any person acting under the authority or employed in the execution of this Act, shall be liable for every such offence to a penalty not exceeding five pounds.

Penalty on Occupier obstructing Owner.

XXXVII. If the occupier of any premises prevent the owner thereof from obeying or carrying into effect the provisions of this Act, any justice to whom application is made in this behalf shall by order in writing require such occupier to desist from such prevention, or to permit the execution of the works required to be executed, provided that such works appear to such justice to be necessary for the purpose of obeying or carrying into effect the provisions of this Act; and if within twenty-four hours after the service of such order the occupier against whom it is made do not comply therewith, he shall be liable to a penalty not exceeding five pounds for every day afterwards during the continuance of such noncompliance.

Penalties and Expenses recoverable under 11 & 12 Vict. c. 43.

XXXVIII. Penalties imposed by this Act for offences committed and sums of money ordered to be paid under this Act may be recovered by persons thereto competent in *England* according to the provisions of the Act of the eleventh and twelfth years of the present reign, chapter forty-three; and all penalties recovered by the local authority under this Act shall be paid to them, to be by them applied in aid of their expenses under this Act.

Proceedings not to be quashed for Want of Form.

XXXIX. No order, nor any other proceeding, matter, or thing done or transacted in or relating to the execution of this Act, shall be vacated, quashed, or set aside for want of form, nor shall any order, nor any other proceeding, matter, or thing done or transacted in relation to the execution of this Act, be removed or removable by certiorari, or by any other writ or process whatsoever, into any of the superior courts; and proceedings under this Act against several persons included in one complaint shall not abate by reason of the death of any among the persons so included, but all such proceedings may be carried on as if the deceased person had not been originally so included.

Appeals under this Act to be to Quarter Sessions.

XL. Appeals under this Act shall be to the court of quarter sessions held next after the making of the order appealed against; but the appellant shall not be heard in support of the appeal unless within fourteen days after the making of the order appealed against he give to the local authority notice in writing stating his intention to bring such appeal, together with a statement in writing of the grounds of appeal, and shall within two days of giving such notice enter into a recognizance before some justice of the peace, with sufficient securities, conditioned to try such appeal at the said court, and to abide the order of and pay such costs as shall be awarded by the justices at such court or any adjournment thereof; and the said court, upon hearing and finally determining the matter of the appeal, may, according to its discretion, award such costs to the party appealing or appealed against as they shall think proper, and its determination in or concerning the premises shall be conclusive and binding on all persons to all intents or purposes whatsoever: Provided always, that if there be not time to give such notice and enter into such recognizance as aforesaid, then such appeal may be made to, and such notice, statement, and recognizance be given and entered into for, the next sessions at which the appeal can be heard; provided also, that on the hearing of the appeal no grounds of appeal shall be gone into or entertained other than those set forth in such statement as aforesaid; provided also, that in any case of appeal the court of quarter sessions may, if they think fit, state the facts specially for the determination of Her Majesty Court of Queen's Bench, in which case it shall be lawful to remove the proceedings, by writ of certiorari or otherwise, into the said Court of Queen's Bench.

Forms to be used as in Schedule.

XLI. The forms contained in the schedule to this Act annexed, or any forms to the like effect, varied as circumstances may require, may be used for instruments under this Act, and shall be sufficient for the purpose intended.

As to Protection of Local Authority; and its Officers.

XLII. The local authority, and any officer or person acting under the authority and in execution or intended execution of this Act, shall be entitled to such protection and privilege in actions and suits, and such exemption from personal liability, as are granted to local boards of health and their officers by the law in force for the time being.

Act not to impair Jurisdiction of Sewers Commissioners, or Common Law Remedies for Nuisance, nor Jurisdiction of Local Authority as to the Nuisances referred to in this Act.

XLIII. Nothing in this Act shall be construed to affect the provisions of any local Act as to matters included in this Act, nor to impair, abridge, or take away any power, jurisdiction, or authority which may at any time be vested in any commissioners of sewers or of drainage, or to take away or interfere with any course of proceedings which might be resorted to or adopted by such commissioners if this Act had not passed, nor to impair any power of abating nuisances at common law; nor any jurisdiction in respect of nuisances that may be possessed by any authority under the Act intituled " An Act to abate the Nuisances arising from the Smoke of Furnaces in the Metropolis, and from Steam Vessels above *London Bridge*," or the Common Lodging Houses Acts, the Act for the Regulation of Municipal Corporations, the Public Health Act, or any Improvement Act respectively, or any Acts incorporated with such Acts, and authorities may respectively proceed for the abatement of nuisances or in respect of any other matter or thing hereinbefore provided or referred to either under the Acts mentioned in this section or any other Act conferring jurisdiction in respect of the nuisances referred to in this Act, or any byelaws framed under any such Act, as they may think fit; and the local authorities constituted under and for the purposes of the Common Lodging House Acts, 1851 and 1853, shall for the purposes of those Acts have all the powers o local authorities under this Act.

Act not to affect Navigation of Rivers or Canals.

XLIV. Nothing herein contained shall enable any local authority, surveyor of highways, or other person, either with or without any order of justices, to

B B

injuriously affect the navigation of any river or canal, or to divert or diminish any supply of water of right belonging to any such river or canal; and the provisions of this Act shall not extend or be construed to extend to mines of different descriptions so as to interfere with or obstruct the efficient working of the same, or to the smelting of ores and minerals or to the manufacturing of the produce of such ores and minerals.

Saving as to Rights of Millowners, &c.

XLV. No power given by this Act shall be exercised in such manner as to injuriously affect the supply, quality, or fall of water contained in any reservoir or stream, or any feeders of such reservoir or stream, belonging to or supplying any waterwork established by Act of Parliament, or in cases where any company or individual are entitled for their own benefit to the use of such reservoir or stream, or to the supply of water contained in such feeders, without the consent in writing of the company or corporation in whom such waterworks may be vested, or of the parties so entitled to the use of such reservoirs, streams, and feeders, and also of the owners thereof in cases where the owners and parties so entitled are not the same person.

Short Title.

XLVI. In citing this Act in other Acts of Parliament, and in legal instruments and other proceedings, it shall be sufficient to use the words "The Nuisances Removal Act for *England*, 1855."

SCHEDULE OF FORMS.

FORM (A.)

Order of Justices for Admission of Officer of Local Authority to inspect private Premises.

WHEREAS [*describe the local authority*] have by their officer [*naming him*] made application to me *A. B.*, one of Her Majesty's justices of the peace having jurisdiction in and for [*describe the place*], and the said officer has made oath to me of his belief that a nuisance, within the meaning of the Nuisances Removal Act for England, 1855, viz. [*describe nuisance*], exists on private premises at [*describe situation of premises so as to identify them*], within my jurisdiction, and demand of admission to such premises for the inspection thereof has been duly made under the said Act, and refused :

Now, therefore, I, the said *A.B.*, do hereby require you to admit the said [*name the local authority*], [or the officer of the said (*local authority*)], for the purpose of inspecting the said premises.

Dated this   day of   18 .

                                                                          A.B.

FORM (B.)

Notice of Nuisance.

To the local authority (*describing it*).

I [*or* we], the person aggrieved by the nuisance herein-after described [*or* the undersigned and described inhabitant householders, sanitary inspector, *or other officer* (*describing him*)], do hereby give you notice, that there exists in or upon the [dwelling house, yard, *etc.*, *as the case may be*], situate at [*giving such description as may be sufficient to identify the premises*], in the parish of   in your district, under the Nuisances Removal Act, 1855, the following nuisance, videlicet [*describing the nuisance, as the case may be; for instance*, a dwelling house or building a nuisance or injurious to health for want of a privy or drain *or* sufficient means of ventilation, *or* so dilapidated *or* so filthy as to be a nuisance or injurious to health, *or*, *for further instance*, a ditch *or* drain so foul as to be a nuisance or injurious to health, *or* an accumulation of

a nuisance or injurious to health, &c., or swine so kept as to be a nuisance or injurious to health]; and that such nuisance is caused by [*naming the person by whose act or default the nuisance is caused, or by some person unknown*].
Dated this   day of   in the year of our Lord one thousand eight hundred and   .

[*Signed by complainant under Section* 10.]

FORM (C.)

Notice to Owner or Occupier of Entry for Examination.

To the owner [*or occupier, as the case may be,*] of [*describe the premises*] situate at [*insert a description sufficient to identify the premises*].

Take notice, That, under the Nuisances Removal Act for England, 1855, the [*local authority, naming it,*] in whose district under the said Act the above premises are situate, have received a notice from [*name complainant*], stating that in or upon the said premises [*insert the cause of nuisance as set forth in the notice*].

And further take notice, That after the expiration of twenty-four hours from the service of this notice the [*local authority*] will cause the said premises to be entered and examined under the provisions of the said Act, and if the cause of nuisance aforesaid be found still existing, or, though removed or discontinued, be likely to be repeated, a summons will be issued requiring your attendance to answer a complaint which will be made to the justices for enforcing the removal of the same, and prohibiting a repetition thereof, and for recovering the costs and penalties that may be incurred thereby.

Dated this   day of   in the year of our Lord one thousand eight hundred and   .

A.B.,
the officer appointed by the [*local authority*] to take proceedings under the Nuisances Removal Act for England, 1855.

FORM (D.)

Summons.

To the owner or occupier of [*describe premises*] situate at [*insert such a description as may be sufficient to identify the premises*], or to *A.B.* of

County of   ⎫   You are required to appear before two of Her Majesty's
[or borough of   ⎪   justices of the peace [*or* one of the magistrates of the
&c. or district of   ⎬   police courts of the metropolis, or the stipendiary ma-
or as the case may be]   ⎪   gistrate] of the county [*or other jurisdiction*] of
to wit.   ⎭   at the petty sessions [*or court*] holden at
on the   day of   next, at the hour of
in the   noon, to answer the complaint this day made to me by
[or by   on behalf of] [*naming the local authority, as the case may be*], that in or upon the premises above mentioned [or in or upon certain premises situate at No.   in the   street in the parish of
            or such other description or reference as may be sufficient to identify the premises], in their district, under the Nuisances Removal Act for England, 1855, the following nuisance exists [*describing it, as the case may be*], and that the said nuisance is caused by the act or default of the occupier [*or owner*] of the said premises, or by you *A.B.* [*or in case the nuisance be discontinued, but likely to be repeated, say*], there existed recently, to wit, on or about the   day of   on the premises, the following nuisance [*describe the nuisance*], and that the said nuisance was caused [*&c.*], and although the same has since the said last-mentioned day been removed or discontinued there is reasonable

ground to consider that the same or the like nuisance is likely to recur on the said premises].

Given under the hand of me, J.P., Esquire, one of Her Majesty's justices of the peace acting in and for the [*jurisdiction*] stated in the margin, or one of the magistrates of the police courts of the metropolis, *or* stipendiary magistrate of   day of   in the year of our Lord one thousand eight hundred and

FORM (E.)

Order of Justices for Removal of Nuisances by Owner, &c.

To the owner [*or* occupier] of [*describe the premises*] situate [*give such description as may be sufficient to identify the premises*], or to A.B. of or to [*giving name of the local authority*], or to their servants or agents, and to all whom it may concern.

County of   WHEREAS on the   day of   com-
[*or* borough, *&c.* of   plaint was made before   Esquire,
           or   one of Her Majesty's justices of the peace acting in and
district of   for the county [*or other jurisdiction*] stated in the margin,
or as the case may   [*or* before the undersigned, one of the magistrates of the
be].   police courts of the metropolis, *or as the case may be*] by
[*or by*   on behalf of] [*the local authority, naming it, as the case may be*], that in or upon certain premises situate at   in the district under the Nuisances Removal Act for England, 1855, of the complainants above named, the following nuisance then existed [*describing it*]; and that the said nuisance was caused by the act or default of the owner [*or* occupier] of the said premises [or was caused by *A.B.*] (*If the nuisance have been removed, say,* the following nuisance existed on or about [*the day the nuisance was ascertained to exist*], and that the said nuisance was caused, *&c.*, and although the same is now removed, the same or the like nuisance is likely to recur on the same premises.)

And whereas   the owner [*or* occupier] within the meaning of the said Nuisances Removal Act, 1855, [*or* the said *A.B.,*] hath this day appeared before us justices, being two of Her Majesty's justices in and for   , sitting in petty sessions at their usual place of meeting [or before me, the said magistrate of the police courts of the metropolis, *or as the case may be*], to answer the matter of the said complaint [*Or in case the party charged do not appear, say,* And whereas it hath been this day proved to our [*or* my] satisfaction that a true copy of a summons requiring the owner [*or* occupier] of the said premises [*or* the said *A.B.*] to appear this day before us [*or* me]   hath been duly served according to the said Act :]

Now upon proof here had before us [*or* me] that the nuisance so complained of doth exist on the said premises, and that the same is caused by the act or default of the owner [*or* occupier] of the said premises [*or* by the said *A.B.*], we [*or* I], in pursuance of the said Act, do order the said owner [*or* occupier, *or A.B.*] within [*specify the time*] from the service of this order or a true copy thereof according to the said Act [*here specify the works to be done, as, for instance,* to cleanse, whitewash, purify, and disinfect the said dwelling house ; *or, for further instance,* to construct a privy *or* drain, *&c.* ; *or, for further instance,* to cleanse *or* to cover *or* to fill up the said cesspool, *&c.*], so that the same shall no longer be a nuisance or injurious to health as aforesaid.

[*And if it appear to the justices that the nuisance is likely to recur on the premises say* [And we] [*or* I] being satisfied that, notwithstanding the said cause or causes of nuisances may be removed under this order, the same is or are likely to recur, do therefore prohibit the said owner [*or* occupier, *or A.B.*] from [*here insert the matter of the prohibition, as, for instance,* from using the said house or building for human habitation until the same, in our judgment, is rendered fit for that purpose]

And if the above order for abatement be not complied with [*or* if the above order of prohibition be infringed], then we [or I] do authorize and require you the said [local authority, naming it], from time to time to enter upon the said premises, and to do all such works, matters, and things as may be necessary for carrying this order into full execution according to the Act aforesaid.

In case the nuisance were removed before complaint, say, [Now, upon proof here had before us that at or recently before the time of making the said complaint, to wit, on   as aforesaid, the cause of nuisance complained of did exist on the said premises, but that the same hath since been removed, yet, notwithstanding such removal, we [or I] being satisfied that it is likely that the same or the like nuisance will recur on the said premises, do hereby prohibit [*order of prohibition*] ; and if this order of prohibition be infringed, then we [or I] *order on local authority to do works*].

Given under the hands and seals of us, two of Her Majesty's justices of the peace in and for   [or the hand and seal of me, one of the magistrates of the police courts of the metropolis, *or as the case may be*], this   day of   in the year of our Lord one thousand eight hundred and   .

Form (F.)

Order of Justices for Removal of Nuisance by Local Authority.

To the town council, &c., *as the case may be.*

County, &c. } Whereas [recite complaint of nuisance as in last form].
to wit.

And whereas it hath been now proved to our [or my] satisfaction that such nuisance exists, but that no owner or occupier of the premises, or person causing the nuisance, is known *or* can be found [*as the case may be*]: Now we [or I], in pursuance of the said Act, do order the said [*local authority, naming it,*] forthwith to [*here specify the works to be done*].

Given, &c.

Form (G.)

Order to permit Execution of Works by Owners.

County of [or borough of or metropolitan police district, *or as the case may be,*] to wit. } Whereas complaint hath been made to me, E.F., Esquire, one of Her Majesty's justices of the peace in and for the county [or borough, &c.] of   [or one of the magistrates of the police courts of the metropolis, *or as the case may be*, or one of Her Majesty's justices of the peace, *as the case may be*, of the county of   ], by A.B., owner within the meaning of the "Nuisances Removal Act for England, 1855," of certain premises, to wit, a dwelling house [or building, *or as the case may be*], situate at [*insert such a description of the premises as may be sufficient to identify them*], in the parish of   in the said county [*or borough, &c.*], that C.D., the occupier of the said premises, doth prevent the said A.B. from obeying and carrying into effect the provisions of the said Act, in this, to wit, that he the said C.D. [*here describe the act of prevention generally according to the circumstances; for instance thus,* doth refuse to quit the said house, the same having by the order of justices been declared unfit for human habitation, *or* doth prevent the said A.B. from cleansing *or* whitewashing *or* purifying the said dwelling house, *or* erecting a privy *or* drain, *or* breaking an aperture for ventilation, *or* cleansing a drain, ditch, gutter, watercourse, privy, urinal, cesspool, *or* ashpit which is a nuisance or injurious to health]: And whereas the said C.D. has been summoned to answer the said complaint, and has not shown sufficient cause against the same, and it appears to me that [*describe the act or works to be done*] is necessary for the purpose of enabling the said A.B. to obey and carry into effect the provisions of the said Act, I do hereby order that the said C.D. do permit the said A.B. [*describe the act or works to be done*] in the manner required by the said Act.

Given under my hand and seal this   day of
          in the year of our Lord one thousand eight hundred and
                                                                E.F.   (L.S.)

FORM (H.)

Summons for Nonpayment of Costs, Expenses, or Penalties. Sec. 20.

To   [*describe
          the person from whom the costs, expenses, and penalties are due*].

County of   ⎱ You are required to appear before two of Her Majesty's
or borough of   ⎰ justices of the peace [*or* one of the magistrates of the
or district of   ⎱ police courts of the metropolis, *or* the stipendiary
to wit.   ⎰ magistrates] of the county [*or other jurisdiction*] of
          at the petty sessions [*or* court] holden at   on the
          day of   next, at the hour of   in the
          noon, to answer the complaint this day made to me by
[*or* by   on behalf of] [*naming the local authority*],
that the sum of   pounds, being costs and expenses incurred
by you under and in relation to a certain complaint touching [*describe the nuisance*],
and an order of [*describe the person making the order*] duly made in pursuance of
the Nuisances Removal Act for England, 1855 [*if penalties are due, add*, and also
the sum of   being the amount of penalties payable by you for disobedience of the said order], remains unpaid and due from you.

Given under the hand of me, J.P., Esquire, one of Her Majesty's justices of the
          peace acting in and for the [*jurisdiction stated in the margin*] [*or* one of the
          magistrates of the police courts of the metropolis, *or* stipendiary magistrate
          of   ] the   day of   in the year of our Lord
          one thousand eight hundred and   .

FORM (I.)

Order for Payment of Costs, Expenses, and Penalties. Sec. 20.

To   [*name the person on whom the order is made*].

County, &c., ⎱ WHEREAS complaint has been made before us [*or* me] for that
to wit.   ⎰ [*recite cause of complaint*].

And whereas the said [*naming the person against whom the complaint is made*] has this day appeared before us the said justices [*or* before me the said magistrate of the police courts of the metropolis, *or as the case may be*], to answer this matter of the said complaint : [*Or, in case the party charged do not appear, say*],

And whereas it has been this day satisfactorily proved to us [*or* me] that a true copy of the summons requiring the said [*naming person charged*] to appear before us [*or* me] this day hath been duly served according to the said Act : Now, having heard the matter of the said complaint, we [*or* I] do adjudge the said [*naming the person charged*] to pay forthwith [*or* by instalments of   payable respectively on or before the   ] to the said [*naming the person or local authority to whom the costs adjudged are payable*], the sum of   for costs in this behalf, and to [*naming the person or authority to whom the expenses are payable*], the sum of   for expenses in this behalf, [*if penalties are due*, add, and the sum of   for penalties incurred in relation to the premises,] together with the sum of   being the charges attending the application for this order and proceedings thereon ; and if the said several sums, amounting in the whole to   [*or if any one of the said instalments*] be not paid within fourteen days after the same is due as aforesaid, we [*or* I] hereby order that the same be levied by distress and sale of the goods and chattels of the said   and in default of sufficient distress in that behalf adjudge the said   to be imprisoned in the common gaol [*or* House of Correction, *as the case may be*,] at   in the said county [*or as the case may be*] for the space of such time, not exceeding three calendar months, as the justices may think fit, unless the said several sums [*or* sum] and all costs and charges of the said distress [and of the commitment and carrying of the said   to the said House of Correction *or* common gaol, *or as the case may be,*] shall be sooner paid.

Given under our [*or* my] hands, this   day of   in the year of our Lord one thousand eight hundred and   at in the [county, *or as the case may be,*] aforesaid.

Form (K.)

Warrant of Distress. Sec. 20.

To the constable of   and to all other peace officers in the said county [*or as the case may*].

WHEREAS on   last past complaint was made before the undersigned, two of Her Majesty's justices of the peace in and for the said county of [*or as the case may be*] [*or* a magistrate of the police courts of the metropolis *or* stipendiary magistrate, *as the case may be*] for that [*&c. as in the order*] ; and thereupon having considered the matter of the said complaint, we [*or* I] adjudged the said [*set out from Form K. the adjudication of payment, and the order for distress and for imprisonment in default of distress*] : And whereas the time in and by the said order appointed for the payment of the said several sums of   and hath elapsed, but the said   hath not paid the same or any part thereof within fourteen days after the date fixed by the order for such payment, but therein hath made default : These are therefore to command you in Her Majesty's name forthwith to make distress of the goods and chattels of the said *A. B.*, and if within the space of   days after the making of such distress the said last-mentioned sums, together with the reasonable charges of taking and keeping the said distress, shall not be paid, that then you do sell the said goods and chattels so by you distrained, and do pay the money arising from such sale over to the clerk of the justices of the peace for the division of   in the said [county, *or as the case may be*], that he may pay and apply the same as by law directed, and may render the overplus, if any, on demand, to the said   ; and if no such distress can be found, then that you certify the same unto me, to the end that such proceedings may be had therein as to the law doth appertain.

Given under our [*or* my] hands and seal, this   day of   in the year of our Lord one thousand eight hundred and   at in the [county] aforesaid.

A. B.
C. D.

(L. S.)

Form (L.)

Return of Proceedings under Nuisances Removal Act, 1855, *by the* [name the local authority at length].

From 25th March 1855 *to 25th March* 1856.

Date of Notice.	By whom given.	Nature of Nuisance.	Proceedings taken.	Remarks :—With any special Work done under the Acts without any Notice.
16 April	The Inspector	Foul Drainage from House.	Owner put down good Drain, on Summons, without Justices' Order.	Several Houses being in a like Position, the Highway Surveyor laid down a Sewer in the old Watercourse, and each House was charged a proportionate Sum for the same, of which the highest Sum was 10s.
18 April	Two Neighbours	Offensive Cesspool.	Abated by Local Authority.	Renewed once; but Penalty recovered, and no subsequent Renewal attempted.

Dated this 26th day of March, 1856. [*To be signed by the chairman of the local authority.*]

APPENDIX.

ANNO VICESIMO TERTIO & VICESIMO QUARTO VICTORIÆ REGINÆ.

CAP. LXXVII.

An Act to amend the Acts for the Removal of Nuisances and the Prevention of Diseases. [6th *August* 1860.]

18 & 19 *Vict.* cc. 121. and 116.

WHEREAS the provisions of "The Nuisances Removal Act for *England*, 1855," and "The Diseases Prevention Act, 1855," concerning the local authority for the execution of the said Acts, are defective, and it is expedient that the said Acts should be amended as herein-after mentioned: Be it enacted by the Queen's most excellent Majesty, by and with the advice and consent of the Lords Spiritual and Temporal, and Commons, in this present Parliament assembled, and by the authority of the same, as follows:

NUISANCES REMOVAL.

Sections 3, 6, 7, and 9 of 18 & 19 Vict. c. 121. repealed.

I. Section three, section six, section seven, and section nine of the said "Nuisances Removal Act for *England*, 1855," shall be repealed: Provided always, that such repeal as aforesaid shall not extend to any charges or expenses already incurred, but the same may be defrayed and recovered, and all proceedings commenced or taken under the said Act, and not yet completed, may be proceeded with, and all contracts under the said Act shall continue and be as effectual, as if this Act had not been passed.

Local Authority to execute the Nuisances Removal Act.

II. The following bodies shall respectively be the local authority to execute the said Nuisances Removal Act in the districts hereunder stated in *England:*
In any place within which the Public Health Act is or shall be in force, the local board of health:
In any other place wherein a council exists or shall exist, the mayor, aldermen, and burgesses by the council, except in the city of *London* and the liberties thereof, where the local authority shall be the commissioners of sewers for the time being, and except in the city of *Oxford* and borough of *Cambridge*, where the local authority shall be the commissioners acting in execution of the local Improvement Acts in force respectively in the said city and borough:
In any place in which there is no local board of health or council, and where there are or shall be trustees or commissioners under an Improvement Act, such trustees or commissioners:
In any place within which there is no such local board of health, council, body of trustees, or commissioners, if there be a board of guardians of the poor for such place, or for any parish or union within which such place is situate, such board of guardians, and, if there be no such board of guardians, the overseers of the poor for such place, or for the parish of which such place forms part.

Highway Board or Nuisances Removal Committees now subsisting may be continued so long as they employ Sanitary Inspectors.

III. Provided, That in any place where a highway board or "the nuisances removal committee" chosen by the vestry in pursuance of the said Act is subsisting, and at the time of the passing of this Act employs or joins with other local authorities in employing a sanitary inspector or inspectors, such highway board or nuisances removal committee may continue to act, and a like committee may be annually chosen by the vestry for such place in the same manner as if this Act had not been passed; but in case in any year the nuisances removal committee be not chosen for such place in manner provided by the said Act, or if the highway board or committee now subsisting or hereafter chosen fail for two months in any year to appoint or employ a sanitary inspector or inspectors, the authority of such highway board or committee shall cease, and no like committee shall be chosen for such place, and the same body or persons shall thenceforth be the local authority for the place as if no such highway board or committee had been appointed therein.

How Expenses of Local Authority to be defrayed.

IV. All charges and expenses incurred by the local authority in executing the said Nuisances Removal Act, and not recovered as therein provided, shall be defrayed as follows; to wit,

Out of general district rates where the local authority is a local board of health:

Out of the borough fund or borough rate where the local authority is the mayor, aldermen, and burgesses by the council:

Provided always, that in the city of *Oxford* and borough of *Cambridge* such expenses shall be deemed annual charges and expenses of cleansing the streets of the said city and borough respectively, and shall be so payable:

Out of the rates levied for purposes of improvement under any Improvement Act, where the local authority is a body of trustees or commissioners acting in execution of the powers of such an Act.

Where a board of guardians for a union is such local authority for the whole of such union, such charges and expenses shall be defrayed by means of an addition to be made to the rate for the relief of the poor of the parish or parishes for which the expense has been incurred, and be raised and paid in like manner, as money expended for the relief of the poor:

Where the board of guardians for a union is such local authority for two or more places maintaining their own poor, but not for all such places in such union, such charges and expenses shall be paid out of the poor rates of the places aforesaid for which the board is the local authority:

Where the board of guardians for a union is under this Act the local authority for a single place maintaining its own poor, and where the board of guardians for any such single place or the overseers of any such place, or "the nuisances removal committee," continued or chosen as herein-before provided in any such place, are under this Act the local authority for such place, such charges and expenses shall be defrayed out of the rates for the relief of the poor thereof:

Where the board of guardians for a union is under this Act the local authority for part only of any place maintaining its own poor, together with the whole of any other such place or part of any other such place such board shall apportion such charges and expenses between or among any or every such part and any or every such place; and so much of such charges and expenses as may be apportioned to any or every such place for the whole of which such board is the local authority shall be defrayed out of the rates or funds applicable to the relief of the poor thereof:

So much of any such charges and expenses as may be apportioned to part of a place maintaining its own poor, and any such charges and expenses incurred by any board of guardians or overseers, where such board or overseers are the local authority for part of any such place only, shall be defrayed by means of an addition to be made to the rate for the relief of the poor thereof, and be raised and paid in like manner as money expended for the relief of the poor.

Board of Guardians may appoint Committees for particular Parishes.

V. Provided, That the board of guardians for a union may appoint a committee or committees of their own body, under section five of the said Nuisances Removal Act, to act in and for one or more of the parishes or places for which the board is the local authority; and every committee so appointed shall have the full power of executing the said Act in all respects, within the specified place or places for which it is appointed, unless its power be expressly limited by the terms of its appointment; and the board of guardians shall cause the charges and expenses of every such committee to be paid out of the poor rates of the place or places for which such committee is appointed; and where a committee is so appointed for any such place or places the charges and expenses of the board as local authority for or in respect of the place or places for which a committee is not appointed shall be paid or contributed by such last-mentioned place or places in like manner as the expenses of a committee : Provided that where any one such committee is appointed for all the places for which the board is the local authority its charges and expenses shall be contributed and paid in like manner as the charges and expenses of the board would have been contributed and paid if such committee had not been appointed.

Saving for the Vestries and District Boards of the Metropolis.

VI. Provided also, That as regards the metropolis, the vestries and district boards under the Act of the Session holden in the eighteenth and nineteenth years of Her Majesty, chapter one hundred and twenty, within their respective parishes and districts, shall continue and be the local authorities for the execution of the said Nuisances Removal Act, and their charges and expenses shall be defrayed as if this Act had not been passed.

Wells, &c. belonging to any Place vested in Local Authority, &c.

VII. All wells, fountains, and pumps provided under section fifty of the Public Health Act, 1848, or otherwise, for the use of the inhabitants of any place and not being the property of or vested in any person or corporation other than officers of such place, shall be vested in the local authority under this Act for such place, who shall from time to time cause to be kept in good repair and condition and free from pollution all wells, fountains, and pumps vested in them under this Act, and may also keep in good repair and condition and free from pollution other wells, fountains, and pumps dedicated to or open to the use of the inhabitants of such place.

Penalty for fouling Water.

VIII. If any person do any act whatsoever whereby any fountain or pump is wilfully or maliciously damaged, or the water of any well, fountain, or pump is polluted or fouled, he shall, upon summary conviction of such offence before two justices, forfeit a sum not exceeding five pounds for such offence, and a further sum not exceeding twenty shillings for every day during which such offence is continued after written notice from the local authority in relation thereto ; but nothing herein contained shall extend to any offence provided against by section twenty-three of the said Nuisances Removal Act.

Appointment of Inspectors of Nuisances.

IX. Local authorities under this Act may, for the purposes of the Act, severally appoint or employ inspectors of nuisances, and make such payments as they see fit for the remuneration and expenses of such inspectors.

DISEASES PREVENTION.

Sections 2 and 3 of 18 & 19 Vict. c. 116. repealed.

X. Sections two and three of the Diseases Prevention Act, 1855, and every other enactment constituting a local authority for the execution of the same Act, or providing for the expenses of the execution thereof, except those contained in the eighteenth and nineteenth of *Victoria*, chapter one hundred and twenty, the Metropolis Local Management Act, shall be repealed.

Guardians and Overseers of the Poor to be the Local Authorities for executing Diseases Prevention Act.

XI. The board of guardians for every union, or parish not within an union, in *England* shall be the local authority for executing the said Diseases Prevention Act in every place within their respective unions and parishes, and in every parish and place in *England* not within a union, and for which there is no board of guardians, the overseers of the poor shall be the local authority to execute the same Act; and the expenses incurred in the execution of such Act by the board of guardians for a union shall be defrayed out of the common fund thereof, and the expenses of the board of guardians or overseers of the poor of any single parish or place shall be defrayed out of the rates for the relief of the poor of such parish or place; provided that every such board of guardians shall, for the execution of the said Act for the Prevention of Diseases, have the like powers of appointing committees, with the like authority, and where any such committee is appointed the expenses thereof and of the board shall be paid in the same manner, as herein-before provided where such a board is the local authority for the execution of the said Nuisances Removal Act; provided also, that any expenses already incurred by any local authority in the execution of the said Act shall be defrayed as if this Act had not been passed; provided, moreover, that in respect of any place where, under this Act, the local authority for executing the Nuisances Removal Act is any other body than the board of guardians or the overseers of the poor, the Privy Council, if it see fit, may, in the manner provided for the exercise of its powers under the Public Health Act, 1858, authorize such other body to be, instead of the board of guardians or the overseers of the poor, the local authority for executing the Diseases Prevention Act; provided also, that as regards the metropolis the vestries and district boards under the Act of the Session holden in the eighteenth and nineteenth years of Her Majesty, chapter one hundred and twenty, within their respective parishes and districts, shall continue to be the local authorities for the execution of the said Diseases Prevention Act, 1855, and their charges and expenses shall be defrayed as if this Act had not been passed.

Local Authorities may provide Carriages for Conveyance of infected Persons.

XII. It shall be lawful for the local authority for executing the said Diseases Prevention Act to provide and maintain a carriage or carriages suitable for the conveyance of persons suffering under any contagious or infectious disease, and to convey such sick and diseased persons as may be residing within such locality to any hospital or other place of destination, and the expense thereof shall be deemed to be an expense incurred in executing the said Act.

Justices, on the Application of Householders, may order the Removal of Nuisances.

XIII. Upon complaint before a justice of the peace by any inhabitant of any parish or place of the existence of any nuisance on any private premises in the same parish or place, such justice shall issue a summons requiring the person by whose act, default, permission, or sufferance the nuisance arises, or if such person cannot be found or ascertained, the owner or occupier of the premises on which the nuisance arises, to appear before two justices in petty sessions assembled at their usual place of meeting, who shall proceed to inquire into the said complaint, and act in relation thereto as in cases where complaint is made by a local authority under section twelve of the said Nuisances Removal Act, and as if the person making the complaint were such local authority: Provided always, that it shall be lawful for the said justices, if they see fit, to adjourn the hearing or further hearing of such summons for an examination of the premises where the nuisance is alleged to exist, and to require the admission or authorize the entry into such premises of any constable or other person or persons, and thereupon the person or persons authorized by the order of the justices may enter and act as the local authority might under a like order made by any justice under section eleven of the said Act: Provided also, that the costs in the case of every such application shall be in the discretion of the justices, and payment thereof may be ordered and enforced as in other cases of summary adjudication by justices: Any order made by justices under this enactment shall be attended with the like penalties and consequences for disobedience thereof and subject to the like appeal as any order made under section twelve of the said Nuisances Removal Act, and the justices making such order may thereby authorize any constable or other person or persons to do all acts for removing or abating the nuisance condemned or prohibited, and for executing such order, in like manner as a local authority obtaining the like order might do under the said Act, and to charge the costs to the person on whom the order is made, as is provided in the case where a like order is obtained and executed by such local authority.

Guardians may procure sanitary Reports and pay for the same.

XIV. The guardians of any union, or parish not within a union, may at any time employ one of their medical officers to make inquiry and report upon the sanitary state of their union or parish, or any part thereof, and pay a reasonable compensation for the same out of their common fund.

Interpretation of Terms.

XV. The several words used in this Act shall be construed in the same manner as is declared with reference to the same words in the above-cited Act, termed "The Nuisances Removal Act for *England*, 1855," and all the provisions therein, and in "The Diseases Prevention Act, 1855," contained, shall respectively be applicable to this Act, except so far as they shall be hereby repealed, or be inconsistent with anything herein provided.

Justices not incapable of acting by being Members of Bodies to execute Nuisances Removal Act.

XVI. No justice of the peace shall, unless objected to at the hearing of any complaint or charge, be deemed incapable of acting in cases other than appeals arising under the said Nuisances Removal Act by reason of his being a member of any body hereby declared to be the local authority to execute the said Act, or by reason of his being a contributor, or liable to contribute, to any rate or fund out of which it is hereby provided that all charges and expenses incurred in executing the said Act, and not recovered as therein provided, shall be defrayed.

ANNO VICESIMO NONO & TRICESIMO VICTORIÆ REGINÆ.

CAP. XC.

An Act to amend the Law relating to the Public Health.

[7th *August* 1866.]

WHEREAS it is expedient to amend the law relating to public health: Be it enacted by the Queen's most excellent Majesty, by and with the advice and consent of the Lords Spiritual and Temporal, and Commons, in this present Parliament assembled, and by the authority of the same, as follows :

PRELIMINARY.

Short Title of Act.

1. This Act may be cited for all purposes as the Sanitary Act, 1866.

PART I.

AMENDMENT OF THE SEWAGE UTILIZATION ACT, 1865.

Definition of "Sewer Authority:" "Lord Lieutenant in Council."

2. "Sewer authority" in this Act shall have the same meaning as it has in the Sewage Utilization Act, 1865.

The words "Lord Lieutenant in Council" shall mean in this Act the Lord Lieutenant or any chief governor or chief governors in *Ireland* acting by and with the consent of Her Majesty's Privy Council in *Ireland*.

This Part to be construed with 28 & 29 *Vict. c.* 75.

3. This part of this Act shall be construed as one with the Sewage Utilization Act, 1865, and the expression "The Sewage Utilization Act, 1865," as used in this or any other Act of Parliament or other document, shall mean the said Sewage Utilization Act, 1865, as amended by this Act.

Power to Sewer Authority to form Committee of its own Members and others.

4. Any sewer authority may from time to time, at any meeting specially convened for the purpose, form one or more committee or committees consisting wholly of its own members, or partly of its own members and partly of such other persons contributing to the rate or fund out of which the expenses incurred by such authority are paid, and qualified in such other manner as the sewer authority may determine, and may delegate, with or without conditions or restrictions, to any committee so formed, all or any powers of such sewer authority, and may from time to time revoke, add to, or alter any powers so given to a committee.

A committee may elect a chairman of its meetings. If no chairman is elected, or if the chairman elected is not present at the time appointed for holding the same, the members present shall choose one of their number to be chairman of such meeting. A committee may meet and adjourn as it thinks proper. The quorum of a committee shall consist of such number of members as may be prescribed by the sewer authority that appointed it, or, if no number be prescribed, of three members. Every question at a meeting shall be determined by a majority of votes of the members present, and voting on that question; and in case of an equal division of votes the chairman shall have a second or casting vote.

The proceedings of a committee shall not be invalidated by any vacancy or vacancies amongst its members.

A sewer authority may from time to time add to or diminish the number of the members or otherwise alter the constitution of any committee formed by it, or dissolve any committee.

A committee of the sewer authority shall be deemed to be the agents of that authority, and the appointment of such committee shall not relieve the sewer authority from any obligation imposed on it by Act of Parliament or otherwise.

Formation of Special Drainage District.

5. Where the sewer authority of a district is a vestry, select vestry, or other body of persons acting by virtue of any Act of Parliament, prescription, custom, or otherwise as or instead of a vestry or select vestry, it may, by resolution at any meeting convened for the purpose after twenty-one c'ear days' notice affixed to the places where parochial notices are usually affixed in its district, form any part of such district into a special drainage district for the purposes of the Sewage Utilization Act, and thereupon such special drainage district shall, for the purposes of the Sewage Utilization Act, 1865, and the powers therein conferred, be deemed to be a parish in which a rate is levied for the maintenance of the poor, and of which a vestry is the sewer authority, subject, as respects any meeting of the inhabitants thereof in vestry, to the Act of the fifty-eighth year of the reign of King *George* III., chapter sixty-nine, and the Acts amending the same; and any officer or officers who may from time to time be appointed by the sewer authority of such special drainage district for the purpose shall have within that district all the powers of levying a rate for the purpose of defraying the expense of carrying the said Sewage Utilization Act into effect that they would have if such district were such parish as aforesaid, and such rate were a rate for the relief of the poor, and they were duly appointed overseers of such parish.

Appeal against Constitution of Special Drainage District.

6. Where the sewer authority of any place has formed a special drainage district in pursuance of this Act, if any number of the inhabitants of such place, not being less than twenty, feel aggrieved by the formation of such district, or desire any modification in its boundaries, they may, by petition in writing under their

hands, bring their case under the consideration of one of Her Majesty's Principal Secretaries of State, and the said Secretary of State may after due investigation annul the formation of the special drainage district or modify its boundaries as he thinks just.

Evidence of Formation of Special Drainage District.

7. A copy of the resolution of a sewer authority forming a special drainage district shall be published by affixing a notice thereof to the church door of the parish in which the district is situate, or of the adjoining parish if there be no church in the said parish, and by advertising notice thereof in some newspaper published or circulating in the county in which such district is situate; and the production of a newspaper containing such advertisement, or a certificate under the hand of the clerk or other officer performing the duties of clerk for the time being of the sewer authority which passed the resolution forming the district, shall be evidence of the formation of such district, and after the expiration of three months from the date of the resolution forming the district such district shall be presumed to have been duly formed, and no objection to the formation thereof shall be entertained in any legal proceedings whatever.

Power to drain into Sewers of Sewer Authority.

8. Any owner or occupier of premises within the district of a sewer authority shall be entitled to cause his drains to empty into the sewers of that authority on condition of his giving such notice as may be required by that authority of his intention so to do, and of complying with the regulations of that authority in respect of the mode in which the communications between such drains and sewers are to be made, and subject to the control of any person who may be appointed by the sewer authority to superintend the making of such communications; but any person causing any drain to empty into any sewer of a sewer authority without complying with the provisions of this section shall incur a penalty not exceeding twenty pounds, and it shall be lawful for the sewer authority to close any communication between a drain and sewer made in contravention of this section, and to recover in a summary manner from the person so offending any expenses incurred by them under this section.

Use of Sewers by Persons beyond District.

9. Any owner or occupier of premises beyond the limits of the district of a sewer authority may cause any sewer or drain from such premises to communicate with any sewer of the sewer authority upon such terms and conditions as may be agreed upon between such owner or occupier, and such sewer authority, or in case of dispute may, at the option of the owner or occupier, be settled by two justices or by arbitration in manner provided by the Public Health Act, 1848, in respect of matters by that Act authorized or directed to be settled by arbitration.

As to the Drainage of Houses.

10. If a dwelling house within the district of a sewer authority is without a drain or without such drain as is sufficient for effectual drainage, the sewer authority may by notice require the owner of such house within a reasonable time therein specified to make a sufficient drain emptying into any sewer which the sewer authority is entitled to use, and with which the owner is entitled to make a communication, so that such sewer be not more than one hundred feet from the site of the house of such owner; but if no such means of drainage are within that distance then emptying into such covered cesspool or other place not being under any house, as the sewer authority directs; and if the person on whom such notice is served fails to comply with the same, the sewer authority may itself, at the expiration of the time specified in the notice, do the work required, and the expenses incurred by it in so doing may be recovered from such owner in a summary manner.

Supply of Water to District of Sewer Authority.

11. A sewer authority within its district shall have the same powers in relation to the supply of water that a local board has within its district, and the provisions

of the sections herein-after mentioned shall apply accordingly in the same manner as if in such provisions "sewer authority" were substituted for "local board of health" or "local board," and the district in such provisions mentioned were the district of the sewer authority and not the district of the local board; that is to say, the sections numbered from seventy-five to eighty, both inclusive, of the Public Health Act, 1848, sections fifty-one, fifty-two, and fifty-three of the Local Government Act, 1858, and section twenty of the Local Government Act, 1858, Amendment Act, 1861.

The sewer authority may, if it think it expedient so to do, provide a supply of water for the use of the inhabitants of the district by

        (1.) Digging wells;
        (2.) Making and maintaining reservoirs;
        (3.) Doing any other necessary acts;

and they may themselves furnish the same, or contract with any other persons or companies to furnish the same: Provided always, that no land be purchased or taken under this clause except by agreement or in manner provided by the Local Government Act, 1858.

Expenses of Sewer Authority in supplying Water.

12. Any expenses incurred by a sewer authority in or about the supply of water to its district, and in carrying into effect the provisions herein-before in that behalf mentioned, shall be deemed to be expenses incurred by that authority in carrying into effect the Sewage Utilization Act, 1865, and be payable accordingly.

Wells, &c. belonging to any Place vested in Sewer Authority, &c.
23 & 24 Vict. c. 77. s. 7.

13. All property in wells, fountains, and pumps, and powers in relation thereto, vested in the nuisance authority by the seventh section of the Act passed in the Session of the twenty-third and twenty-fourth years of the reign of Her present Majesty, chapter seventy-seven, shall vest in the sewer authority, where the sewer authority supplies water to its district.

PART II.

AMENDMENT OF THE NUISANCES REMOVAL ACTS.

Definition of "Nuisances Removal Acts."

14. The expression "Nuisances Removal Acts" shall mean the Acts passed in the years following of the reign of Her present Majesty, that is to say, the one in the Session of the eighteenth and nineteenth years, chapter one hundred and twenty-one, and the other in the Session of the twenty-third and twenty-fourth years, chapter seventy-seven, as amended by this part of this Act; and this part of this Act shall be construed as one with the said Acts, and all expenses incurred by a nuisance authority in carrying into effect any of the provisions of this part of this Act shall be deemed to be expenses incurred by it in carrying into effect the Nuisances Removal Acts.

Definition of "Nuisance Authority."

15. "Nuisance authority" shall mean any authority empowered to execute the Nuisances Removal Acts.

Power of Police with respect to Nuisances.

16. In any place within the jurisdiction of a nuisance authority the chief officer of police within that place, by and under the directions of one of Her Majesty's Principal Secretaries of State, on its being proved to his satisfaction that the nuisance authority has made default in doing its duty, may institute any proceeding which the nuisance authority of such place might institute with respect to the removal of nuisances: Provided always, that no officer of police shall be at liberty to enter any

house or part of a house used as the dwelling of any person without such person's consent, or without the warrant of the justice of the peace, for the purpose of carrying into effect this Act.

Sect. 3. of 23 & 24 Vict. c. 77. repealed. 18 & 19 Vict. c. 120.

17. The third section of the said Act of the Session of the twenty-third and twenty-fourth years of the reign of Her present Majesty, chapter seventy-seven, shall be repealed, and all powers vested in any highway board or "nuisance removal committee" under the Nuisances Removal Acts shall determine, and all property belonging to them for the purposes of the said Nuisances Removal Acts shall, subject to any debts or liabilities affecting the same, be transferred to or vested in the nuisance authority under the said Acts : Provided always, that this section shall not extend to any vestry or district board, under the Act of the Session of eighteenth and nineteenth years of the reign of Her present Majesty, chapter one hundred and twenty, intituled "An Act for the better Local Management of the Metropolis," or to any committee appointed by such vestry or district board for the purpose of carrying into effect the Nuisances Removal Acts or any of them.

Requisition of Ten Inhabitants equivalent to Certificate of Medical Officer.

18. A requisition in writing under the hands of any ten inhabitants of a place shall for the purposes of the twenty-seventh section of "The Nuisances Removal Act for *England*, 1855," be deemed to be equivalent to the certificate of the medical officer or medical practitioners therein mentioned, and the said section shall be enforced accordingly.

Addition to Definition of Nuisance.

19. The word "nuisances" under the Nuisance Removal Acts shall include,
1. Any house or part of a house so overcrowded as to be dangerous or prejudicial to the health of the inmates :
2. Any factory, workshop, or workplace not already under the operation of any general Act for the regulation of factories or bakehouses, not kept in a cleanly state, or not ventilated in such a manner as to render harmless as far as practicable any gases, vapours, dust, or other impurities generated in the course of the work carried on therein that are a nuisance or injurious or dangerous to health, or so overcrowded while work is carried on as to be dangerous or prejudicial to the health of those employed therein :
3. Any fireplace or furnace which does not as far as practicable consume the smoke arising from the combustible used in such fireplace or furnace, and is used within the district of a nuisance authority for working engines by steam, or in any mill, factory, dyehouse, brewery, bakehouse, or gaswork, or in any manufactory or trade process whatsoever :

Any chimney (not being the chimney of a private dwelling house) sending forth black smoke in such a quantity as to be a nuisance :
Provided, first, that in places where at the time of the passing of this Act no enactment is in force compelling fireplaces or furnaces to consume their own smoke, the foregoing enactment as to fireplaces and furnaces consuming their own smoke shall not come into operation until the expiration of one year from the date of the passing of this Act :
Secondly, that where a person is summoned before the justices in respect of a nuisance arising from a fireplace or furnace which does not consume the smoke arising from the combustible used in such fireplace or furnace, the justices may hold that no nuisance is created within the meaning of this Act, and dismiss the complaint, if they are satisfied that such fireplace or furnace is constructed in such a manner as to consume as far as practicable, having regard to the nature of the manufacture or trade, all smoke arising therefrom, and that such fireplace or furnace has been carefully attended to by the person having the charge thereof.

Duties of Nuisance Authorities as to Inspection of Nuisances, &c.

20. It shall be the duty of the nuisance authority to make from time to time either by itself or its officers, inspection of the district, with a view to ascertain what nuisances exist calling for abatement under the powers of the Nuisance Removal Acts, and to enforce the provisions of the said Acts in order to cause the abatement thereof, also to enforce the provisions of any Act that may be in force within its district requiring fireplaces and furnaces to consume their own smoke ; and any justice upon complaint upon oath may make an order to admit the nuisance authority or their officers for these purposes, as well as to ground proceedings under the eleventh section of the Nuisances Removal Act, 1855.

As to Proceedings of Nuisance Authority under Sect. 12. *of* 18 *&* 19 *Vict. c.* 121.

21. The nuisance authority or chief officer of police shall, previous to taking proceedings before a justice under the twelfth section of the Nuisances Removal Act, 1855, serve a notice on the person by whose act, default, or sufferance the nuisance arises or continues, or, if such person cannot be found or ascertained, on the owner or occupier of the premises on which the nuisance arises, to abate the same, and for that purpose to execute such works and to do all such things as may be necessary within a time to be specified in the notice : Provided,

> First, that where the nuisance arises from the want or defective construction of any structural convenience, or where there is no occupier of the premises, notice under this section shall be served on the owner :
>
> Secondly, that where the person causing the nuisance cannot be found, and it is clear that the nuisance does not arise or continue by the act, default, or sufferance of the owner or occupier of the premises, then the nuisance authority may itself abate the same without further order, and the cost of so doing shall be part of the costs of executing the Nuisances Removal Acts, and borne accordingly.

Power to cause Premises to be cleansed or otherwise disinfected.

22. If the nuisance authority shall be of opinion, upon the certificate of any legally qualified medical practitioner, that the cleansing and disinfecting of any house or part thereof, and of any articles therein likely to retain infection, would tend to prevent or check infectious or contagious disease, it shall be the duty of the nuisance authority to give notice in writing requiring the owner or occupier of such house or part thereof to cleanse and disinfect the same as the case may require ; and if the person to whom notice is so given fail to comply therewith within the time specified in the notice, he shall be liable to a penalty of not less than one shilling and not exceeding ten shillings for every day during which he continues to make default ; and the nuisance authority shall cause such house or part thereof to be cleansed and disinfected, and may recover the expenses incurred from the owner or occupier in default in a summary manner ; when the owner or occupier of any such house or part thereof as is referred to in this section is from poverty or otherwise unable, in the opinion of the nuisance authority, effectually to carry out the requirements of this section, such authority may, without enforcing such requirements on such owner or occupier, with his consent, at its own expense, cleanse and disinfect such house or part thereof, and any articles therein likely to retain infection.

Power to provide Means of Disinfection.

23. The nuisance authority in each district may provide a proper place, with all necessary apparatus and attendants, for the disinfection of woollen articles, clothing, or bedding which have become infected, and they may cause any articles brought for disinfection to be disinfected free of charge.

Nuisance Authorities may provide Carriages for Conveyance of infected Persons.

24. It shall be lawful at all times for the nuisance authority to provide and maintain a carriage or carriages suitable for the conveyance of persons suffering

under any contagious or infectious disease, and to pay the expense of conveying any person therein to a hospital or place for the reception of the sick or to his own home.

Penalty on Person suffering from infectious Disorder entering public Conveyance without notifying to Driver that he is so suffering.

25. If any person suffering from any dangerous infectious disorder shall enter any public conveyance without previously notifying to the owner or driver thereof that he is so suffering, he shall on conviction thereof before any justice be liable to a penalty not exceeding five pounds, and shall also be ordered by such justice to pay to such owner and driver all the losses and expenses they may suffer in carrying into effect the provisions of this Act; and no owner or driver of any public conveyance shall be required to convey any person so suffering until they shall have been first paid a sum sufficient to cover all such losses and expenses.

Removal of Persons sick of infectious Disorders, and without proper Lodging, in any District.

26. Where a hospital or place for the reception of the sick is provided within the district of a nuisance authority, any justice may, with the consent of the superintending body of such hospital or place, by order on a certificate signed by a legally qualified medical practitioner, direct the removal to such hospital or place for the reception of the sick, at the cost of the nuisance authority, of any person suffering from any dangerous contagious or infectious disorder, being without proper lodging or accommodation, or lodged in a room occupied by more than one family, or being on board any ship or vessel.

Places for the Reception of dead Bodies may be provided at the Public Expense.

27. Any nuisance authority may provide a proper place for the reception of dead bodies, and where any such place has been provided, and any dead body of one who has died of any infectious disease is detained in a room in which persons live or sleep, or any dead body which is in such a state as to endanger the health of the inmates of the same house or room is retained in such house or room, any justice may, on a certificate signed by a legally qualified medical practitioner, order the body to be removed to such proper place of reception at the cost of the nuisance authority, and direct the same to be buried within a time to be limited in such order; and unless the friends or relations of the deceased undertake to bury the body within the time so limited, and do bury the same, it shall be the duty of the relieving officer to bury such body at the expense of the poor rate, but any expense so incurred may be recovered by the relieving officer in a summary manner from any person legally liable to pay the expense of such burial.

Places for Reception of dead Bodies during Time required for post-mortem Examination may be provided.

28. Any nuisance authority may provide a proper place (otherwise than at a workhouse or at a mortuary house as lastly herein-before provided for) for the reception of dead bodies for and during the time required to conduct any *post-mortem* examination ordered by the coroner of the district or other constituted authority, and may make such regulations as they may deem fit for the maintenance, support, and management of such place; and where any such place has been provided, any coroner or other constituted authority may order the removal of the body for carrying out such *post-mortem* examination and the re-removal of such body, such costs of removal and re-removal to be paid in the same manner and out of the same fund as the cost and fees for *post-mortem* examinations when ordered by the coroner.

Power to remove to Hospital sick Persons brought by Ships.

29. Any nuisance authority may, with the sanction of the Privy Council signified in manner provided by the Public Health Act, 1858, lay down rules for the

removal to any hospital to which such authority is entitled to remove patients, and for keeping in such hospital so long as may be necessary any persons brought within their district by any ship or boat who are infected with a dangerous and infectious disorder, and they may by such rules impose any penalty not exceeding five pounds on any person committing any offence against the same.

Provision as to District of Nuisance Authority extending to Places where Ships are lying.

30. For the purposes of this Act, any ship, vessel, or boat that is in a place not within the district of a nuisance authority shall be deemed to be within the district of such nuisance authority as may be prescribed by the Privy Council, and until a nuisance authority has been prescribed then of the nuisance authority whose district nearest adjoins the place where such ship, vessel, or boat is lying, the distance being measured in a straight line, but nothing in this Act contained shall enable any nuisance authority to interfere with any ship, vessel, or boat that is not in *British* waters.

Power of Entry to Nuisance Authority or their Officer under Sect. 11. of 18 & 19 Vict. c. 121.

31. The power of entry given to the authorities by the eleventh section of the Nuisances Removal Act, 1855, may be exercised at any hour when the business in respect of which the nuisance arises is in progress or is usually carried on.

And any justices' order once issued under the said section shall continue in force until the nuisance has been abated, or the work for which the entry was necessary has been done.

Provision as to Ships within the Jurisdiction of Nuisance Authority.

32. Any ship or vessel lying in any river, harbour, or other water shall be subject to the jurisdiction of the nuisance authority of the district within which such river, harbour, or other water is, and be within the provisions of the Nuisances Removal Acts, in the same manner as if it were a house within such jurisdiction, and the master or other officer in charge of such ship shall be deemed for the purposes of the Nuisances Removal Acts to be the occupier of such ship or vessel ; but this section shall not apply to any ship or vessel belonging to Her Majesty or to any foreign Government.

Provision for raising Money in divided Parishes.

33. Where the guardians are the nuisance authority for part of any parish only, and shall require to expend money on account of such part in execution of the provisions of the said Acts, the overseers of the parish shall, upon receipt of an order from the said guardians, raise the requisite amount from the persons liable to be assessed to the poor rate therein by a rate to be made in like manner as a poor rate, and shall have all the same powers of making and recovering the same, and of paying the expense of collecting the rate when made, and shall account to the auditor of the district for receipt and disbursement of the same, in like manner, and with the same consequences, as in the case of the poor rate made by them.

Nuisance Authority may require Payment of Costs or Expenses from Owner or Occupier, and Occupier paying to deduct from Rent.

34. That it shall be lawful for the nuisance authority, at their discretion, to require the payment of any costs or expenses which the owner of any premises may be liable to pay under the said Nuisances Removal Acts or this Act, either from the owner or from any person who then or at any time thereafter occupies such premises, and such owner or occupier shall be liable to pay the same, and the same shall be recovered in manner authorized by the Nuisance Removal Acts, and the owner shall allow such occupier to deduct the sums of money which he so pays out of the rent from time to time becoming due in respect of the said premises, as if the

same had been actually paid to such owner as part of such rent : Provided always, that no such occupier shall be required to pay any further sum than the amount of rent for the time being due from him, or which, after such demand of such costs or expenses from such occupier, and after notice not to pay his landlord any rent without first deducting the amount of such costs or expenses, becomes payable by such occupier, unless he refuse, on application being made to him for that purpose by or on behalf of the nuisance authority, truly to disclose the amount of his rent and the name and address of the person to whom such rent is payable, but the burden of proof that the sum demanded from any such occupier is greater than the rent due by him at the time of such notice, or which has since accrued, shall lie upon such occupier ; provided also, that nothing herein contained shall be taken to affect any contract made or to be made between any owner or occupier of any house, building, or other property whereof it is or may be agreed that the occupier shall pay or discharge all rates, dues, and sums of money payable in respect of such house, building, or other property, or to affect any contract whatsoever between landlord or tenant.

PART III.

MISCELLANEOUS.

In Cities, Boroughs, or Towns, Secretary of State, on Application of Nuisance Authority, may empower them to make Regulations as to Lodging Houses.

35. On application to one of Her Majesty's Principal Secretaries of State by the nuisance authority of the city of *London*, or any district or parish included within the Act for the better local government of the metropolis, or of any municipal borough, or of any place under the Local Government Act, 1858, or any local Improvement Act, or of any city or town containing, according to the census for the time being in force, a population of not less than five thousand inhabitants, the Secretary of State may, as he may think fit, by notice to be published in the *London Gazette*, declare the following enactment to be in force in the district of such nuisance authority, and from and after the publication of such notice the nuisance authority shall be empowered to make regulations for the following matters ; that is to say,

1. For fixing the number of persons who may occupy a house or part of a house which is let in lodgings or occupied by members of more than one family :
2. For the registration of houses thus let or occupied in lodgings :
3. For the inspection of such houses, and the keeping the same in a cleanly and wholesome state :
4. For enforcing therein the provision of privy accommodation and other appliances and means of cleanliness in proportion to the number of lodgings and occupiers, and the cleansing and ventilation of the common passages and staircases :
5. For the cleansing and lime-whiting at stated times of such premises :

The nuisance authority may provide for the enforcement of the above regulations by penalties not exceeding forty shillings for any one offence, with an additional penalty not exceeding twenty shillings for every day during which a default in obeying such regulations may continue ; but such regulations shall not be of any validity unless and until they shall have been confirmed by the Secretary of State.

But this section shall not apply to common lodging houses within the provisions of the Common Lodging Houses Act, 1851, or any Act amending the same.

Cases in which two Convictions have occurred within Three Months.

36. Where two convictions against the provisions of any Act relating to the overcrowding of a house, or the occupation of a cellar as a separate dwelling place, shall have taken place within the period of three months, whether the persons so convicted were or were not the same, it shall be lawful for any two justices to

direct the closing of such premises for such time as they may deem necessary, and in the case of cellars occupied as aforesaid, to empower the nuisance authority to permanently close the same, in such manner as they may deem fit, at their own cost.

Power to provide Hospitals.

37. The sewer authority, or in the metropolis the nuisance authority, may provide for the use of the inhabitants within its district hospitals or temporary places for the reception of the sick.

Such authority may itself build such hospitals or places of reception, or make contracts for the use of any existing hospital or part of a hospital, or for the temporary use of any place for the reception of the sick.

It may enter into any agreement with any person or body of persons having the management of any hospital for the reception of the sick inhabitants of its district, on payment by the sewer authority of such annual or other sum as may be agreed upon.

The carrying into effect this section shall in the case of a sewer authority be deemed to be one of the purposes of the said Sewage Utilization Act, 1865, and all the provisions of the said Act shall apply according.

Two or more authorities having respectively the power to provide separate hospitals may combine in providing a common hospital, and all expenses incurred by such authorities in providing such hospital shall be deemed to be expenses incurred by them respectively in carrying into effect the purposes of this Act.

Penalty on any Person, with infectious Disorder, exposing himself, or on any Person in charge of such Sufferer causing such Exposure.

38. Any person suffering from any dangerous infectious disorder who wilfully exposes himself, without proper precaution against spreading the said disorder, in any street, public place, or public conveyance, and any person in charge of one so suffering who so exposes the sufferer, and any owner or driver of a public conveyance who does not immediately provide for the disinfection of his conveyance after it has, with the knowledge of such owner or driver, conveyed any such sufferer, and any person who without previous disinfection gives, lends, sells, transmits, or exposes any bedding, clothing, rags, or other things which have been exposed to infection from such disorders, shall, on conviction of such offence before any justice, be liable to a penalty not exceeding five pounds: Provided that no proceedings under this section shall be taken against persons transmitting with proper precautions any such bedding, clothing, rags, or other things for the purpose of having the same disinfected.

Penalty on Persons letting Houses in which infected Persons have been lodging.

39. If any person knowingly lets any house, room, or part of a house in which any person suffering from any dangerous infectious disorder has been to any other person without having such house, room, or part of a house, and all articles therein liable to retain infection, disinfected to the satisfaction of a qualified medical practitioner as testified by a certificate given by him, such person shall be liable to a penalty not exceeding twenty pounds. For the purposes of this section the keeper of an inn shall be deemed to let part of a house to any person admitted as a guest into such inn.

Guardians, &c. of the Poor to be the Local Authorities for executing Diseases Prevention Act.

40. Where in any place two or more boards of guardians or local authorities have jurisdiction, the Privy Council may, by any order made under the Diseases Prevention Act, 1855, authorize or require such boards to act together for the purposes of that Act, and may prescribe the mode of such joint action and of defraying the costs thereof.

APPENDIX.

Evidence of Family in case of overcrowded Houses.

41. In any proceedings under the Common Lodging Houses Act, 1851, if the inmates of any house or part of a house allege that they are members of the same family, the burden of proving such allegation shall lie on the persons making it.

Extension to the whole of England and Ireland of Sect. 67. of 11 & 12 Vict. c. 63.

42. The sixty-seventh section of the Public Health Act, 1848, relating to cellar dwellings, shall apply to every place in *England* and *Ireland* where such dwellings are not regulated by any other Act of Parliament, and in applying that section to places where it is not in force at the time of the passing of this Act the expression "this Act" shall be construed to mean the Sanitary Act, 1866, and not the said Public Health Act, 1848. In construing the said sixty-seventh section as applied by this Act, nuisance authority shall be substituted for the local board.

Local Board in certain Cases may adopt Baths and Wash-houses Acts.

43. Local boards acting in execution of the Local Government Act, 1858, may adopt the Act to encourage the establishment of public baths and wash-houses, and any Act amending the same, for districts in which those Acts are not already in force, and when they have adopted the said Acts they shall have all the powers, duties, and rights of commissioners under the said Acts ; and all expenses incurred by any local board in carrying into execution the Acts referred to in this section shall be defrayed out of the general district rates, and all receipts by them under the said Acts shall be carried to the district fund account.

Power to Burial Boards in certain Cases to transfer their Powers to Local Board.

44. When the district of a burial board is conterminous with the district of a local board of health, the burial board may, by resolution of the vestry, and by agreement of the burial board and local board, transfer to the local board all their estate, property, rights, powers, duties, and liabilities, and from and after such transfer the local board shall have all such estate, property, rights, powers, duties, and liabilities as if the local board had been appointed a burial board by Order in Council under the fourth section of the Act of the Session of the twentieth and twenty-first years of the reign of Her present Majesty, chapter eighty-one.

Penalty for wilful Damage of Works.

45. If any person wilfully damages any works or property belonging to any local board, sewer authority, or nuisance authority, he shall be liable to a penalty not exceeding five pounds.

Incorporation of Sanitary Authorities.

46. The following bodies, that is to say, local boards, sewer authorities, and nuisance authorities, if not already incorporated, shall respectively be bodies corporate designated by such names as they may usually bear or adopt, with power to sue and be sued in such names, and to hold lands for the purposes of the several Acts conferring powers on such bodies respectively in their several characters of local boards, sewer authorities, or nuisance authorities.

Extent of Authority to make Provisional Orders respecting Lands under Sect. 75. of 21 & 22 Vict. s. 98.

47. The authority conferred on one of Her Majesty's Principal Secretaries of State by section seventy-five of the Local Government Act, 1858, to empower by provisional order a local board to put in force, with reference to the land referred to in such order, the powers of the Lands Clauses Consolidation Act, 1845, with respect to the purchase and taking of lands otherwise than by agreement, shall extend and apply and shall be deemed to have always extended and applied to

every case in which, by the Public Health Act, 1848, and the Local Government Act, 1858, or either of them, or any Act extending or amending those Acts, or either of them, a local board are authorized to purchase, provide, use, or take lands or premises for any of the purposes of the said Acts or either of them, or of any such Act as aforesaid ; and sections seventy-three and eighty-four of the Public Health Act, 1848, shall be construed as if the words "by agreement" therein respectively used had been expressly repealed by section seventy-five of the Local Government Act, 1858.

Appearance of Local Authorities in legal Proceedings.

48. Any local board, sewer authority, or nuisance authority may appear before any justice or justices, or in any legal proceeding, by its clerk or by any officer or member authorized generally or in respect of any special proceeding by resolution of such board or authority, and such person being so authorized shall be at liberty to institute and carry on any proceeding which the nuisance authority is authorized to institute and carry on under the Nuisance Removal Acts or this Act.

Mode of Proceeding where Sewer Authority has made default in providing sufficient Sewers, &c.

49. Where complaint is made to one of Her Majesty's Principal Secretaries of State that a sewer authority or local board of health has made default in providing its district with sufficient sewers, or in the maintenance of existing sewers, or in providing its district with a supply of water, in cases where danger arises to the health of the inhabitants from the insufficiency or unwholesomeness of the existing supply of water, and a proper supply can be got at a reasonable cost, or that a nuisance authority has made default in enforcing the provisions of the Nuisance Removal Acts, or that a local board has made default in enforcing the provisions of the Local Government Act, the said Secretary of State, if satisfied, after due inquiry made by him, that the authority has been guilty of the alleged default, shall make an order limiting a time for the performance of its duty in the matter of such complaint ; and if such duty is not performed by the time limited in the order, the said Secretary of State shall appoint some person to perform the same, and shall by order direct that the expenses of performing the same, together with a reasonable remuneration to the person appointed for superintending such performance, and amounting to a sum specified in the order, together with the costs of the proceedings, shall be paid by the authority in default : and any order made for the payment of such costs and expenses may be removed into the Court of Queen's Bench, and be enforced in the same manner as if the same were an order of such court.

Recovery of certain Expenses of Water Supply.

50. All expenses incurred by a sewers authority or local board in giving a supply of water to premises under the provisions of the seventy-sixth section of the Public Health Act, 1848, or the fifty-first section of the Local Government Act, 1858, and recoverable from the owners of the premises supplied, may be recovered in a summary manner.

Power to reduce Penalties imposed by 6 G. 4. c. 78.

51. All penalties imposed by the Act of the sixth year of King George IV., chapter seventy-eight, intituled "An Act to repeal the several Laws relating to Quarantine, and to make other Provisions in lieu thereof," may be reduced by the justices or court having jurisdiction in respect of such penalties to such sum as the justices or court think just.

Description of Vessels within Provisions of 6 G. 4. c. 78.

52. Every vessel having on board any person affected with a dangerous or infectious disorder shall be deemed to be within the provisions of the Act of the sixth

year of King *George* IV., chapter seventy-eight, although such vessel has not commenced her voyage, or has come from or is bound for some place in the United Kingdom; and the Lords and others of Her Majesty's most honourable Privy Council, or any three or more of them (the Lord President of the Council or one of Her Majesty's Principal Secretaries of State being one), may, by order or orders to be by them from time to time made, make such rules, orders, and regulations as to them shall seem fit, and every such order shall be certified under the hand of the clerk in ordinary of Her Majesty's Privy Council, and shall be published in the *London Gazette*, and such publication shall be conclusive evidence of such order to all intents and purposes; and such orders shall be binding and be carried into effect as soon as the same shall have been so published, or at such other time as shall be fixed by such orders, with a view to the treatment of persons affected with cholera, and epidemic, endemic, and contagious disease, and preventing the spread of cholera and such other diseases, as well on the seas, rivers, and waters of the United Kingdom, and on the high seas within three miles of the coasts thereof, as on land; and to declare and determine by what nuisance authority or authorities such orders, rules, and regulations shall be enforced and executed; and any expenses incurred by such nuisance authority or authorities shall be deemed to be expenses incurred by it or them in carrying into effect the Nuisances Removal Acts.

Periodical Removal of Manure in Mews, &c.

53. Where notice has been given by the nuisance authority, or their officer or officers, for the periodical removal of manure or other refuse matter from mews, stables, or other premises (whether such notice shall be by public announcement in the locality or otherwise), and subsequent to such notice the person or persons to whom the manure or other refuse matter belongs shall not so remove the same, or shall permit a further accumulation, and shall not continue such periodical removal at such intervals as the nuisance authority, or their officer or officers, shall direct, he or they shall be liable, without further notice, to a penalty of twenty shillings *per* day for every day during which such manure or other refuse matter shall be permitted to accumulate, such penalty to be recovered in a summary manner: Provided always, that this section shall not apply to any place where the board of guardians or overseers of the poor are the nuisance authority.

Recovery of Penalties.

54. Penalties under this Act, and expenses directed to be recovered in a summary manner, may be recovered before two justices in manner directed by an Act passed in the Session holden in the eleventh and twelfth years of the reign of Her Majesty Queen *Victoria*, chapter forty-three, intituled "An Act to facilitate the Performance of the Duties of Justices of the Peace out of Sessions within *England* and *Wales* with respect to summary Convictions and Orders," or any Act amending the same.

Powers of Act cumulative.

55. All powers given by this Act shall be deemed to be in addition to and not in derogation of any other powers conferred on any local authority by Act of Parliament, law, or custom, and such authority may exercise such other powers in the same manner as if this Act had not passed.

PART IV.

APPLICATION OF ACT TO IRELAND.

Modifications necessary for Application of Part I. to Ireland.

56. In applying the First Part of this Act to *Ireland* the following changes shall be observed:

(1.) The provisions of the sections numbered from seventy-five to eighty, both included, of the Public Health Act, 1848, and sections fifty-one, fifty-two, and fifty-three of the Local Government Act, 1858, and section twenty of the Local Government Act, 1858, Amendment Act, 1861, referred to in the First Part of this Act, shall for all purposes connected with the execution of this Act be extended to *Ireland:*

(2.) The Sewage Utilization Act, 1865, shall be amended by substituting in *Ireland* the sewer authority, as defined by the first schedule to this Act, for the sewers authority as defined by said Act.

Modifications necessary for Application of Part II. to Ireland.

57. The Nuisance Removal Acts as amended by the Second Part of this Act shall apply to *Ireland;* provided, however, that in such application the following changes shall be observed :

(1.) Sewer authority as defined by the Sewage Utilization Act, 1865, and amended by this Act, shall in *Ireland* be the nuisance authority for executing the Nuisance Removal Acts :

(2.) The expenses of executing the Nuisance Removal Acts shall be defrayed out of the funds herein-after provided :

(3.) The penalties shall be recovered in the manner herein-after provided :

(4.) The expressions "mayor, aldermen, and burgesses," "council," "borough rate," borough fund," and "town rate," shall in the first schedule hereto have respectively the same meaning as in the Acts for the regulation of municipal corporations in *Ireland:*

(5.) For the purposes of the twenty-second section of the Nuisance Removal Act, 1855, the nuisance authority shall in *Ireland* have the power of entering land conferred by the Sewage Utilization Act, 1865, and shall have the same power of levying assessments under the said section that they have of levying any other rates they are authorized by law to impose.

How Expenses to be defrayed in Ireland when Nuisance Authority not a Board of Guardians.

58. In *Ireland,* the nuisance authority, not being the guardians of the poor, shall pay all expenses incurred by them in carrying the Nuisance Removal Acts into effect out of the fund in the first schedule in that behalf mentioned, and where such fund arises wholly or in part from rates shall have, in addition to their existing powers of rating, all such powers for making and levying any extra rate, if necessary, respectively, as in the case of any rate authorized to be made under the provisions of the respective Acts of Parliament under which the nuisance authorities are constituted or authorized to levy rates ; and all provisions of such Acts respectively shall be applicable in respect thereof; provided that when the rates to be assessed by such authority are limited by law to a certain rateable amount, such limitation shall not apply or extend to expenses incurred in carrying this Act into execution ; and it shall be lawful for such authority to assess the expenses under this Act in addition to such limited assessment.

When Board of Guardians is Nuisance Authority, how Expenses to be defrayed in Ireland.

59. In *Ireland,* a nuisance authority, being guardians of the poor, shall pay all expenses incurred by them in carrying this Act into effect out of the poor rates of the union, and charge the same to the union, or any electoral division or electoral divisions thereof, in such manner as the Poor Law Commissioners shall from time to time, by general orders applicable to classes of cases or by order in any particular case, direct.

Recovery of Penalties in Ireland.

60. In *Ireland,* penalties under this Act and expenses or compensation directed to be recovered in a summary manner, and nuisances and other offences liable to be prosecuted summarily, shall be recovered and prosecuted in manner directed by the Petty Sessions (*Ireland*) Act, 1851, or any Act amending the same ; and all

penalties recovered by any authority under this Act shall be paid to them respectively, and by them applied in aid of their expenses under this Act.

Any order authorized to be made by justices under this Act shall be deemed to be an order made upon a complaint on which justices are authorized to make orders under the last-mentioned Act.

Modifications necessary for Application of Part III. to Ireland.

61. In applying the provisions of Part III. of this Act to *Ireland* the following changes shall be observed :
   (1.) Applications for power to make regulations as to lodging houses may be made by any nuisance authority, except a board of guardians, and shall be made to the Lord Lieutenant in Council, and the said Lord Lieutenant in Council shall have the power of declaring the enactments as to lodging houses in the Third Part of this Act to be in force in any nuisance district :
   (2.) The said Lord Lieutenant in Council shall have and exercise the power, in respect of boards of guardians acting together, vested in the Privy Council by the said Third Part of this Act.
   (3.) In *Ireland*, any nuisance authority, except a board of guardians, may exercise the powers conferred on local boards acting in the execution of the Local Government Act, 1858, by the said Third Part of this Act :
   (4.) Sewer and nuisance authorities in *Ireland* shall be incorporated for the purposes of this Act by the names set forth in the said first schedule hereto ; and such sewer or nuisance authorities may hold lands by such names for the purposes of Burial Ground (*Ireland*) Act, 1856 :
   (5.) The penalties under the Third Part of this Act shall be recovered in like manner as herein-before provided with respect to penalties under the Second Part of this Act.

Modifications necessary for Application of Disease Prevention Act to Ireland.

62. The Diseases Prevention Act, 1855, as amended by the Nuisance Removal and Disease Prevention Amendment Act, 1860, and this Act, shall extend to *Ireland :* Provided, however, that in such application the following changes shall be observed :
   (1.) The Lord Lieutenant in Council shall have the power with respect to *Ireland* which the Privy Council has under such provisions for prevention of disease in *England :*
   (2.) The commissioners for administering the laws for the relief of the poor in *Ireland,* hereinafter called the Poor Law Commissioners, shall be the authority in *Ireland* for issuing regulations to carry the provisions of the said Act into effect :
   (3.) The regulations of the Poor Law Commissioners shall be authenticated in like manner as orders of theirs under the Dispensary Act, 1851, stat. 14 & 15 *Vict.* c. 68. sect. 8. :
   (4.) In defraying the expenses of the prevention of disease out of the poor rate of the union under this Act the guardians of the poor shall charge the same to the union or any dispensary district or electoral division or divisions thereof, in such manner as the Poor Law Commissioners shall from time to time, by general orders applicable to classes of cases, or by orders in particular cases, direct.

Committee and Officers under Dispensaries Act to aid Local Authority in execution of this Act.

63. In *Ireland*, all committees, inspectors, medical officers, and other persons appointed or employed under the powers of statute fourteenth and fifteenth *Victoria*, chapter sixty-eight (the Dispensaries Act, 1851), shall and they are hereby required within their respective districts to aid the local authority, and such officers or persons as they shall appoint or employ, in the superintendence and execution of any directions and regulations which may at any time be issued by the Poor Law Commissioners for the time being under the authority and by virtue of this Act.

APPENDIX.

The Provisions of 14 & 15 Vict. c. 68. as to Duties and Appointment of Medical Inspectors in Ireland incorporated with this Act.

64. In *Ireland*, the provisions of the Dispensary Act, 1851 (statute 14 & 15 Vict. c. 68.), with respect to the duties and appointment of medical inspectors, shall be incorporated with this Act, and the prevention of disease and inquiry into public health under this Act shall be deemed one of the purposes for which such medical inspectors have been or may be appointed, in like manner as if its provisions had been referred to in the said Act of 1851, instead of the provisions of the said Nuisance Removal and Diseases Prevention Act of 1848.

Remuneration to Medical Practitioners for Services under the Directions and Regulations of the Poor Law Commissioners in Ireland.

65. In *Ireland*, whenever in compliance with any direction or regulation of the Poor Law Commissioners which they may be empowered to make under the laws for the time being as to the public health, any medical officer of a union or dispensary district, or any other medical practitioner specially employed by the guardians for the purpose, shall perform any extra medical service in any union or part of a union, it shall and may be lawful for the guardians of the union to determine, subject to the approval of the said commissioners, and if they shall not approve the amount determined by the guardians, for the said commissioners to fix by order under their seal such remuneration, proportioned to the nature and extent of such services as aforesaid, as to them shall appear just and reasonable ; and the amount of such remuneration shall be paid to such medical officer or other medical practitioner by the guardians of the union out of the rates raised for the relief of the poor, and shall be charged either to the union at large, or to such part or parts of the union, according to the nature of the case, as the said commissioners shall in each case direct.

Poor Law Commissioners to make Inquiries as to Public Health in Ireland.

66. The Lord Lieutenant in Council may from time to time direct the Poor Law Commissioners to cause to be made such inquiries as the Lord Lieutenant in Council see fit in relation to any matters concerning the public health in any place or places in *Ireland*, and the Poor Law Commissioners shall report the result of such inquiries to the Lord Lieutenant in Council.

Publication in Ireland to be made in Dublin Gazette.

67. Publication shall be made in the *Dublin Gazette* in any case in *Ireland* where publication in the *London Gazette* is required in *England*.

Powers in Secretary of State in England to be exercised in Ireland by the Lord Lieutenant in Council.

68. All powers relating to the execution of this Act in *England*, and by this Act vested in one of Her Majesty's Principal Secretaries of State, shall, with regard to the execution of this Act in *Ireland*, in all cases not herein-before expressly provided for, be vested in the Lord Lieutenant or other chief governor or governors of *Ireland* ; and all powers relating to the execution of this Act in *England*, and by this Act vested in the Privy Council in *England*, shall, with regard to the execution of this Act in *Ireland*, in all cases not herein-before expressly provided for, be vested in the Lord Lieutenant in Council in *Ireland*.

Repeal of Statutes applicable to Ireland.

69. From and after the passing of this Act, the Acts set forth in the second schedule hereto shall be repealed, so far as they are still in force : Provided always, that all proceedings commenced or taken under the said Acts and not yet completed may be proceeded with under said Acts, and that all contracts and works undertaken by virtue of said Acts shall continue and be effective as if said Acts had not been repealed.

SCHEDULES.

FIRST SCHEDULE.

APPLICATION TO IRELAND.

Description of Sewers and Nuisance Authority in Ireland.	Description of Sewers and Nuisance District in Ireland.	Corporate Name, for the Purpose of suing or being sued, or holding Property, under the Provisions of this Act.	Rate or Fund out of which Expenses incurred by Sewers or Nuisance Authority under this Act to be defrayed.
The Right Honourable the Lord Mayor, Aldermen, and Burgesses, acting by the Town Council.	The City of Dublin.	The Right Honourable the Lord Mayor, Aldermen, and Burgesses of the City of Dublin.	The Borough Rate or Borough Fund.
The Mayor, Aldermen, and Burgesses, acting by the Town Council.	Towns Corporate, with Exception of Dublin.	The Mayor, Aldermen, and Burgesses of the City or Town of ——.	The Borough Rate or Borough Fund.
The Town Commissioners.	Towns having Town Commissioners, under the Towns Improvement (Ireland) Act, 1854 (17 & 18 Vict. c. 113), or under any Local Act.	The Town Commissioners of ——.	Any Rate levied by the Commissioners.
The Township Commissioners.	Townships having Commissioners under Local Acts.	The Township Commissioners of ——.	
The Commissioners appointed by virtue of an Act made in the 9th Year of the Reign of George the Fourth, intituled "An Act to make Provision for the lighting, cleansing, and watching of Cities and Towns Corporate and Market Towns in Ireland in certain Cases."	Towns under such Commissioners.	The Lighting and Cleansing Commissioners of the Town of ——.	
The Municipal Commissioners.	Towns having Municipal Commissioners, under 3 & 4 Vict. c. 108.	The Municipal Commissioners of ——.	The Town Fund.
The Guardians of the Poor of each Union.	Such Part of each Union as is not under another Sewer or Nuisance Authority.	The Guardians of the Poor of the —— Union.	The Poor Rate of Union.

SECOND SCHEDULE.

Statutes repealed.

Local Boards of Health Act for Ireland, 1818; statute 58 Geo. 3. c. 47. ss. 10 to 15 inclusive.
Officers of Health Act for Ireland, 1819; statute 59 Geo. 3. c. 41.
Nuisance Removal and Disease Prevention Act, 1848.
Nuisance Removal and Disease Prevention Act, 1849.

ANNO VICESIMO PRIMO & VICESIMO SECUNDO VICTORIÆ REGINÆ.

CAP. XC.

An Act to regulate the Qualifications of Practitioners in Medicine and Surgery. [2nd *August* 1858.]

WHEREAS it is expedient that persons requiring medical aid should be enabled to distinguish qualified from unqualified practitioners : Be it therefore enacted by the Queen's most excellent Majesty, by and with the advice and consent of the Lords Spiritual and Temporal, and Commons, in this present Parliament assembled; and by the authority of the same, as follows :

Short Title.

I. This Act may for all purposes be cited as "The Medical Act."

Commencement of Act.

II. This Act shall commence and take effect from the first day of *October* one thousand eight hundred and fifty-eight.

Medical Council.

III. A council which shall be styled "The General Council of Medical Education and Registration of the United Kingdom," herein-after referred to as the general council, shall be established, and branch councils for *England*, *Scotland*, and *Ireland* respectively formed thereout as herein-after mentioned.

Members of Council.

IV. The general council shall consist of one person chosen from time to time by each of the following bodies : (that is to say,)
  The Royal College of Physicians :
  The Royal College of Surgeons of *England:*
  The Apothecaries' Society of *London:*
  The University of *Oxford:*
  The University of *Cambridge:*
  The University of *Durham:*
  The University of *London:*
  The College of Physicians of *Edinburgh:*
  The College of Surgeons of *Edinburgh:*
  The Faculty of Physicians and Surgeons of *Glasgow:*
  One person chosen from time to time by the University of *Edinburgh* and the two universities of *Aberdeen* collectively :
  One person chosen from time to time by the University of *Glasgow* and the University of *Saint Andrew's* collectively :
  One person chosen from time to time by each of the following bodies :
  The King and Queen's College of Physicians in *Ireland:*
  The Royal College of Surgeons in *Ireland:*
  The Apothecaries' Hall of *Ireland:*
  The University of *Dublin:*
  The Queen's University in *Ireland:*
and six persons to be nominated by Her Majesty with the advice of Her Privy Council, four of whom shall be appointed for *England*, one for *Scotland*, and one for *Ireland* ; and of a president to be elected by the general council.

Provision in case the Universities of Glasgow, Aberdeen, and Saint Andrew's fail to appoint a Person to represent them.

V. If the said Universities of *Edinburgh* and *Aberdeen*, of *Glasgow* and *Saint Andrew's* respectively, shall not be able to agree upon some one person to represent them in the council, it shall be lawful for each one of the said universities to select one person; and thereupon it shall be lawful for Her Majesty, with the advice of Her Privy Council, to appoint one of the persons so selected to be a member of the said council for the said universities.

Branches of the Council for England, Scotland, and Ireland.

VI. The members chosen by the medical corporations and universities of *England, Scotland*, and *Ireland* respectively, and the members nominated by Her Majesty, with the advice of Her Privy Council, for such parts respectively of the United Kingdom, shall be the branch councils for such parts respectively of the United Kingdom, to which branch councils shall be delegated such of the powers and duties vested in the council as the council may see fit other than the power to make representations to Her Majesty in Council as herein-after mentioned : the president shall be a member of all the branch councils.

Qualification.

VII. Members of the general council representing the medical corporations must be qualified to be registered under this Act.

Resignation or Death of Member of General Council.

VIII. The members of the general council shall be chosen and nominated for a term not exceeding five years, and shall be capable of re-appointment, and any member may at any time resign his appointment by letter addressed to the president of the said council, and upon the death or resignation of any member of the said council, some other person shall be constituted a member of the said council, in his place in manner herein-before provided; but it shall be lawful for the council during such vacancy to exercise the powers herein-after mentioned.

Time and Place of Meeting of the General Council.

IX. The general council shall hold their first meeting within three months from the commencement of this Act, in such place and at such time as one of Her Majesty's Principal Secretaries of State shall appoint, and shall make such rules and regulations as to the times and places of the meetings of the general council, and the mode of summoning the same, as to them shall seem expedient, which rules and regulations shall remain in force until altered at any subsequent meeting; and in the absence of any rule or regulation as to the summoning a meeting of the general council, it shall be lawful for the president to summon a meeting at such time and place as to him shall seem expedient by letter addressed to each member; and at every meeting, in the absence of the president, some other member to be chosen from the members present shall act as president; and all Acts of the general council shall be decided by the votes of the majority of the members present at any meeting, the whole number present not being less than eight, and at all such meetings the president for the time being shall, in addition to his vote as a member of the council, having a casting vote, in case of an equality of votes; and the general council shall have power to appoint an executive committee out of their own body, of which the quorum shall not be less than three, and to delegate to such committee such of the powers and duties vested in the council as the council may see fit, other than the power of making representations to Her Majesty in Council as herein-after mentioned.

Appointment of Registrars and other Officers.

X. The general council shall appoint a registrar, who shall act as secretary of the general council, and who may also act as treasurer, unless the council shall appoint another person or other persons as treasurer or treasurers; and the person or persons so appointed shall likewise act as registrar for *England*, and as secretary and treasurer or treasurers, as the case may be, for the branch council for *England;* the general council and branch council for *England* shall also appoint so many clerks and servants as shall be necessary for the purposes of this Act; and every person so appointed by any council shall be removable at the pleasure of that council, and shall be paid such salary as the council by which he was appointed shall think fit.

Appointment of Registrars and other Officers by Branch Councils.

XI. The branch councils for *Scotland* and *Ireland* shall each respectively in like manner appoint a registrar and other officers and clerks, who shall be paid such salaries as such branch councils respectively shall think fit, and be removable at the pleasure of the council by which they were appointed; and the person appointed registrar shall also act as secretary to the branch council, and may also act as treasurer, unless the council shall appoint some other person or persons as treasurer or treasurers.

Fees for Attendance at Councils.

XII. There shall be paid to the members of the councils such fees for attendance and such reasonable travelling expenses as shall from time to time be allowed by the general council and approved by the Commissioners of Her Majesty's Treasury.

Expenses of the Councils.

XIII. All monies payable to the respective councils shall be paid to the treasurers of such councils respectively, and shall be applied to defray the expenses of carrying this Act into execution in manner following; that is to say, separate accounts shall be kept of the expenses of the general council, and of those of the branch councils; and the expenses of the general council, including those of keeping, printing, and publishing the register for the United Kingdom, shall be defrayed, under the direction of the general council, by means of an equal percentage rate upon all monies received by the several branch councils; returns shall be made by the treasurers of the respective branch councils, at such times as the general council shall direct, of all monies received by them; and the necessary percentage having been computed by the general council, the respective contributions shall be paid by the treasurers of such branch councils to the treasurer or treasurers of the general council; and the expenses of the branch councils shall be defrayed, under the direction of those councils respectively, out of the residue of the monies so received as aforesaid.

Duty of Registrar to keep the Register correct.

XIV. It shall be the duty of the registrars to keep their respective registers correct in accordance with the provisions of this Act, and the orders and regulations of the general council, and to erase the names of all registered persons who shall have died, and shall from time to time make the necessary alterations in the addresses or qualifications of the persons registered under this Act; and to enable the respective registrars duly to fulfil the duties imposed upon them it shall be lawful for the registrar to write a letter to any registered person, addressed to him according to his address on the register, to inquire whether he has ceased to practice, or has changed his residence, and if no answer shall be returned to such letter within the period of six months from the sending of the letter it shall be lawful to erase the name of such person from the register; provided always, that the same may be restored by direction of the general council should they think fit to make an order to that effect.

Registration of Persons now qualified, and of Persons hereafter becoming qualified.

XV. Every person now possessed, and (subject to the provisions herein-after contained) every person hereafter becoming possessed, of any one or more of the qualifications described in the Schedule (A.) to this Act, shall, on payment of a fee, not exceeding two pounds in respect of qualifications obtained before the first day of *January*, one thousand eight hundred and fifty-nine, and not exceeding five pounds in respect of qualifications obtained on or after that day, be entitled to be registered on producing to the registrar of the branch council for *England, Scotland,* or *Ireland* the document conferring or evidencing the qualification or each of the qualifications in respect whereof he seeks to be so registered, or upon transmitting by post to such registrar information of his name and address, and evidence of the qualification or qualifications in respect whereof he seeks to be registered, and of the time or times at which the same was or were respectively obtained : Provided always, that it shall be lawful for the several colleges and other bodies mentioned in the said Schedule (A.) to transmit from time to time to the said registrar lists certified under their respective seals of the several persons who, in respect of qualifications granted by such colleges and bodies respectively, are for the time being entitled to be registered under this Act, stating the respective qualifications and places of residence of such persons : and it shall be lawful for the registrar thereupon, and upon payment of such fee as aforesaid in respect of each person to be registered, to enter in the register the persons mentioned in such lists, with their qualifications and places of residence as therein dated, without other application in relation thereto.

Council to make Orders for regulating Registers to be kept.

XVI. The general council shall, with all convenient speed after the passing of this Act, and from time to time as occasion may require, make orders for regulating the registers to be kept under this Act as nearly as conveniently may be in accordance with the form set forth in Schedule (D.) to this Act, or to the like effect.

Persons practising in England before 1st August, 1815, entitled to be registered.

XVII. Any person who was actually practising medicine in *England* before the first day of *August*, one thousand eight hundred and fifteen, shall, on payment of a fee to be fixed by the general council, be entitled to be registered on producing to the registrar of the branch council for *England, Scotland,* or *Ireland* a declaration according to the form in the Schedule (B.) to this Act, signed by him; or upon transmitting to such registrar information of his name and address, and enclosing such declaration as aforesaid.

Council may require Information as to Course of Study, &c. required for obtaining Qualifications.

XVIII. The several colleges and bodies in the United Kingdom mentioned in Schedule (A.) to this Act, shall from time to time, when required by the general council, furnish such council with such information as they may require as to the courses of study and examinations to be gone through in order to obtain the respective qualifications mentioned in Schedule (A.) to this Act, and the ages at which such courses of study and examination are required to be gone through, and such qualifications are conferred, and generally as to the requisites for obtaining such qualifications ; and any member or members of the general council, or any person or persons deputed for this purpose by such council, or by any branch council, may attend and be present at any such examinations.

Colleges may unite in conducting Examinations.

XIX. Any two or more of the colleges and bodies in the United Kingdom mentioned in Schedule (A.) to this Act may, with the sanction and under the directions of the general council, unite or co-operate in conducting the examinations required for qualifications to be registered under this Act.

Defects in the Course of Study or Examinations may be represented by General Council to Privy Council.

XX. In case it appear to the general council that the course of study and examinations to be gone through in order to obtain any such qualification from any such college or body are not such as to secure the possession by persons obtaining such qualification of the requisite knowledge and skill for the efficient practice of their profession, it shall be lawful for such general council to represent the same to Her Majesty's most Honourable Privy Council.

APPENDIX. 401

Privy Council may suspend the Right of Registration in respect of Qualifications granted by College, &c. in default, but may be revoked.

XXI. It shall be lawful for the Privy Council, upon any such representation as aforesaid, if it see fit, to order that any qualification granted by such college or body, after such time as may be mentioned in the order, shall not confer any right to be registered under this Act; Provided always, that it shall be lawful for Her Majesty, with the advice of Her Privy Council, when it is made to appear to Her, upon further representation from the general council or otherwise, that such college or body has made effectual provision, to the satisfaction of such general council, for the improvement of such course of study or examinations, or the mode of conducting such examinations, to revoke any such order.

Persons not to be registered in respect of Qualifications granted by the College or Body before Revocation.

XXII. After the time mentioned in this behalf in any such order in Council no person shall be entitled to be registered under this Act in respect of any such qualification as in such order mentioned granted by the college or body to which such order relates, after the time therein mentioned, and the revocation of any such order shall not entitle any person to be registered in respect of any qualification granted before such revocation.

Privy Council may prohibit Attempts to impose Restrictions as to any Theory of Medicine or Surgery by Bodies entitled to grant Certificates.

XXIII. In case it shall appear to the general council that an attempt has been made by any body, entitled under this Act to grant qualifications, to impose upon any candidate offering himself for examination an obligation to adopt or refrain from adopting the practice of any particular theory of medicine or surgery as a test or condition of admitting him to examination, or of granting a certificate, it shall be lawful for the said council to represent the same to Her Majesty's most Honourable Privy Council, and the said Privy Council may thereupon issue an injunction to such body so acting, directing them to desist from such practice; and in the event of their not complying therewith, then to order that such body shall cease to have the power of conferring any right to be registered under this Act so long as they shall continue such practice.

As to the making and Authentication of Orders, &c.

XXIV. All powers vested in the Privy Council by this Act may be exercised by any three or more of the Lords and others of the Privy Council, the Vice-President of the Committee of the said Privy Council on Education being one of them; and all orders and acts of the Privy Council under this Act shall be sufficiently made and signified by a written or printed document, signed by one of the clerks of the Privy Council, or such officer as may be appointed by the Privy Council in this behalf; and all orders and acts made or signified by any written or printed document purporting to be so signed shall be deemed to have been duly made, issued, and done by the Privy Council; and every such document shall be received in evidence in all courts, and before all justices and others, without proof of the authority or signature of such clerk or other officer or other proof whatsoever, until it be shown that such document was not duly signed by the authority of the Privy Council.

As to Registration by Branch Registrars.

XXV. Where any person entitled to be registered under this Act applies to the registrar of any of the said branch councils for that purpose, such registrar shall forthwith enter in a local register in the form set forth in Schedule (D.) to this Act, or to the like effect, to be kept by him for that purpose, the name and place of residence, and the qualification or several qualifications in respect of which the person is so entitled, and the date of the registration, and shall, in the case of

D D

the registrar of the branch council for *Scotland* or *Ireland*, with all convenient speed send to the registrar of the general council a copy, certified under the hand of the registrar, of the entry so made, and the registrar of the general council shall forthwith cause the same to be entered in the general register ; and such registrar shall also forthwith cause all entries made in the local register for *England* to be entered in the general register ; and the entry on the general register shall bear date from the local register.

Evidence of Qualification to be given before Registration.

XXVI. No qualification shall be entered on the register, either on the first registration or by way of addition to a registered name, unless the registrar be satisfied by the proper evidence that the person claiming is entitled to it ; and any appeal from the decision of the registrar may be decided by the general council, or by the council for *England, Scotland,* or *Ireland* (as the case may be) ; and any entry which shall be proved to the satisfaction of such general council or branch council to have been fraudulently or incorrectly made may be erased from the register by order in writing of such general council or branch council.

Register to be published.

XXVII. The registrar of the general council shall in every year cause to be printed, published, and sold, under the direction of such council, a correct register of the names in alphabetical order according to the surnames, with the respective residences, in the form set forth in Schedule (D.) to this Act, or to the like effect, and medical titles, diplomas, and qualifications conferred by any corporation or university, or by doctorate of the Archbishop of *Canterbury,* with the dates thereof, of all persons appearing on the general register as existing on the first day of *January* in every year ; and such register shall be called " the medical register ;" and a copy of the medical register for the time being, purporting to be so printed and published as aforesaid, shall be evidence in all courts and before all justices of the peace and others that the persons therein specified are registered according to the provisions of this Act ; and the absence of the name of any person from such copy shall be evidence, until the contrary be made to appear, that such person is not registered according to the provisions of this Act : Provided always, that in the case of any person whose name does not appear in such copy, a certified copy, under the hand of the registrar of the general council or of any branch council, of the entry of the name of such person on the general or local register, shall be evidence that such person is registered under the provisions of this Act.

Names of Members struck off from List of College, &c. to be signified to General Council.

XXVIII. If any of the said colleges or the said bodies at any time exercise any power they possess by law of striking off from the list of such college or body the name of any one of their members, such college or body shall signify to the general council the name of the member so struck off; and the general council may, if they see fit, direct the registrar to erase forthwith from the register the qualification derived from such college or body in respect of which such member was registered, and the registrar shall note the same therein : Provided always, that the name of no person shall be erased from the register on the ground of his having adopted any theory of medicine or surgery.

Medical Practitioners convicted of Felony may be struck off the Register.

XXIX. If any registered medical practitioner shall be convicted in *England* or *Ireland* of any felony or misdemeanor, or in *Scotland* of any crime or offence, or shall after due inquiry be judged by the general council to have been guilty of infamous conduct in any professional respect, the general council may, if they see fit, direct the registrar to erase the name of such medical practitioner from the register.

APPENDIX.

Registered Persons may have subsequent Qualifications inserted in the Register.

XXX. Every person registered under this Act who may have obtained any higher degree or any qualification other than the qualification in respect of which he may have been registered, shall be entitled to have such higher degree or additional qualification inserted in the register in substitution for or in addition to the qualification previously registered, on payment of such fee as the council may appoint.

Privileges of registered Persons.

XXXI. Every person registered under this Act shall be entitled according to his qualification or qualifications to practise medicine or surgery, or medicine and surgery, as the case may be, in any part of Her Majesty's dominions, and to demand and recover in any court of law, with full costs of suit, reasonable charges for professional aid, advice, and visits, and the costs of any medicines or other medical or surgical appliances rendered or supplied by him to his patients: Provided always, that it shall be lawful for any college of physicians to pass a byelaw to the effect that no one of their fellows or members shall be entitled to sue in manner aforesaid in any court of law, and thereupon such byelaw may be pleaded in bar to any action for the purposes aforesaid commenced by any fellow or member of such college.

None but registered Persons to recover Charges.

XXXII. After the first day of *January* one thousand eight hundred and fifty-nine, no person shall be entitled to recover any charge in any court of law for any medical or surgical advice, attendance, or for the performance of any operation, or for any medicine which he shall have both prescribed and supplied, unless he shall prove upon the trial that he is registered under this Act.

Poor Law Medical Officers not disqualified if registered within Six Months of passing of Act.

XXXIII. Provided also, that no person who on the first of *October* one thousand eight hundred and fifty-eight shall be acting as medical officer under an order of the Poor Law Commissioners or Poor Law Board shall be disqualified to hold such office by reason of his not being registered as herein required, unless he shall have failed to be registered within six months from the passing of this Act.

Meaning of Terms "legally qualified Medical Practitioner," &c.

XXXIV. After the first day of *January* one thousand eight hundred and fifty-nine, the words "legally qualified medical practitioner" or "duly qualified medical practitioner," or any words importing a person recognised by law as a medical practitioner, or member of the medical profession, when used in any Act of Parliament, shall be construed to mean a person registered under this Act.

Registered Persons exempted from serving on Juries, &c.

XXXV. Every person who shall be registered under the provisions of this Act shall be exempt, if he shall so desire, from serving on all juries and inquests whatsoever, and from serving all corporate, parochial, ward, hundred, and township offices, and from serving in the militia, and the name of such person shall not be returned in any list of persons liable to serve in the militia, or in any such office as aforesaid.

Unregistered Persons not to hold certain Appointments.

XXXVI. After the first day of *January* one thousand eight hundred and fifty-nine, no person shall hold any appointment as a physician, surgeon, or other medical officer either in the military or naval service, or in emigrant or other vessels, or in any hospital, infirmary, dispensary, or lying-in hospital, not supported wholly by voluntary contributions, or in any lunatic asylum, gaol, penitentiary, house of correction, house of industry, parochial or union workhouse or poorhouse, parish union, or other public establishment, body, or institution, or to any friendly or other society for affording mutual relief in sickness, infirmity, or old age, or as a medical officer of health, unless he be registered under this Act: Provided always, that nothing in this Act contained shall extend to repeal or alter any of the provisions of the Passengers Act, 1855.

No Certificate to be valid unless Person signing be registered.

XXXVII. After the first day of *January* one thousand eight hundred and fifty-nine, no certificate required by any Act now in force or that may hereafter be passed from any physician, surgeon, licentiate in medicine and surgery, or other medical practitioner, shall be valid unless the person signing the same be registered under this Act.

Penalty on wilful Falsification of Register.

XXXVIII. Any registrar who shall wilfully make or cause to be made any falsification in any matters relating to the register shall be deemed guilty of a misdemeanor in *England* or *Ireland*, and in *Scotland* of a crime or offence punishable by fine or imprisonment, and shall, on conviction thereof, be imprisoned for any term not exceeding twelve months.

Penalty for obtaining Registration by false Representations.

XXXIX. If any person shall wilfully procure or attempt to procure himself to be registered under this Act, by making or producing or causing to be made or produced any false or fraudulent representation or declaration, either verbally or in writing, every such person so offending, and every person aiding and assisting him therein, shall be deemed guilty of a misdemeanor in *England* and *Ireland*, and in *Scotland* of a crime or offence punishable by fine or imprisonment, and shall, on conviction thereof, be sentenced to be imprisoned for any term not exceeding twelve months.

Penalty for falsely pretending to be a registered Person.

XL. Any person who shall wilfully and falsely pretend to be or take or use the name or title of a physician, doctor of medicine, licentiate in medicine and surgery, bachelor of medicine, surgeon, general practitioner or apothecary, or any name, title, addition, or description implying that he is registered under this Act, or that he is recognised by law as a physician, or surgeon, or licentiate in medicine and surgery, or a practitioner in medicine, or an apothecary, shall, upon a summary conviction for any such offence, pay a sum not exceeding twenty pounds.

Recovery of Penalties.

XLI. Any penalty to which under this Act any person is liable on summary conviction of any offence may be recovered as follows; (that is to say,) in *England*, in manner directed by the Act of the Session holden in the eleventh and twelfth years of Her Majesty, chapter forty-three, and in *Ireland* in manner directed by "The Petty Sessions (*Ireland*) Act, 1851," or any other Act for the time being in force in *England* and *Ireland* respectively for the like purpose; and any such penalty may in *Scotland* be recovered by the procurator fiscal of the county, or by any other person before the sheriff or two justices, who may proceed in a summary way and grant warrant for bringing the party complained against before him or them, or issue an order requiring such party to appear on a day and at a time and place to be named in such order, and every such order shall be served on the party by delivering to him in person or by leaving at his usual place of abode a copy of such order and of the complaint whereupon the same has proceeded, and upon the appearance or default to appear of the party, it shall be lawful for the sheriff or justices to proceed to the hearing of the complaint, and upon proof on oath or confession of the offence, the sheriff or justices shall without any written pleadings or record of evidence commit the offender and decern him to pay the penalty named as well as such expenses as the sheriff or justices shall think fit, and failing payment shall grant warrant for recovery thereof by poinding and imprisonment, such imprisonment to be for such period as the discretion of the sheriff or justices may direct, not exceeding three calendar months, and to cease on payment of the penalty and expenses.

Application of Penalties.

XLII. Any sum or sums of money arising from conviction and recovery of penalties as aforesaid shall be paid to the treasurer of the general council.

Application of Monies received by Treasurer.

XLIII. All monies received by any treasurer arising from fees to be paid on registration, from the sale of registers, from penalties, or otherwise, shall be applied for expenses of registration and of the execution of this Act.

Accounts to be published.

XLIV. The treasurers of the general and branch councils shall enter in books to be kept for that purpose a true account of all sums of money by them received and paid, and such accounts shall be submitted by them to the respective general council and branch councils at such times as the councils shall require; and the said accounts shall be published annually, and such accounts shall be laid before both Houses in the month of *March* in every year, if Parliament be sitting, or, if Parliament be not sitting, then within one month after the next meeting of Parliament.

Notice of Death of Medical Practitioners to be given by Registrars.

XLV. Every registrar of deaths in the United Kingdom on receiving notice of the death of any medical practitioner shall forthwith transmit by post to the registrar of the general council and to the registrar of the branch council a certificate under his own hand of such death, with the particulars of time and place of death, and may charge the cost of such certificate and transmission as an expense of his office, and on the receipt of such certificate the medical registrar shall erase the name of such deceased medical practitioner from the register.

Provision for Persons practising in the Colonies and elsewhere, and for Students.

XLVI. It shall be lawful for the general council by special orders to dispense with such provisions of this Act or with such part of any regulations made by its authority as to them shall seem fit in favour of persons not practising medicine or surgery in any part of Her Majesty's dominions other than *Great Britain* and *Ireland* by virtue of any of the qualifications described in Schedule (A.); and also in favour of persons practising medicine or surgery within the United Kingdom on foreign or colonial diplomas or degrees before the passing of this Act; and also in favour of any persons who have held appointments as surgeons or assistant surgeons in the army, navy, or militia, or in the service of the *East India* Company, or are acting as surgeons in the public service, or in the service of any charitable institutions, and also, so far as to the council shall seem expedient, in favour of medical students who shall have commenced their professional studies before the passing of this Act.

New Charter may be granted to the College of Physicians of London.

XLVII. It shall be lawful for Her Majesty to grant to the corporation of the Royal College of Physicians of *London* a new charter, and thereby to give to such corporation the name of " The Royal College of Physicians of *England*," and to make such alterations in the constitution of the same corporation as to Her Majesty may seem expedient; and it shall be lawful for the said corporation to accept such charter under their common seal, and such acceptance shall operate as a surrender of all charters heretofore granted to the said corporation, except the charter granted by King *Henry* the Eighth, and shall also operate as a surrender of such charter, and of any rights, powers, or privileges conferred by or enjoyed under an Act of the Session holden in the fourteenth and fifteenth years of King *Henry* the Eighth, chapter five, confirming the same, as far as such charter and Act respectively may be inconsistent with such new charter: Provided nevertheless, that within twelve months after the granting of such charter to the College of Physicians of *London*, any fellow, member, or licentiate of the Royal College of Physicians of *Edinburgh*, or of the Queen's College of Physicians of *Ireland*, who may be in practice as a physician in any part of the United Kingdom called *England*, and who may be desirous of becoming a member of such College of Physicians of *England*, shall be at liberty to do so, and be entitled to receive the diploma of the said college, and to be admitted to all the rights and privileges thereunto appertaining, on the payment of a registration fee of two pounds to the said college.

Her Majesty may grant Power to College of Surgeons to institute Examinations, &c., for Dentists.

XLVIII. It shall, notwithstanding anything herein contained, be lawful for Her Majesty, by charter, to grant to the Royal College of Surgeons of *England* power to institute and hold examinations for the purpose of testing the fitness of persons to practise as dentists who may be desirous of being so examined, and to grant certificates of such fitness.

New Charter may be granted to College of Physicians of Edinburgh.

XLIX. It shall be lawful for Her Majesty to grant to the corporation of the Royal College of Physicians of *Edinburgh* a new charter, and thereby to give to the said College of Physicians the name of "The Royal College of Physicians of *Scotland*," and it shall be lawful for the said Royal College of Physicians, under their common seal, to accept such new charter and such acceptance shall operate as a surrender of all charters heretofore granted to the said corporation.

The Faculty at Glasgow may be amalgamated.

L. If at any future period the Royal College of Surgeons of *Edinburgh* and Faculty of Physicians and Surgeons of *Glasgow* agree to amalgamate, so as to form one united corporation, under the name of "The Royal College of Surgeons of *Scotland*," it shall be lawful for Her Majesty to grant, and for such college and faculty under their respective common seals to accept, such new charter or charters as may be necessary for effecting such union, and such acceptance shall operate as a surrender of all charters heretofore granted to such college and faculty; and in the event of such union it shall be competent for the said college and faculty to make such arrangements as to the time and place of their examinations as they may agree upon, these arrangements being in conformity with the provisions of this Act, and subject to the approval of the general council.

New Charter may be granted to the King and Queen's College of Physicians in Ireland.

LI. It shall be lawful for Her Majesty to grant to the corporation of the King and Queen's College of Physicians in *Ireland* a new charter, and thereby to give to such corporation the name of "The Royal College of Physicians of *Ireland*," and to make such alterations in the constitution of the said corporation as to Her Majesty may seem expedient; and it shall be lawful for the said corporation to accept such charter under their common seal, and such acceptance shall operate as a surrender of the charter granted by King *William* and Queen *Mary*, so far as it may be inconsistent with such new charter.

Charters not to contain new Restrictions in the Practice of Medicine or Surgery.

LII. Provided always, that nothing herein contained shall extend to authorize Her Majesty to create any new restriction in the practice of medicine or surgery, or to grant to any of the said corporations any powers or privileges contrary to the common law of the land or to the provisions of this Act, and that no such new charter shall in anywise prejudice, affect, or annul any of the existing statutes or byelaws of the corporations to which the same shall be granted, further than shall be necessary for giving full effect to the alterations which shall be intended to be effected by such new charters and by this Act in the constitution of such corporation.

Provisions of 17 & 18 Vict. c. 114, as to University of London to continue in force.

LIII. The enactments and provisions of the University of *London* Medical Graduates Act, 1854, shall be deemed and construed to have applied and shall apply to the University of *London* for the time being, notwithstanding the surrender or determination of the therein-recited charter, and the granting or acceptance of the now existing charter of the University of *London*, or the future determination of the present or any future charter of the said university, and the granting of any new charter to the said university ; and that every bachelor of medicine and doctor of medicine of the University of *London* for the time being shall be deemed to have been and to be entitled and shall be entitled to the privileges conferred by the said Act, in the same manner and to the same extent as if the charter recited in the said Act remained in force, subject nevertheless to the provisions of this Act.

British Pharmacopœia to be published.

LIV. The general council shall cause to be published under their direction a book containing a list of medicines and compounds, and the manner of preparing them, together with the true weights and measures by which they are to be prepared and mixed, and containing such other matter and things relating thereto as the general council shall think fit, to be called "*British* Pharmacopœia ;" and the general council shall cause to be altered, amended, and republished such pharmacopœia as often as they shall deem it necessary.

Chemists, &c. not to be affected.

LV. Nothing in this Act contained shall extend or be construed to extend to prejudice or in any way to affect the lawful occupation, trade, or business of chemists and druggists and dentists, or the rights, privileges, or employment of duly licensed apothecaries in *Ireland*, so far as the same extend to selling, compounding, or dispensing medicines.

SCHEDULE (A.)

1. Fellow, licentiate, or extra licentiate of the Royal College of Physicians of London.
2. Fellow or licentiate of the Royal College of Physicians of Edinburgh.
3. Fellow or licentiate of the King's and Queen's College of Physicians of Ireland.
4. Fellow or member or licentiate in midwifery of the Royal College of Surgeons of England.
5. Fellow or licentiate of the Royal College of Surgeons of Edinburgh.
6. Fellow or licentiate of the Faculty of Physicians and Surgeons of Glasgow.
7. Fellow or licentiate of the Royal College of Surgeons in Ireland.
8. Licentiate of the Society of Apothecaries, London.
9. Licentiate of the Apothecaries' Hall, Dublin.
10. Doctor, or bachelor, or licentiate of medicine, or master in surgery of any university of the United Kingdom ; or doctor of medicine by doctorate granted prior to passing of this Act by the Archbishop of Canterbury.
11. Doctor of medicine of any foreign or colonial university or college, practising as a physician in the United Kingdom before the first day of October, 1858, who shall produce certificates to the satisfaction of the council of his having taken his degree of doctor of medicine after regular examination, or who shall satisfy the council under section forty-five of this Act, that there is sufficient reason for admitting him to be registered.

SCHEDULE (B.)

DECLARATION required of a person who claims to be registered as a medical practitioner upon the ground that he was in practice as a medical practitioner in England or Wales before the first day of August, 1815:

To the registrar of the medical council.

I,   residing at   in the county of   hereby declare that I was practising as a medical practitioner at   in the county of   before the first day of August, 1815.

                        (Signed)   [*Name.*]

Dated this   day of   185 .

SCHEDULE (D.)

Name.	Residence.	Qualification.	Title.
A.B.	London - -	Fellow of the Royal College of Physicians of	
C.D.	Edinburgh -	Fellow and Member of the Royal College of Surgeons of	
E.F.	Dublin - -	Graduate in Medicine of University of	
G.H.	Bristol - -	Licentiate of the Society of Apothecaries.	
I.K.	London - -	Member of College of Surgeons and Licentiate of the Society of Apothecaries.	

ANNO VICESIMO SECUNDO VICTORIÆ REGINÆ.

CAP. XXI.

An Act to amend the Medical Act (1858).

[19th *April* 1859.]

21 & 22 *Vict.* c. 90.

WHEREAS by an Act passed in the last Session of Parliament, chapter ninety, "The Medical Act," provision is made for the registration of members of the medical profession, and certain disabilities are imposed, after the first day of *January* one thousand eight hundred and fifty-nine, on members of that profession who are not then registered: And whereas, by reason of the time required for the collection and examination of the proper evidence on the first formation of "the medical register," it is expedient to amend the said Act as herein-after mentioned: And whereas it is expedient that Schedule D. of the aforesaid Act should be amended: And whereas in sections thirty-one and forty-seven of the Medical Act (1858) the terms "fellow" and "member" of the Royal Colleges of Physicians of *London* and *Edinburgh* are made use of, whilst in Schedule A. in the same Act "fellows," "licentiates," and "extra licentiates" of the said colleges are alone entitled to be registered: Be it enacted by the Queen's most excellent Majesty, by and with the advice and consent of the Lords Spiritual and Temporal, and Commons, in this present Parliament assembled, and by the authority of the same, as follows:

1st July 1859 to be substituted in ss. 32, 34, 36, and 37, of recited Act for 1st Jan. 1859.

I. The first day of *July* one thousand eight hundred and fifty-nine shall be substituted, in sections thirty-two, thirty-four, thirty-six, and thirty-seven respectively of the said Act, for the first day of *January* one thousand eight hundred and fifty-nine; and the said several sections, and all provisions of the said Act having reference thereto, shall be construed and take effect as if the word *July* had been originally inserted in each of the said sections instead of the word *January*.

Section 33 of recited Act repealed.

II. Section thirty-three of the said Act shall be repealed, and no person shall by reason of the said Act be or be deemed to have been disqualified to hold such office as mentioned in the said section thirty-three, or any appointment mentioned in the said section thirty-six, unless he shall have failed to be registered on or before the first day of *July* one thousand eight hundred and fifty-nine.

Fourth Column of Schedule D. repealed.

III. The fourth column of Schedule D. of the said Act with its heading shall be repealed and omitted.

The Term "Member" to be added in First and Second Heads of Schedule A.

IV. The term "member" shall be added after the term "fellow" to the qualifications described in the first and second heads of Schedule A.

The Words "Forty-six" to be substituted for Forty-five in Schedule A.

V. And whereas in Schedule A. of the said Act there is a reference to section "forty-five," but the word "five" is there inserted by mistake: now it is hereby enacted, that the words "forty-six" shall be deemed to be substituted in this schedule in the place of the words "forty-five."

Any Person not a British Subject having obtained his Degree or Diploma may act as Resident Physician, &c. of any Hospital exclusively for Foreigners.

VI. Nothing in the said Act contained shall prevent any person not a *British* subject who shall have obtained from any foreign university a degree or diploma of doctor in medicine, and who shall have passed the regular examinations entitling him to practise medicine in his own country, from being and acting as the resident physician or medical officer of any hospital established exclusively for the relief of foreigners in sickness: Provided always, that such person is engaged in no medical practice except as such resident physician or medical officer.

ANNO VICESIMO TERTIO VICTORIÆ REGINÆ.

CAP. VII.

An Act to amend the Medical Acts.

[23d *March* 1860.]

21 & 22 *Vict.* c. 90.   22 *Vict.* c. 21.

WHEREAS by an Act passed in the twenty-first and twenty-second years of the reign of Her Majesty, chapter ninety, intituled " The Medical Act," provision is made for the registration of members of the medical profession, and the said Act was amended by an Act passed in the twenty-second year of the reign of Her Majesty, chapter twenty-one; and certain disabilities are imposed by the said Acts, after a period mentioned therein, on members of that profession who are not then registered: And whereas it is expedient that the said recited Acts should be amended as herein-after mentioned: Be it therefore enacted, by the Queen's most excellent Majesty, by and with the advice and consent of the Lords Spiritual and Temporal, and Commons, in this present Parliament assembled, and by the authority of the same, as follows:

Licentiates in Surgery of any University in Ireland entitled to be registered under first-recited Act in like Manner as Masters in Surgery.

I. From and after the passing of this Act the diploma or licence in surgery, granted by any university of that part of the United Kingdom called *Ireland*, legally authorized to grant the same, shall be considered a sufficient qualification to practise under the said first-recited Act, and every person to whom such diploma or licence has been granted shall be entitled to be registered under the provisions of the said first-recited Act, in the like manner, and with the like effect, and subject to the like provisions as are prescribed by the said first-recited Act in respect of the registration of any master in surgery of any university of the United Kingdom.

Certain Powers given to Medical Council extended to this Act.

II. The powers given to the medical council in the said first-recited Act with respect to the studies and examinations required for obtaining a qualification under the said Act shall be extended to the studies and examinations required for a qualification under this Act.

1st January 1861 to be substituted in Sections 32. 34. 36. and 37. of first-recited Act, for 1st July 1859, so far as relates to Persons authorized to be registered under this Act.

III. The first day of *January* one thousand eight hundred and sixty-one shall be deemed to be substituted in sections thirty-two, thirty-four, thirty-six, and thirty-seven respectively of the said first-recited Act, as the same are amended by the said second-recited Act, for the first day of *July* one thousand eight hundred and fifty-nine, so far as the same relate to any person authorized to be registered under this Act; and the said several sections, as so amended, and all the provisions of the said Act having reference thereto, shall, with respect to any such Person so authorized to be registered under this Act, be construed and take effect as if the words "the first day of *January* one thousand eight hundred and sixty-one" had been originally inserted in each of the said sections instead of the words "the first day of *July* one thousand eight hundred and fifty-nine."

No Person authorized to be registered under this Act disqualified to hold certain Offices, unless he has failed to be registered.

IV. No person authorized to be registered under this Act who shall be acting as medical officer under an order of the Poor Law Commissioners, or Poor Law Board, shall by reason of the said recited Acts, or either of them, be or be deemed to have been disqualified to hold such office, or any appointment mentioned in section thirty-six of the said first-recited Act, unless he shall have failed to be registered on or before the first day of *January* one thousand eight hundred and sixty-one.

Recited Acts and this Act to be as one.

V. The said recited Acts and this Act shall be construed together as one Act.

Short Title.

VI. This Act may for all purposes be cited as "The Medical Acts Amendment Act, 1860."

APPENDIX.   411

ANNO VICESIMO TERTIO & VICESIMO QUARTO VICTORIÆ REGINÆ.

CAP. LXVI.

An Act to amend the Medical Act (1858).

[6th *August* 1860.]

21 & 22 *Vict.* c. 90.   14 & 15 *Hen.* 8. c. 5.

WHEREAS by "The Medical Act, 1858," it is provided that it shall be lawful for Her Majesty to grant to the corporation of the Royal College of Physicians of *London* a new charter, and thereby to give to such corporation the name of "The Royal College of Physicians of *England*," and to grant to the corporation of the Royal College of Physicians of *Edinburgh* a new charter, and thereby to give to the said College of Physicians the name of "The Royal College of Physicians of *Scotland*," and to grant to the corporation of the King and Queen's College of Physicians in *Ireland* a new charter, and thereby to give to such corporation the name of "The Royal College of Physicians of *Ireland*;" but provision is not made by the said Act for reserving to the said colleges, and the presidents and censors, fellows, members, licentiates, and extra licentiates thereof respectively, by their said new names, the powers, privileges, liberties, and immunities to which they are respectively entitled by their existing names, and doubts have arisen whether, in case of the acceptance by these colleges respectively of new charters under such altered names respectively, the said powers, privileges, liberties, and immunities would legally attach and be preserved to them, and it is expedient that such doubts should be removed: And whereas by an Act passed in the fourteenth and fifteenth years of the reign of King *Henry* VIII., intituled "The Privileges and Authorities of Physicians in *London*," certain letters patent, dated the twenty-third day of *September*, in the tenth year of the reign of His said Majesty, whereby certain physicians in *London* therein named were incorporated by the name of "The President and College or Commonalty of the Faculty of Physic in *London*," were ratified and confirmed; and by the said Act it was enacted, that the six persons named in the said letters patent, and two more of the said commonalty to be chosen by them, should be called elects, and that the said elects should yearly choose one of them to be president of the said commonalty, and that as oft as any of the places of the said elects should become void the survivors should choose and admit one or more, as need should require, of the said faculty to supply the number of eight persons, and that no person should from thenceforth be suffered to practise in physic through *England* until he be examined by the said president and three of the said elects, and have from them letters testimonial, except he be a graduate of *Oxford* or *Cambridge*: And whereas the main function of the said elects, viz., that of examining and granting letters testimonial, has been virtually superseded by the said Medical Act, and they have ceased to grant letters testimonial in accordance with the provisions contained in the last-recited Act; and it is therefore expedient that the before-recited provisions should be repealed: Be it enacted by the Queen's most excellent Majesty, by and with the advice and consent of the Lords Spiritual and Temporal, and Commons, in this present Parliament assembled, and by the authority of the same, as follows:

Interpretation of Terms.

I. The expression in the Medical Act and this Act "The Corporation of the Royal College of Physicians of *London*," or "The Royal College of Physicians of *London*," shall be taken to denote the corporation of "The President and College or Commonalty of the Faculty of Physic in *London*."

New Charters may be granted to the Colleges.

II. Any new charter which, under the provisions of the Medical Act, shall be granted to the corporation of the Royal College of Physicians of *London*, may be granted to them either by and in the name of the Royal College of Physicians of London, or, as provided by that Act, by and in the name of the Royal College of Physicians of *England;* and any such new charter granted to the corporation of the Royal College of Physicians of *Edinburgh* may be granted to that college either by and in its present name, or, as provided by the Medical Act, by and in the name of the Royal College of Physicians of *Scotland;* and any such new charter granted to the corporation of the King and Queen's College of Physicians in *Ireland* may be granted to that college either by and in its present name, or, as provided by the Medical Act, by and in the name of the Royal College of Physicians of *Ireland.*

Colleges to retain all existing Rights, notwithstanding Change of Name.

III. The granting of new charters to the said corporations respectively by and in the altered names and styles respectively, as provided in the Medical Act, shall not, in respect of such alteration of name or style merely, alter or affect in any way the rights, powers, authorities, qualifications, liberties, exemptions, immunities, duties, and obligations granted, conferred, or imposed to or upon, or continued and preserved to the said corporations respectively, and the respective presidents, censors, fellows, members, and licentiates thereof, by the respective charters and Acts of Parliament relating to the said corporations respectively, or by the Medical Act, the Act to amend the Medical Act, the Medical Acts Amendment Act, 1860, and this Act respectively; but the said corporations respectively, and the respective presidents, censors, fellows, members, and licentiates thereof, shall, notwithstanding any such change of name and style, have and retain all such and the same rights, powers, authorities, qualifications, liberties, exemptions, and immunities, and be subject to all such and the same duties and obligations, as if such new charters respectively had been granted to them by and in their respective names and styles as then existed.

Colleges to hold Property notwithstanding Change of Name.

IV. Each of the said corporations shall also, notwithstanding any such alteration of name or style, have, hold, and enjoy, and continue to have, hold, and enjoy, all lands and other real and personal, heritable and moveable property belonging to such corporation, either beneficially or in trust, at the date of the granting of such new charter, and may execute and perform any use or trust for the time being vested or reposed in such corporation.

Provisions in 14 & 15 Hen. 8. c. 5. as to the Elects repealed.

V. So much of the Act of the fourteenth and fifteenth *Henry* VIII., chapter five, as relates to the elects of the said Royal College of Physicians of *London*, and their powers and functions, shall be and the same is hereby repealed, but this repeal shall not prejudice or affect the rights and privileges of any persons to whom the said president and elects may have granted letters testimonial; and all trusts which by any deed, gift, devise, or bequest are vested in, or to be executed or performed by the elects, or some defined number of them, shall vest in and accrue to, and be executed and performed by the censors of the said college for the time being as if the name of the censors had in such instruments respectively been used instead of that of the elects, and the office and name of elects of the said college shall henceforth wholly cease and determine.

Election of the President of the Royal College of Physicians of London.

VI. The office of president of "The Royal College of Physicians of *London*," shall be an annual office; and *Thomas Mayo*, doctor of physic, the now president of the said corporation, shall remain such president until the day next after *Palm Sunday* in the year one thousand eight hundred and sixty-one, when he shall go out of office; and the fellows of the said corporation shall, at a meeting to be holden by them for that purpose, on the same day, and on the same day in every subsequent year, elect some one of the fellows of the said corporation in such manner as shall be provided by any byelaw or byelaws made in that behalf by the said corporation, and for the time being in force, to be president of the said corporation, but the retiring president shall always be capable of being re-elected, and every president shall remain in office until the actual election of a new president; or in case of the death, resignation, or other avoidance of any such president before the expiration of his year of office, the said fellows shall, at a meeting to be holden by them for that purpose, as soon as conveniently may be (of which due notice shall be given), elect one other of the fellows of the said corporation in such manner as aforesaid to be president for the remainder of the year in which such death, resignation, or other avoidance shall happen, and until such election the duties of president shall be performed by the senior censor for the time being.

ANNO TRICESIMO PRIMO VICTORIÆ REGINÆ.

CAP. XXIX.

An Act to amend the Law relating to Medical Practitioners in the Colonies. [29th *May* 1868.]

WHEREAS by the thirty-first section of the Medical Act, passed in the Session holden in the twenty-first and twenty-second years of Her Majesty, chapter ninety, it is enacted as follows: "Every person registered under this Act shall be entitled, according to his qualification or qualifications, to practise medicine or surgery, or medicine and surgery, as the case may be, in any part of Her Majesty's dominions, and to demand and recover in any court of law, with full costs of suit, reasonable charges for professional aid, advice, and visits, and the cost of any medicines or other medical or surgical appliances rendered or supplied by him to his patients:" And whereas it is expedient to amend the said enactment: Be it enacted by the Queen's most excellent Majesty, by and with the advice and consent of the Lords Spiritual and Temporal, and Commons, in this present Parliament assembled, and by the authority of the same, as follows:

Short Title.

1. This Act may be cited as "The Medical Act Amendment Act, 1868."

Interpretation of Act.

2. The term "colony" shall in this Act include all of Her Majesty's possessions abroad in which there shall exist a legislature as herein-after defined, except the *Channel Islands* and the *Isle of Man.*

The term "colonial legislature" shall signify the authority other than the Imperial Parliament or Her Majesty in Council competent to make laws for any colony.

Power to Colonial Legislatures to enforce Registration of Persons registered under the Medical Act.

3. Every colonial legislature shall have full power from time to time to make laws for the purpose of enforcing the registration within its jurisdiction of persons who have been registered under the Medical Act, anything in the said Act to the contrary notwithstanding: Provided, however, that any person who has been duly registered under the Medical Act shall be entitled to be registered in any colony, upon payment of the fees (if any) required for such registration, and upon proof, in such manner as the said colonial legislature shall direct, of his registration under the said Act.

INDEX.

Accoucheur, 50, 71, 72
Æsculapius, 1, 8
Affidavit, 101, 102
Alchymists, 6, 10 (see Ancient Orders of the Medical Profession).
Anatomy, 3, 6
,, professor of, Cambridge, 5
,, schools of, 63
Ancient Orders of the Medical Profession, 1
Antimony, 125
Apothecaries, 5, 9, 15, 24, 25, 28, 31, 32, 34, 36, 37, 38, 42, 46
Apothecaries, Society of, 16, 24, 31, 37
Apprentice, 74
Arsenic, poisoning by, 123
Assistant, 70, 73, 74
Astrologers, 6, 10 (see Ancient Orders of the Medical Profession).
Attorney, 72

Bachelor of Medicine, 6, 25, 26, 33, 155
Bacon, Roger, 10
Barbers, 8, 9, 10, 146
,, Company, 8
Barber-surgeons, 9
Barrister, 45
Births, marriages, and deaths, registration of, 126
Bishops, 5
Boerhaave, 12

Cemeteries, 124
Certificates, medical, with forms, 93, 94, 95, 96, 97, 100, 110—112, 133, 134, 136—142
Character, 43, 44, 46, 50, 52
Chemists and druggists, 14, 15, 42, 143
Contagious diseases, 34, 82, 84 (see Public Health).
Copyright, law of, 53
Coroners' inquests, 114, 127

Defamation, 43
Dentists, 15, 146
Diagnosis, 3
Dietetics, 3

Disabilities of unregistered practitioners, 35, 36
Dislocations, 12, 34
Dissections, 65, 66
Doctors of medicine, 17, 25, 26, 27, 33, 36, 153, 155
Doctors of medicine, by Doctorate, 26, 33, 36
Doctors of medicine, foreign and colonial, 27, 33, 36
Doctors of philosophy, 151
,, of physic, 4, 18, 153
,, quack (see Quack).
Druids, 12, 14

Empirics, 6, 149
Ethics, medical, 147
Examination (Conjoint Medical Board), 152

Factory and Workshop Regulation Acts, 86—88
Fees, 28, 31, 32, 33, 45, 79, 87, 91, 116
Fellows of College of Physicians, London, 5, 16, 17, 18, 29, 32, 155
Fellows of College of Physicians, Edinburgh, 23, 32
Fellows of King's and Queen's College of Physicians, Ireland, 23, 32
Fellows of College of Surgeons, England, 5, 23, 32
Fellows of College of Surgeons, Edinburgh, 23, 32
Fellows of Faculty of Physicians and Surgeons, Glasgow, 24, 32
Fellows of College of Surgeons, Ireland, 24, 32
Fellow of Trinity College, Cambridge, 5
Fractures, 34

Galen, 2
,, works of (Kühn's edition), 3
Gregory the Great, Pope, 3

Herbalists, 6
Hippocrates, 1

INDEX.   415

Homœopathists, 47
Hygiene, 3

IDIOTS, law relating to, 93
Ignorance, crass, 37, 38, 41, 42
Immoral, offence, 44, 46
Immunities of duly registered practitioners (see Rights, Privileges, &c.), 29
Infectious diseases, 34, 44, 45 (and see Public Health).
Interment, 64, 66

JEWS, 11
Judges, answers of, to House of Lords, 98, 99, 100

KING'S evil, touching for the, 12
Knights Hospitallers, 8

LATERAN, Council of, 4
Lepers, 8
Leprosy, 82, 83
Liabilities of legally qualified medical practitioners, 36, 38
Libel (see Defamation).
Licentiates of College of Physicians, London, 16, 19, 20, 32, 155
Licentiates, extra, 16, 17, 32
Licentiates of College of Physicians, Edinburgh, 23, 32
Licentiates of King's and Queen's College of Physicians, Ireland, 23, 32
Licentiates in midwifery of the College of Surgeons, England, 23, 32
Licentiates of the College of Surgeons, Edinburgh, 23, 32
Licentiates of the Faculty of Physicians and Surgeons, Glasgow, 24, 32
Licentiates of the College of Surgeons, Ireland, 24, 32
Licentiates of the Society of Apothecaries, London, 24, 32
Licentiates of Apothecaries' Hall, Dublin, 24, 32
Life assurance, the law of, 78

MALPRACTICE, or mala-praxis, 22, 39
Manslaughter, 38, 40, 41, 42
Masters in surgery, 25, 33
Medical adviser, 80
　,,　　man, 80
　,,　　practitioner, 29, 35, 36, 38, 47, 54
　,,　　practitioner, legally qualified, 36, 85
　,,　　referees, 78, 79
　,,　　"usual attendant," 80, 81
　,,　　Witnesses Act, 36
Members of the College of Physicians, London, 16, 17, 20, 21, 29, 32, 34, 155

Members of the College of Physicians, Edinburgh, 23, 32
Members of the College of Surgeons of England, 23, 32
Midwife, 4, 6
　,,　　man, 42
Midwifery, 23, 26, 32, 143
Monastery, 4
Monastic, 4, 5
Monks, 4, 5, 8
Murder, 40

NEGLIGENCE, 37, 38

ORDERS, holy, 4, 6
　,,　ancient, of the medical profession
　　　1 Scholars,
　　　2 Surgeons,
　　　3 Grocers or poticaries,
　　　4 Empirics,   } 6—10
　　　5 Alchymists,
　　　6 Sorcerers,
　　　7 Witches,
　　　8 Astrologers,
Orders, modern, of the medical profession, 15—28

PARTNERSHIP, law of, 68
Pathology, 3
Pestilence, 34
Pharmaceutical chemists, 143, 144
　,,　　Society, 143, 144
Pharmacopœia, 2, 62, 145, 146
Pharmacy, 3, 15, 16, 17, 23, 24, 26, 29, 32, 33
Pharmacy Acts, 143
Physicians, 6, 7, 14—23, 29, 32, 38, 40, 45, 46, 47, 153, 154, 155
Physicians, monkish, 7
　,,　　clerical, 4, 5, 6
Physic, 4, 6, 11, 15, 17, 25, 26, 34
Physiology, 3, 6
Plague, 82, 83
Poor-law Legislation, 130
Post-mortem examinations, 36, 116
Privileged communications, 47, 50, 107, 125
Prussic acid, poisoning by, 119—121
Public health, 82—92
Pupil, 70, 71

QUACK, 39, 44, 45, 84

RELIGION, hostile to, 55 (see Defamation).
Restraint, 112, 134, 137
Rheims, Council of, 4
Rights, Privileges, &c., 29

St. Lazarus, order of, 8
Salerno, College of, 4, 6
Semeiology, 3
Skill, want of, 38, 40, 41, 42
Small-pox, 34, 83 (see Public Health).
Solicitor (see Attorney).
Sorcerers (see Ancient Orders of the Medical Profession).
Sorcery, 66
Surgeons, 4, 5, 6, 7, 8, 9, 11, 15, 23, 24, 25, 26, 29, 30, 32, 34, 35, 37, 38, 39, 40, 41, 46, 55, 63
Surgery, 3, 8, 9, 11, 15, 16, 23, 24, 25, 26, 32, 33, 155
Syphilis, 34, 82

Tetanus, 122, 123
Therapeutics, 3
Tours, Council of, 4
Tumours, 34
Typhus fever, 34

Ulcers, 34
Universities, 25
  ,, English, 6
  ,, foreign and colonial, 27, 33

Universities, French, 6
  ,, of the United Kingdom, 33
University of Cambridge, 5, 6, 18. 26, 33, 153
  ,, Durham, 33
  ,, Edinburgh, 18
  ,, London, 26, 33
  ,, Oxford, 6, 18, 26, 33, 153 154
Unqualified medical practitioners, 39, 71
Unregistered medical practitioners, 36

Vaccination, 132 (see Public Health, Section III.)
Vectis or lever, improper use of, 42

Witches (see Ancient Orders of the Medical Profession).
Witchcraft, 34
Women, medical practitioners, 11
  ,, attempt to prohibit practising, 7, 12
Workhouse (see Poor-law Legislation).
Workshops, 86

THE END.

www.ingramcontent.com/pod-product-compliance
Lightning Source LLC
Chambersburg PA
CBHW051724300426
44115CB00007B/456